# De Gruyter Studies in Mathematical Physics 27

*Edited by*
Michael Efroimsky, Bethesda, Maryland, USA
Leonard Gamberg, Reading, Pennsylvania, USA
Dmitry Gitman, São Paulo, Brazil
Alexander Lazarian, Madison, Wisconsin, USA
Boris Smirnov, Moscow, Russia

Ivan A. Lukovsky

# Nonlinear Dynamics

## Mathematical Models for Rigid Bodies with a Liquid

De Gruyter

*Physics and Astronomy Classification Scheme 2010:*
02.30.Xx, 45.20.Jj, 45.40.-f, 47.10.A-, 47.20.Ky, 47.35.Lf, 47.52.+j

*Author*
Prof. Dr. Ivan A. Lukovsky
National Academy of Sciences of Ukraine
Institute of Mathematics
Tereschenkivska, 3 str
Kiev, 01601
Ukraine

Translated by Peter V. Malyshev

ISBN 978-3-11-055536-3
e-ISBN (PDF) 978-3-11-031657-5
e-ISBN (EPUB) 978-3-11-038973-9
Set-ISBN 978-3-11-031658-2
ISSN 2194-3532

*Library of Congress Cataloging-in-Publication Data*
A CIP catalog record for this book has been applied for at the Library of Congress.

*Bibliographic information published by the Deutsche Nationalbibliothek*
The Deutsche Nationalbibliothek lists this publication in the Deutsche Nationalbibliografie; detailed bibliographic data are availableon the Internet at http://dnb.dnb.de.


Dieser Band ist text- und seitenidentisch mit der 2015 erschienenen gebundenen Ausgabe.
Printing and binding: CPI books GmbH, Leck
♾ Printed on acid-free paper
Printed in Germany

www.degruyter.com

# Foreword to English Edition

The Russian Edition of the present book was published in 2010 as an extended and elaborated version of the earlier author's monograph "Introduction to Nonlinear Dynamics of a Rigid Body with Cavities Partially Filled with Liquid" (Naukova Dumka, 1990). The book deals with analytic approaches to nonlinear sloshing problems and the coupled "rigid body–contained liquid" dynamics focusing on the results developed by the author starting from the early 1970s. The majority of these approaches are associated with the so-called multimodal method. The book presents various theoretical aspects of the multimodal method as they appeared driven by author's collaboration with spacecraft designers and the inner logics of the research. This partly explains a special emphasis of the book on axisymmetric containers quite common for the spacecraft applications. Furthermore, the Russian Edition basically concentrates on the applied mathematical tools without giving practically important numerical examples that were mainly obtained by the author within the framework of the above-mentioned collaboration and are not yet available for the readers.

In preparing the English Edition, the book underwent serious improvements and additions. Thus, new Introduction and Chapter 3 were written. In Introduction, we present a deeper historical review of the multimodal method and other related topics associated with the dynamics of rigid body containing a liquid. In Chapter 3, one can find an applied mathematical background of the dynamics of floating ships with partially filled tanks presented in the same way as in the other chapters, where the Bateman–Luke variational principle plays the role of the main analytic instrument. As for the improvements and revisions made in the other chapters, it is necessary, first of all, to mention new author's results obtained in 2012 on the general Narimanov–Moiseev-type nonlinear modal equations for an upright circular cylindrical tank. These results are presented in Chapter 4. The infinite-dimensional equations can be regarded as a breakthrough in the nonlinear multimodal method. They are deduced by using a special analytic technique based on a computer algebra.

Chapter 5 is devoted to axisymmetric tank shapes with nonvertical walls. The multimodal method is combined here with the so-called nonconformal mapping technique proposed by the author in 1975. Among the general aspects of the combined approach, the Russian Edition presented a five-dimensional nonlinear modal equations deduced in 2005 guaranteeing the possibility of accurate description of resonant sloshing in V-shaped conical containers excited by sway and surge with a forcing frequency close to the lowest natural sloshing frequency. In this connection, new author's results of 2013 for a truncated conical tank are added to the English Edition. They can be regarded as important examples of application of the combined approach. The corresponding seven-dimensional nonlinear modal system is extensively used in Chapter 8 to study the steady-state resonant wave modes in tapered conical tanks. New examples illustrating the nonlinear sloshing phenomena are also included in Chapter 8 with a focus on the visualization of nonlinear wave patterns in upright circular cylindrical tanks. The visualization was performed by using a specially developed software and the five-dimensional nonlinear modal model from Chapter 4. Finally, the author corrected some misprints found in the Russian Edition and improved the presentation by adding more discussions and remarks.

Kiev, Ukraine,
March, 2014

# Foreword to Russian Edition

The book is devoted to the nonlinear theory of motion of rigid bodies with cavities containing a liquid, which is an important branch of the mechanics of deformable systems.

In the last decades, a significant progress has been attained in the nonlinear dynamics of rigid bodies partially filled with liquids in connection with the solution of specific fluid-structure interaction problems. As the most complicated problems, we can mention the problems of establishing the laws governing the dynamics of space motion of coupled body-liquid systems and finding the interaction forces and moments acting between the rigid body and contained liquids in the case where external loads are known. As examples, we can mention motion of tanks, the problems of safety of containers with ecologically dangerous liquids, especially, in seismically dangerous regions, the dynamics, stability, and control of aircrafts, spacecrafts, and sea-going vessels carrying significant masses of liquids.

The physical processes encountered in the investigation of these applied problems are strongly nonlinear and, hence, their study naturally involves the nonlinear governing equations of the dynamics of rigid body and the nonlinear equations of the free-surface of a bounded volume of the contained liquid.

The efforts of various researchers and scientific groups working with mechanical problems of this kind were mainly concentrated on the overcoming the fundamental theoretical difficulty connected with the necessity of description of coupling of two different mechanical systems, a rigid body and a liquid, in the nonlinear statement, when the motion of the body is described by nonlinear ordinary differential equations, whereas the dynamics of sloshing liquid is governed by the corresponding nonlinear initial-boundary-value problem for the free surface.

As early as in the 1950s and 1960s, the development of linear mathematical models used to describe the liquid-body dynamics was motivated by the possibility of solution of the applied problems of spacecraft engineering. This development was largely based on the use of the so-called multimodal approach. Specifically, this approach reduces the investigation of the hybrid mechanical system to the analysis of an infinite-dimensional system of ordinary differential equations. The procedures of getting the indicated systems of equations are based on the application of the well-known results of analytic mechanics, including the variational principles, as well as on the use of eigensolutions of special spectral boundary problems for the natural modes and frequencies of sloshing.

The multimodal approach remains applicable in both linear and nonlinear cases. It is based on the Lagrange variational formalism and the methods of perturbation theory. In our monograph, main attention is given to the variational methods used to construct finite-dimensional nonlinear modal systems aimed at the description of coupled space motions of a rigid body and a liquid. The dimension of these finite-dimensional models is determined by the necessity of taking into account the perturbations of a certain number of natural sloshing modes. Unlike the linear theory in which an infinite system of ordinary differential equations decomposes into independent second-order differential equations (linear oscillators), the nonlinear case deals with an infinite system of nonlinearly coupled ordinary differential equations.

As a result, the existing practical possibilities to construct the indicated nonlinear models (even if we use computer technologies) and their realizations for the solution of specific practical problems are strongly restricted. Most efficient and well-tested are the so-called nonlinear models of the third order of smallness obtained in the general theory of space motions of a rigid body with cylindrical cavities under certain restrictions imposed on the orders of smallness of deviations of the free surface. Similar models were also constructed for some other shapes of the tanks. Testing these models in the course of the solution of a series of applied problems of nonlinear dynamics revealed their high efficiency.

The survey of the available scientific works in this field and their critical analysis shows that the nonlinear theory of motion of a rigid body with cavities partially filled with liquid is far from being complete and remains in the stage of active development. However, it is possible to state that this theory has already passed through the period of paradoxes of the linear theory. It was theoretically and experimentally established that the elevations of the free surface are always bounded for various perturbations applied to the body. The nodal lines on the free surface are always mobile. For the forced liquid motions, the height of the "hump" on the free surface is much (sometimes twice) larger than the depth of the trough. Moreover, the steady-state modes of the free surface motion change in the vicinity of the main resonance (we observe bifurcations, e.g., associated with swirling) .

Numerous specific nonlinear physical phenomena are also caused by high-frequency perturbations (in the case where the forcing frequency is much higher than the lowest natural sloshing frequency and the amplitude of excitations is low). All these nonlinear hydrodynamic processes affect, in a certain way, the resulting hydrodynamic forces and moments. This leads to the formation of qualitatively new motions of the entire body-liquid system.

At the same time, it is worth noting that the nonlinear analytic theory of coupled dynamics is not widely known for engineers. This can be partly explained, by the laboriousness of the developed analytic methods and a sharp increase in the dimensionalities of the proposed mathematical models in realistic cases. Our experience of application of the multimodal method to the coupled dynamics of a rigid body containing an upright circular cylindrical tank partially filled with a liquid shows that the indicated nonlinear phenomena can be satisfactorily described (with an accuracy to within 2% for the hydrodynamic forces and to within 5% for free-surface elevations) with a system of nonlinear ordinary differential equations of the 11th order (six degrees of freedom for the absolutely rigid body and five natural sloshing modes). However, the number of tanks in realistic applications can sometimes be as large as several dozens (even for medium-tonnage tankers). Moreover, they are characterized by different performances, which strongly affects the actual dimension of the mathematical model of the entire mechanical system.

In the present monograph, a special emphasis is made on the mathematical background required to deduce approximate analytic mathematical models of hybrid mechanical systems carrying liquids with free surfaces as well as on the analytic approaches to the investigation of these models. The monograph is based on the previous book of the author "Introduction to the nonlinear dynamics of rigid bodies with cavities filled with liquid" and recent author's publications written together with his colleagues who continue the development of the analytic ideology proposed in the cited book. Among the essential points, we can especially mention the analytic technique aimed at improving the construction of high-dimensional models, efficient approximate methods for the evaluation of the kinematic and dynamic characteristics of the rigid body containing a liquid, the comparative analysis depending on the dimension of mathematical models, and the development of analytic methods facilitating the efficient investigation of the stability of steady-state wave motions.

As an important achievement of the recent period in the nonlinear sloshing dynamics, we can mention the generalization of the multimodal approach to the case of dissipative ("viscous") incompressible liquid developed by the author and discussed in the present monograph. A special chapter contains the results of investigations related to a practically important problem of construction of the equivalent mechanical systems (analogs of nonlinear sloshing). The presented invariant form of the governing equations of equivalent mechanical systems can be quite useful for some dynamic problems, e.g., in the vibration mechanics.

The main topics discussed in the present book are split into several chapters.

First, we present the conservation laws for the problems of continuous media, formulate the basic governing equations, and discuss the boundary and initial conditions for trapped and external surface wave. Separately, we consider the nonlinear sloshing problem in a rigid tank participating in the space motion. The problem is posed both in Cartesian and in special curvilinear coordinate systems.

In the second chapter, a variational principle is formulated for the liquid sloshing dynamics, starting with the case of a motionless container and finishing with arbitrary space motions of the rigid body with liquid caused by the external loads applied to the body. For the first dynamical problem (the body performs a prescribed space motion), according to the variational principle, the variational problems are formulated for the domains with variable boundaries. The variational problem is further used to develop a series of direct (projective) approximate methods. By virtue of the equivalence of the variational statement and the original nonlinear free-surface problem, these approximate methods give an approximate solution of the boundary-value problem. As a result, we deduce a system of nonlinear ordinary differential equations for the description of the coupled motions of a rigid body and a liquid.

The examples of practical applications of the proposed methods can be found in the fourth chapter. Quite comprehensively, we analyze the problem of motion of a rigid body whose liquid-filled cavity is formed by coaxial cylindrical surfaces. The developed mathematical model includes five generalized coordinates characterizing the free-surface motions caused by the forced excitations in the vicinity of the main resonance. In the case of space motions, the model includes, in addition, six quasivelocities $v_{0i}$ and $\omega_i$, $i = 1, 2, 3$. The explicit formulas for computing the added masses and moments of inertia due to the liquid sloshing are given. All coefficients of the system of nonlinear equations are tabulated for a broad range of input geometric parameters of the cylindrical cavities.

A similar mathematical model is constructed for an upright circular cylindrical container whose bottom has an arbitrary geometric shape. The possibilities of the method for more complicated tank shapes are illustrated for the case of liquid sloshing in a cylindrical container of the elliptic cross section. The theory of nonlinear sloshing of dissipative ("viscous") liquids is formulated for the cases of free and forced liquid motions in an upright cylindrical container. A five-dimensional mathematical model is constructed. In this chapter, we also present the nonlinear equations of perturbed motion of a rigid body with cylindrical cavity partially filled with liquid.

The fifth chapter is devoted to the variational algorithm of construction of the mathematical models used to describe the liquid sloshing in axisymmetric tanks. With the same level of completeness as in the case of upright cylindrical containers, we consider sloshing in a conic tank which performs a translational oscillatory motion. An example of the corresponding modal equations is given. The hydrodynamic coefficients are computed and partly tabulated for the containers of conical, ellipsoidal, and spherical shapes. All theoretical results are illustrated by numerical examples.

In Chapter 6, we briefly outline the methods of perturbation theory in the liquid sloshing dynamics. Despite the fact that this problem is considered in numerous works, we also discuss it in the proposed monograph for several reasons. First, the method developed by G.S. Narimanov and the variational method lead (under the same basic conditions) to the approximate models of the same structure, and the numerical values of their hydrodynamic coefficients are practically identical. From the methodological viewpoint, the joint analysis of these two alternative methods is quite desirable. Second, the available literature sources dealing with this perturbation method contain a large number of misprints and algebraic errors. The most complete information obtained by the perturbation method proposed by G.S. Narimanov is presented for a class of upright cylindrical cavities with circular and annular cross sections. The approximate nonlinear (modal) equations (models) aimed at the description of the space motions of the rigid body filled with a liquid are obtained under the assumptions accepted for the variational method. The scalar form of these equations and the formulas for their hydrodynamic coefficients are compared with the results obtained in the earlier publications. To make our presentation self-contained, we also present a generalization of the method to the case of complex tank shapes.

In Chapter 7, we consider the case of nonlinear sloshing in a cylindrical reservoir in the state of translational motion. An infinite system of nonlinear ordinary differential equations coupling the generalized coordinates responsible for the free-surface elevations is reduced (by a special transformation of the generalized coordinates) to equivalent nonlinear equations governing an equivalent mechanical system of the pendulum type. We discuss how to choose the parameters of the corresponding mathematical pendulum systems (masses, lengths, points of support, etc.) and propose a physical interpretation of nonlinear oscillators of the hydrodynamic type in terms of the analytic mechanics. Practical advantages of the proposed invariant form of the governing equations of the body-liquid system are discussed.

In the final chapter, we present analytic results on various dynamic properties of liquid sloshing in motionless and spatially moving containers. The well-known problem of forced sloshing induced by translation harmonic excitations is studied by using the two-or-five dimensional nonlinear mathematical models for the case of an upright circular cylindrical tank. This enables us to describe various nonlinear phenomena. A satisfactory agreement of the theoretical results with experimental data is reported. We also consider the problem of forced nonlinear sloshing caused by angular harmonic oscillations. Specifically, the analysis of the steady-state sloshing solutions is performed on the basis of the Bubnov–Galerkin method whose accuracy is controlled both by alternative analytic approximate methods and by the Runge–Kutta numerical integration. The problem of dynamic stability of the steady-state modes is stated and solved by using the first Lyapunov method. This leads to the construction of equations in variations whose solutions are studied by using the Floquet–Lyapunov theory. The response curves (for the free-surface elevations and the hydrodynamic forces) are analyzed in detail. The presented nonlinear modal equations (models) confirm their applicability to the description of the basic hydrodynamic phenomena discovered in model tests, both qualitatively and quantitatively, with satisfactory accuracy.

In writing the book, the author attempted to present the nonlinear theory of motion of rigid bodies partially filled with liquid and its mathematical background in the form convenient for ordinary readers. The author emphasizes, as frequently as possible, the links between the theory and the experimental data. The major part of the results presented in the book are substantiated on the level required for the purposes of applied mathematics. Therefore, many problems mentioned in the book remain open for subsequent discussions from the viewpoint of pure mathematics. Thus, the author hopes that the methods developed in the present book will become a challenge for pure mathematicians.

I want to express my deep gratitude to my colleagues Prof. A. N. Timokha and Drs. A. M. Pil'kevich, A. V. Solodun, and D. V. Ovchinnikov for fruitful scientific collaboration, which favored appearance of the book. I am also especially thankful to the researchers of the Department of Dynamics and Stability of Multidimensional Systems at the Institute of Mathematics of the Ukrainian Academy of Sciences, Dr. G. A. Shvets and Mrs. N. M. Kuz'menko, who prepared the manuscript for publication. The author is also grateful to V. D. Kubenko, Academician of the Ukrainian National Academy of Sciences, and Prof. V. A. Trotsenko for their valuable remarks made, while reading the manuscript.

# Contents

# Introduction

The coupled "rigid tank–contained liquid" dynamics is connected with the investigation of so-called *hybrid mechanical systems* formed by two coupled subsystems of different mechanical and mathematical nature. The first subsystem, i.e., the rigid tank, can move with six degrees of freedom and, hence, its dynamic equations form a six-dimensional system of ordinary differential equations. The hydromechanical subsystem, i.e., the contained liquid, is described by the solutions of a free-surface boundary problem and, therefore, has infinitely many degrees of freedom.

The theoretical foundations of the analysis of these hybrid mechanical systems were laid by G. Stokes, H. Helmholtz, W. Lamb, J. Neumann, and N. E. Zhukovsky as early as in the 19th century. They considered the case of cavities in the body completely filled with ideal incompressible liquids with irrotational flows in the absence of the free surface (interface). However, even in this case, the problem remains quite complicated and requires the analysis of boundary-value problems coupled with the partial differential equations used to describe the motions of the body via the Neumann-type boundary conditions. N. E. Zhukovsky studied the most general statement of the coupled dynamics and showed that (i) the assumption of "frozen" liquid is a wrong way but (ii) the hybrid mechanical system can be associated with an auxiliary rigid body in which the contained liquid is replaced by an artificial solid medium [83] different from the "frozen" liquid. N. E. Zhukovsky introduced the so-called Stokes–Zhukovsky potentials to find the inertia tensor for the artificial solid media.

The presence of an interface (free surface) between the liquid and ullage gas in the cavity of the body (rigid tank) leads to the so-called problems of liquid sloshing [56]. These sloshing problems look mathematically similar to the free-surface problems in the *water wave theory*. This theory was created by Isaak Newton (Book II, Prop. XLV of Principia, 1687, [175]), Leonhard Euler [50, 49, 51], Pierre-Simon Laplace [188], Joseph Louis Lagrange [95], Augustin-Louis Cauchy [30], and many other prominent scientists. An extensive historical survey of their contributions was given by Craik [38]. We refer the interested readers to this survey as well as to [39], where Craik discussed the invaluable contribution to the water wave theory made by George Gabriel Stokes.

The water wave theory normally assumes that the liquid (water) is (i) ideal, (ii) incompressible, and (iii) with irrotational flows and that (iv) the bottom, beach, and floating structures are rigid bodies. For the gravitational water waves, the surface tension is neglected.

The assumptions (i)–(iv) were used for the construction of the sloshing theory for a heavy liquid in rigid mobile/immobile tanks. Under these assumptions, the classical water-wave problem and the problem of liquid sloshing in a motionless tank have the same mathematical structure described in the form of the following free-surface problem:

$$\nabla^2\varphi(x, y, z, t) = 0, \quad \boldsymbol{r} \in Q, \tag{1a}$$

$$\frac{\partial\varphi}{\partial\nu} = 0, \quad \boldsymbol{r} \in S, \tag{1b}$$

$$\frac{\partial\varphi}{\partial\nu} = -\frac{\zeta_t}{\sqrt{(\nabla\zeta)^2}}, \quad \boldsymbol{r} \in \Sigma, \tag{1c}$$

$$\frac{\partial\varphi}{\partial t} + \frac{1}{2}(\nabla\varphi)^2 + \mathrm{g}x = 0, \quad \boldsymbol{r} \in \Sigma. \tag{1d}$$

Here, $\boldsymbol{r}$ is the radius vector, $\varphi(x, y, z, t)$ is the velocity potential of the liquid occupying the domain $Q$ variable with time, $\boldsymbol{\nu}$ is the outer unit normal, $S$ is the wetted surface of the container, $\Sigma$ is the free surface described, in the most general sloshing statement, by the equation $\zeta(x, y, z, t) = 0$, and $\mathrm{g}$ is the gravitational acceleration. The free-surface problem (1) should be equipped with appropriate initial conditions. Together with the Bernoulli equation

$$\frac{\partial\varphi}{\partial t} + \frac{1}{2}(\nabla\varphi)^2 + \mathrm{g}x + \frac{p}{\rho} = 0, \tag{2}$$

this problem enables us to determine all hydrodynamic characteristics interesting for practical purposes.

Despite the fact that the free-surface problem (1) deals with the Laplace equation in the liquid domain $Q$, i.e., it can be formally regarded as an elliptic boundary-value problem with respect to the velocity potential, the presence of the time derivatives in the kinematic (1c) and dynamic (1d) boundary conditions on the free surface turns the entire free-surface problem into a problem of hyperbolic type. This becomes especially clear in the linear statement when, combining the kinematic and dynamic boundary conditions on the mean free surface $\Sigma_0$, we get the following boundary condition of hyperbolic type:

$$\frac{\partial^2\varphi}{\partial t^2} + \mathrm{g}\frac{\partial\varphi}{\partial z} = 0 \text{ on } \Sigma_0.$$

The water-wave theory became an independent part of hydromechanics as early as at the beginning of the 19th century (for its development, Cauchy was awarded a Prize of the Paris Academy of Sciences in 1815). The pioneer investigations of this theory dealt with the linearized case with a special focus on the propagation of low-amplitude waves under fairly general assumptions about initial perturbations. The most important results in the water-wave theory were then obtained under the assumptions that the liquid flow is stationary and two-dimensional. The proof of the existence theorems for two-dimensional periodic progressive waves with finite (not small) amplitudes and infinite depth of water was first proposed by A. I. Nekrasov [174] and then by T. Levi-Civita. It was also based on the reduction of the problem to the investigation of steady processes. Numerous results have been recently obtained in the exact water-wave theory of unsteady wave motions in the form of existence and uniqueness theorems for the solution of the Cauchy–Poisson problem in the two- and three-dimensional statements. In fact, these results deal with the nonlinear wave patterns either on the short-time scale or under the stationary conditions. A good survey of these results can be found in the book [180].

Stokes was the first who used the perturbation (asymptotic) technique to solve the problem of steady surface waves under the assumption that the free-surface elevation and velocity potential can be represented in the form of Fourier series

$$x = \xi(y) = a_1 \cos y + a_2 \cos 2y + \ \ldots \tag{3}$$

and

$$\varphi(x, y) = cy + b_1 \operatorname{ch}(x + x_0) \sin y + b_2 \operatorname{ch} 2(x + x_0) \sin 2y \ \ldots, \tag{4}$$

respectively. The coefficients $a_n$, $b_n$, and $c$ are functions of the depth of liquid. It is assumed that these coefficients can be expanded in power series in the parameter $a_1$ and the leading orders of the quantities $a_n$ and $b_n$ are $O\left(a_1^n\right)$. Substituting (3) and (4) in the original free-surface problem, we get an ordered system of equations from which we can successively find the coefficients of power series. Stokes solved the problem for the case of deep water with an accuracy to within $O\left(a_1^5\right)$ and obtained similar results for the case of finite-depth water to within $O\left(a_1^3\right)$. At present, the problem is solved by the Stokes method with an accuracy to within $O\left(a_1^{10}\right)$. Stokes also showed that the wave with the maximum height must have an inscribed angle of $120^o$ at the crest. However, the results of contemporary investigations demonstrate that this is not true. Despite the commonly accepted ideas, it turns out that the integral quantities, such as the wave velocity, momentum, potential energy, and kinetic energy attain their maximum values prior to the time of attainment of the maximum wave height.

At present, unlike the case of steady waves, there are no rigorous proofs of the existence theorems for the problem on standing waves with finite amplitudes. In generalizing the Stokes idea by his method, J. Rayleigh [194] solved the problem of plane standing waves for an infinite depth of liquid with the help of the Taylor expansion up to the third order. W. Penny and A. Price [181] obtained a fifth-order asymptotic solution of the problem. They showed that the maximum steepness of a wave is equal to 0.22. They also assumed that the inscribed angle at the crest of the highest wave is equal to $90^o$. The experiments carried out by G. Taylor [218] and D. Edge and G. Walters [46] confirmed that the angle at the crest is close to $90^o$. In [217], this method was simplified and developed for the case of a finite depth. L. Schwartz and A. Whitney [197] recently obtained the 25th order asymptotic solution by using the technique of time-dependent conformal mapping. It was discovered that the very steep waves have several inflection points near the crest as in the case of steady waves. For this solution, the highest steepness of the waves was approximately equal to 0.208 with an inscribed angle of about $90^o$. The free surface is never planar and has no nodal points. The frequency of free oscillations decreases as their amplitude increases, whereas the frequency of shallow standing waves increases with amplitude. The critical value of the ratio of the mean depth of liquid to the wavelength for which the soft-to-hard-spring changes is equal to 0.17 [217]. This fact was experimentally confirmed in [62].

The ideas of the Stokes method were also used for the solution of the problems of three-dimensional progressive and three-dimensional standing waves [198]. The main results in this field were fairly completely reviewed in [227].

It is worth noting that serious mathematical difficulties encountered in the solution of the problems of exact theory of waves in water motivated the appearance of various approximate nonlinear theories. In this connection, we should especially mention numerous versions of the asymptotic approximate equations for shallow and deep-water waves.

As the most frequently used equations/models, we can indicate the models proposed by J. Boussinesq [26], D. Korteweg and G. de Vries [90], K. Friedrichs [61], and V. E. Zakharov [228]. The contemporary soliton theory is also largely based on the analysis of the classical solutions obtained within the framework of approximate models of the water wave theory. The problem of substantiation of approximate water waves theories in two- and three-dimensional statements was considered by L.V. Ovsyannikov, N.I. Makarov, V.I. Nalimov, and other researchers. A survey of these results can be found in [180].

The practically important problems connected with coupled "rigid body–contained liquid" motions are mathematically and mechanically complicated. For the same hydrodynamic model of inviscid potential flows with prescribed motions of the body, sloshing is normally considered in the tank-fixed inertialess coordinate system $Oxyz$ such that $O$ coincides with the geometric center of the mean flat (unperturbed) free surface $\Sigma_0$ ($x = 0$) and the $Ox$-axis is directed upward when both the tank and the liquid are in the state of rest. To describe the motions of the tank with six degrees of freedom, it is customary to introduce an absolute (inertial) coordinate system $O'x'y'z'$. This can, e.g., be the Earth-fixed coordinate system or a reference system for a ship tank moving with a steady speed (of the ship). The notation is schematically illustrated in Fig. 1.

Physically, the *six degrees of freedom*, $\eta_i(t)$, $i = 1, \ldots, 6$, of motions of the rigid body ($Oxyz$-system) relative to the $O'x'y'z'$ system are associated with

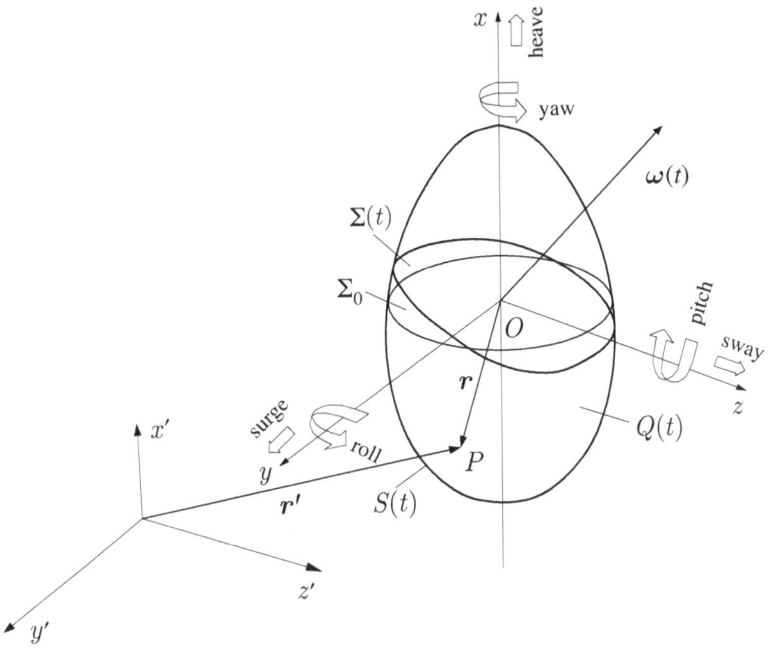


**Figure 1.** Schematic diagram of a moving rigid tank partly filled with liquid. The free surface $\Sigma(t)$ confines the liquid domain $Q(t)$ and the wetted surface of the tank $S(t)$. In the hydrostatic state, the unperturbed free surface $\Sigma_0$ is perpendicular to the $Ox$-axis of a tank-fixed inertialess coordinate system $Oxyz$ such that $O$ coincides with the geometric center of $\Sigma_0$. The motions of the rigid tank are realized relative to an inertial coordinate system $O'x'y'z'$ and associated with the instant translational motions, velocity of the origin $O$, $\boldsymbol{v}_O(t)$, and the instant angular velocity $\boldsymbol{\omega}(t)$. These vectors are normally expressed via the projections of the $Oxyz$ unit vectors: $\boldsymbol{e}_1(t)$, $\boldsymbol{e}_2(t)$, and $\boldsymbol{e}_3(t)$. The six degrees of freedom of the body are often called, sway, surge, and heave (translational motions) and roll, pitch, and yaw (angular motions). For low-amplitude motions, the angular degrees of freedom are the same as the Euler angles.

the surge, sway, heave, roll, pitch, and yaw generalized coordinates, as shown in Fig. 1. On the other hand, it can be demonstrated that the mathematical description of the liquid sloshing dynamics requires the *absolute* translational velocity $\boldsymbol{v}_0(t)$ and the instant angular velocity $\boldsymbol{\omega}(t)$. The vectors $\boldsymbol{v}_O(t)$ and $\boldsymbol{\omega}(t)$ are expressed via the projections onto the unit vectors $\boldsymbol{e}_i$, $i = 1, 2, 3$, of the $Oxyz$-coordinate system, i.e.,

$$\begin{aligned} \boldsymbol{v}_O(t) &= v_{O1}(t)\boldsymbol{e}_1(t) + v_{O2}(t)\boldsymbol{e}_2(t) + v_{O3}(t)\boldsymbol{e}_3(t), \\ \boldsymbol{\omega}(t) &= \omega_1(t)\boldsymbol{e}_1(t) + \omega_2(t)\boldsymbol{e}_2(t) + \omega_3(t)\boldsymbol{e}_3(t). \end{aligned} \tag{5}$$

One must explicitly know the relations connecting the generalized coordinates $\boldsymbol{\eta} = \{\eta_i,\ i = 1, \dots, 6\}$ and $\boldsymbol{v}_0(t)$, $\boldsymbol{\omega}(t)$. These relations are discussed in Sections 2.3 of [56]. If the oscillatory motions of the rigid body have small amplitudes, then the corresponding relations can be linearized as follows:

$$v_{O1} = \dot{\eta}_1; \quad v_{O2} = \dot{\eta}_2; \quad v_{O3} = \dot{\eta}_3; \quad \omega_1 = \dot{\eta}_4; \quad \omega_2 = \dot{\eta}_5; \quad \omega_3 = \dot{\eta}_6. \tag{6}$$

The gravity potential can be represented in the form

$$U_{\mathrm{g}}(x, y, z, t) = U_{\mathrm{g}}(x, y, z; \eta_i(t)) = -\mathbf{g} \cdot \boldsymbol{r}', \quad \boldsymbol{r}' = \boldsymbol{r}'_0 + \boldsymbol{r},$$

where $\boldsymbol{r}'$ is the radius vector of a point of the body–liquid system with respect to $O'$, $\boldsymbol{r}'_0$ is the radius vector of $O$ with respect to $O'$, $\boldsymbol{r}$ is the radius vector with respect to $O$, and $\mathbf{g}$ is the gravitational acceleration. In Fig. 1, we show the liquid domain $Q(t)$ bounded by the free surface $\Sigma(t)$ and the wetted surface of the tank $S(t)$. If the tank and liquid are in the state of rest, then the flat unperturbed liquid surface $\Sigma_0$ should cut the mean liquid domain $Q_0$ and the mean wetted tank surface $S_0$.

Solving the sloshing problem is connected with finding the *absolute velocity potential* $\Phi(x, y, z, t)$ and the free surface $\Sigma(t)$ implicitly given by the equation $\zeta(x, y, z, t) = 0$. We also determine the absolute (with respect to the $O'x'y'z'$ system) velocity field in the *projections onto the $Oxyz$-axes* but $\Phi$ is a function of $x, y, z$, and $t$.

The pressure field in $Q(t)$ is computed by using the Bernoulli equation rewritten in the inertialess $Oxyz$-system as follows:

$$p - p_0 = -\rho\left(\frac{\partial\Phi}{\partial t} + \tfrac{1}{2}(\nabla\Phi)^2 - \nabla\Phi \cdot [\boldsymbol{v}_0(t) + \boldsymbol{\omega}_0 \times \boldsymbol{r}] + U_g\right), \tag{7}$$

where $p_0$ is the ullage (atmospheric) pressure. By using (7), we compute the internal hydrodynamic loads including the resulting hydrodynamic force $\boldsymbol{P}(t)$ and the moment $\boldsymbol{N}_O(t)$ (relative to $O$).

Under the assumption of *prescribed* tank motions, we conclude that the six generalized coordinates $\eta_i(t)$, $i = 1, \ldots, 6$, are known and, hence, the two vectors $\boldsymbol{v}_0(t) = (v_{O1}(t), v_{O2}(t), v_{O3}(t))$ and $\boldsymbol{\omega}(t) = (\omega_1(t), \omega_2(t), \omega_3(t))$ are the known input parameters. The problem is normally reduced to finding the unknown functions $\Phi$ and $\zeta$ and, in addition, the hydrodynamic loads, including the resulting hydrodynamic force $\boldsymbol{P}(t)$ and the moment $\boldsymbol{N}_O(t)$. The velocity field and the hydrodynamic loads are functions of the six input generalized coordinates and initial perturbations of the contained liquid.

*Coupling* the tank and liquid motions means that the six generalized coordinates $\boldsymbol{\eta} = \{\eta_i(t),\ i = 1, \ldots, 6\}$ are unknown. In this case, one can symbolically represent the dynamic equations for the rigid tank in the form

$$M(\boldsymbol{\eta})\ddot{\boldsymbol{\eta}} + \boldsymbol{S}(\boldsymbol{\eta}, \dot{\boldsymbol{\eta}}) = \boldsymbol{F}_h(\boldsymbol{\eta}) + \boldsymbol{F}_e(t), \tag{8}$$

where $\boldsymbol{F}_h(t)$ is the hydrodynamic force and moment caused by sloshing, $\boldsymbol{F}_e(t)$ is the external force and moment (relative to $O$) applied to the rigid tank and ordered as follows: $\boldsymbol{F}(t) = (F_1(t), F_2(t), F_3(t), M_{O1}(t), M_{O2}(t), M_{O3}(t))$, and the explicit form of the mass matrix $M(\boldsymbol{\eta})$ and the function $\boldsymbol{S}$ depend on the shape of the body. The dynamic equations (8) contain the resulting hydrodynamic force and moment which are, in turn, functions of the six generalized coordinates $\eta_i(t)$, $i = 1, \ldots, 6$. As long as we have an analytically-oriented method which enables us to explicitly find $\boldsymbol{F}_h(t)$ as functions of the generalized coordinates, we can immediately deduce the dynamic equations (8) in the analytic form and, thereby, effectively study the coupled body–liquid dynamics.

Finding $\Phi$ and $\zeta$ as functions of the six generalized coordinates $\eta_i$, $i = 1, \ldots, 6$, yields the solution of the following free-surface problem [119, 56]:

$$\nabla^2\Phi = 0 \quad \text{in } Q(t); \quad \frac{\partial \Phi}{\partial \nu} = \boldsymbol{v}_O \cdot \boldsymbol{\nu} + \boldsymbol{\omega} \cdot [\boldsymbol{r} \times \boldsymbol{\nu}] \quad \text{on } S(t),$$

$$\frac{\partial \Phi}{\partial \nu} = \boldsymbol{v}_O \cdot \boldsymbol{\nu} + \boldsymbol{\omega} \cdot [\boldsymbol{r} \times \boldsymbol{\nu}] + \frac{\partial \zeta / \partial t}{|\nabla \zeta|} \quad \text{on } \Sigma(t), \quad \int\limits_{Q(t)} \mathrm{d}Q = \text{const}, \tag{9a}$$

$$\frac{\partial \Phi}{\partial t} + \tfrac{1}{2}(\nabla \Phi)^2 - \nabla\Phi \cdot (\boldsymbol{v}_O + \boldsymbol{\omega} \times \boldsymbol{r}) + U_g = 0 \quad \text{on } \Sigma(t), \tag{9b}$$

where $\boldsymbol{\nu}$ is the outer normal.

The major part of analytic approaches to the coupled body–liquid dynamics is based on the ideas of replacement of the hybrid mechanical systems by their *mathematical models* in the form of infinite-dimensional systems of ordinary differential equations. As an example of approach of this kind, we can mention the so-called multimodal method developed in the present book. Based on the free-surface problem or its Lagrange variational formulation, the multimodal method enables one to explicitly select the hydrodynamic generalized coordinates and get a system of ordinary differential equations which plays the role of a system of approximate dynamic equations for the hydromechanical subsystem.

The etymology of the word *"multimodal"* in sloshing problems is connected, most likely, with the notion of "nonlinear *multimodal* analysis" appearing in the title of the paper [53]. However, the multimodal method was originated not in 2000 but more than 40 years before, in the 1950-60s, when the aircraft, spacecraft, and marine applications were regarded as a great challenge for mathematicians and engineers involved in the investigation of the dynamics of vehicles carrying partly filled containers. The author was one of the founders of the proposed method.

In the 1950-60s, all experimental, computational, and theoretical aspects of this circle of problems were of great interest. The enthusiastic atmosphere of this period is well described in the memoir article by H. Abramson [4] who was a leader of the corresponding NASA research programs. The scientific results of these investigations were partly presented in the fundamental books and monographs published mainly in the USA [3, 60, 80, 208, 216, 215] and in the Soviet Union [1, 59, 86, 149, 150, 162, 163, 164, 191].

The model tests made it possible to establish the input parameters for which the free-surface nonlinearity is significant. The experiments performed in the USA are well reviewed in the NASA Report [3]. At the same time, the systematic experiments of Soviet scientists were carried out in Moscow, Kiev, Dnipropetrovsk, and Tomsk but their results are less known from the available (open) literature. In Fig. 2, we present some unpublished photos made in the 1960s by the Kiev research group (the author was one of the members of this group). These photos reveal the increasing importance of the free-surface nonlinearity, which becomes visible due to the presence of steep-wave patterns, breaking, overturning, bubbling, and free-surface fragmentations.

Summarizing the main results obtained in the 1950-60s, we especially mention, among the above-mentioned books and reports, the papers [178, 170, 189] and the books [59, 149, 164] in which the concept of *linear multimodal method* was originated and developed for an ideal incompressible contained liquid with irrotational flow. According to this concept, the coupled "rigid body–contained liquid" dynamics is interpreted as a conservative mechanical system with infinitely many degrees of freedom.

The linear multimodal method gives an infinite-dimensional system of linear ordinary differential equations with respect to the indicated six generalized coordinates governing the low-amplitude motions of the rigid body and an infinite set of *hydrodynamic* generalized coordinates responsible for the linearly perturbed (relative to the hydrostatic shape of the liquid) natural sloshing modes. The indicated hydrodynamic-type subsystem is called the *linear modal system.* In the derivations, it is necessary to know the natural sloshing modes and frequencies and the so-called Stokes–Zhukovsky potentials given in the explicit analytic form. The Stokes–Zhukovsky potentials were introduced by Nikolay Zhukovsky, "father of the Russian aviation," in 1885 [83]. He studied a body moving in the space and containing a cavity completely filled with an ideal incompressible liquid. He formulated the corresponding Neumann boundary-value problems. Finding the Stokes–Zhukovsky potentials in the case of linear sloshing is reduced to the solution of these Neumann problems in the mean (hydrostatic) liquid domain.

The spectral boundary-value problem for the natural sloshing modes and frequencies can also be formulated in the mean liquid domain so that the spectral parameter $\varkappa$ appears in the mean free-surface boundary condition and specifies the natural sloshing frequencies as $\sigma = \sqrt{\varkappa g}$, where g is the gravitational acceleration.

The natural sloshing modes determine the standing surface wave patterns. The standing waves were first described for an upright circular basin by M. Ostrogradskii (his manuscript [179] was submitted to the Paris Academy of Sciences in 1826). For some other basins, these standing waves were also studied by Poisson and Rayleigh. However, a rigorous mathematical theory of the spectral boundary problems for natural sloshing modes and frequencies was constructed

**Figure 2.** A series of photos made in the 1960s by the Kiev research group illustrating the set of instant free-surface patterns caused by different types of resonant excitations of the tank and revealing the effect of free-surface nonlinearity. The photos (a)-(d) illustrate a weakly nonlinear sloshing when the lowest [degenerating] modes [(a) and (b)] are predominant. The second and third antisymmetric modes are nonlinearly excited in the cases (c) and (d) but a stronger excitation of these modes is shown in (e), (f), and (g). The nonlinearity leads to the formation of very steep wave profiles even in the case where the contained liquid is almost in the state of rest (h). The pictures (i-l) show the breaking waves and the phenomena of overturning and bubbling (k), as well as the other strongly nonlinear phenomena leading to the fragmentation of the free surface.

only in the 1960s. The spectral theorems reveal the pure positive point spectrum formed by the eigenvalues with a single limiting point at infinity. This differs from the case of water-wave theory in which the continuous spectrum exists. As representative mathematical publications, we can especially mention [44] and Ch. VI in [59].

S. Krein [92] generalized the spectral theorems for a viscous incompressible contained liquid. At the same time, N. Kopachevskii proved the corresponding theorems for a capillary liquid (see Part II in [167] and [88]).

Coupling of the linear modal equations with the dynamic equations of a rigid container turns into a mathematically simple procedure if we use the so-called linearized Lukovsky formulas to express the resulting hydrodynamic force and moment via the introduced hydrodynamic generalized coordinates. The hydrodynamic coefficients in the linear modal system and in the Lukovsky formulas are integrals of the Stokes–Zhukovsky potentials and the natural sloshing modes. After adding the corresponding initial conditions responsible for the initial position and velocity of the contained liquid and the body, we apply a Runge–Kutta-type simulation of the coupled linear ordinary differential equations in order to describe the linear transient "rigid body–contained liquid" dynamics. Under harmonic loads applied to the rigid tank, we can also analytically find the periodic (steady-state) wave solution.

The canonical description of the linear multimodal method is presented in [59, 124]. Faltinsen & Timokha [56], Chap. 5, gave a contemporary treatment of the method as of the 2000-10s. Since the possibility of getting the hydrodynamic coefficients in the linear modal equations depends solely on whether the natural sloshing modes and the Stokes–Zhukovsky potentials are known, the major part of the results obtained in the 1950-80s and dealing with the linear multimodal method were devoted to the construction of approximate solutions of the spectral boundary problem for natural sloshing modes and the Neumann boundary-value problem for the Stokes–Zhukovsky potentials. The efficient ideas on the possibility of construction of the analytic approximate solutions are collected in [124]. Starting from the 1990s, the numerical solutions of these spectral and Neumann boundary-value problems are found by using various open and/or commercial solvers based on, e.g., the finite-element method. By using these and Runge–Kutta solvers, one can state that the multimodal method can always give a solution of the linear coupled "rigid tank–contained liquid" problem.

The readers interested in the linear modal theory and, especially, in the historical context, as well as in the other hydrodynamic aspects of the linear sloshing problem in the coupled dynamics of rigid bodies containing liquids are referred to books by N. N. Moiseev & V. V. Rumyantsev, G. N. Mikishev & B. I. Rabinovich, K. S. Kolesnikov, I. A. Lukovsky, M. Ya. Barnyak & A. N. Komarenko, H. N. Abramson, and many other authors [24, 86, 124, 147, 149, 150, 162, 164, 163, 191, 59, 3, 37]. These books and the surveys [134, 110, 2, 37, 82] contain a comprehensive bibliography related to various aspects of the linear sloshing theory. However, numerous phenomena observed in practice cannot be explained from the positions of this theory even qualitatively. The linear theory contradicts the experimental data if the amplitude of liquid begins to increase in the process of motion of carrying bodies.

The practical necessity of the investigation of sloshing with finite (not small) elevations is connected with the analysis of the limits of applicability of the linear sloshing theory. It was experimentally [5, 6, 21, 147, 148, 149, 58, 84, 146, 193, 196, 215, 218, 223] and theoretically established that the linear sloshing theory satisfactorily describes the interaction of a body with liquid when the wave elevations do not exceed 0.15 of the linear size of the mean free surface (e.g., the radius of the cylindrical container). In the experiments, the researchers revealed the set of nonlinear sloshing phenomena. These are, first of all, the dependence of the sloshing frequency on the wave amplitude; the boundedness of the liquid sloshing amplitudes under the resonance conditions; the asymmetry of wave patterns for standing waves; the mobility of nodal lines; complicated spatial wave motions caused by harmonic longitudinal excitations of the container, and other features. Some interesting nonlinear effects also appear as a result of the interaction between the elastic thin-walled shell and the liquid with free surface [2, 36, 40]. It is also established that all nonlinear processes are explainable within the framework of the nonlinear theory of standing waves in an ideal incompressible liquid in the case of vortex-free flows. The construction of this theory is connected with serious fundamental and technical difficulties.

The most important directions in the nonlinear sloshing analysis were originated in the 1950-60s. Under certain conditions, these directions can be associated with the original works by N.N. Moiseev [161], G.S. Narimanov [172], and L. Perko [165, 182].

Moiseev [161] constructed an asymptotic steady-state wave (periodic) solution of the nonlinear free-surface problem for an upright tank in the process of prescribed horizontal and/or angular harmonic motions with a forcing frequency $\sigma$ close to the lowest natural sloshing frequency $\sigma_1$. He assumed that the depth of an ideal incompressible liquid with irrotational flows is finite. It was proved that if the dimensionless forcing amplitude is a small parameter of the order $\epsilon \ll 1$, then the amplification of primary excited mode(s) is characterized by the order $\epsilon^{1/3}$ and the matching asymptotics (the so-called Moiseev detuning) is $|\sigma^2 - \sigma_1^2|/\sigma_1^2 = O(\epsilon^{2/3})$.

Moiseev also implicitly assumed that the so-called *secondary resonances* are absent. The Moiseev asymptotic steady-state solutions were analytically constructed for two-dimensional rectangular tanks [52, 176] and some other [207, 144, 80, 14] shapes of the tank. They were based on the solutions of auxiliary boundary-value problems found in terms of the natural sloshing modes known for the inertialess (or analytically approximate) forms analytically expandable over the mean free surface (together with the higher-order derivatives) for these shapes of the tank. The construction of these solutions is connected with cumbersome transformations.

Note that Moiseev focused his attention solely on the resonant steady-state sloshing caused by the prescribed harmonic excitations of the tank. At the same time, G.S. Narimanov [172], bearing in mind the description of both steady-state and transient sloshing, proposed a perturbation technique for getting *weakly nonlinear* analogs of the linear modal equations. The weakly nonlinear modal equations also facilitate the analysis of the coupled "tank-liquid" dynamics. To deduce the weakly nonlinear modal equations, Narimanov [172] *postulated* a set of asymptotic relations between the hydrodynamic generalized coordinates. These coordinates are introduced in the same way as in the linear case. Note that Narimanov did not know Moiseev's results because they were published one year later. Nevertheless the applied asymptotic relations between the hydrodynamic generalized coordinates coincide, in fact, with the relations following from Moiseev's analysis if the Moiseev periodic solution is represented in terms of the Fourier solutions for the natural sloshing modes. The Narimanov perturbation technique is outlined in Chapter 7. The Narimanov-type multimodal method was first developed for an upright circular cylindrical tank but V. Stolbetsov [212, 211, 210, 213] and the author [112, 173] extended it for the other shapes of the tanks. At present, the Narimanov-type modal systems are derived for upright tanks with circular, annular, and rectangular cross sections, conical and spherical tanks, and upright circular cylindrical tanks with rigid annular baffles [65, 66, 41, 111, 112, 128, 119, 173, 211, 210, 212, 72].

It is worth noting that the original Narimanov expressions contain some algebraic errors. The corrected expressions can be found in the author's works published after 1975 (see, e.g., [112, 119, 173]). Unfortunately, as indicated in the Moiseev case, the procedure of construction of the Narimanov-type weakly nonlinear modal equations leads to difficult and cumbersome transformations; moreover, these difficulties drastically increase if we introduce a large set of hydrodynamic generalized coordinates. As a result, the dimensions of all existing Narimanov-type modal systems are low and these systems connect at most 2–5 generalized coordinates.

The phenomenon of nonlinear liquid sloshing was studied in [161, 163], where the algorithm of solution was based on the Lyapunov–Poincaré method. This method was realized, in particular, for the problem of free sloshing between two vertical walls [163] and for the problem of subharmonic sloshing in a container moving in the direction perpendicular to the unperturbed free surface [35].

Due to the absence of the required computer facilities and suitable numerical methods, the data of engineering computations based on the space-and-time discretization of the free-surface sloshing problem are poorly presented in the literature of the 1960-70s. They became common only in the 1990s. As a unique exception, we can mention the so-called Perko method [165, 182], which can be regarded both as a numerical version of the multimodal method and as a Computational Fluid Dynamics solver. The Perko method uses, on the one hand, the Galerkin projective scheme and, on the other hand, the natural sloshing modes in the Fourier-type representation of the velocity potential. The Perko method was forgotten until the 2000s when it was combined (see [94, 200]) with the Bateman–Luke variational formalism. The Perko method serves as an origin of some other numerical and semiinertialess techniques.

Lukovsky and J. W. Miles [151, 119, 113] used the Bateman–Luke variational formalism [10, 107] to derive the fully nonlinear modal equations. Moreover, the Narimanov–Moiseev asymptotic relations were used to get their approximate weakly nonlinear forms. Miles [152, 153] generalized the Moiseev results to study the amplitude modulation of weakly nonlinear nearly steady-state sloshing (almost periodic) solutions for the case of harmonically excited tanks (also by using the Bateman-Luke variational formalism). Finally, O. S. Limarchenko [98, 100, 99, 101] proposed a weakly nonlinear version of the Perko method and combined it with the classical Lagrange variational principle and the Galerkin projective scheme to solve the kinematic part of the sloshing problem. These are the origins of different versions of the multimodal method.

In 1976, Lukovsky and Miles independently used the Bateman–Luke variational formalism to get a fully nonlinear modal system. In their original papers [151, 119], the authors studied prescribed harmonic translational motions of the tank but later [113, 53] generalized these results for arbitrary prescribed motions of the tank and proposed the so-called technique of nonconformal mappings [112, 135, 57] aimed at getting fully nonlinear modal equations for tanks with nonvertical walls. The so-called fully nonlinear Lukovsky formulas were also derived for the hydrodynamic force and moment in [119] (Chap. 7 of [56] gives an alternative procedure). Furthermore, it was shown how to use the Bateman–Luke formalism to deduce the dynamic equations for the coupled "rigid tank–contained liquid" mechanical system [119].

Since both kinematic and dynamic relations of the free-surface problem naturally follow from the Bateman–Luke variational formulation, these modal equations correspond to the kinematic and dynamic subsystems obtained as the first-order infinite-dimensional systems of ordinary differential equations coupling the hydrodynamic-type generalized coordinates and velocities. In order to get an approximate finite-dimensional system of the second-order modal equations, as in the Narimanov case, Lukovsky and some other authors used the Narimanov–Moiseev asymptotics for the generalized coordinates and velocities. They deduced a set of weakly nonlinear modal systems and applied these systems to the inertialess investigation of the steady-state resonant waves and transients.

In the Lukovsky–Miles method, it is assumed that we know analytic expressions for the approximate natural sloshing modes and the Stokes–Zhukovsky potentials are analytically defined over the mean free surface. This is a serious restriction of the method preventing its generalization to tanks of arbitrary shapes. Thus, it becomes clear why the extensive study of fully nonlinear modal equations and their finite-dimensional weakly nonlinear versions (performed by using the rich collection of inertialess methods leading to the approximate natural sloshing modes which were developed as early as in the 1970-90s) was originated only in 2000.

Bearing in mind the generalization of Moiseev's results on resonant steady-state wave modes, Miles [152, 153] deduced the so-called *Miles equations* governing the slow-time variations of the predominant amplitudes of almost periodic sloshing caused by low-amplitude horizontal harmonic excitations of the upright circular cylindrical tank with a forcing frequency close to the lowest natural sloshing frequency. He adopted the Moiseev asymptotic ordering, the Moiseev detuning, and the technique of multiple time scales. The separation of the fast and slow time scales was performed directly in the Bateman–Luke action. According to the Narimanov–Moiseev asymptotics, there are four (or less) independent slowly varying amplitudes for upright cylindrical tanks. They are of the $O(\epsilon^{1/3})$-order ($\epsilon$ is the dimensionless forcing amplitude). The Miles equations were later derived for upright tanks with rectangular cross sections.

The application of Miles equations is a fairly popular approach in the applied mathematical investigations aimed at the classification of nearly steady-state sloshing, detection of periodic orbits, and clarifying chaos in the hydrodynamic systems. Both horizontal and vertical (Faraday waves) harmonic excitations were studied in [79, 63, 64, 153, 155, 154, 78, 156, 157, 158]. Krasnopolskaya and Shvets [91, 203] extended the Miles technique to the case of "rigid tank–contained liquid" mechanical systems with limited power supply of forcing.

In the 1970s, the Perko method was used quite rarely. As an exception, we can mention the papers [45, 166] in which the Galerkin-type numerical schemes were quite similar to the schemes proposed by Perko. However, this method served, most probably, as the key idea for the weakly nonlinear "computational" versions of the multimodal method [23, 43, 98, 99]. They were created for the simulation of weakly nonlinear transient sloshing and the related tank-liquid dynamics.

The numerical multimodal methods are well exemplified by Limarchenko [98, 99, 100, 101, 102] who combined the Perko method with the classical Lagrange variational formulation. This variational formulation leads solely to the dynamic boundary condition (pressure balance on the free surface) and the kinematic relations should be regarded as a constraint. Adopting a modal-type solution, Limarchenko solved the kinematic constraint by the Galerkin projective method (similar to the Perko method) but the dynamic modal equations were obtained in the variational formulation as in the works by Lukovsky and Miles. Furthermore, the kinematic and dynamic equations are recombined to get a finite-dimensional system of second-order ordinary differential equations. These are (i) weakly nonlinear equations containing only second- and third-order polynomial nonlinearities; (ii) based on the *a priori* postulated predominant and higher-order generalized coordinates; (iii) full of the zero-order hydrodynamic coefficients which can hardly be obtained in the inertialess way and, therefore, (iv) inapplicable to the inertialess studies but used for simulating transients by the Runge-Kutta solver. In other words, the Perko type weakly nonlinear modal equations form a specific computer package, which is not directly available for the interested readers.

The Computational Fluid Dynamics (CFD) simulations of nonlinear sloshing actually started only in the 1970-80s. The primary focus was made on the finite-difference (maker-and-cell, etc.) and finite-element methods [169, 7, 48, 204, 214, 192, 177]. Parallel with publishing open-source algorithms, the commercial FLOW-3D was founded to become one of the leading Navier–Stokes commercial solvers in the 1990-2000s. A collection of FLOW-3D simulations was presented by F. Solaas [206] who discussed its advantages and drawbacks.

In 2000, the papers [94, 200] gave a new life to the Perko method by combining it with the Bateman-Luke variational formalism. In fact, the authors of these papers used the truncated fully nonlinear Lukovsky–Miles modal system for time-step integrations with appropriate initial conditions. A drawback of this approach is that this system is unrealistically stiff and, hence, special artificial damping terms should be introduced to suppress the formation of parasitic higher harmonics. This problem was also discussed in [45].

The 1990-2010s opened a new computational era in simulating the nonlinear sloshing of viscous liquids. The Volume of Fluid (VoF), Smoothed Partitions Hydromechanics (SPH), and their modifications made it possible, by using parallel computations, to perform a fairly accurate and efficient analysis of the transient nonlinear sloshing and coupled liquid-tank dynamics. Interested readers are referred to the paper [27], which outlines the state-of-the-art of the 1990s.

The recent results of the numerical analysis of the sloshing problem are presented in [195, 75, 222]. As advantages of the contemporary CFD, we can mention the facts that it is normally based on viscous and fully nonlinear statement and enables one to model specific free-surface phenomena associated with (i) the free-surface fragmentation; (ii) wave breaking; (iii) overturning (typically at the walls); (iv) roof and wall impacts, flip-through.

These phenomena are partly illustrated in Fig. 2. As compared with these features, the existing modal systems look rather poor. They (i) are weakly nonlinear and based on ideal potential liquid motions; (ii) unable to describe the above-mentioned specific free-surface phenomena; (iii) require the extensive experimental validation due to numerous physical and mathematical assumptions made in getting these systems. This most likely explains why the multimodal method was forgotten in the 1990s when the main emphasis was made on the construction of efficient CFD solvers. One can say that the multimodal method lost the competition with the CFD in the engineering computations.

A new era of the nonlinear multimodal method, now as an inertialess tool, was originated in 2000 [53]. The weakly nonlinear modal equations obtained by the multimodal method play the same role as weakly nonlinear water-wave theories proposed by Korteweg–de Vries, Boussinesq, and many other researchers. The method proves to be efficient in getting fundamental theoretical results on the dynamics and stability of liquid sloshing, classification of steady-state wave modes, and identifying chaos, as well as in the parametric studies suggesting, in view of the CFD, exhaustive testing of numerous initial scenarios and input parameters. In addition, the multimodal method is of great importance for the investigation of novel sloshing problems for which the CFD remains less applicable.

The present book focuses on the nonlinear multimodal method and the associated weakly nonlinear modal systems. We mainly analyze the case of so-called heavy liquids when the surface tension and viscosity are neglected. Interested readers are referred to [18, 19] and to the fundamental monograph [88] which gives the state-of-the-art of the problem of sloshing affected by surface tension.

As already indicated, the contemporary CFD methods prove to be quite efficient in the investigation of the problems of coupled rigid-body–liquid dynamics in the case of a viscous liquid. However, the literature on the inertialess studies of viscous liquid flows in mobile rigid tanks is not empty. Parallel with the fundamental mathematical results in this field obtained by S. Krein and N. Kopachevsky [92, 167, 89, 8] and F. Chernous'ko [32, 33, 31], there are special research topics in which the coupled rigid-body–viscous-liquid dynamics is of especial interest. As an example, we can mention the planetary dynamics [47] when it comes, e.g., to the phenomenon of near-separatrix deceleration (the so-called call lingering effect). This phenomenon was first mentioned in the book by Chernous'ko [31] who studied the free precession of a tank filled with viscous liquid and proved that, despite the apparent trap, the separatrix is crossed within a finite period of time.

Finally, the book does not consider parametric (Faraday) waves in the tanks caused by their vertical excitations, i.e., when the excitation is perpendicular to the mean free surface. Numerous works were devoted to the nonlinear Faraday waves; see, e.g., [97, 201, 12, 17, 35, 40, 52, 53, 55, 60, 71, 72, 74, 105, 169, 205, 221, 223, 224, 226]. In these works, one can find the theoretical description of some interesting nonlinear phenomena experimentally discovered by Faraday [58], Matthiessen [146], and Rayleigh [194] as early as in the 19th century. In particular, these are the subharmonic waves and the phenomena of harmonic and superharmonic sloshing which can be predicted in the linear approximation [15] but their quantitative description requires nonlinear sloshing models.

Chapter 1

# Governing equations and boundary conditions in the dynamics of a bounded volume of liquid

## 1.1 Conservation laws. Governing equations

The motion of continua is most frequently described by using two approaches developed by Lagrange and Euler. The Lagrange approach is based on the use of the basic theorems of mechanics for the direct description of every individual liquid particle by tracking its position with the help of a function of time, whereas the Euler equations deal with the distribution of the velocity field in the analyzed liquid domain. In what follows, we consider, following Euler, the velocity field with components $v_x$, $v_y$, and $v_z$ as functions of the space variables and time. Then the trajectories of individual particles can be obtained by the integration of the system of ordinary differential equations $\dot{x} = v_x$, $\dot{y} = v_y$, $\dot{z} = v_z$, where the dots over $x$, $y$, and $z$ denote the operation differentiation with respect to the time preformed along the motion of an individual particle.

We now present some auxiliary formulas frequently used in what follows to deduce the equations of motion and for some other purposes. It is important to be able to perform the mathematical operation of evaluation of the time derivatives of various hydrodynamical quantities related to a given liquid particle in the course of its motion.

Let $A(x, y, z, t)$ be a hydrodynamic quantity related to a particle moving along the trajectory $x = x(t),\ y = y(t),\ z = z(t)$. By definition, the quantity

$$\dot{\boldsymbol{r}} = \dot{x}(t)\boldsymbol{i} + \dot{y}(t)\boldsymbol{j} + \dot{z}(t)\boldsymbol{k} = v_x\boldsymbol{i} + v_y\boldsymbol{j} + v_z\boldsymbol{k}$$

is the velocity vector related to the particle. For this particle, the variables $x, y, z$ of the function $A(x, y, z, t)$ are also functions of $t$ and characterize the motion of the particle. According to the rule of differentiation of a composite function, we get

$$\frac{dA}{dt} = \frac{\partial A}{\partial t} + v_x\frac{\partial A}{\partial x} + v_y\frac{\partial A}{\partial y} + v_z\frac{\partial A}{\partial z}. \tag{1.1.1}$$

This formula for the total derivative can also be represented in the form

$$\frac{dA}{dt} = \frac{\partial A}{\partial t} + \boldsymbol{v} \cdot \nabla A. \tag{1.1.2}$$

Another frequently used formula related to the hydrodynamic quantity $A(x, y, z, t)$ is the time derivative of the expression $J = \int_\tau A d\tau$, where $\tau$ is a time-dependent liquid domain containing the same particles for the entire period of motion.

The value of the integral $J$ varies in the course of time due to the variations of the integrand $A$ and the liquid domain $\tau$. For two close times, we get the following formula:

$$\Delta J = \int_{\tau} A d\tau - \int_{\tau'} A' d\tau = \int_{\tau} (A - A') d\tau + \int_{\tau-\tau'} A' d\bar{\tau}. \tag{1.1.3}$$

Here, $d\bar{\tau}$ is an element of the domain $\tau - \tau'$. To within the terms of higher orders in $\Delta t$, we obtain

$$A - A' = \frac{\partial A}{\partial t} \Delta t, \quad d\bar{\tau} = v_\nu dS \Delta t,$$

where $S$ is the surface of $\tau$ and $v_\nu$ is the external normal velocity of $S$. Passing in (1.1.3) to the limit as $\Delta t \to 0$, we find [209]

$$\frac{dJ}{dt} = \int_{\tau} \frac{\partial A}{\partial t} d\tau + \int_{S} A v_\nu dS. \tag{1.1.4}$$

This formula is often called the Reynolds transport theorem.

We now transform the surface integral in (1.1.4) into the volume integral with the help of the Gauss theorem [202]. As a result, we get

$$\frac{dJ}{dt} = \int_{\tau} \left( \frac{\partial A}{\partial t} + \operatorname{div} A \boldsymbol{v} \right) d\tau. \tag{1.1.5}$$

In view of the definition of the total derivative (1.1.2), we can represent relation (1.1.5) in the form

$$\frac{dJ}{dt} = \int_{\tau} \left( \frac{\partial A}{\partial t} + A \operatorname{div} \boldsymbol{v} \right) d\tau. \tag{1.1.6}$$

The equations of liquid motions can be obtained by using the general conservation laws. We now separate an elementary volume $d\tau$ with elementary mass $dm$ and introduce, by definition, the density $\rho$ by the formula $dm = \rho d\tau$. Then the mass $m$ and momentum $\boldsymbol{K}$ in the liquid domain $\tau$ are given by the formulas

$$m = \int_{\tau} \rho d\tau \quad \text{and} \quad \boldsymbol{K} = \int_{\tau} \rho \boldsymbol{v} d\tau, \tag{1.1.7}$$

respectively. By using the law of conservation of the mass of liquid in the domain $\tau$, we obtain

$$\frac{d}{dt} \int_{\tau} \rho d\tau = 0. \tag{1.1.8}$$

In view of (1.1.5), equality (1.1.8) written for an arbitrary domain $\tau$ takes the form

$$\int_{\tau} \left[ \frac{\partial \rho}{\partial t} + \operatorname{div}(\rho \boldsymbol{v}) \right] d\tau = 0. \tag{1.1.9}$$

This implies that the equation

$$\frac{\partial \rho}{\partial t} + \operatorname{div}(\rho \boldsymbol{v}) = 0 \tag{1.1.10}$$

is satisfied everywhere in $\tau$. The mass conservation law in the form (1.1.10) is usually called the equation of continuity.

It is known that the theorem on changes in the momentum $\boldsymbol{K}$ can be written in the form

$$\frac{d\boldsymbol{K}}{dt} = \boldsymbol{F}^{(m)} + \boldsymbol{F}^{(S)}, \tag{1.1.11}$$

where $\boldsymbol{F}^{(m)}$ and $\boldsymbol{F}^{(S)}$ are, respectively, the principal vectors of external mass forces and surface forces. We now introduce the vector of force $\boldsymbol{F}$ related to a unit mass as

$$\boldsymbol{F} = \lim_{\tau \to 0} \frac{\boldsymbol{F}^{(m)}}{m},$$

and the stress vector $\boldsymbol{p}_\nu$ of surface forces acting at a given point of the elementary surface $\Delta S$ with normal $\boldsymbol{\nu}$ and satisfying the relation

$$\boldsymbol{p}_\nu = \lim_{\Delta S \to 0} \frac{\Delta \boldsymbol{F}_\nu^{(S)}}{\Delta S}.$$

Then the principal vectors of mass and surface forces can be represented in the form

$$\boldsymbol{F}^{(m)} = \int_{\tau} \rho \boldsymbol{F} d\tau \text{ and } \boldsymbol{F}^{(S)} = \int_{S} \boldsymbol{p}_\nu dS.$$

Equality (1.1.11) can be now rewritten as

$$\frac{d}{dt} \int_{\tau} \rho \boldsymbol{v} d\tau = \int_{\tau} \rho \boldsymbol{F} d\tau + \int_{S} \boldsymbol{p}_\nu dS \tag{1.1.12}$$

or, in view of relation (1.1.6),

$$\int_{\tau} \left[ \frac{d(\rho \boldsymbol{v})}{dt} + \rho \boldsymbol{v} \operatorname{div} \boldsymbol{v} - \rho \boldsymbol{F} \right] d\tau = \int_{S} \boldsymbol{p}_\nu dS. \tag{1.1.13}$$

Relation (1.1.13) is the integral formulation of the theorem on conservation of momentum valid for any continuum media. To obtain the appropriate governing equation in the differential form, we deduce the Cauchy formula for the stress $\boldsymbol{p}_\nu$ acting at a point of an arbitrary zone with normal $\boldsymbol{\nu}$ via the known stresses $\boldsymbol{p}_x$, $\boldsymbol{p}_y$ and $\boldsymbol{p}_z$ acting at the same point in the zones with normals directed along the coordinate axes. First, we note that the relation

$$\int_\Sigma (\boldsymbol{p}_\nu + \boldsymbol{p}_{-\nu}) dS = 0 \tag{1.1.14}$$

holds in the plane $\Sigma$ splitting any domain $\tau$ occupied by the continuum into two domains $\tau_1$ and $\tau_2$. This relation is established by applying (1.1.13) to the domains $\tau_1$, $\tau_2$, and $\tau$. Adding the first two equations (for $\tau_1$ and $\tau_2$) and subtracting the third equation, we obtain (1.1.14). In view of the arbitrariness of the domains, this yields

$$\boldsymbol{p}_\nu = -\boldsymbol{p}_{-\nu}. \tag{1.1.15}$$

The classical procedure of getting the Cauchy formula

$$\boldsymbol{p}_\nu = \boldsymbol{p}_x \cos(n, x) + \boldsymbol{p}_y \cos(n, y) + \boldsymbol{p}_z \cos(n, z) \tag{1.1.16}$$

is based on the following reasoning: We rewrite (1.1.13) for a small 3D region $\tau$ in the form of a tetrahedron with vertex at the origin cut out by the plane with normal $\boldsymbol{\nu}$ in the first octant:

$$\int_\tau \left[ \frac{d(\rho \boldsymbol{v})}{dt} + \rho \boldsymbol{v} \operatorname{div} \boldsymbol{v} - \rho \boldsymbol{F} \right] d\tau = \int_{S_\nu} \boldsymbol{p}_\nu dS + \int_{S_x} \boldsymbol{p}_{-x} dS + \int_{S_y} \boldsymbol{p}_{-y} dS + \int_{S_z} \boldsymbol{p}_{-z} dS. \tag{1.1.17}$$

We denote the height of tetrahedron dropped from the origin by $h$ and replace the integrands in (1.1.17) by their mean values. This yields

$$d\tau = \frac{1}{3} h\, dS, \quad dS_x = dS \cos(\nu, x), \quad dS_y = dS \cos(\nu, y), \quad dS_z = dS \cos(\nu, z),$$
$$\boldsymbol{p}_\nu + \boldsymbol{p}_{-x} \cos(\nu, x) + \boldsymbol{p}_{-y} \cos(\nu, y) + \boldsymbol{p}_{-z} \cos(\nu, z) = \frac{1}{3} h \left[ \frac{d(\rho \boldsymbol{v})}{dt} + \rho \boldsymbol{v} \operatorname{div} \boldsymbol{v} - \rho \boldsymbol{F} \right]. \tag{1.1.18}$$

In view of relation (1.1.15), relation (1.1.18) yields the well-known Cauchy formula (1.1.16) as $h \to 0$.

By using the Gauss theorem, the surface integral in (1.1.13) can be transformed into the integral over the domain:

$$\int_S \boldsymbol{p}_\nu dS = \int_\tau \left( \frac{\partial \boldsymbol{p}_x}{\partial x} + \frac{\partial \boldsymbol{p}_y}{\partial y} + \frac{\partial \boldsymbol{p}_z}{\partial z} \right) d\tau.$$

As a result, relation (1.1.13) takes the form

$$\int_\tau \left[ \frac{d(\rho \boldsymbol{v})}{dt} + \rho \boldsymbol{v} \operatorname{div} \boldsymbol{v} - \rho \boldsymbol{F} - \frac{\partial \boldsymbol{p}_x}{\partial x} - \frac{\partial \boldsymbol{p}_y}{\partial y} - \frac{\partial \boldsymbol{p}_z}{\partial z} \right] d\tau = 0. \tag{1.1.19}$$

We now exclude $d\rho/dt$ from (1.1.19) by using the equation of continuity. Hence, in view of the arbitrariness $\tau$, relation (1.1.19) yields the following equation for momentum in the differential form:

$$\rho \frac{d\boldsymbol{v}}{dt} = \rho \boldsymbol{F} + \frac{\partial \boldsymbol{p}_x}{\partial x} + \frac{\partial \boldsymbol{p}_y}{\partial y} + \frac{\partial \boldsymbol{p}_z}{\partial z}. \tag{1.1.20}$$

This equation is also true for any continuum.

We now decompose the vectors $\boldsymbol{p}_x = \boldsymbol{p}_1, \boldsymbol{p}_y = \boldsymbol{p}_2, \boldsymbol{p}_z = \boldsymbol{p}_3$ in the basis vectors $\boldsymbol{e}_1 = \boldsymbol{i}$, $\boldsymbol{e}_2 = \boldsymbol{j}$, $\boldsymbol{e}_3 = \boldsymbol{k}$ of the Cartesian system of coordinates, $\boldsymbol{p}_i = p_{ij}\boldsymbol{e}_j$, and introduce a matrix formed by nine elements $p_{ij}$

$$\boldsymbol{P} = \left\| \begin{matrix} p_{11} & p_{12} & p_{13} \\ p_{21} & p_{22} & p_{23} \\ p_{31} & p_{32} & p_{33} \end{matrix} \right\| = \|p_{ij}\|. \tag{1.1.21}$$

Then the vector equation (1.1.20) can be represented in terms of projections onto the axes of the Cartesian coordinate system in the form

$$\begin{aligned} \rho \frac{dv_x}{dt} &= \rho F_x + \frac{\partial p_{11}}{\partial x} + \frac{\partial p_{12}}{\partial y} + \frac{\partial p_{13}}{\partial z}, \\ \rho \frac{dv_y}{dt} &= \rho F_y + \frac{\partial p_{21}}{\partial x} + \frac{\partial p_{22}}{\partial y} + \frac{\partial p_{23}}{\partial z}, \\ \rho \frac{dv_z}{dt} &= \rho F_z + \frac{\partial p_{31}}{\partial x} + \frac{\partial p_{32}}{\partial y} + \frac{\partial p_{33}}{\partial z}, \end{aligned} \tag{1.1.22}$$

where $\boldsymbol{P}$ is called the stress tensor.

Equations (1.1.22), together with the equation of continuity (1.1.10), form a system of four equations with 13 unknown functions: density $\rho$, the components of the velocity $v_x$, $v_y$, and $v_z$, and the nine components of the internal surface stresses $p_{ij}$. This system is incomplete.

Among the other universal equations of motion of the continuum, we distinguish the equation for the angular momenta. Under the ordinary assumptions of absence of the internal angular momenta and the external bulk and surface distributed interacting couples, this equation takes the following form:

$$\frac{d}{dt} \int_\tau \boldsymbol{r} \times \rho \boldsymbol{v} d\tau = \int_\tau \boldsymbol{r} \times \rho \boldsymbol{F} d\tau + \int_S \boldsymbol{r} \times \boldsymbol{p}_\nu dS. \tag{1.1.23}$$

By using the Gauss theorem, we transform the integral over the surface into the integral over the volume:

$$\int_S \boldsymbol{r} \times \boldsymbol{p}_\nu dS = \int_\tau \left[\frac{\partial}{\partial x}(\boldsymbol{r} \times \boldsymbol{p}_x) + \frac{\partial}{\partial y}(\boldsymbol{r} \times \boldsymbol{p}_y) + \frac{\partial}{\partial z}(\boldsymbol{r} \times \boldsymbol{p}_z)\right] d\tau.$$

Hence, relation, (1.1.23) can be represented in the form

$$\int_\tau \boldsymbol{r} \times \left(\rho\frac{d\boldsymbol{v}}{dt} - \rho\boldsymbol{F} - \frac{\partial \boldsymbol{p}_x}{\partial x} - \frac{\partial \boldsymbol{p}_y}{\partial y} - \frac{\partial \boldsymbol{p}_z}{\partial z}\right) d\tau$$

$$= \int_\tau \left[\frac{\partial \boldsymbol{r}}{\partial x} \times \boldsymbol{p}_x + \frac{\partial \boldsymbol{r}}{\partial y} \times \boldsymbol{p}_y + \frac{\partial \boldsymbol{r}}{\partial z} \times \boldsymbol{p}_z\right] d\tau. \quad (1.1.24)$$

By virtue of the momentum equation (1.1.10), the left-hand side of (1.1.24) vanishes. Since

$$\frac{\partial \boldsymbol{r}}{\partial x} = \boldsymbol{i}, \ \frac{\partial \boldsymbol{r}}{\partial y} = \boldsymbol{j}, \ \frac{\partial \boldsymbol{r}}{\partial z} = \boldsymbol{k},$$

and the domain $\tau$ in relation (1.1.24) is arbitrary, we find

$$\boldsymbol{i} \times \boldsymbol{p}_x + \boldsymbol{j} \times \boldsymbol{p}_y + \boldsymbol{k} \times \boldsymbol{p}_z = 0. \quad (1.1.25)$$

This yields the equalities

$$p_{xy} = p_{yx}, \ p_{xz} = p_{zx}, \ p_{yz} = p_{zy}.$$

Thus, in the analyzed case, the momentum equation has the following important consequence: The stress tensor is symmetric. The three additional equations decrease the total number of independent components of the stress tensor to six.

For moving continua, it is important to study the energy balance. Let $\tau$ be a domain occupied by a liquid and bounded by the "geometric" surface $S$, which can be mobile or immobile relative to the liquid. The total energy $U$ of a liquid domain $\tau$ consists of the kinetic energy $T$ of liquid particles located in $\tau$, their potential energy $\Pi$ caused, as a rule, by the gravity field, and the internal energy $\mathcal{E}$. Assuming that the direction of the $Ox$-axis is opposite to the direction of the vector of gravity forces **g**, we can represent the energy of the liquid domain $\tau$ (with free surface) in the form

$$U = \int_\tau \rho\left(\frac{1}{2}v^2 + \mathrm{g}x + E\right) d\tau. \quad (1.1.26)$$

Here, $\rho E d\tau$ is the internal energy energy accumulated in an elementary liquid domain with mass $dm = \rho d\tau$ and $E$ is assumed to be an explicitly given function such that the total internal energy

$$\mathcal{E} = \int_{\tau} \rho E d\tau. \tag{1.1.27}$$

The general relation for the law of conservation of energy has the form

$$dU = dA^{(e)} + dQ^{(e)} + dQ^*, \tag{1.1.28}$$

where $dA^{(e)}$ is an elementary work of external macroscopic forces, $dQ^{(e)}$ is an elementary inflow of heat into the body, and $dQ^*$ is an elementary inflow of the other (nonthermal) kinds of energy.

In finding the inflow of heat, we usually distinguish the heat flow $Q_S$ through the wetted surface of the body and the heat flow $Q_\tau$ from the heat sources distributed over the liquid domain. We denote the density of heat flow through the surface by $q_\nu$ and the rate of bulk heat release by $\varepsilon$. As a result, we arrive at the following relations:

$$Q_S = \int_{t_1}^{t_2} dt \int_S q_\nu dS \quad \text{and} \quad Q_\tau = \int_{t_1}^{t_2} dt \int_\tau \varepsilon d\tau. \tag{1.1.29}$$

Numerous isotropic media obey the Fourier law of heat conduction. According to this law, the density of heat flow $q_\nu$ and the vector of heat flow $\boldsymbol{q}$ can be written in the form

$$q_\nu = k\frac{\partial T}{\partial \nu} \quad \text{and} \quad \boldsymbol{q} = k \operatorname{grad} T,$$

where $k$ is the heat-conduction coefficient.

We now represent the energy equation in the integral form for the case where relation (1.1.28) includes only the work of external forces and the heat flows are given by relations (1.1.29):

$$\frac{dU}{dt} = \int_\tau \rho(\boldsymbol{F} \cdot \boldsymbol{v}) d\tau + \int_S (\boldsymbol{p}_\nu \cdot \boldsymbol{v}) dS + \int_S q_\nu dS + \int_\tau \varepsilon d\tau. \tag{1.1.30}$$

In (1.1.30), we transform the surface integrals into the volume integrals as indicated above. By virtue of the equations of motion (1.1.20), we obtain the following energy equation in the differential form:

$$\rho\frac{dE}{dt} = \boldsymbol{p}_x\frac{\partial \boldsymbol{v}}{\partial x} + \boldsymbol{p}_y\frac{\partial \boldsymbol{v}}{\partial y} + \boldsymbol{p}_z\frac{\partial \boldsymbol{v}}{\partial z} + \operatorname{div} \boldsymbol{q} + \varepsilon. \tag{1.1.31}$$

The presented system of equations containing the equation of continuity (1.1.10), the momentum equation (1.1.20), the equation for angular momenta (1.1.25), and the energy equation (1.1.31) is also not complete. In order to construct a complete system of governing equations for the motion of continua, it is necessary, first of all, to establish additional relations between the parameters of the analyzed system in specific cases.

## 1.2 Stress tensors, strain rates, and their links

The Cauchy formula (1.1.16) establishes the links between the stresses $\boldsymbol{p}_\nu$ acting in an arbitrarily oriented area and the stresses $\boldsymbol{p}_1$, $\boldsymbol{p}_2$, $\boldsymbol{p}_3$ acting in the coordinate areas and enables one to introduce a symmetric tensor $\boldsymbol{P}$ of the second rank with the components $p_{ij}$ (1.1.21). It is called the stress tensor. This tensor is associated with a symmetric quadratic form

$$2\Phi(x_1,\ x_2,\ x_3) = p_{ij}x_ix_j,$$

which describes the second-order surface

$$2\Phi(x_1,\ x_2,\ x_3) = \text{const}$$

(called the tensor surface of the stress tensor) in the coordinate system $Ox_1x_2x_3$. It is always possible to find a coordinate system $O\overline{x}_1\overline{x}_2\overline{x}_3$ in which

$$2\Phi(x_1,\ x_2,\ x_3) = p_1\overline{x}_1^2 + p_2\overline{x}_2^2 + p_3\overline{x}_3^2, \tag{1.2.1}$$

and the stress tensor has only diagonal components

$$\boldsymbol{P} = \begin{Vmatrix} p_1 & 0 & 0 \\ 0 & p_2 & 0 \\ 0 & 0 & p_3 \end{Vmatrix}.$$

It is known that a quadratic form can be reduced to the canonical form (1.2.1) (in any Cartesian coordinate system) with the help of a nondegenerate linear transformation. In this case, the quantities $p_1$, $p_2$, and $p_3$, i.e., the so-called principal stresses, are defined as the roots of the secular equation

$$\begin{vmatrix} p_{11}-\lambda & p_{12} & p_{13} \\ p_{21} & p_{22}-\lambda & p_{23} \\ p_{31} & p_{32} & p_{33}-\lambda \end{vmatrix} = 0 \tag{1.2.2}$$

or, in the expanded form,

$$-\lambda^3 + I_1\lambda^2 - I_2\lambda + I_3 = 0, \tag{1.2.3}$$

where

$$I_1 = p_{11} + p_{22} + p_{33}, \quad I_3 = \det \| p_{ij} \|,$$

$$I_2 = \begin{vmatrix} p_{22} & p_{23} \\ p_{32} & p_{33} \end{vmatrix} + \begin{vmatrix} p_{33} & p_{31} \\ p_{13} & p_{11} \end{vmatrix} + \begin{vmatrix} p_{11} & p_{12} \\ p_{21} & p_{22} \end{vmatrix}. \tag{1.2.4}$$

Three real roots of the secular equation determine the stresses in the zones orthogonal to the principal directions:

$$\lambda_1 = p_{\nu_1} = p_1, \quad \lambda_2 = p_{\nu_2} = p_2, \quad \lambda_3 = p_{\nu_3} = p_3.$$

These stresses are called the principal components of the stress tensor. For the components of the stress tensor along the principal axes, we find

$$p_{ii} = \lambda_i = p_i, \quad p_{ki} = p_{ik} = 0 \quad \text{for} \quad k \neq i.$$

The coefficients of the secular equation (1.2.3) are invariants of the stress tensor. They can be represented in terms of the roots of the secular equation as follows:

$$I_1 = p_1 + p_2 + p_3, \quad I_2 = p_2 p_3 + p_1 p_2 + p_1 p_3, \quad I_3 = p_1 p_2 p_3. \tag{1.2.5}$$

We now consider some notions from the kinematics of continua in the vicinity of a given point related, mainly, to the fundamental Cauchy–Helmholtz theorem on the decomposition of the velocities of points of an infinitely small particle of the medium. As usual, an infinitely small particle is understood as the set of points of the medium located at infinitely small distances $\boldsymbol{\rho}$ from a fixed point. The velocity field $\boldsymbol{v}(\boldsymbol{r})$ is regarded as a continuous function of the coordinates of these points.

In the general case of deformable medium, its motion can be decomposed, as in the case of a perfectly rigid body, into certain components. Thus, in particular, we can separate the translational and rotational motions. However, in this case, the vector of angular velocity is not identical at all points of the medium. Therefore, it is necessary to consider the decomposition of motion of the medium only in a certain small elementary volume in the vicinity of a given point $O$ (Fig. 1.1) with coordinates $x$, $y$, $z$.

We now consider two consecutive positions of a particle at times $t$ and $t+dt$, where $dt$ is an infinitely small time interval. For the first position of the particle, we fix two arbitrary points $O$ and $A$ and denote their absolute radius vectors relative to a point $\overline{O}$ of the space by $\boldsymbol{r}_0$ and $\boldsymbol{r}$. By $\boldsymbol{\rho}$, we denote the directed segment $\overline{OA}$. For the second position of the particle at the time $t+\Delta t$, the same quantities are denoted by $\boldsymbol{r}'_0$, $\boldsymbol{r}'$, and $\boldsymbol{\rho}'$. In what follows, we consider the point $O$ as a basic point (pole). Following the generally accepted ideas, we define the elementary displacements of the points $O$ and $A$ as

$$d\boldsymbol{r}_0 = \boldsymbol{r}'_0 - \boldsymbol{r}_0 \quad \text{and} \quad d\boldsymbol{r} = \boldsymbol{r}' - \boldsymbol{r}.$$

The difference $\boldsymbol{\rho}' - \boldsymbol{\rho}$ is called an elementary relative displacement of the point $A$ relative to $O$. It is denoted by $d\boldsymbol{\rho}$ and, hence,

$$d\boldsymbol{\rho} = \boldsymbol{\rho}' - \boldsymbol{\rho}.$$

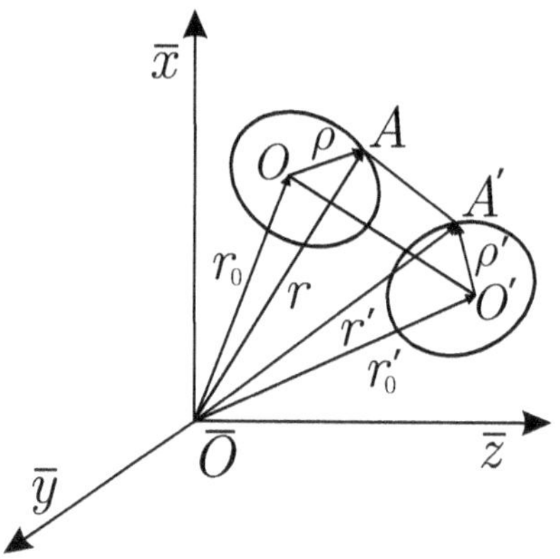


**Figure 1.1**

It follows from the obvious geometric considerations that

$$\boldsymbol{\rho}' = \boldsymbol{r}' - \boldsymbol{r}_0, \quad \boldsymbol{\rho} = \boldsymbol{r} - \boldsymbol{r}_0, \quad d\boldsymbol{\rho} = d\boldsymbol{r} - d\boldsymbol{r}_0 = (\boldsymbol{v} - \boldsymbol{v}_0)dt, \tag{1.2.6}$$

where $\boldsymbol{v}$ and $\boldsymbol{v}_0$ are the velocities of the points $A$ and $O$ at time $t$.

At the same time, it is possible to find $d\boldsymbol{\rho}$ by assuming that the velocity is a function of position of the point, i.e.,

$$\boldsymbol{v} = \boldsymbol{v}(\boldsymbol{r}), \quad \boldsymbol{v}_0 = \boldsymbol{v}_0(\boldsymbol{r}_0),$$

and analyzing the field of velocity of the liquid at time $t$. Setting $\boldsymbol{r} = \boldsymbol{r}_0 + \boldsymbol{\rho}$ in the vicinity of the point $O$ and using the Taylor formula to within the terms of the second order of smallness, we obtain

$$\boldsymbol{v}(\boldsymbol{r}) = \boldsymbol{v}(\boldsymbol{r}_0) + \left(\frac{\partial \boldsymbol{v}}{\partial x}\right)_0 \xi + \left(\frac{\partial \boldsymbol{v}}{\partial y}\right)_0 \eta + \left(\frac{\partial \boldsymbol{v}}{\partial z}\right)_0 \zeta\,. \tag{1.2.7}$$

Hence,

$$d\boldsymbol{\rho} = (\boldsymbol{\rho} \cdot \nabla)\boldsymbol{v}dt\,. \tag{1.2.8}$$

In relation (1.2.7), the derivatives of $\boldsymbol{v}$ with respect to $\xi$, $\eta$, $\zeta$ are found at the point $O$. In terms of the projections onto the immobile axes $\bar{O}\,\bar{x}\,\bar{y}\,\bar{z}$, relation (1.2.8) can be rewritten in the form

$$\begin{aligned} d\xi &= \left(\frac{\partial v_x}{\partial x}\xi + \frac{\partial v_x}{\partial y}\eta + \frac{\partial v_x}{\partial z}\zeta\right) dt, \\ d\eta &= \left(\frac{\partial v_y}{\partial x}\xi + \frac{\partial v_y}{\partial y}\eta + \frac{\partial v_y}{\partial z}\zeta\right) dt, \\ d\zeta &= \left(\frac{\partial v_z}{\partial x}\xi + \frac{\partial v_z}{\partial y}\eta + \frac{\partial v_z}{\partial z}\zeta\right) dt. \end{aligned} \tag{1.2.9}$$

In (1.2.9), we now realize the following identical replacement of the derivatives proposed by Cauchy:

$$\frac{\partial v_x}{\partial x} \equiv \frac{\partial v_x}{\partial x}; \quad \frac{\partial v_x}{\partial y} \equiv \frac{1}{2}\left(\frac{\partial v_x}{\partial y} + \frac{\partial v_y}{\partial x}\right) + \frac{1}{2}\left(\frac{\partial v_x}{\partial y} - \frac{\partial v_y}{\partial x}\right),$$
$$\frac{\partial v_x}{\partial z} \equiv \frac{1}{2}\left(\frac{\partial v_x}{\partial z} + \frac{\partial v_z}{\partial x}\right) + \frac{1}{2}\left(\frac{\partial v_x}{\partial z} - \frac{\partial v_z}{\partial x}\right),$$
$$\frac{\partial v_y}{\partial x} \equiv \frac{1}{2}\left(\frac{\partial v_y}{\partial x} + \frac{\partial v_x}{\partial y}\right) + \frac{1}{2}\left(\frac{\partial v_y}{\partial x} - \frac{\partial v_x}{\partial y}\right), \quad \frac{\partial v_y}{\partial y} \equiv \frac{\partial v_y}{\partial y}, \tag{1.2.10}$$
$$\frac{\partial v_y}{\partial z} \equiv \frac{1}{2}\left(\frac{\partial v_y}{\partial z} + \frac{\partial v_z}{\partial y}\right) + \frac{1}{2}\left(\frac{\partial v_y}{\partial z} - \frac{\partial v_z}{\partial y}\right),$$
$$\frac{\partial v_z}{\partial x} \equiv \frac{1}{2}\left(\frac{\partial v_z}{\partial x} + \frac{\partial v_x}{\partial z}\right) + \frac{1}{2}\left(\frac{\partial v_z}{\partial x} - \frac{\partial v_x}{\partial z}\right),$$
$$\frac{\partial v_z}{\partial y} \equiv \frac{1}{2}\left(\frac{\partial v_z}{\partial y} + \frac{\partial v_y}{\partial z}\right) + \frac{1}{2}\left(\frac{\partial v_z}{\partial y} - \frac{\partial v_y}{\partial z}\right), \quad \frac{\partial v_z}{\partial z} \equiv \frac{\partial v_z}{\partial z}.$$

Denote

$$\dot\varepsilon_{xx} = \frac{\partial v_x}{\partial x}, \quad \dot\varepsilon_{xy} = \dot\varepsilon_{yx} = \frac{1}{2}\left(\frac{\partial v_x}{\partial y} + \frac{\partial v_y}{\partial x}\right),$$
$$\dot\varepsilon_{yy} = \frac{\partial v_y}{\partial y}, \quad \dot\varepsilon_{yz} = \dot\varepsilon_{zy} = \frac{1}{2}\left(\frac{\partial v_y}{\partial z} + \frac{\partial v_z}{\partial y}\right), \tag{1.2.11}$$
$$\dot\varepsilon_{zz} = \frac{\partial v_z}{\partial z}, \quad \dot\varepsilon_{zx} = \dot\varepsilon_{xz} = \frac{1}{2}\left(\frac{\partial v_z}{\partial x} + \frac{\partial v_x}{\partial z}\right);$$

$$\Omega_x = \frac{1}{2}\left(\frac{\partial v_z}{\partial y} - \frac{\partial v_y}{\partial z}\right), \quad \Omega_y = \frac{1}{2}\left(\frac{\partial v_x}{\partial z} - \frac{\partial v_z}{\partial x}\right);$$
$$\Omega_z = \frac{1}{2}\left(\frac{\partial v_y}{\partial x} - \frac{\partial v_x}{\partial y}\right). \tag{1.2.12}$$

This enables us to rewrite relations (1.2.9) in the form

$$d\xi = \frac{\partial \Phi}{\partial \xi}dt + (\Omega_y\zeta - \Omega_z\eta)dt,$$
$$d\eta = \frac{\partial \Phi}{\partial \eta}dt + (\Omega_z\xi - \Omega_x\zeta)dt, \tag{1.2.13}$$
$$d\zeta = \frac{\partial \Phi}{\partial \zeta}dt + (\Omega_x\eta - \Omega_y\xi)dt,$$

where

$$2\Phi = \dot\varepsilon_{xx}\xi^2 + \dot\varepsilon_{yy}\eta^2 + \dot\varepsilon_{zz}\zeta^2 + 2\dot\varepsilon_{xy}\xi\eta + 2\dot\varepsilon_{yz}\eta\zeta + 2\dot\varepsilon_{zx}\xi\zeta. \tag{1.2.14}$$

Thus, the elementary relative displacement $d\boldsymbol{\rho}$ is the geometric sum of two terms, namely, of the vector $\mathbf{grad}\,\Phi dt$ and the vector $(\boldsymbol{\Omega} \times \boldsymbol{\rho})dt$. The latter describes an elementary rotational displacement of the point $A$ relative to the pole $O$ in the case where the particle becomes rigid and is in the state of rotational motion with instantaneous angular velocity

$$\boldsymbol{\Omega} = \tfrac{1}{2}\,\mathrm{rot}\,\boldsymbol{v}.$$

By virtue of the relation $d\boldsymbol{r} = d\boldsymbol{r}_0 + d\boldsymbol{\rho}$, we can state that an elementary displacement of any point of the liquid particle has three components: translational, rotational, and deformation displacements. The relation

$$\boldsymbol{v}_A = \boldsymbol{v}_0 + \boldsymbol{\Omega} \times \boldsymbol{\rho} + \mathbf{grad}\,\boldsymbol{\Phi} \tag{1.2.15}$$

reveals the main meaning of the Cauchy–Helmholtz theorem on the distribution of the velocity of points of an infinitely small particle: the velocity of any point of an infinitely small liquid particle is a superposition of the velocities of translational $\boldsymbol{v}_0$ and rotational $\boldsymbol{\Omega} \times \boldsymbol{\rho}$ motions of this particle regarded as an absolutely rigid body and the pure strain rate $\boldsymbol{v}^* = \mathbf{grad}\,\boldsymbol{\Phi}$. For comparison, we recall that the corresponding theorem for the rigid body takes the form

$$\boldsymbol{v} = \boldsymbol{v}_0 + \boldsymbol{\omega} \times (\boldsymbol{r} - \boldsymbol{r}_0). \tag{1.2.16}$$

In this case, the angular velocity $\boldsymbol{\omega}$ can be expressed via the velocity $\boldsymbol{v}$ at a given point of the body

$$\boldsymbol{\omega} = \tfrac{1}{2}\,\mathrm{rot}\,\boldsymbol{v}.$$

However, unlike the deformable medium, the vectors of angular velocity are identical for all points of the body because the length $\rho$ remains constant in the course of motion of the body.

The deformation component of the motion of continua is described by nine parameters (1.2.11) (with only six different values). These parameters form a symmetric tensor $\dot{\boldsymbol{S}}$ of the second rank, which is called the strain-rate tensor [104]:

$$\dot{\boldsymbol{S}} = \begin{Vmatrix} \dot{\varepsilon}_{xx} & \dot{\varepsilon}_{xy} & \dot{\varepsilon}_{xz} \\ \dot{\varepsilon}_{yx} & \dot{\varepsilon}_{yy} & \dot{\varepsilon}_{yz} \\ \dot{\varepsilon}_{zx} & \dot{\varepsilon}_{zy} & \dot{\varepsilon}_{zz} \end{Vmatrix}. \tag{1.2.17}$$

The tensor $\dot{\boldsymbol{S}}dt = \boldsymbol{S}$ whose components can be represented via the projections of infinitely small translations $v_x dt$, $v_y dt$, and $v_z dt$, is called the strain tensor. It is extensively used in the elasticity theory. The strain-rate tensor $\dot{\boldsymbol{S}}$ is connected with the symmetric quadratic form $\boldsymbol{\Phi}$ (1.2.14). On the principal coordinate axes $\bar{\xi}$, $\bar{\eta}$, $\bar{\zeta}$, this expression takes the following canonical form:

$$2\boldsymbol{\Phi} = \dot{\varepsilon}_1\bar{\xi}^2 + \dot{\varepsilon}_2\bar{\eta}^2 + \dot{\varepsilon}_3\bar{\zeta}^2, \tag{1.2.18}$$

and the strain-rate tensor reduces to the form

$$\dot{\boldsymbol{S}} = \begin{Vmatrix} \dot{\varepsilon}_1 & 0 & 0 \\ 0 & \dot{\varepsilon}_2 & 0 \\ 0 & 0 & \dot{\varepsilon}_3 \end{Vmatrix}. \tag{1.2.19}$$

The principal strain rates $\dot{\varepsilon}_1$, $\dot{\varepsilon}_2$, and $\dot{\varepsilon}_3$ are defined as the roots of the secular equation

$$\begin{vmatrix} \dot{\varepsilon}_{11} - \lambda & \dot{\varepsilon}_{12} & \dot{\varepsilon}_{13} \\ \dot{\varepsilon}_{21} & \dot{\varepsilon}_{22} - \lambda & \dot{\varepsilon}_{23} \\ \dot{\varepsilon}_{31} & \dot{\varepsilon}_{32} & \dot{\varepsilon}_{33} - \lambda \end{vmatrix} = 0.$$

This equation can also be represented in the form (1.2.3), by introducing three invariants of the strain-rate tensor, namely, the linear $J_1$, quadratic $J_2$, and cubic $J_3$ invariants. In particular, the first invariant of the strain-rate tensor is given by the formula

$$J_1 = \dot{\varepsilon}_{11} + \dot{\varepsilon}_{22} + \dot{\varepsilon}_{33} = \dot{\varepsilon}_1 + \dot{\varepsilon}_2 + \dot{\varepsilon}_3 = \operatorname{div} \boldsymbol{v}. \tag{1.2.20}$$

The stress $\boldsymbol{P}$ and strain-rate $\dot{\boldsymbol{S}}$ tensors characterizing the stress–strain state at a given point of continuum are fundamental notions of the continuum mechanics. The subsequent construction of adequate mathematical models for the description of motion of the analyzed medium on the basis of the equations of continuity and the momentum, angular momentum, and energy equations can be performed by using additional hypotheses reflecting the relationships between these tensors and additional relations between the introduced parameters, which are called the equations of state.

Thus, the model of viscous liquid is based on the following assumptions: (a) the liquid is isotropic, i.e., its properties are identical in all directions; (b) if the liquid is at rest or moves as a rigid body, then only normal stresses are formed in the liquid; and (c) the components of the stress tensor are linear functions of the components of strain rate. These assumptions correspond to a more general relationship between the tensors $\boldsymbol{P}$ and $\dot{\boldsymbol{S}}$, which is called the generalized Newton law:

$$\boldsymbol{P} = a\dot{\boldsymbol{S}} + b\mathcal{E}. \tag{1.2.21}$$

Here, $a$ and $b$ are scalars and $\mathcal{E}$ is a unit tensor with the components

$$\varepsilon_{ij} = \begin{cases} 0, & \text{for } i \neq j, \\ 1, & \text{for } i = j, \quad i, j = 1,\ 2,\ 3. \end{cases}$$

In the case of linear relationship between the tensors $\boldsymbol{P}$ and $\dot{\boldsymbol{S}}$, the scalar $a$ should be independent of their components. The Newton law for the viscous liquid

$$\tau = \mu \frac{\partial v}{\partial \nu} \tag{1.2.22}$$

($\tau$ is the tangential stress acting in the planes of contact of the liquid layers, $v$ is the velocity of longitudinal laminar motion of the liquid, $\mu$ is the dynamic viscosity coefficient, and $\nu$ is the normal to the planes) is a special case of (1.2.21). Therefore, we can set that the scalar $a$ is equal to $2\mu$. As already indicated, the linear invariant of the stress tensor is the sum of three normal stresses applied to three mutually perpendicular zones at a given point of the flow, $I_1 = p_{11} + p_{22} + p_{33}$, and the linear invariant of the strain-rate tensor is given by the formula

$$J_1 = \dot{\varepsilon}_{11} + \dot{\varepsilon}_{22} + \dot{\varepsilon}_{33} = \operatorname{div} \boldsymbol{v}.$$

To find the scalar $b$, we equate the linear invariants of the tensors on the left- and right-hand sides of equality (1.2.21) under the conditions of incompressible liquid, i.e., for $\operatorname{div} \boldsymbol{v} = 0$. This yields $p_{11} + p_{22} + p_{33} = 3b$, whence it follows that

$$b = \tfrac{1}{3}\,(p_{11} + p_{22} + p_{33}). \tag{1.2.23}$$

For the Newton incompressible viscous liquid, it is usually assumed that the arithmetic mean of three normal stresses applied to three mutually perpendicular zones at a given point of the medium is the pressure taken with the opposite sign at the indicated point, i.e.,

$$-\tfrac{1}{3}\,(p_{11} + p_{22} + p_{33}) = p. \tag{1.2.24}$$

For the hydrodynamic model of ideal liquid in which the tangential forces in the contact zones of two liquids moving relative to each other are neglected, the pressure is introduced by the equalities

$$p_{11} = p_{22} = p_{33} = -p \tag{1.2.25}$$

reflecting the fundamental property of the ideal liquid, namely, that the normal stress at a given point is independent of the orientation of the zone to which it is applied. According to (1.2.21), (1.2.23), and (1.2.24) for incompressible viscous liquids, we finally obtain

$$\boldsymbol{P} = 2\mu\dot{\boldsymbol{S}} - p\mathcal{E} \tag{1.2.26}$$

or, in the componentwise form,

$$p_{ij} = \begin{cases} \mu\left(\dfrac{\partial v_i}{\partial x_j} + \dfrac{\partial v_j}{\partial x_i}\right) & \text{for } j \neq i, \\[2ex] -p + 2\mu\dfrac{\partial v_i}{\partial x_j} & \text{for } i = j. \end{cases} \tag{1.2.27}$$

Equality (1.2.26) is usually called the rheological equation of the Newton incompressible viscous liquid. In a more detailed form, the formulas of the accepted relationships in the Cartesian coordinate system can be represented as follows:

$$p_{xx} = -p + 2\mu\frac{\partial v_x}{\partial x}, \quad p_{yy} = -p + 2\mu\frac{\partial v_y}{\partial y}, \quad p_{zz} = -p + 2\mu\frac{\partial v_z}{\partial z},$$

$$p_{xy} = p_{yx} = \mu\left(\frac{\partial v_x}{\partial y} + \frac{\partial v_y}{\partial x}\right), \quad p_{yz} = p_{zy} = \mu\left(\frac{\partial v_y}{\partial z} + \frac{\partial v_z}{\partial y}\right), \tag{1.2.28}$$

$$p_{zx} = p_{xz} = \mu\left(\frac{\partial v_z}{\partial x} + \frac{\partial v_x}{\partial z}\right).$$

In the elasticity theory, the models based on Hooke's law are constructed in the same way.

For the compressible medium, the scalar coefficient $b$ in (1.2.21) may depend not only on the linear invariant of the stress tensor $p_{11} + p_{22} + p_{33}$ but also on the linear invariant of the strain-rate tensor $\operatorname{div} \boldsymbol{v}$, which is not equal to zero. As above, we set $a = 2\mu$. Thus, to determine $b$, we equate the linear invariants of both sides of equality (1.2.21) and obtain

$$p_{11} + p_{22} + p_{33} = 2\mu \operatorname{div} \boldsymbol{v} + 3b. \tag{1.2.29}$$

Generalizing the definition of pressure presented above for the case of incompressible liquid, we set

$$\tfrac{1}{3}(p_{11} + p_{22} + p_{33}) = -p + \mu' \operatorname{div} \boldsymbol{v}, \tag{1.2.30}$$

where $\mu'$ is the coefficient of bulk viscosity or the second viscosity coefficient.

Relations (1.2.29) and (1.2.30) imply that

$$b = -p - \left(\tfrac{2}{3}\mu - \mu'\right) \operatorname{div} \boldsymbol{v}. \tag{1.2.31}$$

Thus the generalized Newton's law (1.2.21) takes the form

$$\boldsymbol{P} = 2\mu\dot{\boldsymbol{S}} + \left[-p - \left(\tfrac{2}{3}\mu - \mu'\right) \operatorname{div} \boldsymbol{v}\right] \mathcal{E}. \tag{1.2.32}$$

The law of motion of a viscous liquid in the form (1.2.32) is connected with the names of C.-L. Navier (1843) and G. Stokes (1845).

The value of the coefficient $\mu'$ noticeably differs from zero only in exceptional cases (in the problems of mechanics with rapidly developing processes, such as explosions, in the problems of passing of a medium through a densification jump, etc.). Setting $\mu' = 0$ in (1.2.32), we obtain

$$\boldsymbol{P} = 2\mu\dot{\boldsymbol{S}} - \left(p + \tfrac{2}{3}\mu \operatorname{div} \boldsymbol{v}\right) \mathcal{E} \tag{1.2.33}$$

or, in the matrix form,

$$\begin{pmatrix} p_{xx} & p_{yx} & p_{zx} \\ p_{xy} & p_{yy} & p_{zy} \\ p_{xz} & p_{yz} & p_{zz} \end{pmatrix}$$

$$= 2\mu \begin{bmatrix} \dfrac{\partial v_x}{\partial x} & \dfrac{1}{2}\left(\dfrac{\partial v_x}{\partial y} + \dfrac{\partial v_y}{\partial x}\right) & \dfrac{1}{2}\left(\dfrac{\partial v_x}{\partial z} + \dfrac{\partial v_z}{\partial x}\right) \\ \dfrac{1}{2}\left(\dfrac{\partial v_y}{\partial x} + \dfrac{\partial v_x}{\partial y}\right) & \dfrac{\partial v_y}{\partial y} & \dfrac{1}{2}\left(\dfrac{\partial v_y}{\partial z} + \dfrac{\partial v_z}{\partial y}\right) \\ \dfrac{1}{2}\left(\dfrac{\partial v_z}{\partial x} + \dfrac{\partial v_x}{\partial z}\right) & \dfrac{1}{2}\left(\dfrac{\partial v_z}{\partial y} + \dfrac{\partial v_y}{\partial z}\right) & \dfrac{\partial v_z}{\partial z} \end{bmatrix}$$

$$- \left(p + \frac{2}{3}\mu \operatorname{div} \boldsymbol{v}\right) \begin{pmatrix} 1 & 0 & 0 \\ 0 & 1 & 0 \\ 0 & 0 & 1 \end{pmatrix}. \tag{1.2.34}$$

Equation (1.2.32) for the stress tensor and the strain-rate tensor contains new scalar quantities $p$, $\mu$, $\mu'$, which are not constant in the general case. According to their physical nature, these quantities are the parameters specifying the state of a system including, e.g., temperature, phase characteristics of the substance, the heat-conduction coefficients, elasticity moduli, etc. The collection of these parameters forms the space of states $\mu^i$ (phase space). These parameters satisfy some relations called the equations of state. In what follows, we present some of these equations.

For compressible media, their density depends on pressure and temperature, and the equation of state takes the form

$$f(p, \rho, T) = 0. \tag{1.2.35}$$

It is assumed that the internal energy is completely defined by the pressure $p$ and temperature $T$ of the medium:

$$E = E(T, p).$$

Thus, for a compressible medium called an ideal gas, the equation of state takes the form of the Clapeyron equation

$$p = R\rho T. \tag{1.2.36}$$

This equation is true for relatively low pressures $p$ and temperatures $T$. The constant $R$ can be represented in the form $R = km^{-1}$, where $k$ is the Boltzmann constant and $m$ is the mean mass of a molecule in grams. The ideal gas obeys the equation

$$E = \int c_V dT,$$

where $c_V$ is the specific heat capacity at constant volume.

For the real gases, $c_V = \text{const}$ for very high values of $T$. If this condition is satisfied, then

$$E = c_V T + E_0. \tag{1.2.37}$$

Parallel with $c_V$, we also use a constant quantity

$$c_p = c_V + R$$

called the specific heat capacity at constant pressure and a quantity

$$h = c_p T + h_0 \tag{1.2.38}$$

called enthalpy.

The temperature dependence of the viscosity coefficient of gases $\mu$ is described by the Sutherland formula

$$\mu = \frac{\text{const}}{T + T_0} T^{\frac{3}{2}}, \quad T_0 \approx 114^o K,$$

or by the formula

$$\frac{\mu}{\mu_0} = \left(\frac{T}{T_0}\right)^n.$$

For the incompressible liquids without heat conduction, the equations of state take the form:

$$\rho = \text{const}, \ \mu = \text{const}.$$

In this case, the internal energy is a linear function of temperature,

$$E = cT + \text{const}.$$

The outlined theory of motion of a viscous liquid is based on the Newton hypothesis (1.2.22) and its generalizations of the form (1.2.21). This theory proves to be quite efficient for the solution of some external problems of hydrodynamics in which the laminar flows are, as a rule, predominant. As for the application of the Navier–Stokes mathematical models to the solution of the problems of internal hydrodynamics, e.g., about the oscillatory motions of liquids in bounded domains, it is worth noting that, for various reasons, these models turned out to be inefficient (at least, quantitatively). For this class of problems, the theory of surface waves based on the other Newton hypothesis is proposed in Section 5.2.

## 1.3 Mathematical and physical models

The process of construction of various hydrodynamical models is reduced to getting complete systems of differential equations and equations of state capable of the adequate description of the actual liquid motion. For the viscous heat-conducting liquid, the role of these equations is played by the equation of continuity (1.1.10), the momentum equation (1.1.20), the energy equation (1.1.31), the relationships between the stress tensor and the strain-rate tensor in the form of the Navier–Stokes equation (1.2.32), the relationship between the heat flow and the temperature of the flow in the form of the Fourier law, the equations of state of the form (1.2.35), the relation for the internal energy $E$ via the parameters of state $T$ and $p$, and the relations specifying the parameters $\mu$ and $k$.

To construct the dynamic equation of motion of a viscous compressible liquid, we use Eq. (1.1.20) in terms of stresses:

$$\rho\frac{d\boldsymbol{v}}{dt} = \rho\boldsymbol{F} + \operatorname{div}\boldsymbol{P}.$$

Substituting $\boldsymbol{P}$ from the generalized Newton law (1.2.33), we arrive at the following main Navier–Stokes equation of the dynamics of viscous compressible liquid:

$$\rho\frac{d\boldsymbol{v}}{dt} = \rho\boldsymbol{F} - \operatorname{grad}\left(p + \frac{2}{3}\mu\operatorname{div}\boldsymbol{v}\right) + 2\operatorname{div}(\mu\,\dot{\boldsymbol{S}}). \tag{1.3.1}$$

In the projections onto the axes of a Cartesian coordinate system, it takes the form

$$\begin{aligned}
\rho\frac{dv_x}{dt} &= \rho F_x - \frac{\partial p}{\partial x} + 2\frac{\partial}{\partial x}\left(\mu\frac{\partial v_x}{\partial x}\right) + \frac{\partial}{\partial y}\left[\mu\left(\frac{\partial v_x}{\partial y} + \frac{\partial v_y}{\partial x}\right)\right] \\
&\quad + \frac{\partial}{\partial z}\left[\mu\left(\frac{\partial v_x}{\partial z} + \frac{\partial v_z}{\partial x}\right)\right] - \frac{2}{3}\frac{\partial}{\partial x}(\mu\operatorname{div}\boldsymbol{v}), \\
\rho\frac{dv_y}{dt} &= \rho F_y - \frac{\partial p}{\partial y} + \frac{\partial}{\partial x}\left[\mu\left(\frac{\partial v_x}{\partial y} + \frac{\partial v_y}{\partial x}\right)\right] + 2\frac{\partial}{\partial y}\left(\mu\frac{\partial v_y}{\partial y}\right) \\
&\quad + \frac{\partial}{\partial z}\left[\mu\left(\frac{\partial v_y}{\partial z} + \frac{\partial v_z}{\partial y}\right)\right] - \frac{2}{3}\frac{\partial}{\partial y}(\mu\operatorname{div}\boldsymbol{v}), \qquad (1.3.2) \\
\rho\frac{dv_z}{dt} &= \rho F_z - \frac{\partial p}{\partial z} + \frac{\partial}{\partial x}\left[\mu\left(\frac{\partial v_x}{\partial z} + \frac{\partial v_z}{\partial x}\right)\right] + \frac{\partial}{\partial y}\left[\mu\left(\frac{\partial v_y}{\partial z} + \frac{\partial v_z}{\partial y}\right)\right] \\
&\quad + 2\frac{\partial}{\partial z}\left(\mu\frac{\partial v_z}{\partial z}\right) - \frac{2}{3}\frac{\partial}{\partial z}(\mu\operatorname{div}\boldsymbol{v}).
\end{aligned}$$

For the incompressible liquid with $\mu = \text{const}$, the vector form of the equation of motion is as follows:

$$\rho\frac{d\boldsymbol{v}}{dt} = \rho\boldsymbol{F} - \nabla p + \mu\Delta\boldsymbol{v}. \tag{1.3.3}$$

By using the generalized Newton law (1.2.33), we can exclude $\boldsymbol{p}_x$, $\boldsymbol{p}_y$, and $\boldsymbol{p}_z$ from the energy equation (1.1.31)

$$\boldsymbol{p}_x\frac{\partial\boldsymbol{v}}{\partial x} + \boldsymbol{p}_y\frac{\partial\boldsymbol{v}}{\partial y} + \boldsymbol{p}_z\frac{\partial\boldsymbol{v}}{\partial z} = -p\operatorname{div}\boldsymbol{v} + \Phi,$$

$$\Phi = \mu \left[ 2\left(\frac{\partial v_x}{\partial x}\right)^2 + 2\left(\frac{\partial v_y}{\partial y}\right)^2 + 2\left(\frac{\partial v_z}{\partial z}\right)^2 + \left(\frac{\partial v_x}{\partial y} + \frac{\partial v_y}{\partial x}\right)^2 \right. \tag{1.3.4}$$
$$\left. + \left(\frac{\partial v_y}{\partial z} + \frac{\partial v_z}{\partial y}\right)^2 + \left(\frac{\partial v_x}{\partial z} + \frac{\partial v_z}{\partial x}\right)^2 \right] + \lambda(\operatorname{div} \boldsymbol{v})^2,$$

where $\Phi$ is called the dissipative function ($\Phi \geqslant 0$ for $\mu \geqslant 0$). In view of (1.3.4), the energy equation (1.1.31) takes the form

$$\rho\frac{dE}{dt} + p \operatorname{div} \boldsymbol{v} = \operatorname{div}(k \operatorname{grad} T) + \Phi + \varepsilon. \tag{1.3.5}$$

We now present the most frequently used mathematical (physical) models obtained as special cases of the models presented above.

1. *Viscous ideal gas.* The closed system of equations for the unknown quantities $\boldsymbol{v}$, $p$, $\rho$, and $T$ includes the Navier–Stokes equation (1.3.2), the energy equations (1.3.5), the equation of continuity (1.1.10), and the following relations:

$$p = R\rho T, \quad E = c_V T + E_0, \quad \mu = \mu(T), \quad k = k(T).$$

By using relations (1.2.36) and (1.2.38), the energy equation can be transformed as follows:

$$\rho\frac{dh}{dt} - \frac{dp}{dt} = \operatorname{div}(k \operatorname{grad} T) + \Phi + \varepsilon. \tag{1.3.6}$$

Since

$$p \operatorname{div} \boldsymbol{v} = -\frac{1}{\rho} p \frac{d\rho}{dt} = -\frac{dp}{dt} + \rho \frac{dT R}{dt},$$

we get the following equation:

$$\rho \frac{dE}{dt} + p \operatorname{div} \boldsymbol{v} = \rho \frac{dh}{dt} - \frac{dp}{dt}. \tag{1.3.7}$$

2. *Incompressible viscous liquid.* For a liquid of this kind, we have $\rho = \text{const}$. Taking the relations

$$k = \text{const}, \quad \mu = \text{const}, \quad E = cT$$

as the equations of state, we obtain the following closed system of equations with respect to $\boldsymbol{v}, p$, and $T$:

$$\rho\frac{d\boldsymbol{v}}{dt} = \rho\boldsymbol{F} - \nabla p + \mu\Delta\boldsymbol{v}, \quad \operatorname{div} \boldsymbol{v} = 0, \quad c\rho\frac{dT}{dt} = k\Delta T + \Phi + \varepsilon. \tag{1.3.8}$$

In this case, the field of flows of the liquid is defined independently of the equation for the temperature field. If $\mu = \mu(T)$ and $k = k(T)$, then the velocity field is defined by the system of equations (1.3.2) and the distribution of temperature is described by the energy equation

$$\rho c \frac{dT}{dt} = \frac{\partial}{\partial x}\left(k\frac{\partial T}{\partial x}\right) + \frac{\partial}{\partial y}\left(k\frac{\partial T}{\partial y}\right) + \frac{\partial}{\partial z}\left(k\frac{\partial T}{\partial z}\right) + \boldsymbol{\Phi} + \varepsilon. \tag{1.3.9}$$

Hence, in this case, the systems of equations for the velocities and temperature field must be solved together.

3. *Incompressible ideal liquid.* The Navier–Stokes equations for an incompressible ideal liquid (with $\mu = 0$, $\rho = \text{const}$, and $k = 0$) are transformed into the Euler equation

$$\rho \frac{d\boldsymbol{v}}{dt} = \rho \boldsymbol{F} - \nabla p. \tag{1.3.10}$$

Together with the equation of continuity $\operatorname{div} \boldsymbol{v} = 0$ this equation forms a closed system of equations for the projections of the velocities $v_x, v_y$, and $v_z$ and pressure $p$.

We now apply the well-known vector transformation

$$\tfrac{1}{2} \operatorname{grad} \boldsymbol{v}^2 = (\boldsymbol{v} \cdot \nabla)\boldsymbol{v} + \boldsymbol{v} \times \operatorname{rot} \boldsymbol{v}. \tag{1.3.11}$$

This enables us to represent Eq. (1.3.10) in the form

$$\frac{\partial \boldsymbol{v}}{\partial t} + \nabla\left(\tfrac{1}{2} v^2\right) - \boldsymbol{v} \times \operatorname{rot} \boldsymbol{v} = \boldsymbol{F} - \frac{1}{\rho}\nabla p. \tag{1.3.12}$$

Consider the case where $\operatorname{rot} \boldsymbol{v} = 0$ at every point of the field and, hence, the velocity $\boldsymbol{v}$ is a potential vector, i.e.,

$$\boldsymbol{v} = \operatorname{grad} \varphi.$$

Then the basic equation of liquid motion (1.3.12) is simplified

$$\boldsymbol{F} = \operatorname{grad}\left(\frac{\partial \varphi}{\partial t} + \tfrac{1}{2} v^2 + \frac{p}{\rho}\right),$$

i.e., the mass forces must have a potential. Denoting the potential of forces by $U$, we transform the equations of motion (1.3.12) as follows:

$$\frac{p}{\rho} + \frac{\partial \varphi}{\partial t} + \tfrac{1}{2} v^2 + U = f(t), \tag{1.3.13}$$

where $f(t)$ is a function of time.

If we assume that the direction of the $Ox$-axis is opposite to the direction of the vector of gravitational acceleration $\mathbf{g}$, then $U = \mathrm{g}x$. Equation (1.3.13), which is an integral of the Euler equations, is called the Bernoulli equation.

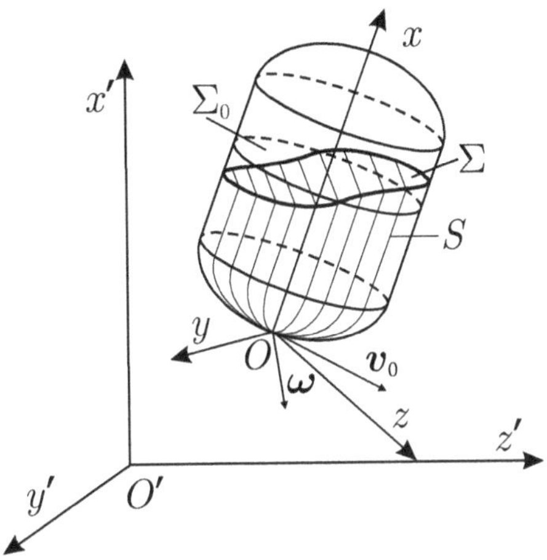


**Figure 1.2**

For a scalar function $\varphi$ depending on the space coordinates and time, the equation of continuity $\operatorname{div} \boldsymbol{v} = 0$ yields the Laplace equation

$$\Delta\varphi \equiv \frac{\partial^2\varphi}{\partial x^2} + \frac{\partial^2\varphi}{\partial y^2} + \frac{\partial^2\varphi}{\partial z^2} = 0. \tag{1.3.14}$$

We now consider the equations of motion of a liquid in the case where its bounded domain $Q$ moves in the space and this motion is described by the vector of translational velocity $\boldsymbol{v}_0$ and the vector of instantaneous angular velocity $\boldsymbol{\omega}$ relative to an immobile coordinate system $O'x'y'z'$ (Fig. 1.2). The motion of liquid is studied in a inertialess coordinate system $Oxyz$ rigidly connected with the tank. It is assumed that, at the initial time, both coordinate systems coincide. We denote the radius vector by $\boldsymbol{r}$ and the vectors of relative velocity and relative acceleration of liquid particles in the inertialess coordinate system by $\boldsymbol{v}_r$ and $\boldsymbol{w}_r$, respectively. By using the well-known relations of kinematics of the relative motions, we obtain

$$\boldsymbol{v}_a = \boldsymbol{v}_r + \boldsymbol{v}_e, \quad \boldsymbol{w}_a = \boldsymbol{w}_r + \boldsymbol{w}_e + \boldsymbol{w}_C,$$

where the vectors of transport velocity $\boldsymbol{v}_e$, transport acceleration $\boldsymbol{w}_e$, and Coriolis acceleration $\boldsymbol{w}_C$ of liquid particles are given by the formulas

$$\begin{gathered} \boldsymbol{v}_e = \boldsymbol{v}_0 + \boldsymbol{\omega} \times \boldsymbol{r}, \\ \boldsymbol{w}_e = \frac{d\boldsymbol{v}_0}{dt} + \frac{d\boldsymbol{\omega}}{dt} \times \boldsymbol{r} + \boldsymbol{\omega} \times (\boldsymbol{\omega} \times \boldsymbol{r}), \\ \boldsymbol{w}_C = 2(\boldsymbol{\omega} \times \boldsymbol{v}_r). \end{gathered} \tag{1.3.15}$$

In view of (1.3.15), the equation of relative liquid motions takes the form

$$\boldsymbol{w}_r = \boldsymbol{F} - \frac{1}{\rho}\operatorname{grad} p - \boldsymbol{w}_e - 2(\boldsymbol{\omega} \times \boldsymbol{v}_r). \tag{1.3.16}$$

Since

$$\boldsymbol{w}_r = \frac{\partial' \boldsymbol{v}_r}{\partial t} + \text{grad}\left(\tfrac{1}{2}v_r^2\right) - (\boldsymbol{v}_r \times \text{rot}\,\boldsymbol{v}_r),$$

we can represent (1.3.16) in the form

$$\frac{\partial' \boldsymbol{v}_r}{\partial t} + \text{grad}\left(\tfrac{1}{2}v_r^2\right) - \boldsymbol{v}_r \times \text{rot}\,\boldsymbol{v}_r + 2(\boldsymbol{\omega} \times \boldsymbol{v}_r) = \boldsymbol{F} - \frac{1}{\rho}\,\text{grad}\,p - \boldsymbol{w}_e. \quad (1.3.17)$$

In what follows, we study the absolute liquid motions by using a mobile (inertialess) coordinate system. The corresponding equation can be obtained from (1.3.17) and the identity that follows from this equation for $\boldsymbol{v}_r = \boldsymbol{v}_e$, i.e., in the case where liquid is in the state of absolute rest: $\boldsymbol{v}_a = 0$, $\boldsymbol{F} = 0$, and $p = \text{const}$. After simple transformations, the equation of absolute liquid motions in the mobile coordinate system takes the form [104]

$$\frac{\partial' \boldsymbol{v}_a}{\partial t} + \text{grad}\left(\frac{v_a^2}{2} - \boldsymbol{v}_a \cdot \boldsymbol{v}_e\right) - (\boldsymbol{v}_a - \boldsymbol{v}_e) \times \text{rot}\,\boldsymbol{v}_a = \boldsymbol{F} - \frac{1}{\rho}\,\text{grad}\,p, \quad (1.3.18)$$

where the prime at the time derivative shows that the operation of differentiation is performed in the mobile coordinate system. For the potential motions, we get

$$\boldsymbol{v}_a = \text{grad}\,\Phi. \quad (1.3.19)$$

By analogy with (1.3.13), we get the following primitive of the equations of motion (1.3.18):

$$\frac{\partial' \Phi}{\partial t} + \frac{1}{2}(\nabla\Phi)^2 - \nabla\Phi \cdot \boldsymbol{v}_e + U + \frac{p}{\rho} = f(t), \quad (1.3.20)$$

where $U$ is the potential of mass forces. This is the Bernoulli equation in the inertialess coordinate system. For $\rho = \text{const}$, the equation of continuity implies that the velocity potential $\Phi(x,\, y,\, z,\, t)$ satisfies the Laplace equation $\triangle\Phi = 0$ in the liquid domain.

Another procedure of construction of the equations of absolute liquid motions in the mobile coordinate system is based on the direct application of the formulas of transformation of the coordinate systems $O'x'y'z'$ and $Oxyz$.

## 1.4 Boundary and initial conditions

In other to find a solution of the equations of motion for a viscous (or ideal) liquid, it is necessary to assume that this solution satisfies the corresponding boundary and initial conditions. In this connection, it is customary to distinguish the statements of internal and external problems of aerohydrodynamics. In the internal problem, a mobile (or quiescent) finite liquid domain $Q(t)$ is bounded, in the general case, by a given rigid surface $S$ and and a free surface $\Sigma$ unknown in advance. If the free surface is absent, then the domain $Q$ is bounded solely by the closed rigid surface $S$.

For the Navier–Stokes equations, the no-slip boundary conditions are usually imposed on $S$. According to these conditions, the velocity of liquid particles located near the surface of the body is equal to the velocity $\boldsymbol{v}_S$ of the corresponding points of the surface:

$$\boldsymbol{v} = \boldsymbol{v}_S, \quad \boldsymbol{r} \in S. \tag{1.4.1}$$

For the immobile surface $S$, the no-slip condition takes the following simple form:

$$\boldsymbol{v} = 0, \quad \boldsymbol{r} \in S. \tag{1.4.2}$$

In this case, the normal and tangential components of the velocity are set equal to zero.

The boundary conditions more complicated than (1.4.1) or (1.4.2) are used in the cases where the surface $S$ is permeable or deformable and in the case where the slip of liquid particles is possible.

Among the important problems of the dynamics of bounded volumes of liquid, we can especially mention the problems with free surface and the problems of the dynamics of liquid with interface. If the interface $S_p$ is present, then it is necessary to formulate the conjugation conditions

$$\boldsymbol{v}_1 = \boldsymbol{v}_2, \quad \boldsymbol{r} \in S_p, \tag{1.4.3}$$

and

$$\boldsymbol{p}_{\nu_1} = \boldsymbol{p}_{\nu_2}, \quad \boldsymbol{r} \in S_p, \tag{1.4.4}$$

which enables one to find continuous solutions in the domains adjacent to the interface. The kinematic condition (1.4.3) expresses the equality of velocities of touching liquid particles, whereas the dynamic condition (1.4.4) reflects the equality of stresses acting on the sides of the first and second liquids. Condition (1.4.4) can be deduced by using a procedure used to obtain relation (1.1.16).

The boundary conditions on the free surface are quite important for our subsequent presentation. If the surface is specified by the equation

$$F(x,\, y,\, z,\, t) = 0, \tag{1.4.5}$$

then kinematic boundary condition can be written in the form

$$\frac{\partial F}{\partial t} + v_x \frac{\partial F}{\partial x} + v_y \frac{\partial F}{\partial y} + v_z \frac{\partial F}{\partial z} = 0, \quad \boldsymbol{r} \in F. \tag{1.4.6}$$

In deducing relations (1.4.6), it is assumed that the free surface has the following property: any liquid particle located on this surface stays on it for the entire period of motion.

The dynamic boundary condition on the free surface is established under the assumption that the pressure is constant on this surface. In this case, it follows from relation (1.4.4) that

$$\boldsymbol{p}_{\nu_1} = p_0 \boldsymbol{\nu}, \quad \boldsymbol{r} \in F, \tag{1.4.7}$$

where $\boldsymbol{\nu}$ is the unit vector of the outer normal to the free surface.

The boundary conditions (1.4.1), (1.4.2), (1.4.6), and (1.4.7) are used in the case of motion of a viscous liquid satisfying the Navier–Stokes equations and the equation of continuity.

For an ideal incompressible liquid, the boundary condition of impermeability

$$\frac{\partial \varphi}{\partial \nu} = \boldsymbol{v}_S \cdot \boldsymbol{\nu}, \quad \boldsymbol{r} \in S, \tag{1.4.8}$$

is usually used as a kinematic condition to obtain the solution of the Laplace equation on the wetted rigid surface. For the immobile boundary $S$, we find

$$\frac{\partial \varphi}{\partial \nu} = 0, \quad \boldsymbol{r} \in S. \tag{1.4.9}$$

Since $\boldsymbol{v} = \operatorname{grad} \varphi$ and the vector $\{F_x, F_y, F_z\}$ is normal to the surface $F$, the kinematic condition (1.4.6) on the free surface takes the form

$$\frac{\partial \varphi}{\partial \nu} = -\frac{F_t}{\sqrt{|\nabla F|^2}} = u_\nu, \tag{1.4.10}$$

where $u_\nu$ is the normal velocity of liquid. The dynamical condition on the free surface of an ideal liquid is formulated by using the Bernoulli equation (1.3.13) under the assumption that the pressure is constant on the free surface and equal to $p_0$. Note that the function $f(t)$ in (1.3.13) can be omitted without loss of generality because, instead of $\varphi$, one can introduce the function

$$\varphi_1 := \varphi - \int\limits_0^t f(t)dt$$

such that

$$\frac{\partial \varphi_1}{\partial t} = \frac{\partial \varphi}{\partial t} + f(t).$$

Therefore, equation (1.3.13) can be rewritten in the form

$$\frac{\partial \varphi}{\partial t} + \frac{1}{2}(\nabla \varphi)^2 + \mathrm{g}x + \frac{p_0}{\rho} = 0. \tag{1.4.11}$$

For simplicity, the term $p_0/\rho$ is included in the derivative $\partial\varphi/\partial t$, by setting $\varphi_2 = \varphi + p_0 t/\rho$. Thus, Eq. (1.3.13) takes the form (the index of the function $\varphi_2$ is omitted)

$$\frac{\partial \varphi}{\partial t} + \frac{1}{2}(\nabla \varphi)^2 + \mathrm{g}x + \frac{p - p_0}{\rho} = 0. \tag{1.4.12}$$

For $p = p_0$, relation (1.4.12) implies the following dynamic boundary condition on the free surface:

$$\frac{\partial \varphi}{\partial t} + \frac{1}{2}(\nabla \varphi)^2 + \mathrm{g}x = 0. \tag{1.4.13}$$

By the similar reasoning, the Bernoulli equation (1.3.20) yields the following dynamic condition on the free surface also for the general case of space motions of the bounded liquid volume:

$$\frac{\partial' \Phi}{\partial t} + \frac{1}{2}(\nabla \Phi)^2 - \nabla \Phi \cdot \boldsymbol{v}_e + U = 0. \tag{1.4.14}$$

Here, $U$ is the potential of the mass (gravity) forces.

The equations of motion of viscous and ideal liquids contain the time derivatives of the required quantities. Hence, in addition to the above-mentioned boundary conditions, it is also necessary to specify, in the general case, the initial distributions of these quantities at time $t = 0$ in the liquid domain and on its boundary.

In the statement of the external problems of hydroaeromechanics connected, e.g., with the circumfluence of finite-size bodies by uniform spatially unbounded flows, it is also necessary to specify one or another condition of boundedness of solutions at infinitely remote points in addition to the boundary and initial conditions introduced above.

## 1.5 Formulation of the main boundary-value problem in a curvilinear coordinate system

In our subsequent presentation, we deal with nonlinear free-surface problems of description of motions of an ideal incompressible liquid (with irrotational flows) partially filling a vessel participating, in the general case, in 3D motions. On the basis of Eqs. (1.3.14), (1.3.19), (1.3.20), (1.4.8), (1.4.10), and (1.4.14), the indicated problem of sloshing can be represented in the following form:

$$\begin{gathered}
\nabla^2 \Phi(x,\, y,\, z,\, t) = 0, \quad \boldsymbol{r} \in Q, \\
\frac{\partial \Phi}{\partial \nu} = \boldsymbol{v}_0 \cdot \boldsymbol{\nu} + \boldsymbol{\omega} \cdot (\boldsymbol{r} \times \boldsymbol{\nu}), \quad \boldsymbol{r} \in S, \\
\frac{\partial \Phi}{\partial \nu} = \boldsymbol{v}_0 \cdot \boldsymbol{\nu} + \boldsymbol{\omega} \cdot (\boldsymbol{r} \times \boldsymbol{\nu}) + u_\nu, \quad \boldsymbol{r} \in \Sigma, \\
\frac{\partial \Phi}{\partial t} + \frac{1}{2}(\nabla \Phi)^2 - \nabla \Phi \cdot (\boldsymbol{v}_0 + \boldsymbol{\omega} \times \boldsymbol{r}) + U = 0, \quad \boldsymbol{r} \in \Sigma.
\end{gathered} \tag{1.5.1}$$

Here, $Q(t)$ is a liquid domain, $S$ and $\Sigma$ are the rigid wetted wall and the free surface, respectively, $\boldsymbol{\nu}$ is the unit vector of the outer normal to the surface of liquid, and $\boldsymbol{v}_0$ and $\boldsymbol{\omega}$ are the vectors of translational and angular velocities of the rigid vessel.

If we know the solution of the free-boundary problem (1.5.1), then we can find the pressure $p$ by using the Bernoulli equation

$$\frac{\partial \Phi}{\partial t} + \frac{1}{2}(\nabla\Phi)^2 - \nabla\Phi \cdot (\boldsymbol{v}_0 + \boldsymbol{\omega} \times \boldsymbol{r}) + U + \frac{p}{\rho} = 0, \quad \boldsymbol{r} \in Q. \tag{1.5.2}$$

Note that the derivative $\partial\Phi/\partial t$ is computed in the mobile coordinate system, i.e., at a point $M$ with coordinates $x, y, z$ rigidly fixed in the indicated mobile system.

The equations obtained above can be represented in the tensor form as follows:

$$\begin{gathered}\nabla\varphi = \boldsymbol{e}_j q^{ij}\frac{\partial\varphi}{\partial x^j}, \quad \frac{\partial\varphi}{\partial\nu} = \nabla\varphi \cdot \boldsymbol{\nu} = q^{ij}\frac{\partial\varphi}{\partial x^j}\nu_i, \\ \Delta\varphi = \nabla^2\varphi = q^{ij}\left(\frac{\partial^2\varphi}{\partial x^i \partial x^j} - \Gamma^k_{ij}\frac{\partial\varphi}{\partial x^k}\right), \\ \Gamma^k_{ij} = \frac{1}{2}q^{\alpha k}\left(\frac{\partial q_{i\alpha}}{\partial x^j} + \frac{\partial q_{i\alpha}}{\partial x^i} - \frac{\partial q_{ij}}{\partial x^\alpha}\right),\end{gathered} \tag{1.5.3}$$

where $q^{ij}$ are the components of the metric tensor and $\Gamma^k_{ij}$ are the Christoffel symbols of the second kind.

We now construct a special parametrization of a domain occupied by the liquid convenient for the formulation and solution of the main boundary-value problem (1.5.1). According to the general idea of introduction of curvilinear coordinates, which is called the nonconformal mapping technique, we consider two spaces referred to a Cartesian coordinate system $xyz$ and to a curvilinear coordinate system $x^1x^2x^3$. In these spaces, we consider two closed domains $\tau$ and $\tau^*$ bounded by piecewise smooth surfaces $S$ and $S^*$, respectively. If these domains are connected by a bijective continuous relationship realized by the formulas

$$x = x(x^1, x^2, x^3), \quad y = y(x^1, x^2, x^3), \quad z = z(x^1, x^2, x^3), \tag{1.5.4}$$

then every inner point $(x_0^1, x_0^2, x_0^3)$ of the domain $\tau^*$ is associated, by virtue of (1.5.4), with the inner point $(x_0, y_0, z_0)$ of the domain $\tau$. Conversely, every inner point of the domain $\tau$ always corresponds to an inner point of the domain $\tau^*$. In this case, the points of the surface $S^*$ correspond to the points of the surface $S$, and vice versa. It is known that this is a consequence of the fact that the Jacobian of the transformation

$$J = \frac{D(x, y, z)}{D(x^1, x^2, x^3)}$$

in the domain $\tau^*$ is nonzero. Setting fixed variables $x^1, x^2, x^3$ in the domain $\tau^*$, we uniquely determine a point of the domain $\tau$. This enables us to say that

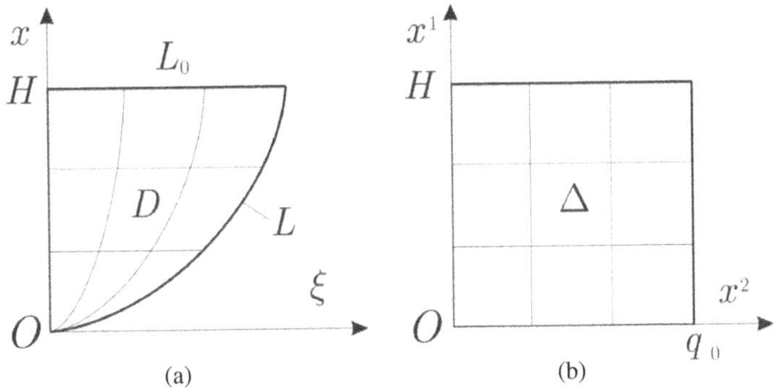


**Figure 1.3**

the numbers $x^1, x^2, x^3$ are (curvilinear) coordinates of points of the domain $\tau$. The points of the space $xyz$ for which one of these coordinates is constant form the corresponding coordinate surface. If a piecewise smooth surface in the domain $\tau^*$ is described by the equations

$$x^1 = x^1(u, v), \; x^2 = x^2(u, v), \; x^3 = x^3(u, v), \tag{1.5.5}$$

then relations (1.5.4) transform it into in a piecewise smooth surface of the domain $\tau$. In this case, the variables $u$ and $v$ vary in a certain domain $E$ of the plane $uv$.

We now construct a transformation of the form (1.5.4) leading to a domain $\tau^*$ of the simplest possible shape. In this case, we also add necessary quantities, such as the components of the metric tensor, which enable us to construct the differential expressions of the boundary-value problem (1.5.1).

In what follows, we restrict ourselves to the investigation of a class of vessels formed by the rotation of a domain $D$ of the plane $x\xi$ about the $x$-axis (Fig. 1.3a). We choose a curvilinear coordinate system in the following way: In the domain $D$ of the plane $x\xi$, we consider two families of curves each of which depends on a single parameter and fills the entire domain under consideration. As a result of the intersection of two neighboring curves in each pair, we obtain a curvilinear quadrangle depicted in Fig. 1.3b.

These two families of curves form a network of coordinate lines and the parameters representing these families are regarded as curvilinear coordinates. As the first family of coordinate lines, we choose a system of straight lines parallel to the $O\xi$-axis:

$$x = p \quad (0 \leqslant p \leqslant H). \tag{1.5.6}$$

We also assume that the system of coordinate lines of the second family whose limiting positions coincide with the $Ox$-axis and with the boundary $L$ of the domain $D$ is characterized by the parameter $q$ $(0 \leqslant q \leqslant q_0)$. Thus, by choosing the parameters $p$ and $q$ and the polar angle $\eta$ ($0 \leqslant \eta \leqslant 2\pi$) as curvilinear coordinates $x^1 = p$, $x^2 = q$, and $x^3 = \eta$, we get the following relationship between the Cartesian and curvilinear coordinates [117]:

$$x = x^1, \quad y = \xi(x^1, x^2)\cos x^3, \quad z = \xi(x^1, x^2)\sin x^3; \tag{1.5.7}$$

$$x^1 = x, \quad x^2 = x^2(x, y, z), \quad x^3 = \operatorname{arctg}\frac{z}{y}. \tag{1.5.8}$$

The domain $\triangle$ of variation of curvilinear coordinates is presented in Fig. 1.3b and the corresponding 3D domain $\tau^*$ is obtained by the rotation of the domain $\Delta$ about the axis $Ox^1$. In this case, the Jacobian of transformation takes the form

$$J = \begin{vmatrix} 1 & 0 & 0 \\ \dfrac{\partial \xi}{\partial x^1}\cos x^3 & \dfrac{\partial \xi}{\partial x^2}\cos x^3 & -\xi\sin x^3 \\ \dfrac{\partial \xi}{\partial x^1}\sin x^3 & \dfrac{\partial \xi}{\partial x^2}\sin x^3 & \xi\cos x^3 \end{vmatrix} = \xi\frac{\partial \xi}{\partial x^2}. \tag{1.5.9}$$

Thus, the bijective correspondence between the points of the domains $D$ and $\Delta$ can be violated only at points for which $\xi = 0$ or $\partial\xi/\partial x^2 = 0$. However, it is known that this does not make the application of the corresponding formulas for curvilinear coordinates impossible provided that the Jacobian remains bounded in the domain $\Delta$. Thus, the volume of the body of revolution in the analyzed coordinate system $x^1x^2x^3$ is given by the formula

$$V = \int_0^H\int_0^{q_0}\int_0^{2\pi} \xi\frac{\partial \xi}{\partial x^2}\,dx^1dx^2dx^3. \tag{1.5.10}$$

By virtue of the well-known relation

$$q_{ij} = \frac{\partial x}{\partial x^i}\frac{\partial x}{\partial x^j} + \frac{\partial y}{\partial x^i}\frac{\partial y}{\partial x^j} + \frac{\partial z}{\partial x^i}\frac{\partial z}{\partial x^j}, \tag{1.5.11}$$

the components of the metric tensor take the form

$$q_{11} = 1 + \left(\frac{\partial \xi}{\partial x^1}\right)^2, \quad q_{12} = q_{21} = \frac{\partial \xi}{\partial x^1}\frac{\partial \xi}{\partial x^2}, \quad q_{13} = q_{31} = 0,$$
$$q_{22} = \left(\frac{\partial \xi}{\partial x^2}\right)^2, \quad q_{23} = q_{32} = 0, \quad q_{33} = \xi^2. \tag{1.5.12}$$

We now introduce the quantity

$$q = \det(q_{ij}) = \xi^2\left(\frac{\partial \xi}{\partial x^2}\right)^2 \tag{1.5.13}$$

and find the components of the mutual metric tensor:

$$q^{11} = 1, \quad q^{12} = q^{21} = -\frac{\partial \xi}{\partial x^1} \Big/ \frac{\partial \xi}{\partial x^2}, \quad q^{13} = q^{31} = 0,$$

$$q^{22} = \frac{1 + (\partial \xi / \partial x^1)^2}{(\partial \xi / \partial x^2)^2}, \quad q^{23} = q^{32} = 0, \quad q^{33} = \frac{1}{\xi^2}. \tag{1.5.14}$$

The elements of the surfaces $x^1 = \text{const}$, $x^2 = \text{const}$, and $x^3 = \text{const}$ are given by the formulas

$$d\sigma^{(1)} = \xi \frac{\partial \xi}{\partial x^2} dx^2 dx^3, \quad d\sigma^{(2)} = \xi \sqrt{1 + \left(\frac{\partial \xi}{\partial x^1}\right)^2} dx^3 dx^1,$$

$$d\sigma^{(3)} = \frac{\partial \xi}{\partial x^2} dx^1 dx^2. \tag{1.5.15}$$

We now consider the problem of representation of an elementary area of a curvilinear surface given in the domain $\tau^*$ by the equation $F(x^1,\ x^2,\ x^3) = 0$. Assume that the equation of this surface has the form

$$x^1 = f(x^2,\ x^3). \tag{1.5.16}$$

By using the parametric representation of the given surface (1.5.5), we choose $u = x^2$ and $v = x^3$ as parameters. The surface element $dS$ is defined via the Gaussian coefficients in the following way:

$$dS = \sqrt{EG - F^2 du\, dv}, \tag{1.5.17}$$

where

$$E = x_u^2 + y_u^2 + z_u^2, \quad F = x_u x_v + y_u y_v + z_u z_v, \quad G = x_v^2 + y_v^2 + z_v^2. \tag{1.5.18}$$

By using the matrix of derivatives

$$\begin{pmatrix} \dfrac{\partial f}{\partial x^2} & \left(\dfrac{\partial \xi}{\partial x^1}\dfrac{\partial f}{\partial x^2} + \dfrac{\partial \xi}{\partial x^2}\right)\cos x^3 & \left(\dfrac{\partial \xi}{\partial x^1}\dfrac{\partial f}{\partial x^2} + \dfrac{\partial \xi}{\partial x^2}\right)\sin x^3 \\ \dfrac{\partial f}{\partial x^3} & \dfrac{\partial \xi}{\partial x^1}\dfrac{\partial f}{\partial x^3}\cos x^3 - \xi \sin x^3 & \dfrac{\partial \xi}{\partial x^1}\dfrac{\partial f}{\partial x^3}\sin x^3 + \xi \cos x^3, \end{pmatrix} \tag{1.5.19}$$

we find

$$E = \left(\frac{\partial f}{\partial x^2}\right)^2 + \left(\frac{\partial \xi}{\partial x^1}\frac{\partial f}{\partial x^2} + \frac{\partial \xi}{\partial x^2}\right)^2,$$

$$G = \left(\frac{\partial f}{\partial x^3}\right)^2 + \left(\frac{\partial \xi}{\partial x^1}\frac{\partial f}{\partial x^3}\right)^2 + \xi^2, \tag{1.5.20}$$

$$F = \frac{\partial f}{\partial x^2}\frac{\partial f}{\partial x^3} + \left(\frac{\partial \xi}{\partial x^1}\right)^2 \frac{\partial f}{\partial x^2}\frac{\partial f}{\partial x^3} + \frac{\partial \xi}{\partial x^1}\frac{\partial f}{\partial x^3}\frac{\partial \xi}{\partial x^2}.$$

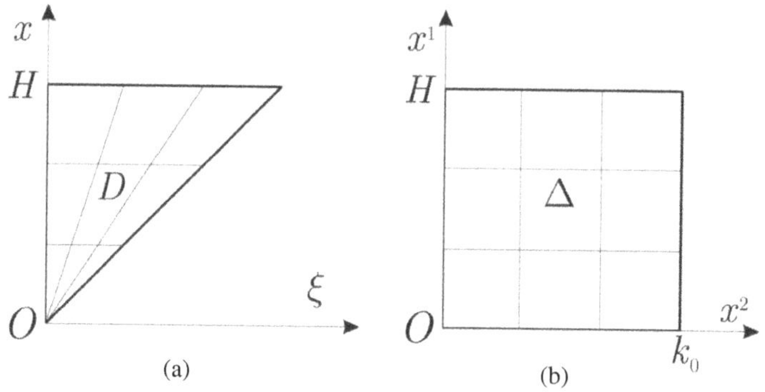


**Figure 1.4**

We now define an elementary area in the curvilinear coordinate system by relation (1.5.17). The indicated relation takes the form

$$dS = \sqrt{1 + q^{22}\left(\frac{\partial f}{\partial x^2}\right)^2 + q^{33}\left(\frac{\partial f}{\partial x^3}\right)^2 - 2q^{12}\frac{\partial f}{\partial x^2}\,\xi\frac{\partial \xi}{\partial x^2}}\,dx^2 dx^3, \qquad (1.5.21)$$

where the components of the mutual metric tensor and the other functions are determined for the value $x^1 = f(x^2, x^3)$.

Furthermore, we present two examples illustrating the curvilinear coordinates introduced above.

1. *Circular cone.* In this case, the domain $D$ has the shape depicted in Fig. 1.4a. As the first family of coordinate lines, we choose a family of straight lines $x = p$ $(0 \leqslant p \leqslant H)$ parallel to the $O\xi$-axis. As the second family, it is convenient to choose a bundle of straight lines going from the coordinate origin and described by the equation $\xi = kx$ $(k = \tan\alpha)$. For $k_0 = \tan\alpha_0$ and $k = 0$, we get the boundary of the cone and the straight line going along the $Ox$-axis, respectively. If we take the parameters $p$ and $k$ as curvilinear coordinates in the domain $D$, i.e., if we set $x^1 = p$, $x^2 = k$, then the formulas of transformations take the form

$$\begin{gathered} x = x^1, \quad y = x^1 x^2 \cos x^3, \quad z = x^1 x^2 \sin x^3, \\ x^1 = x, \quad x^2 = \frac{\sqrt{y^2 + z^2}}{x}, \quad x^3 = \operatorname{arctg}\frac{z}{y}. \end{gathered} \qquad (1.5.22)$$

The domain $\Delta$ of variation of the coordinates $x^1$ and $x^2$ is shown in Fig. 1.4b.

In this case, the components of the metric tensor $q_{ij}$ are given by the formulas

$$q_{11} = 1 + (x^2)^2, \quad q_{12} = q_{21} = x^1 x^2, \quad q_{13} = q_{31} = 0,$$

$$q_{22} = (x^1)^2, \quad q_{23} = q_{32} = 0, \quad q_{33} = (x^1 x^2)^2.$$

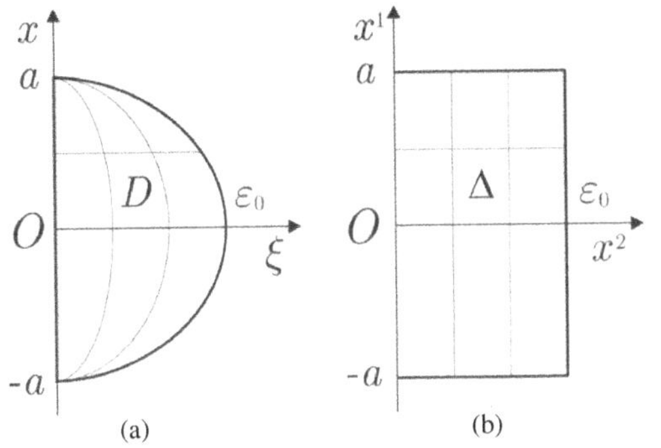


**Figure 1.5**

For the components of the mutual metric tensor, we get

$$q^{11}=1,\quad q^{12}=q^{21}=-\frac{x^2}{x^1},\quad q^{13}=q^{31}=0,$$

$$q^{22}=\frac{1+(x^2)^2}{(x^1)^2},\quad q^{23}=q^{32}=0,\quad q^{33}=\frac{1}{(x^1x^2)^2}. \tag{1.5.23}$$

The Jacobian of transformation takes the form

$$J=\sqrt{q}=(x^1)^2x^2$$

and, hence, the volume of the cone in the curvilinear coordinates $x^1$, $x^2$, $x^3$ is given by the integral

$$V=\int_0^H\int_0^{k_0}\int_0^{2\pi}(x^1)^2x^2dx^1dx^2dx^3=\frac{1}{3}\pi H^3k_0^2, \tag{1.5.24}$$

where $k_0=r_0/H$ and $r_0$ is the base radius of the cone.

2. *Ellipsoid of rotation.* Consider an ellipsoid formed as a result of rotation of the domain $D$ (Fig. 1.5a) about the $Ox$-axis. We denote the major semiaxis of the ellipsoid by $a$ and the minor semiaxis by $b$. Then we form a grid of coordinate lines with the help of the straight lines $x=p$ and ellipses $\xi=\varepsilon\sqrt{a^2-x^2}$ $\left(0\leqslant\varepsilon\leqslant\varepsilon_0=\frac{b}{a}\right)$ constructed on the major axis of the ellipsoid. Thus, by setting $x^1=p$, $x^2=\varepsilon$, we present the formulas of transformation in the form

$$x=x^1,\quad y=x^2\sqrt{a^2-(x^1)^2}\cos x^3,\quad z=x^2\sqrt{a^2-(x^1)^2}\sin x^3,$$

$$x^1=x,\quad x^2=\frac{\sqrt{y^2+z^2}}{\sqrt{a^2-x^2}},\quad x^3=\operatorname{arctg}\frac{z}{y}. \tag{1.5.25}$$

The domain $\Delta$ is shown in Fig. 1.5b. The Jacobian of transformation is given by the formula

$$J = \sqrt{q} = x^2(a^2 - x^1)^2,$$

and the components of the metric tensors $q_{ij}$ and $q^{ij}$ take the form

$$q_{11} = 1 + \frac{(x^1x^2)^2}{a^2 - (x^1)^2}, \quad q_{12} = q_{21} = -x^1x^2, \quad q_{13} = q_{31} = 0,$$

$$q_{22} = a^2 - (x^1)^2, \quad q_{23} = q_{32} = 0, \quad q_{33} = (x^2)^2(a^2 - (x^1)^2);$$

$$q^{11} = 1, \quad q^{12} = q^{21} = \frac{x^1x^2}{a^2 - (x^1)^2}, \quad q^{23} = q^{32} = q^{13} = q^{31} = 0,$$

$$q^{22} = \frac{1}{a^2 - (x^1)^2} + \frac{(x^1/x^2)^2}{(a^2 - (x^1)^2)^2}, \quad q^{33} = \frac{1}{(x^2)^2(a^2 - (x^1)^2)}.$$

The transformed domain $\tau^*$ (and, hence, also the domain $Q(t)$ occupied by the liquid) has some advantages as compared with the initial domain. One of these advantages is connected with the fact that the perturbed free surface can be projected at any time onto a plane domain which does not vary in time and coincides with the unperturbed surface $\Sigma_0^*$. This circumstance plays an important role in the development of approximate methods for the solution of the main nonlinear boundary-value problem (1.5.1) in cavities of complex geometric shapes.

A special place in the nonlinear water-wave theory is occupied by the problem of sloshing of a liquid in a basin. This problem follows from problem (1.5.1) for $\boldsymbol{v}_0 = 0$ and $\boldsymbol{\omega} = 0$. Under the assumption that the equation of perturbed free surface $\Sigma$ can be written as follows:

$$x = f(y, z, t),$$

this problem takes the following form

$$\nabla^2\varphi(x,\ y,\ z,\ t) = 0, \quad \boldsymbol{r} \in Q; \tag{1.5.26}$$

$$\frac{\partial f}{\partial t} - \frac{\partial \varphi}{\partial x} + \frac{\partial \varphi}{\partial y}\frac{\partial f}{\partial y} + \frac{\partial \varphi}{\partial z}\frac{\partial f}{\partial z} = 0, \quad x \in f; \tag{1.5.27}$$

$$\frac{\partial \varphi}{\partial t} + \frac{1}{2}(\nabla\varphi)^2 + gx = 0, \quad x \in f; \tag{1.5.28}$$

$$\frac{\partial \varphi}{\partial \nu} = 0, \quad \boldsymbol{r} \in S. \tag{1.5.29}$$

Equations (1.5.26)–(1.5.29) should be supplemented with the Bernoulli equation (1.3.13).

In the linear theory, the motion is regarded as sufficiently slow, i.e., such that the term $\frac{1}{2}(\nabla\varphi)^2$ in Eqs. (1.5.28) and (1.3.13) can be neglected. The same argument also enables us to simplify the kinematic boundary condition on the free surface:

$$\frac{\partial\varphi}{\partial x} = \frac{\partial f}{\partial t}. \tag{1.5.30}$$

As a result of differentiation of (1.5.28) with respect to $t$, we can replace the dynamic and kinematic conditions on the free surface by a single combined condition

$$\frac{\partial^2\varphi}{\partial t^2} + \mathrm{g}\frac{\partial\varphi}{\partial x} = 0. \tag{1.5.31}$$

Moreover, in view of the fact that the magnitude of liquid sloshing is regarded as small, condition (1.5.31) should be written for the unperturbed free surface $\Sigma_0$.

Thus, the vortex-free motion of a liquid domain in the gravitational field is described, in the linear approximation, by the following relations:

$$\boldsymbol{v} = \nabla\varphi, \quad \frac{p}{\rho} = -\frac{\partial\varphi}{\partial t} - \mathrm{g}x + \frac{p_0}{\rho},$$

where the function $\varphi$ is the solution of the boundary-value problem

$$\nabla^2\varphi = 0, \ \ \boldsymbol{r} \in Q; \quad \frac{\partial\varphi}{\partial x} + \frac{1}{\mathrm{g}}\frac{\partial^2\varphi}{\partial t^2} = 0, \ \ \boldsymbol{r} \in \Sigma_0; \quad \frac{\partial\varphi}{\partial\nu} = 0, \ \ \boldsymbol{r} \in S. \tag{1.5.32}$$

A special place in the analysis of the problems of liquid sloshing is occupied by the problem of free harmonic oscillations. We assume that

$$\varphi(x,\, y,\, z,\, t) = \cos(\sigma t + \varepsilon)\,\widetilde{\varphi}(x,\, y,\, z), \tag{1.5.33}$$

where $\sigma$ is the natural sloshing frequency and $\widetilde{\varphi}(x, y, z)$ is the natural sloshing mode. In view of (1.5.32), the function $\widetilde{\varphi}(x,\, y,\, z)$ depending solely on the space variables can be found from the spectral boundary problem

$$\Delta\widetilde{\varphi} = 0, \ \ \mathbf{r} \in Q; \quad \frac{\partial\widetilde{\varphi}}{\partial x} = \varkappa\widetilde{\varphi}, \ \ \mathbf{r} \in \Sigma_0; \quad \frac{\partial\widetilde{\varphi}}{\partial\nu} = 0, \ \ \mathbf{r} \in S, \tag{1.5.34}$$

where $\varkappa = \sigma^2/\mathrm{g}$ is the eigenvalue which is usually called the frequency parameter.

Chapter 2

# Direct methods in the nonlinear problems of the dynamics of bodies containing liquids

The present chapter is devoted to the background of construction of mathematical models in the dynamics of rigid bodies with cavities containing incompressible ideal liquids with irrotational flows. The analysis is based on the Bateman–Luke variational principle and its generalizations. This principle differs from the classical Lagrange variational principle by a different Lagrangian [16, 119, 199]. It is shown that the indicated variational principle makes it possible to deduce a complete system of equations of motion, including both kinematic and dynamic nonlinear boundary conditions, which naturally follow from the variational formulation.

On the basis of the variational principle, a direct approximate (multimodal) method is developed for the investigation of the dynamics of body–liquid systems. The proposed method reduces the original free-surface (sloshing) problem to a finite-dimensional system of nonlinear ordinary differential equations. As the first example, we consider the variational principle and the multimodal method for a nonlinear sloshing problem in an immobile basin. Then the results are generalized to the case of a rigid vessel participating in a given 3D motion and to the coupled dynamics of the rigid tank and contained liquid caused by the action of external forces and moments applied to the tank. Our presentation in Chapter 2 is based on the results reported in [113, 115, 116, 109, 118].

## 2.1 Variational formulation of the free-surface problem of liquid sloshing in an immobile basin

We assume that the domain $Q$ is occupied by a liquid and bounded by a free surface $\Sigma$ and a rigid wetted wall $S$ in the form of an upright cylinder in the vicinity of the free surface. Assume that the origin of coordinates $Oxyz$ is located at the geometric center of the unperturbed free surface $\Sigma_0$ and that the direction of the $Ox$-axis is opposite to the direction of the gravitational acceleration $\mathbf{g}$. We present the equation of the perturbed free surface $\Sigma$ in the normal form, i.e., resolved with respect to the vertical space variable:

$$x = f(y, z, t). \tag{2.1.1}$$

As discussed in the previous chapter, the corresponding nonlinear boundary problem of liquid sloshing can be formulated as follows:

$$\nabla^2\varphi(x, y, z, t) = 0, \quad \boldsymbol{r} \in Q; \tag{2.1.2}$$

$$\frac{\partial f}{\partial t} - \frac{\partial \varphi}{\partial x} + \frac{\partial \varphi}{\partial y}\frac{\partial f}{\partial y} + \frac{\partial \varphi}{\partial z}\frac{\partial f}{\partial z} = 0, \quad \boldsymbol{r} \in \Sigma; \tag{2.1.3}$$

$$\frac{\partial \varphi}{\partial t} + \frac{1}{2}(\nabla\varphi)^2 + \mathrm{g}x = 0, \quad \boldsymbol{r} \in \Sigma; \tag{2.1.4}$$

$$\frac{\partial \varphi}{\partial \nu} = 0, \quad \boldsymbol{r} \in S, \tag{2.1.5}$$

where $\varphi(x, y, z, t)$ is the velocity potential, $\boldsymbol{\nu}$ is the unit vector of the outer normal to the liquid surface of the domain. The pressure field in the ideal liquid with irrotational flows is specified by the Bernoulli equation (1.4.12).

In what follows, we show that the boundary problem (2.1.2)–(2.1.5) is a necessary condition for the existence of a stationary point of the action

$$W = \int_{t_1}^{t_2} L dt, \tag{2.1.6}$$

where

$$L = -\rho \int_Q \left[\frac{\partial \varphi}{\partial t} + \frac{1}{2}(\nabla\varphi)^2 + \mathrm{g}x\right] dQ. \tag{2.1.7}$$

The Bateman–Luke variational principle based on the action (2.1.6) can be formulated as follows: For the actual motions specifying the liquid sloshing in the gravitational field, integral (2.1.6) possesses a stationary point, i.e.,

$$\delta W = \delta \int_{t_1}^{t_2} L dt = 0. \tag{2.1.8}$$

Specifically, action (2.1.6) is a function of the velocity potential $\varphi(x, y, z, t)$ and its domain of definition $Q(t)$ (the free surface $\Sigma(t)$). In the variational principle, it is assumed that the volume of liquid is preserved. In the case where the volume of liquid is infinite, the equations of motion (2.1.2)–(2.1.5) were derived from the variational principle (2.1.8) in [107]. The case of a finite volume of liquid was studied, e.g., in [113, 116].

The variational problem obtained from the Bateman–Luke variational principle can be formulated as follows:

For all functions $\varphi(x, y, z, t)$ and $f(y, z, t)$ continuous together with their derivatives with respect to the space variables and time $t$, it is necessary to find a pair of functions for which the integral $W$ takes a stationary value. The "actual motions" are, in this case, defined by the required pair of functions $\varphi(x, y, z, t)$ and $f(y, z, t)$, whereas the comparison motions are defined by the functions $\varphi + \delta\varphi$ and $f + \delta f$. Here, the variations $\delta\varphi$ and $\delta f$ are arbitrary functions with the corresponding domains of definition satisfying the conditions

$$\delta\varphi(x, y, z, t_1) = 0, \quad \delta\varphi(x, y, z, t_2) = 0; \tag{2.1.9}$$

$$\delta f(y, z, t_1) = 0, \quad \delta f(y, z, t_2) = 0. \tag{2.1.10}$$

These conditions reflect the fact that, the initial and final positions of the mechanical system are the same for the actual and comparison motions.

In what follows, we assume that, for comparison motions, the functions $\varphi(x, y, z, t, \alpha)$ and $f(y, z, t, \alpha)$ depend on the parameter $\alpha$ sufficiently close to zero. Moreover, the value $\alpha = 0$ corresponds to the actual motion with $\varphi(x, y, z, t) = \varphi(x, y, z, t, 0) and f(y, z, t) = f(y, z, t, 0)$ for which the first variation (2.1.8) is equal to zero. In agreement with the general definition of the first variation, we get

$$\delta\varphi = \left.\frac{\partial\varphi}{\partial\alpha}\right|_{\alpha=0} \alpha, \quad \delta f = \left.\frac{\partial f}{\partial\alpha}\right|_{\alpha=0} \alpha. \tag{2.1.11}$$

Differentiating (2.1.6) with respect to $\alpha$, setting $\alpha = 0$, and multiplying by $\alpha$, we arrive at the following relations for the first variation:

$$\begin{aligned}
\delta W &= \delta \int_{t_1}^{t_2} \int_{\Sigma_0} \left[ \int_{h(y,z)}^{f(y,z,t)} \left( \varphi_t + \frac{1}{2}(\nabla\varphi)^2 + \text{g}x \right) dx \right] dSdt \\
&= \int_{t_1}^{t_2} \int_{\Sigma_0} \left\{ \left[ \varphi_t + \frac{1}{2}(\nabla\varphi)^2 \text{g}x \right]_{x=f} \delta f \right. \\
&\quad \left. + \int_{h(y,z)}^{f(y,z,t)} (\varphi_x\delta\varphi_x + \varphi_y\delta\varphi_y + \varphi_z\delta\varphi_z + \delta\varphi_t) dx \right\} dSdt = 0.
\end{aligned} \tag{2.1.12}$$

Here, $x = h(x, y)$ is the equation of the bottom of a container . The integrand of the second integral can be represented in the form

$$\varphi_x\delta\varphi_x + \varphi_y\delta\varphi_y + \varphi_z\delta\varphi_z + \delta\varphi_t = \nabla\varphi \cdot \nabla\delta\varphi + \delta\varphi_t. \tag{2.1.13}$$

Hence, by the Green formula, we get

$$\int_Q \nabla\varphi \cdot \nabla(\delta\varphi) dQ = \int_{S+\Sigma} \frac{\partial\varphi}{\partial\nu} \delta\varphi dS - \int_Q \delta\varphi \nabla^2\varphi dQ. \tag{2.1.14}$$

Since the operations of variation and differentiation with respect to time are commutative in the problem under consideration, we obtain

$$\int_Q \delta\varphi_t dQ = \int_Q \frac{\partial}{\partial t}(\delta\varphi) dQ. \tag{2.1.15}$$

The integral on the right-hand side of (2.1.15) can be transformed by using the Reynolds transport theorem (1.1.4) for the domain $Q(t)$ which is not fixed due to the presence of the free surface $\Sigma$. In theorem (1.1.4), $v_\nu$ denotes the normal velocity of the free boundary of the domain $Q$ regarded as positive along the outer normal to $S+\Sigma$. Hence,

$$\int_Q \frac{\partial}{\partial t}(\delta\varphi)dQ = \frac{d}{dt}\int_Q \delta\varphi dQ - \int_\Sigma \delta\varphi \frac{f_t}{\sqrt{1+f_y^2+f_z^2}}\delta f dS. \tag{2.1.16}$$

Substituting (2.1.14) and (2.1.16) in (2.1.12), we find

$$\delta W = \int_{t_1}^{t_2} \Bigg\{ \int_{\Sigma_0} \left[\varphi_t + \frac{1}{2}(\nabla\varphi)^2 + gx\right]_{x=f} \delta f dS + \int_{S+\Sigma} \left(\frac{\partial\varphi}{\partial\nu} - v_\nu\right)\delta\varphi dS$$
$$- \int_Q \nabla^2\varphi\delta\varphi dQ + \frac{d}{dt}\int_Q \delta\varphi dQ \Bigg\} dt = 0. \tag{2.1.17}$$

In view of conditions (1.1.9) and (1.1.10), the variations $\delta\varphi$ and $\delta f$ are equal to zero for $t = t_1$ and $t = t_2$. Therefore, the last term in (2.1.17) also turns to zero.

Passing to the final form of the expression for the first variation of the functional $W$, we see that, for the mobile boundary of the liquid domain described by the equation $\zeta(x, y, z, t) = 0$, the normal velocity $v_\nu$ is given by the following well-known relation:

$$v_\nu = -\frac{\zeta_t}{\sqrt{(\nabla\zeta)^2}}. \tag{2.1.18}$$

Hence, on the free surface $\Sigma$, we get

$$v_\nu = \frac{f_t}{\sqrt{1+f_y^2+f_z^2}}. \tag{2.1.19}$$

Moreover, on the wetted rigid wall $S$, we obtain $v_\nu = 0$. In view of the fact that an elementary area of the perturbed free surface $\Sigma$ is defined as

$$dS = dS_{yz}\sqrt{1+f_y^2+f_z^2}, \tag{2.1.20}$$

and the normal derivative on this surface is given by the formula

$$\frac{\partial\varphi}{\partial\nu} = \frac{1}{\sqrt{1+f_y^2+f_z^2}}\left(\frac{\partial\varphi}{\partial x} - \frac{\partial\varphi}{\partial y}\frac{\partial f}{\partial y} - \frac{\partial\varphi}{\partial z}\frac{\partial f}{\partial z}\right), \tag{2.1.21}$$

we finally find

$$\delta W = \int_{t_1}^{t_2} \left( \int_{\Sigma_0} \left\{ \left[ \varphi_t + \frac{1}{2}(\nabla\varphi)^2 + \mathrm{g}x \right]_{x=f} \delta f \right.\right.$$
$$\left. + \left[ \left( \frac{\partial\varphi}{\partial x} - \frac{\partial\varphi}{\partial y}\frac{\partial f}{\partial y} - \frac{\partial\varphi}{\partial z}\frac{\partial f}{\partial z} - f_t \right) \delta\varphi \right]_{x=f} \right\} dS$$
$$\left. + \int_S \frac{\partial\varphi}{\partial\nu}\delta\varphi dS - \int_Q \Delta\varphi\delta\varphi dQ \right) dt = 0. \tag{2.1.22}$$

In view of the arbitrariness of $\delta f$, $[\delta\varphi]_{x=f}$, and $\delta\varphi$ on $S$, by the standard reasoning, we obtain the nonlinear boundary problem (2.1.2)–(2.1.5) from (2.1.22). It is easy to see that all boundary conditions of the problem are natural for this variational formulation.

We now consider containers of more complicated geometric shapes with the help of suitable curvilinear coordinates and the invariant form of the adopted variational formulation. We restrict ourselves to the case of containers in the form of bodies of revolution and introduce a Cartesian coordinate system $xyz$ and a curvilinear coordinate system $x^1x^2x^3$ connected by the following relations:

$$x = x^1, \quad y = \xi(x^1, x^2)\cos x^3, \quad z = \xi(x^1, x^2)\sin x^3; \tag{2.1.23}$$

$$x^1 = x, \quad x^2 = x^2(x, y, z), \quad x^3 = \operatorname{arctg}\frac{z}{y}. \tag{2.1.24}$$

The function $\xi(x^1, x^2)$ in (2.1.23) depends on the shape of the container and is constructed by using the principle described above, so that the domain $\Delta$ of variation of the parameters $x^1$ and $x^2$ becomes rectangular. The components of the metric tensor $q_{ij}$ in the space referred to the coordinate system $x^1x^2x^3$ are given by the formulas

$$\begin{aligned} q_{11} &= 1 + \left(\frac{\partial\xi}{\partial x^1}\right)^2, \quad q_{12} = q_{21} = \frac{\partial\xi}{\partial x^1}\frac{\partial\xi}{\partial x^2}, \quad q_{13} = q_{31} = 0, \\ q_{22} &= \left(\frac{\partial\xi}{\partial x^2}\right)^2, \quad q_{23} = q_{32} = 0, \quad q_{33} = \xi^2. \end{aligned} \tag{2.1.25}$$

Hence, the chosen curvilinear coordinates are not orthogonal.

The components of the mutual metric tensor are as follows:

$$\begin{aligned} q^{11} &= 1, \quad q^{12} = q^{21} = -\frac{\partial\xi}{\partial x^1}\Big/\frac{\partial\xi}{\partial x^2}, \quad q^{13} = q^{31} = 0; \\ q^{22} &= 1 + \left(\frac{\partial\xi}{\partial x^1}\right)^2 \Big/ \left(\frac{\partial\xi}{\partial x^2}\right), \quad q^{23} = q^{32} = 0, \quad q^{33} = \frac{1}{\xi^2}. \end{aligned} \tag{2.1.26}$$

Relations (2.1.25) and (2.1.26) enable us to find the necessary metric characteristics of the domain $Q$ in the coordinate system $x^1x^2x^3$ and to get the required differential and integral equations in this space. Thus, an elementary volume in the curvilinear coordinate system is given by the formula

$$dQ = \sqrt{q}\, dx^1 dx^2 dx^3, \tag{2.1.27}$$

where

$$q = \det|q_{ij}| = \xi^2 \left(\frac{\partial \xi}{\partial x^2}\right)^2. \tag{2.1.28}$$

For elements of the surfaces $x^1 = \text{const}$, $x^2 = \text{const}$, and $x^3 = \text{const}$, we can write

$$d\sigma^{(1)} = \sqrt{qq^{11}}\, dx^2 dx^3, \quad d\sigma^{(2)} = \sqrt{qq^{22}}\, dx^3 dx^1, \quad d\sigma^{(3)} = \sqrt{qq^{33}}\, dx^1 dx^2. \tag{2.1.29}$$

Assume that the equation of the surface $\zeta(x^1, x^2, x^3) = 0$ can be resolved with respect to $x^1$, i.e., we can write

$$\zeta(x^1,\, x^2,\, x^3) = x^1 - x_0^1 - f(x^2,\, x^3) = 0, \tag{2.1.30}$$

This enables us to compute the elementary area of this surface as follows:

$$dS = \sqrt{1 + q^{22}\left(\frac{\partial f}{\partial x^2}\right)^2 + q^{33}\left(\frac{\partial f}{\partial x^3}\right)^2 - 2q^{12}\frac{\partial f}{\partial x^2}}\, \sqrt{q}\, dx^2 dx^3. \tag{2.1.31}$$

The statement of the free-surface problem (2.1.2)–(2.1.5) in the chosen coordinate system $x^1x^2x^3$ has the following advantage for the analyzed class of containers: the shape of the free surface can be represented in terms of the spatial variable $x^1$ as follows:

$$\zeta(x^1, x^2, x^3, t) \equiv x^1 - x_0^1 - f(x^2, x^3, t) = 0. \tag{2.1.32}$$

Moreover, the domain of definition of the function $f(x^2, x^3, t)$ in the space variables $x^2$ and $x^3$ is independent of time and its shape coincides with the shape of the unperturbed free surface.

In the variational principle formulated above (2.1.8), it is convenient to represent the Lagrangian $L$ in the form

$$L = -\rho \int_D \left( \int_0^{x_0^1+f} \left[\varphi_t + \frac{1}{2}(\nabla\varphi)^2 + \mathrm{g}x^1\right] \sqrt{q}\, dx^1 \right) dx^2 dx^3, \tag{2.1.33}$$

where the upper limit of the internal integral depends on $x^2$, $x^3$, and $t$ according to (2.1.32) and $D$ is the domain of variation of the parameters $x^2$ and $x^3$.

The variational equation (2.1.8) can be reduced to the form

$$\delta W = \int\limits_{t_1}^{t_2} \int\limits_{D} \Bigg\{ \left[ \varphi_t + \frac{1}{2}(\nabla\varphi)^2 + \mathrm{g}x^1 \right] \sqrt{q}\,|_{x^1=x_0^1+f}\, \delta f$$

$$+ \int\limits_{0}^{x_0^1+f} (\nabla\varphi \cdot \nabla\delta\varphi + \delta\varphi_t)\sqrt{q}\, dx^1 \Bigg\} dx^2 dx^3 dt = 0. \qquad (2.1.34)$$

The subsequent transformations can be performed by analogy with the transformations made above by applying the appropriate integral theorems in the tensor form. Finally, relation (2.1.34) takes the form

$$\delta W = \int\limits_{t_1}^{t_2} \Bigg( \int\limits_{D} \Bigg\{ \left[ \varphi_t + \frac{1}{2}(\nabla\varphi)^2 + \mathrm{g}x^1 \right] \sqrt{q}\,|_{x^1=x_0^1+f}\, \delta f$$

$$+ \Big[ (\nabla\varphi \cdot \nabla\zeta - f_t)\sqrt{q}\,\delta\varphi \Big]_{x^1=x_0^1+f} \Bigg\} dx^2 dx^3$$

$$+ \int\limits_{S} \frac{\partial\varphi}{\partial\nu}\delta\varphi dS - \int\limits_{Q} \nabla^2\varphi\delta\varphi dQ \Bigg) dt = 0. \qquad (2.1.35)$$

In view of the arbitrariness of $\delta f$, $[\delta\varphi]_{x^1=x_0^1+f}$, and $\delta\varphi$ on $S$, we get the corresponding free-boundary problem from (2.1.35). It is worth noting that the stationary point of the functional $W$ should be sought in the class of solutions satisfying the condition of conservation of volume. In the analyzed case, this condition takes the following form:

$$\int\limits_{D} dx^2 dx^3 \int\limits_{x_0^1}^{x_0^1+f} \sqrt{q}\, dx^1 = 0. \qquad (2.1.36)$$

Specifying the shape of a container, we explicitly express condition (2.1.36) via the function $f(x^2, x^3, t)$. Thus, for the containers in the form of cones, ellipsoids, or paraboloids of revolution, this dependence takes the form

$$\int\limits_{\Sigma_0} (f + \alpha_1 f^2 + \alpha_2 f^3)\, d\sigma_0^{(1)} = 0. \qquad (2.1.37)$$

The condition of conservation of volume is closely connected with the condition of solvability of the investigated boundary-value problem

$$\int\limits_{\Sigma} \frac{\zeta_t}{\sqrt{(\nabla\zeta)^2}} dS = 0. \qquad (2.1.38)$$

In the relations presented above, we have

$$\nabla\varphi = \boldsymbol{e}_i q^{ij}\frac{\partial\varphi}{\partial x^j}, \quad \frac{\partial\varphi}{\partial\nu} = \nabla\varphi\cdot\boldsymbol{\nu} = q^{ij}\frac{\partial\varphi}{\partial x^j}\nu_i,$$

$$\nabla^2\varphi = q^{ij}\left(\frac{\partial^2\varphi}{\partial x^i\partial x^j} - \Gamma^k_{ij}\frac{\partial\varphi}{\partial x^k}\right), \tag{2.1.39}$$

$$\Gamma^k_{ij} = \frac{1}{2}q^{\alpha k}\left(\frac{\partial q_{i\alpha}}{\partial x^j} + \frac{\partial q_{i\alpha}}{\partial x^i} - \frac{\partial q_{ij}}{\partial x^k}\right).$$

In its physical meaning, the variational principle presented above differs from the classical Lagrange formulation by the choice of a different Lagrangian. In the classical version, the Lagrangian is defined as the difference between the kinetic and potential energies of the contained liquid. In our case, it is equal to pressure. This Lagrangian enables us to get a complete system of governing equations and boundary conditions, including both nonlinear boundary conditions on the free surface. At the same, only dynamic boundary condition follows from the classical Lagrange formulation.

The equivalence of the free-surface problem (2.1.2)–(2.1.5) and the corresponding variational problem for the quadratic functional (2.1.6) established, in a certain sense, earlier is used in what follows for the construction of approximate solutions of the sloshing problem.

In conclusion, we note that the variational formulation based on the classical Lagrange variational principle was used in numerous works (see, e.g., [100, 103, 152]) for the construction of discrete mathematical models of the motion of liquid with free boundary. In this case, the Lagrangian is chosen in the form

$$L_\Gamma = \rho\int_Q\left[\frac{1}{2}(\nabla\varphi)^2 - \mathrm{g}x\right]dQ, \tag{2.1.40}$$

and the variational equation

$$\delta\int_{t_1}^{t_2} L_\Gamma dt = 0 \tag{2.1.41}$$

is equivalent to the boundary problem (2.1.2)–(2.1.5) in the class of harmonic functions satisfying the kinematic boundary conditions (2.1.3) and (2.1.5). This implies that the Lagrangians $L$ and $L_\Gamma$ differ from each other by the quantity

$$\triangle L = L - L_\Gamma = -\rho\int_Q\left[\frac{\partial\varphi}{\partial t} + (\nabla\varphi)^2\right]dQ. \tag{2.1.42}$$

Transforming the first term in (2.1.42) by the Reynolds transport theorem and the second Green formula, we obtain

$$\Delta L = \rho \int\limits_{\Sigma} \left( \frac{\partial \varphi}{\partial x} - \frac{\partial f}{\partial t} - \frac{\partial \varphi}{\partial y}\frac{\partial f}{\partial y} - \frac{\partial \varphi}{\partial z}\frac{\partial f}{\partial z} \right) dS$$
$$+ \int\limits_{S} \frac{\partial \varphi}{\partial n} \varphi dS - \int\limits_{Q} \nabla^2 \varphi \, \varphi dQ - \frac{d}{dt} \int\limits_{Q} \varphi dQ. \tag{2.1.43}$$

This implies that the variational formulations (2.1.17) and (2.1.41) are equivalent in the class of harmonic functions satisfying the kinematic conditions (2.1.3) and (2.1.5).

## 2.2 Multimodal method for an immobile basin

The nonlinear free-boundary problem (2.1.2)–(2.1.5) belongs to the class of fairly complicated boundary-value problems of the continuum mechanics. Its complexity is caused, on the one hand, by the necessity of satisfying two nonlinear boundary conditions on the free surface. On the other hand, the fact that the free surface is unknown in advance and should be found in the process of solution causes much more serious difficulties. The approximate methods [161, 171, 173, 181] proposed earlier for the solution the indicated problem are based on the ideas of perturbation theory and allow one to reduce the solution of the nonlinear problem to a sequence of linear boundary-value problems in the unperturbed liquid domain. These methods are fairly efficient when these linear boundary-value problems admit simple analytic solutions.

The variational principles prove to be an additional important source for the construction of approximate methods aimed at the solution of nonlinear sloshing problem. The methods based on these principles have a series of important advantages typical of the energy methods.

First, we consider the case of cavities of cylindrical shape in the vicinity of the free surface [113]. We specify the equation of perturbed free surface (2.1.1) in the constructive form as the Fourier series

$$x = f(y, z, t) = \sum_{i=1}^{\infty} \beta_i(t) f_i(y, z), \tag{2.2.1}$$

where $f_i(y, z)$ is a complete system of functions orthogonal, together with a constant, and defined on the unperturbed free surface $\Sigma_0$ and $\beta_i(t)$ are the generalized Fourier coefficients depending on time (they have the meaning of generalized coordinates characterizing the deviations of the free surface):

$$\beta_i(t) = \int\limits_{\Sigma_0} f(y, z, t) f_i(y, z) dS. \tag{2.2.2}$$

Since Eq. (2.1.2) and the boundary condition (2.1.5) of (2.1.2)–(2.1.5) are linear, we specify the velocity potential $\varphi(x, y, z, t)$ in the form

$$\varphi(x, y, z, t) = \sum_{n=1}^{\infty} R_n(t)\varphi_n(x, y, z), \tag{2.2.3}$$

where $R_n(t)$ are parameters characterizing the time dependences of the velocity potential and $\varphi_n(x, y, z)$ is the system of harmonic functions in the domain $Q$ satisfying the boundary condition on the wetted surface $S$.

It is known that the construction of the systems of functions $f_i(y, z)$ and $\varphi_i(x, y, z)$ appearing in relations (2.2.2) and (2.2.3) is reduced to the solution of the spectral boundary problem with eigenvalue in the boundary conditions of the form (1.5.34), i.e.

$$\nabla^2\varphi = 0, \quad \boldsymbol{r} \in Q; \quad \frac{\partial\varphi}{\partial x} = \varkappa\varphi, \quad \boldsymbol{r} \in \Sigma_0; \quad \frac{\partial\varphi}{\partial\nu} = 0, \quad \boldsymbol{r} \in S. \tag{2.2.4}$$

The eigensolutions of this problem form a complete system of functions $f_i = \varkappa_i\varphi_i(0, y, z)$ orthogonal, together with a constant, on the surface $\Sigma_0$ [164, 59].

The spectral boundary problem (2.2.4) is the main problem of the linear sloshing theory. In some cases, this problem can be solved by the method of separation of variables. However, in the major part of cases, its solution can be found only approximately. Thus, the energy methods, such as, e.g., the Ritz–Treftz method, are now most extensively used for the solution of problem (2.2.4).

We now substitute (2.2.3) and (2.2.1) in relation (2.1.7). Integrating first with respect to the variable $x$ and then with respect to the variables $y$ and $z$, we arrive at the Lagrangian, which is a function of $\beta_i$, $R_n$, and $\dot{R}_n$. To find $\beta_i(t)$ and $R_n(t)$, we use relation (2.1.8) leading to the following system of nonlinear ordinary differential equations:

$$\frac{\partial L}{\partial\beta_i} = 0, \; i = 1, 2, \dots, \quad \frac{d}{dt}\frac{\partial L}{\partial\dot{R}_n} - \frac{\partial L}{\partial R_n} = 0, \; n = 1, 2, \dots. \tag{2.2.5}$$

In a more detailed form, we get

$$L = -\rho\int\limits_Q \left[\sum_{n=1} \dot{R}_n\varphi_n + \frac{1}{2}\sum_n\sum_k R_nR_k(\nabla\varphi_n, \nabla\varphi_k) + \mathrm{g}x\right] dQ$$

$$= -\left(\sum_n A_n\dot{R}_n + \frac{1}{2}\sum_n\sum_k A_{nk}R_nR_k + \mathrm{g}l_1\right), \tag{2.2.6}$$

where the quantities

$$A_n = \rho\int\limits_Q \varphi_n dQ, \quad A_{nk} = A_{kn} = \rho\int\limits_Q (\nabla\varphi_n, \nabla\varphi_k)dQ = \rho\int\limits_{S+\Sigma} \varphi_n\frac{\partial\varphi_n}{\partial\nu}dS,$$

$$l_1 = \rho \int\limits_Q x dQ \tag{2.2.7}$$

are functions of the parameters $\beta_i(t)$ [by virtue of (2.2.1)].

In view of notation (2.2.7), the system of nonlinear equations (2.2.5) takes the form

$$\frac{d}{dt} A_n - \sum_k A_{nk} R_k = 0, \quad n = 1, 2, \dots, \tag{2.2.8}$$

$$\sum_n \dot{R}_n \frac{\partial A_n}{\partial \beta_i} + \frac{1}{2} \sum_n \sum_k \frac{\partial A_{nk}}{\partial \beta_i} R_n R_k + \mathrm{g} \frac{\partial l_1}{\partial \beta_i} = 0, \quad i = 1, 2, \dots. \tag{2.2.9}$$

Since

$$l_1 = \rho \int\limits_Q x dQ = \frac{1}{2} \rho \int\limits_{\Sigma_0} f^2 dS + C_1, \tag{2.2.10}$$

where $C_1$ is a constant depending on the geometry of the cavity, we obtain

$$\frac{\partial l_1}{\partial \beta_i} = \rho N_i^2 \beta_i = \lambda_{1i} \beta_i, \quad N_i^2 = \int\limits_\Sigma f_i^2 dS. \tag{2.2.11}$$

Another tool for the construction of a system of nonlinear equations in the parameters $R_n$ and $\beta_i$ is the direct application of the formula (2.1.22) for the first variation of the action $W$. In view of (2.1.22) and the restrictions imposed on the system of coordinate functions $\varphi_i(x, y, z)$, we arrive at the following system of nonlinear ordinary differential equations:

$$\sum_n B_n^{(i)} \dot{R}_n + \frac{1}{2} \sum_k \sum_n B_{kn}^{(i)} R_k R_n + \mathrm{g} B^{(i)} \beta_i = 0, \quad i = 1, 2, \dots; \tag{2.2.12}$$

$$\sum_k R_k C_k^{(n)} - \sum_k \sum_i R_k \beta_i C_{ki}^{(n)} - \sum_i \dot{\beta}_i G_i^{(n)} = 0, \quad n = 1, 2, \dots, \tag{2.2.13}$$

where

$$B_n^{(i)} = \rho \int\limits_{\Sigma_0} (\varphi_n f_i)_{x=f}\, dS, \quad B_{kn}^{(i)} = \rho \int\limits_{\Sigma_0} [(\nabla \varphi_n, \nabla \varphi_k) f_i]_{x=f}\, dS,$$

$$B^{(i)} = \rho \int\limits_{\Sigma_0} f_i^2 dS, \quad C_k^{(n)} = \rho \int\limits_{\Sigma_0} \left( \frac{\partial \varphi_k}{\partial x} \varphi_n \right)_{x=f} dS,$$

$$C_{ki}^{(n)} = \rho \int\limits_{\Sigma_0} \left[ \left( \frac{\partial \varphi_k}{\partial y} \frac{\partial f_i}{\partial y} + \frac{\partial \varphi_k}{\partial z} \frac{\partial f_i}{\partial z} \right) \varphi_n \right]_{x=f} dS,$$

$$G_i^{(n)} = \rho \int\limits_{\Sigma_0} (f_i \varphi_n)_{x=f}\, dS. \tag{2.2.14}$$

Comparing the systems of equations (2.2.8), (2.2.9) and (2.2.12), (2.2.13), we establish the following relations between the coefficients of (2.2.12) and (2.2.13)

$$B_k^{(i)} = -G_i^{(k)}, \quad B_{kn}^{(i)} = \frac{\partial}{\partial \beta_i}\left(\sum_i C_{ki}^{(n)} \beta_i - C_k^{(n)}\right), \tag{2.2.15}$$

which are quite useful for the practical purposes.

Both outlined procedures used for the construction of nonlinear modal equations and computing their hydrodynamic coefficients can be used in numerous cases for the independent control of the results of numerical realization of the method.

Note that systems (2.2.8) and (2.2.13) are linear with respect to $R_n(t)$. Therefore, they can be resolved with respect to these parameters. As a result, relations (2.2.9) or (2.2.12) yield a system of nonlinear differential equations of the second order for the parameters $\beta_i(t)$ characterizing the deviations of the free surface from its unperturbed position.

For the system of equations (2.2.8), (2.2.9) or (2.2.12), (2.2.13), we can pose the Cauchy problem by using the shape of the free surface and the distribution of normal velocities over this surface at the initial time as initial conditions of the problem. As an alternative, we can mention the conditions of periodicity specifying the steady-state sloshing.

The investigation of the obtained infinite-dimensional systems of nonlinear ordinary differential equations aimed at the description of liquid sloshing encounters significant mathematical difficulties. Some practically useful results can be obtained if we restrict ourselves to finitely many parameters $R_n(t)$ and $\beta_i(t)$ and select the parameters playing the predominant role in various special cases.

The analytic realization of the proposed procedure is based on imposing the restrictions on the order of smallness of these parameters. However, these restrictions should be introduced only in the final stage of solution of the problem. This gives us the possibility to deduce the equations of motion with regard for the terms of higher orders of smallness with less efforts than, e.g., in the case of application of the method of perturbation theory.

We also note that the coefficients of the nonlinear equations of motion (2.2.8), (2.2.9) and (2.2.12), (2.2.13) are determined by the integral quantities of the form (2.2.7) and (2.2.14). For their evaluation, it is sufficient to have the eigensolutions of Eq. (2.2.4) characterized, together with their first derivatives, solely by the weak convergence. These are, e.g., the corresponding solutions of Eq. (2.2.4) obtained by the Ritz–Treftz method [59].

In what follows, by using a specific example, we show that the coefficients of model equations can be found with the help of these solutions with a sufficiently high accuracy.

It is also worth noting that, in the analyzed method, the system of coordinate functions $\varphi_n$ in the Fourier series (2.2.3) should not necessarily be subordinated to any additional restrictions, except the restrictions required for the realization of the variational problems for functionals with natural boundary conditions.

## 2.3 Modal equations for noncylindrical basins

We now consider noncylindrical containers in the form of the bodies of revolution and introduce a curvilinear coordinate system $x^1x^2x^3$ connected with a Cartesian coordinate system $xyz$ by relations (2.1.23) and (2.1.24). The equation of the perturbed free surface in the space of parameters $x^1x^2x^3$ can be represented in the form

$$x^1 = x_0^1 + f(x^2, x^3, t) = x_0^1 + \beta_0(t) + \sum_{i=1}^{\infty} \beta_i(t) f_i(x^2, x^3), \qquad (2.3.1)$$

where $f_i(x^2, x^3)$ is the complete system of functions orthogonal, with a certain weight, and given in the domain of variation of the variables $x^2x^3$, and $\beta_i(t)$ are the generalized coordinates characterizing the free-surface patterns.

For the velocity potential $\varphi(x^1, x^2, x^3, t)$, we use the Fourier (modal) solution

$$\varphi(x^1, x^2, x^3, t) = \sum_{n=1}^{\infty} R_n(t) \varphi_n(x^1, x^2, x^3), \qquad (2.3.2)$$

where $R_n(t)$ characterize the time variations of the velocity potential and $\varphi_n$ is a complete system of functions in the domain occupied by the vessel.

As above, we relate the choice of the system of functions $f_i(x^2, x^3)$ and $\varphi_i(x^1, x^2, x^3)$ to the eigensolutions of the spectral boundary problem (2.2.4) in the Cartesian coordinates. We substitute (2.3.1) and (2.3.2) in relation (2.1.33) for the function $L$. From the condition of steadiness of the functional $W$, we deduce the system of nonlinear ordinary differential equations similar to (2.2.8) and (2.2.9) and connecting $\beta_i(t)$ and $R_n(t)$. In this case, relations (2.2.7) take the form

$$\begin{aligned}
A_n &= \rho \int_D \left( \int_0^{x_0^1+f} \varphi_n \sqrt{q}\, dx^1 \right) dx^2 dx^3, \\
A_{nk} &= \rho \int_D \left[ \int_0^{x_0^1+f} (\nabla\varphi_n, \nabla\varphi_k) \sqrt{q}\, dx^1 \right] dx^2 dx^3, \qquad (2.3.3) \\
l_1 &= \rho \int_D \left[ \int_0^{x_0^1+f} (x_0^1 + f) \sqrt{q}\, dx^1 \right] dx^2 dx^3,
\end{aligned}$$

where $D$ is the domain of variation of the variables $x^2$ and $x^3$. Relations (2.3.3) are mainly used in the cases where the integrands take the form convenient for the integration with respect to the variable $x^1$.

More universal forms of the equations and relations for their coefficients can be obtained directly from the formula for the first variation of the action $W$ (2.1.35). In what follows, for a fairly general case of dependence of the functions $\varphi_n$ and $\sqrt{q}$ on the variable $x^1$, we assume that the value of a function $F(x^1, x^2, x^3)$ for $x^1 = x_0^1 + f$ appearing in the variational equation (2.1.35) can be obtained with a certain degree of accuracy with the help of the following Taylor series in the tensor form:

$$F(x_0^1 + f, x^2, x^3) = [F]_{x^1=x_0^1} + f[\nabla_1 F]_{x^1=x_0^1} + \frac{1}{2} f^2 [\nabla_{11} F]_{x^1=x_0^1} + \dots . \tag{2.3.4}$$

Here, the covariant derivatives $\nabla_i$ and $\nabla_{ij}$ are associated with the corresponding derivatives by the formulas

$$\nabla_i F = \frac{\partial F}{\partial x^i}, \quad \nabla_{ij} F = \frac{\partial^2 F}{\partial x^i \partial x^j} - \Gamma_{ij}^{\nu} \frac{\partial F}{\partial x^\nu}. \tag{2.3.5}$$

Substituting (2.3.1) and (2.3.2) in the variational equation (2.1.35), we get two systems of nonlinear ordinary differential equations for $\beta_i(t)$ and $R_n(t)$:

$$\sum_n \dot{R}_n B_n^{(i)} + \frac{1}{2} \sum_k \sum_n R_k R_n B_{kn}^{(i)} + g\beta_i B^{(i)} = 0, \quad i = 1, 2, \dots; \tag{2.3.6}$$

$$\sum_n R_n \left( C_n^{(k)} - \sum_i \beta_i C_{ni}^{(k)} \right) - \dot{\beta}_0(t) D^{(k)} - \sum_i \dot{\beta}_i G_i^{(k)} = 0, \; k = 1, \dots , \tag{2.3.7}$$

where

$$B_n^{(i)} = \int_D [\varphi_n \mu]_{x^1=x_0^1+f} \left( \frac{\partial \beta_0}{\partial \beta_i} + f_i \right) dx^2 dx^3,$$

$$B_{kn}^{(i)} = \int_D [(\nabla\varphi_k, \nabla\varphi_n)\mu]_{x^1=x_0^1+f} \left( \frac{\partial \beta_0}{\partial \beta_i} + f_i \right) dx^2 dx^3, \tag{2.3.8a}$$

$$B^{(i)} = \int_D [x^1 \mu]_{x^1=x_0^1+f} \left( \frac{\partial \beta_0}{\partial \beta_i} + f_i \right) dx^2 dx^3;$$

$$C_n^{(k)} = \int_D \left[ \left( q^{11} \frac{\partial \varphi_n}{\partial x^1} + q^{12} \frac{\partial \varphi_n}{\partial x^2} \right) \mu \varphi_k \right]_{x^1=x_0^1+f} dx^2 dx^3,$$

$$C_{ni}^{(k)} = \int_D \left[ \left( q^{12} \frac{\partial \varphi_n}{\partial x^1} \frac{\partial f_i}{\partial x^2} + q^{22} \frac{\partial \varphi_n}{\partial x^2} \frac{\partial f_i}{\partial x^2} \right. \right.$$

$$+ q^{33}\frac{\partial\varphi_n}{\partial x^3}\frac{\partial f_i}{\partial x^3}\bigg)\mu\varphi_k\bigg]_{x^1=x_0^1+f} dx^2dx^3, \qquad (2.3.8b)$$

$$D^{(k)} = \int\limits_D [\mu\varphi_k]_{x^1=x_0^1+f}\, dx^2dx^3, \quad G_i^{(k)} = \int\limits_D [f_i\mu\varphi_k]_{x^1=x_0^1+f}\, dx^2dx^3.$$

Here, $q^{ij}(x^1, x^2)$ stand for components of the mutual metric tensor,

$$\mu(x^1, x^2) = \sqrt{q} = \xi\frac{\partial\xi}{\partial x^2}, \quad \text{and} \quad (\nabla\varphi_k, \nabla\varphi_n) = q^{ij}\frac{\partial\varphi_k}{\partial x^i}\frac{\partial\varphi_n}{\partial x^j}.$$

Note that, by virtue of condition (2.1.36), $\beta_0(t)$ in relation (2.3.1) is a function of the generalized coordinates $\beta_i$.

Preserving the terms up to the third order of smallness in the parameters characterizing the liquid motions in (2.3.6) and (2.3.7) and taking into account (2.3.4), we represent relations (2.3.8a) and (2.3.8b) in the form

$$B_n^{(i)} = a_n^{(i)} + \frac{\partial\beta_0}{\partial\beta_i}\Bigg(a_n + \sum_j \beta_j a_{nj}\Bigg) + a_{1n}^{(i)}\beta_0 + \sum_j \beta_j a_{nj}^{(i)} + \frac{1}{2}\sum_{j,k}\beta_j\beta_k a_{njk}^{(i)},$$

$$B_{kn}^{(i)} = b_{kn}^{(i)} + \sum_j \beta_j b_{kn}^{(i,j)} + \frac{\partial\beta_0}{\partial\beta_i} b_{kn},$$

$$B^{(i)} = s_0^{(i)} + \sum_j \beta_j s_{1j}^{(i)} + s_0\frac{\partial\beta_0}{\partial\beta_i} + s_1^{(i)}\beta_0 + \frac{1}{2}\sum_{j,k}\beta_j\beta_k s_{2jk}^{(i)} + \frac{\partial\beta_0}{\partial\beta_i}\sum_j \beta_j s_{1j}$$

$$+\beta_0\sum_j \beta_j s_{2j} + \frac{1}{2}\frac{\partial\beta_0}{\partial\beta_i}\sum_{j,k} s_{2jk}\beta_j\beta_k + s_1\frac{\partial\beta_0}{\partial\beta_i}\beta_0 + \frac{1}{6}\sum_{j,k,l}\beta_j\beta_k\beta_l s_{3jkl}^{(i)}, \qquad (2.3.9)$$

$$C_n^{(k)} = d_{nk} + d_{1nk}\beta_0 + \sum_j \beta_j d_{nk}^{(j)} + \frac{1}{2}\sum_{j,m}\beta_j\beta_m d_{nk}^{(jm)},$$

$$C_{ni}^{(k)} = e_{ni}^{(k)} + \sum_j e_{nij}^{(k)}\beta_j, \quad D^{(k)} = a_k + \sum_j \beta_j a_{1k}^{(j)},$$

$$G_i^{(k)} = a_k^{(i)} + a_{1k}^{(i)}\beta_0 + \sum_j \beta_j a_{kj}^{(i)} + \frac{1}{2}\sum_{j,m}\beta_j\beta_m a_{kjm}^{(i)}.$$

Since relations (2.3.9) depend only on the parameters $\beta_i(t)$ characterizing the shape of the free surface, the system of equations (2.3.7) is linear in the parameters $R_n(t)$. This fact enables us to consider a system of nonlinear second-order differential equations with respect to the generalized coordinates $\beta_i(t)$ instead of the systems of nonlinear equations (2.3.6) and (2.3.7).

In order to completely determine (2.3.9), we represent the eigenfunctions $\varphi_n(x^1, x^2, x^3)$ of the spectral problem (2.2.4) for a class of cavities of revolution in the form

$$\varphi_n(x^1, x^2, x^3) = \psi_n(x^1, x^2)\omega_n(x^3). \tag{2.3.10}$$

This yields

$$a_n^{(i)} = \int\limits_D \mu\varphi_n f_i dx^2 dx^3, \quad a_n = \int\limits_D \mu\varphi_n dx^2 dx^3,$$

$$a_{nj} = \int\limits_D \nabla_1(\mu\varphi_n) f_j dx^2 dx^3, \quad a_{nj}^{(i)} = \int\limits_D \nabla_1(\mu\varphi_n) f_i f_j dx^2 dx^3,$$

$$a_{1n}^{(i)} = \int\limits_D \nabla_1(\mu\varphi_n) f_i dx^2 dx^3, \quad a_{njk}^{(i)} = \int\limits_D \nabla_{11}(\mu\varphi_n) f_i f_j f_k dx^2 dx^3,$$

$$b_{kn} = \int\limits_D (l_{kn}^{(0)}\omega_n\omega_k + l_{kn}^{(1)}\omega_n'\omega_k') dx^2 dx^3,$$

$$b_{kn}^{(i)} = \int\limits_D (l_{kn}^{(0)}\omega_k\omega_n + l_{kn}^{(1)}\omega_k'\omega_n') f_i dx^2 dx^3,$$

$$b_{kn}^{(ij)} = \int\limits_D (l_{kn}^{(2)}\omega_k\omega_n + l_{kn}^{(3)}\omega_k'\omega_n') f_i f_j dx^2 dx^3,$$

$$s_0 = x_0^1 \int\limits_D \mu dx^2 dx^3, \quad s_0^{(i)} = x_0^1 \int\limits_D \mu f_i dx^2 dx^3, \tag{2.3.11}$$

$$s_1^{(i)} = \int\limits_D \nabla_1(x^1\mu) f_i dx^2 dx^3, \quad s_1 = \int\limits_D \nabla_1(x^1\mu) dx^2 dx^3,$$

$$s_{1j} = \int\limits_D \nabla_1(x^1\mu) f_j dx^2 dx^3, \quad s_{1j}^{(i)} = \int\limits_D \nabla_1(x^1\mu) f_i f_j dx^2 dx^3,$$

$$s_{2j}^{(i)} = \int\limits_D \nabla_{11}(x^1\mu) f_i f_j dx^2 dx^3, \quad s_{2jk} = \int\limits_D \nabla_{11}(x^1\mu) f_j f_k dx^2 dx^3,$$

$$s_{2jk}^{(i)} = \int\limits_D \nabla_{11}(x^1\mu) f_i f_j f_k dx^2 dx^3, \quad s_{3jkl}^{(i)} = \int\limits_D \nabla_{111}(x^1\mu) f_i f_j f_k f_l dx^2 dx^3,$$

$$d_{nk}^{(j)} = \int\limits_D t_{nk}^{(1)} f_j dx^2 dx^3, \quad d_{nk}^{(jm)} = \int\limits_D t_{nk}^{(2)} f_j f_m dx^2 dx^3,$$

$$d_{nk} = \int\limits_D t_{nk}^{(0)} dx^2 dx^3, \quad d_{1nk} = \int\limits_D t_{nk}^{(1)} dx^2 dx^3,$$

$$e_{ni}^{(k)} = \int\limits_D p_{0ni}^{(k)} dx^2 dx^3, \quad e_{nij}^{(k)} = \int\limits_D p_{1ni}^{(k)} f_j dx^2 dx^3,$$

where

$$\nabla_{11}(\mu\varphi_n) = \mu\frac{\partial^2\varphi_n}{\partial x^{1^2}} + 2\frac{\partial\varphi_n}{\partial x^1}\frac{\partial\mu}{\partial x^1} + \varphi_n\frac{\partial^2\mu}{\partial x^{1^2}} - \Gamma_{11}^{(2)}\left(\frac{\partial\varphi_n}{\partial x^2}\mu + \varphi_n\frac{\partial\mu}{\partial x^2}\right),$$

$$l_{kn}^{(0)} = \mu q^{11}\frac{\partial\psi_k}{\partial x^1}\frac{\partial\psi_n}{\partial x^1} + \mu q^{12}\left(\frac{\partial\psi_k}{\partial x^1}\frac{\partial\psi_n}{\partial x^2} + \frac{\partial\psi_k}{\partial x^2}\frac{\partial\psi_n}{\partial x^1}\right) + \mu q^{22}\frac{\partial\psi_k}{\partial x^2}\frac{\partial\psi_n}{\partial x^2},$$

$$l_{kn}^{(1)} = \mu q^{33}\psi_k\psi_n,$$

$$\begin{aligned} l_{kn}^{(2)} &= \frac{\partial}{\partial x^1}(\mu q^{11})\frac{\partial\psi_k}{\partial x^1}\frac{\partial\psi_n}{\partial x^1} + \frac{\partial}{\partial x^1}(\mu q^{12})\left(\frac{\partial\psi_k}{\partial x^1}\frac{\partial\psi_n}{\partial x^2} + \frac{\partial\psi_k}{\partial x^2}\frac{\partial\psi_n}{\partial x^1}\right) \\ &+ \frac{\partial}{\partial x^1}(\mu q^{22})\frac{\partial\psi_k}{\partial x^2}\frac{\partial\psi_n}{\partial x^2} + \mu q^{11}\left(\frac{\partial^2\psi_k}{\partial x^{1^2}}\frac{\partial\psi_n}{\partial x^2} + \frac{\partial\psi_k}{\partial x^1}\frac{\partial^2\psi_n}{\partial x^1\partial x^2}\right) \\ &+ \mu q^{12}\left(\frac{\partial^2\psi_k}{\partial x^{1^2}}\frac{\partial\psi_n}{\partial x^2} + \frac{\partial\psi_k}{\partial x^1}\frac{\partial^2\psi_n}{\partial x^1\partial x^2} + \frac{\partial^2\psi_k}{\partial x^1\partial x^2}\frac{\partial\psi_n}{\partial x^1} + \frac{\partial\psi_k}{\partial x^2}\frac{\partial^2\psi_n}{\partial x^{1^2}}\right) \\ &+ \mu q^{22}\left(\frac{\partial^2\psi_k}{\partial x^1\partial x^2}\frac{\partial\psi_n}{\partial x^2} + \frac{\partial\psi_k}{\partial x^2}\frac{\partial^2\psi_n}{\partial x^1\partial x^2}\right), \end{aligned}$$

$$l_{kn}^{(3)} = \frac{\partial}{\partial x^1}(\mu q^{33})\psi_k\psi_n + \mu q^{33}\left(\frac{\partial\psi_k}{\partial x^1}\psi_n + \frac{\partial\psi_n}{\partial x^1}\psi_k\right), \tag{2.3.12}$$

$$\nabla_{11}(x^1\mu) = 2\frac{\partial\mu}{\partial x^1} + x^1\frac{\partial^2\mu}{\partial x^{1^2}} - \Gamma_{11}^2 x^1\frac{\partial\mu}{\partial x^2},$$

$$\nabla_{111}(x^1\mu) = \frac{\partial^3(x^1\mu)}{\partial x^{1^3}} - \frac{\partial\Gamma_{11}^2}{\partial x^2}\frac{\partial(x^1\mu)}{\partial x^2} - 3\Gamma_{11}^2\frac{\partial^2(x^1\mu)}{\partial x^1\partial x^2} + 2\Gamma_{11}^2\Gamma_{21}^2\frac{\partial(x^1\mu)}{\partial x^2},$$

$$t_{n0}^{(0)} = \left(\mu q^{11}\frac{\partial\varphi_n}{\partial x^1} + \mu q^{12}\frac{\partial\varphi_n}{\partial x^2}\right)\varphi_k,$$

$$\begin{aligned} t_{nk}^{(1)} &= \left[\frac{\partial}{\partial x^1}(\mu q^{11})\frac{\partial\varphi_n}{\partial x^1} + \mu q^{11}\frac{\partial^2\varphi_n}{\partial x^{1^2}} + \frac{\partial}{\partial x^1}(\mu q^{12})\frac{\partial\varphi_n}{\partial x^2} + \mu q^{12}\frac{\partial^2\varphi_n}{\partial x^1\partial x^2}\right]\varphi_k \\ &+ \left(\mu q^{11}\frac{\partial\varphi_n}{\partial x^1} + \mu q^{12}\frac{\partial\varphi_n}{\partial x^2}\right)\frac{\partial\varphi_k}{\partial x^1}, \end{aligned}$$

$$\begin{aligned} t_{nk}^{(2)} &= \varphi_k\frac{\partial^2}{\partial x^{1^2}}\left(\mu q^{11}\frac{\partial\varphi_n}{\partial x^1} + \mu q^{12}\frac{\partial\varphi_n}{\partial x^2}\right) + 2\frac{\partial}{\partial x^1}\left(\mu q^{11}\frac{\partial\varphi_n}{\partial x^1} + \mu q^{12}\frac{\partial\varphi_n}{\partial x^2}\right)\frac{\partial\varphi_k}{\partial x^1} \\ &+ \frac{\partial^2\varphi_k}{\partial x^{1^2}}\left(\mu q^{11}\frac{\partial\varphi_n}{\partial x^1} + \mu q^{12}\frac{\partial\varphi_n}{\partial x^2}\right) - \Gamma_{11}^2\varphi_k\frac{\partial}{\partial x^2}\left(\mu q^{11}\frac{\partial\varphi_n}{\partial x^1} + \mu q^{12}\frac{\partial\varphi_n}{\partial x^2}\right) \\ &- \Gamma_{11}^2\left(\mu q^{11}\frac{\partial\varphi_n}{\partial x^1} + \mu q^{12}\frac{\partial\varphi_n}{\partial x^2}\right)\frac{\partial\varphi_k}{\partial x^2}, \end{aligned}$$

$$p_{0ni}^{(k)} = \left(\mu q^{12}\frac{\partial\varphi_n}{\partial x^1}\frac{\partial f_i}{\partial x^2} + \mu q^{22}\frac{\partial\varphi_n}{\partial x^2}\frac{\partial f_i}{\partial x^2} + \mu q^{33}\frac{\partial\varphi_n}{\partial x^3}\frac{\partial f_i}{\partial x^3}\right)\varphi_k,$$

$$\begin{aligned}p_{1ni}^{(k)} = \varphi_k\Bigg[&\mu q^{12}\frac{\partial^2\varphi_n}{\partial x^{1^2}}\frac{\partial f_i}{\partial x^2} + \mu q^{22}\frac{\partial^2\varphi_n}{\partial x^1\partial x^2}\frac{\partial f_i}{\partial x^2} + \mu q^{33}\frac{\partial^2\varphi_n}{\partial x^1\partial x^3}\frac{\partial f_i}{\partial x^3}\\ &+ \frac{\partial}{\partial x^1}(\mu q^{12})\frac{\partial\varphi_n}{\partial x^1}\frac{\partial f_i}{\partial x^2} + \frac{\partial}{\partial x^1}(\mu q^{22})\frac{\partial\varphi_n}{\partial x^2}\frac{\partial f_i}{\partial x^2}\\ &+ \frac{\partial}{\partial x^1}(\mu q^{33})\frac{\partial\varphi_n}{\partial x^3}\frac{\partial f_i}{\partial x^3}\Bigg] + \frac{\partial\varphi_k}{\partial x^1}\frac{p_{0ni}^{(k)}}{\varphi_k}.\end{aligned}$$

All quantities in relations (2.3.11) and (2.3.12) are computed for $x^1 = x_0^1$. The components of the mutual metric tensor $q^{ij}$ and the Christoffel symbols $\Gamma_{ij}^{\alpha}$ are determined solely by the geometry of the container, i.e., by the function $\xi(x^1, x^2)$.

## 2.4 Linear modal equations for an immobile tank

The theory of low-amplitude waves can be developed as an approximation of the general theory presented above based on the assumption that the velocity of liquid particles, the elevation of the free surface $x = f(y, z, t)$ and their derivatives are small quantities. In addition, the expressions in the square brackets in the variational equation (2.1.22) should be found for $x = 0$. In other words, the boundary conditions of the corresponding boundary-value problem must be satisfied on the initially unperturbed surface $\Sigma_0$.

Under these assumptions, Eq. (2.1.22) takes the form

$$\begin{aligned}\delta W = \int_{t_1}^{t_2}\Bigg(&\int_{\Sigma_0}\left\{\left[\frac{\partial\varphi}{\partial t} + gf\right]_{x=0}\delta f + \left[\left(\frac{\partial\varphi}{\partial x} - \frac{\partial f}{\partial t}\right)\delta\varphi\right]_{x=0}\right\}dS\\ &+ \int_S \frac{\partial\varphi}{\partial\nu}\delta\varphi dS - \int_Q \nabla^2\varphi\delta\varphi\, dQ\Bigg)dt = 0. \qquad (2.4.1)\end{aligned}$$

It is clear that the main boundary-value problem of the linear sloshing theory (1.5.32) can be readily derived from the variational equation (2.4.1).

The performed linearization significantly simplifies the investigation of the sloshing problem. As a result, the original problem becomes linear and the domain in which its solution should be defined becomes fixed. Hence, the variational equation of the form (2.4.1) is suitable for the analysis of the phenomenon of sloshing in containers of any geometric shape.

The linear theory is applicable to a set of problems dealing with the theory of small oscillations of dynamical systems near a static equilibrium. As the simplest of these problems, we can mention the problem of standing waves of the form (1.5.32), i.e., of simple harmonic oscillations in time. It is most frequently solved by the method of separation of variables, which makes it possible to reduce the analyzed problem to finding the natural sloshing modes (2.2.4).

However, the problem of standing waves can also be solved by using the variational method. This is demonstrated in what follows.

As above, we represent the equation of perturbed free surface in the form

$$x = f(y, z, t) = \sum_{i=1}^{\infty} \beta_i(t) f_i(y, z), \tag{2.4.2}$$

where $f_i(y, z)$ is a complete system of functions orthogonal, together with a constant, and defined on the unperturbed free surface $\Sigma_0$ and $\beta_i(t)$ are the unknown generalized Fourier coefficients parametrically depending on time. These coefficients have the meaning of generalized coordinates characterizing the deviations of the free surface from its equilibrium position:

$$\beta_i(t) = \int_{\Sigma_0} f(y, z, t) f_i(y, z) dS. \tag{2.4.3}$$

Similarly, we specify the velocity potential $\varphi(x, y, z, t)$ in the form

$$\varphi(x, y, z, t) = \sum_{n=1}^{\infty} R_n(t) \tilde{\varphi}_n(x, y, z), \tag{2.4.4}$$

where $R_n(t)$ are the parameters and $\tilde{\varphi}_n(x, y, z)$ is a system of harmonic functions in $Q$ satisfying the boundary condition on $S$.

In what follows, we associate the functions $f_i(y, z)$ and $\tilde{\varphi}_n(x, y, z)$ with the eigensolutions of the spectral boundary problem (2.2.4), which possesses the indicated properties and, moreover, $f_i(y, z) = \varkappa_i \tilde{\varphi}_i(0, y, z)$. We now substitute expansions (1.3.2) and (1.3.4) in the variational equation (1.3.1). As a result, we obtain the following infinite system of ordinary differential equations with respect to $\beta_i(t)$ and $R_n(t)$:

$$\begin{gathered}
\sum_{n=1}^{\infty} B_n^{(i)} \dot{R}_n + \mathrm{g} B^{(i)} \beta_i = 0, \quad i = 1, 2, \ldots, \\
\sum_{k=1}^{\infty} C_k^{(n)} R_k - \sum_{i=1}^{\infty} \dot{\beta}_i G_i^{(n)} = 0, \quad n = 1, 2, \ldots,
\end{gathered} \tag{2.4.5}$$

where

$$B_n^{(i)} = \rho \int_{\Sigma_0} (\varphi_n f_i)_{x=0} dS = \begin{cases} \dfrac{\rho N_i^2}{\varkappa_i} \text{ for } n = i, \\ 0 \text{ for } n \neq i, \end{cases}$$

$$B^{(i)} = \rho \int\limits_{\Sigma_0} f_i^2 dS = \rho N_i^2, \quad N_i^2 = \int\limits_{\Sigma_0} f_i^2 dS,$$

$$C_k^{(n)} = \rho \int\limits_{\Sigma_0} \left( \frac{\partial \varphi_k}{\partial x} \varphi_n \right)_{x=0} dS = \begin{cases} \frac{\rho N_n^2}{\varkappa_n} \text{ for } n = k, \\ 0 \text{ for } n \neq k, \end{cases} \tag{2.4.6}$$

$$G_i^{(n)} = \rho \int\limits_{\Sigma_0} (f_i \varphi_n)_{x=0} \, dS = \begin{cases} \frac{\rho N_n^2}{\varkappa_n} \text{ for } n = i, \\ 0 \text{ for } n \neq i. \end{cases}$$

Hence, the system of equations (2.4.5) is decomposed into independent equations of the form

$$\mu_i(\dot{R}_i + \sigma_i^2 \beta_i) = 0, \quad i = 1, 2, \dots, \qquad R_n = \dot{\beta}_n, \quad n = 1, 2, \dots, \tag{2.4.7}$$

where

$$\mu_i = \frac{\rho N_i^2}{\varkappa_i}, \quad \sigma_i^2 = \varkappa_i g. \tag{2.4.8}$$

These equations can be reduced to the following equivalent system of the second-order equations:

$$\mu_i(\ddot{\beta}_i + \sigma_i^2 \beta_i) = 0, \quad i = 1, 2, \dots. \tag{2.4.9}$$

The general solution of (2.4.9) is given by the formula

$$\beta_i(t) = a_i \cos \sigma_i t + b_i \sin \sigma_i t, \tag{2.4.10}$$

where $a_i$ and $b_i$ are unknown constants. The velocity potential and the instant shape of the free surface can be represented in the form

$$\varphi(x, y, z, t) = \sum_{i=1}^{\infty} \sigma_i(-a_i \sin \sigma_i t + b_i \cos \sigma_i t) \tilde{\varphi}_i(x, y, z); \tag{2.4.11a}$$

$$f(y, z, t) = \sum_{i=1}^{\infty} \varkappa_i(a_i \cos \sigma_i t + b_i \sin \sigma_i t) \tilde{\varphi}_i(0, y, z). \tag{2.4.11b}$$

The constants $a_i$ and $b_i$ are found from the initial conditions.

We denote $a_i = A_i \sin \varepsilon_i$ and $b_i = A_i \cos \varepsilon_i$. Then the solution of the analyzed problem takes the form

$$f(y, z, t) = \sum_{i=1}^{\infty} \varkappa_i A_i \tilde{\varphi}_i(0, y, z) \sin(\sigma_i t + \varepsilon_i); \tag{2.4.12}$$

$$\varphi(x, y, z, t) = \sum_{i=1}^{\infty} \sigma_i A_i \tilde{\varphi}_i(x, y, z) \cos(\sigma_i t + \varepsilon_i). \tag{2.4.13}$$

Every term of series (2.4.12) specifying the instant shape of the free surface is a standing wave in which the points of the free surface perform harmonic oscillations with amplitudes $A_i\varphi_i(0, y, z)$ depending on the location of the point, the frequency $\sigma_i = \sqrt{\varkappa_i g}$, and the same phase shift $\varepsilon_i$.

As for the energy balance, we note that the lowest sloshing frequency is predominant. Its contribution to the hydrodynamical forces and moments is of primary importance. Thus, for almost all practical problems of the dynamics of rigid bodies with cavities containing liquids, it is sufficient to consider several lowest natural sloshing frequencies.

In conclusion, we note that the requirement concerning the choice of the functions $f_i(y, z)$ and $\varphi_n(x, y, z)$ can be somewhat weakened in the general case. To apply the variational equation (2.4.1), it suffices to require that these functions belong to the Sobolev space $W_2^1$. However, in this case, Eq. (2.4.9) becomes complicated and the generalized coordinates $\beta_i(t)$ do not, generally speaking, belong to the family of normal coordinates of the system. As usual, in this case, it is necessary to perform some additional transformations reducing the obtained system of equations to the canonical form.

## 2.5 Bateman–Luke principle for the nonlinear sloshing problem in a tank performing prescribed motions

Assume that a container is partially filled with an ideal incompressible liquid and performs translational and rotational motions in the space with a given translational velocity $\boldsymbol{v}_0$ and an instantaneous angular velocity $\boldsymbol{\omega}$. By $Q$, we denote the liquid domain bounded by the free surface $\Sigma$ and the wetted rigid surface of the container $S$. First, we suppose that the shape of the container is cylindrical in the vicinity of the free surface and introduce an absolute coordinate system $O'x'y'z'$ and a coordinate system of $Oxyz$ rigidly fixed with the container. We choose the origin of coordinates $Oxyz$ on the unperturbed free surface and assume that the equation of the perturbed free surface $\zeta(x, y, z, t) = 0$ can be resolved with respect to the vertical space variable, i.e.,

$$x = f(y, z, t). \tag{2.5.1}$$

If we consider the liquid sloshing problem in a potential field of the gravity forces, then we represent the potential of the gravity forces in the form

$$U = -\mathbf{g} \cdot \boldsymbol{r}', \tag{2.5.2}$$

where

$$\boldsymbol{r}' = \boldsymbol{r}_0' + \boldsymbol{r}, \tag{2.5.3}$$

$\boldsymbol{r}'$ is the radius vector of a point of the body–liquid system relative to the origin $O'$ of the absolute coordinate system, $\boldsymbol{r}_0'$ is the radius vector of a point $O$ relative to the immobile point $O'$, $\boldsymbol{r}$ is the radius vector of a point of the system relative to the point $O$, and $\mathbf{g}$ is the gravitational acceleration.

By its nature, the problem under consideration can be regarded as the first problem of the dynamics of body–liquid systems. It is reduced to the determination of the motion of liquid caused by the motion of the container and the action of hydrodynamic loads. In this case, the key problem is to find the velocity potential and the elevation of the free surface .

It is convenient to perform our subsequent analysis in a body-fixed coordinate system $Oxyz$. It is known that, in the potential field of mass forces, the vector of absolute velocity of a particle is specified by a scalar function $\Phi(x, y, z, t)$ as follows:

$$\boldsymbol{v}_a = \nabla\Phi.$$

The absolute velocity potential $\Phi(x, y, z, t)$ used to describe the absolute liquid motion in a mobile (inertialess) coordinate system and the elevation of the perturbed free surface can be found, according to (1.5.1), from the solution of the nonlinear boundary-value problem

$$\begin{aligned}
&\nabla^2\Phi = 0, \quad \boldsymbol{r} \in Q, \\
&\frac{\partial\Phi}{\partial\nu} = \boldsymbol{v}_0 \cdot \boldsymbol{\nu} + \boldsymbol{\omega} \cdot (\boldsymbol{r} \times \boldsymbol{\nu}), \quad \boldsymbol{r} \in S, \\
&\frac{\partial\Phi}{\partial\nu} = \boldsymbol{v}_0 \cdot \boldsymbol{\nu} + \boldsymbol{\omega} \cdot (\boldsymbol{r} \times \boldsymbol{\nu}) + u_\nu, \quad \boldsymbol{r} \in \Sigma, \\
&\frac{\partial\Phi}{\partial t} + \frac{1}{2}(\nabla\Phi)^2 - \nabla\Phi \cdot (\boldsymbol{v}_0 + \boldsymbol{\omega} \times \boldsymbol{r}) + U = 0, \quad \boldsymbol{r} \in \Sigma,
\end{aligned} \tag{2.5.4}$$

where $\boldsymbol{\nu}$ is the unit vector of the outer normal to the surface of the domain $Q$, $\boldsymbol{r}$ is the radius vector of points of the liquid domain $Q$ in the body-fixed system of coordinates, and $u_\nu$ is the relative normal velocity of particles of the free surface given by the formula

$$u_\nu = \frac{f_t}{\sqrt{1 + f_y^2 + f_z^2}}. \tag{2.5.5}$$

The last boundary condition of problem (2.5.4) follows from the Bernoulli equation (1.5.2) considered in the body-fixed (inertialess) coordinate system under the condition of equality of the pressure acting on the free surface to a constant pressure $p_0$ (ullage gas pressure). Note that the derivative $\partial\Phi/\partial t$ is found in the mobile system of coordinates, i.e., at a point $M$ rigidly fixed (with coordinates $x$, $y$, $z$) in the mobile coordinate system.

We now show that the free-surface problem (2.5.4) follows from the Bateman–Luke variational principle in which the role of Lagrangian is played by the expression

$$L = \int_Q p dQ = -\rho \int_Q \left[ \frac{\partial \Phi}{\partial t} + \frac{1}{2}(\nabla \Phi)^2 - \nabla \Phi \cdot (\boldsymbol{v}_0 + \boldsymbol{\omega} \times \boldsymbol{r}) + U \right] dQ$$

$$= -\rho \int_{\Sigma_0} \left( \int_{h(y,z)}^{f(y,z,t)} \left[ \frac{\partial \Phi}{\partial t} + \frac{1}{2}(\nabla \Phi)^2 - \nabla \Phi \cdot (\boldsymbol{v}_0 + \boldsymbol{\omega} \times \boldsymbol{r}) + U \right] dx \right) dS, \tag{2.5.6}$$

where $\rho$ is the density of liquid, $\Sigma_0$ is its unperturbed free surface, and $x = h(y, z)$ is the equation of bottom of the container .

According to the Bateman–Luke variational principle, for the actual motions in the form of wave motions of a liquid with free surface in a moving container, the action

$$W = \int_{t_1}^{t_2} L dt \tag{2.5.7}$$

takes a stationary value, i.e.,

$$\delta W = \delta \int_{t_1}^{t_2} L dt = 0. \tag{2.5.8}$$

In this case, we assume that the actual motions and all motions of comparison simultaneously start at time $t_1$ and simultaneously terminate at time $t_2$. The initial and final positions of the system must be identical for the actual motions and the motions of comparison.

The variational problem obtained from the Bateman–Luke principle is posed as follows: In the collection of all functions $\Phi(x, y, z, t)$ and $f(y, z, t)$ continuous together with their first derivatives with respect to the space variables and time $t$, it is necessary to find a function for which the action $W$ has a stationary point. As above, the actual motions are specified by the functions $\Phi(x, y, z, t)$ and $f(y, z, t)$, whereas the motions of comparison are specified by the functions $\Phi + \delta\Phi$ and $f + \delta f$. Moreover,

$$\delta\Phi(x, y, z, t_1) = \delta\Phi(x, y, z, t_2) = \delta f(y, z, t_1) = \delta f(y, z, t_2) = 0. \tag{2.5.9}$$

In finding the first variation of $W$, we analyze the functions $\Phi(x, y, z, t, \alpha)$ and $f(y, z, t, \alpha)$ for the motions of comparison depending on arbitrary parameters $\alpha$ ($\alpha$ is sufficiently close to zero). For $\alpha = 0$, we obtain an actual motion

$$\Phi(x, y, z, t) = \Phi(x, y, z, t, 0), \quad f(y, z, t) = f(y, z, 0).$$

Passing to the evaluation of $\delta W$, we first transform relation (2.5.6) with regard for (2.5.2) and (2.5.3):

$$\rho\int_Q U dQ = -\rho\int_Q \mathbf{g}\cdot\boldsymbol{r}'\,dQ = -m_1\mathbf{g}\cdot\boldsymbol{r}'_{1C}. \tag{2.5.10}$$

In this case, we have

$$\boldsymbol{r}'_{1C} = \boldsymbol{r}'_0 + \boldsymbol{r}_{1C}, \tag{2.5.11}$$

where $\boldsymbol{r}'_{1C}$ and $\boldsymbol{r}_{1C}$ are the radius vectors of the centers of mass of the liquid relative to the points $O'$ and $O$, respectively, and $m_1$ is the mass of liquid. By using (2.5.6) and (2.5.10), we obtain

$$L = -\rho\int_Q\left[\frac{\partial\Phi}{\partial t} + \frac{1}{2}(\nabla\Phi)^2 - \nabla\Phi\cdot(\boldsymbol{v}_0 + \boldsymbol{\omega}\times\boldsymbol{r})\right]dQ + m_1\mathbf{g}\cdot\boldsymbol{r}'_{1C}. \tag{2.5.12}$$

In view of the variation of the domain of integration in the course of time and the general definition of the first variation, we get

$$\begin{aligned}\delta W = -\rho\int_{t_1}^{t_2}\int_{\Sigma_0}\Bigg\{&\left[\Phi_t + \frac{1}{2}(\nabla\Phi)^2 - \nabla\Phi\cdot(\boldsymbol{v}_0 + \boldsymbol{\omega}\times\boldsymbol{r})\right]_{x=f}\delta f\\ &+\int_h^f\left[\nabla\Phi\cdot\nabla(\delta\Phi) + \delta\Phi_t - \nabla(\delta\Phi)\cdot(\boldsymbol{v}_0 + \boldsymbol{\omega}\times\boldsymbol{r})\right]dx\Bigg\}dS\,dt\\ &+ m_1\int_{t_1}^{t_2}\mathbf{g}\cdot\delta\boldsymbol{r}'_{1C}dt = 0,\end{aligned} \tag{2.5.13}$$

where

$$\delta\Phi = \left.\frac{\partial\Phi}{\partial\alpha}\right|_{\alpha=0}\cdot\alpha,\quad \delta f = \left.\frac{\partial f}{\partial\alpha}\right|_{\alpha=0}\cdot\alpha.$$

The integrals

$$\int_Q\nabla\Phi\cdot\nabla(\delta\Phi)dQ,\quad \int_Q\boldsymbol{v}_0\cdot\nabla(\delta\Phi)dQ,\quad\text{and}\quad \int_Q(\boldsymbol{\omega}\times\boldsymbol{r})\cdot\nabla(\delta\Phi)dQ$$

are transformed by using the Green identity

$$\int_Q\nabla\varphi\cdot\nabla\psi dQ = \int_{S+\Sigma}\psi\frac{\partial\varphi}{\partial\nu}dS - \int_Q\psi\nabla^2\varphi dQ. \tag{2.5.14}$$

We set

$$\nabla\varphi = \boldsymbol{\omega}\times\boldsymbol{r} = \boldsymbol{i}_1(\omega_2 z - \omega_3 y) + \boldsymbol{i}_2(\omega_3 x - \omega_1 z) + \boldsymbol{i}_3(\omega_1 y - \omega_2 x).$$

Since $\boldsymbol{v}_0 = \nabla(\boldsymbol{v}_0 \cdot \boldsymbol{r})$, we find

$$\int\limits_Q \nabla\Phi \cdot \nabla(\delta\Phi) dQ = \int\limits_{S+\Sigma} \delta\Phi \frac{\partial\Phi}{\partial\nu} dS - \int\limits_Q \delta\Phi \nabla^2\Phi dQ,$$

$$\int\limits_Q \boldsymbol{v}_0 \cdot \nabla(\delta\Phi) dQ = \int\limits_Q \nabla(\boldsymbol{v}_0 \cdot \boldsymbol{r}) \cdot \nabla(\delta\Phi) dQ = \int\limits_{S+\Sigma} \delta\Phi \boldsymbol{v}_0 \cdot \boldsymbol{\nu}\, dS, \tag{2.5.15}$$

$$\int\limits_Q (\boldsymbol{\omega} \times \boldsymbol{r}) \cdot \nabla(\delta\Phi) dQ = \int\limits_{S+\Sigma} \delta\Phi(\boldsymbol{\omega} \times \boldsymbol{r}) \cdot \boldsymbol{\nu}\, dS.$$

In view of the fact that time is not varied, we get

$$\int\limits_Q \delta\Phi_t dQ = \int\limits_Q \frac{\partial}{\partial t}(\delta\Phi) dQ \tag{2.5.16}$$

(here, the differentiation with respect to time is carried out in the mobile system of coordinates). Applying formula (1.1.4) to this relation, we arrive at the relation

$$\int\limits_Q \frac{\partial}{\partial t}(\delta\Phi) dQ = \frac{d}{dt} \int\limits_Q \delta\Phi dQ - \int\limits_\Sigma \delta\Phi u_\nu dS. \tag{2.5.17}$$

We now transform the last integral in (2.5.13) by using the relation

$$\delta\boldsymbol{r}' = \delta\boldsymbol{r}_0' + \delta\boldsymbol{\theta} \times \boldsymbol{r} + \delta_1 \boldsymbol{r}, \tag{2.5.18}$$

where $\delta\boldsymbol{r}_0'$ is a virtual displacement of the origin $O$, $\delta\boldsymbol{\theta}$ is a vector of infinitely small rotation of the body, and $\delta_1$ denotes the variation for fixed unit vectors of the system $\boldsymbol{i}_S$ rigidly connected with the body. After simple transformations of the formula for the radius vector of the center-of-mass of the liquid relative to the point $O$

$$m_1 \boldsymbol{r}_{1C} = \rho \int\limits_Q \boldsymbol{r}\, dQ, \tag{2.5.19}$$

we obtain

$$m_1 x_{1C} = \frac{1}{2}\rho \int\limits_{\Sigma_0} f^2 dS + C_1, \quad m_1 y_{1C} = \rho \int\limits_{\Sigma_0} f y\, dS + C_2, \tag{2.5.20}$$

$$m_1 z_{1C} = \rho \int\limits_{\Sigma_0} f z\, dS + C_3,$$

where $C_1$, $C_2$, and $C_3$ are constants of integration depending on the shape of the container.

By using equalities (2.5.20), we find

$$m_1\delta_1\boldsymbol{r}_{1C} = \rho\int\limits_{\Sigma_0}(\boldsymbol{i}_1 f + \boldsymbol{i}_2 y + \boldsymbol{i}_3 z)\delta f\, dS = \rho\int\limits_{\Sigma_0}\boldsymbol{r}|_{x=f}\,\delta f\, dS.$$

Then the last integral in (2.5.13) takes the form

$$m_1\int\limits_{t_1}^{t_2}\mathbf{g}\cdot\delta\boldsymbol{r}'_{1C}\,dt = \int\limits_{t_1}^{t_2}\Big[m_1\mathbf{g}\cdot\delta\boldsymbol{r}'_0 + (m_1\boldsymbol{r}_{1C}\times\mathbf{g})\cdot\delta\boldsymbol{\theta} + \rho\int\limits_{\Sigma_0}\mathbf{g}\cdot\boldsymbol{r}|_{x=f}\,\delta f\, dS\Big]dt. \tag{2.5.21}$$

Since the motion of the container is regarded as known, we conclude that $\delta\boldsymbol{r}'_0 = 0$ and $\delta\boldsymbol{\theta} = 0$.

Thus, in the case under consideration, we get

$$m_1\int\limits_{t_1}^{t_2}\mathbf{g}\cdot\delta\boldsymbol{r}'_{1C}\,dt = \rho\int\limits_{t_1}^{t_2}\int\limits_{\Sigma_0}\mathbf{g}\cdot\boldsymbol{r}\,|_{x=f}\,\delta f\, dSdt. \tag{2.5.22}$$

Substituting relations (2.5.15), (2.5.17), and (2.5.22) in (2.5.13), we obtain

$$\begin{aligned}\delta W = \rho\int\limits_{t_1}^{t_2}\Big\{\int\limits_{\Sigma_0}&\Big[\Phi_t + \frac{1}{2}(\nabla\Phi)^2 - \nabla\Phi\cdot(\boldsymbol{v}_0 + \boldsymbol{\omega}\times\boldsymbol{r}) - \mathbf{g}\cdot\boldsymbol{r}\Big]_{x=f}\delta f\, dS\\ &+\int\limits_{S}\Big[\frac{\partial\Phi}{\partial\nu} - \boldsymbol{v}_0\cdot\boldsymbol{\nu} - (\boldsymbol{\omega}\times\boldsymbol{r})\cdot\boldsymbol{\nu}\Big]\delta\Phi\, dS\\ &+\int\limits_{\Sigma_0}\Big(\Big[\frac{\partial\Phi}{\partial\nu} - \boldsymbol{v}_0\cdot\boldsymbol{\nu} - (\boldsymbol{\omega}\times\boldsymbol{r})\cdot\boldsymbol{\nu} - u_\nu\Big]\delta\Phi\Big)_{x=f} dS\\ &-\int\limits_{Q}\nabla^2\Phi\delta\Phi\, dQ + \frac{d}{dt}\int\limits_{Q}\delta\Phi\, dQ\Big\}dt = 0.\end{aligned} \tag{2.5.23}$$

In view of conditions (2.5.9), the last term in this equation is equal to zero. The variational equation (2.5.23) yields the boundary-value problem (2.5.4) because the variations $\delta f$ and $\delta\Phi$ are independent.

Note that the nonlinear boundary conditions of the free-surface problem (2.5.4) directly follow from the condition of extremum of the action (2.5.8); i.e., play the role of natural boundary conditions. This fact is is decisive for the application of functional (2.5.8) or the variational equation (2.5.23) in the construction of efficient approximate methods for the solution of the nonlinear boundary-value problem (2.5.4). By analogy with the results presented above, the appropriate variational principle can be also formulated for containers with more complicated geometric shapes. To study a class of vessels in the form of bodies of revolution, we introduce, parallel with the system of coordinates $Oxyz$, a curvilinear system of coordinates $Ox^1x^2x^3$, which is also rigidly connected with the container. The $Ox^1$-axis is directed along the axis of symmetry of the vessel and the relationship between the coordinates systems $xyz$ and $x^1x^2x^3$ is described by relations (2.1.23) and (2.1.24). We now represent the equation of the free surface $\zeta(x^1, x^2, x^3, t) = 0$ in the form

$$x^1 = x_0^1 + f(x^2, x^3, t). \tag{2.5.24}$$

This enables us to rewrite the variational equation (2.5.23) in the system of coordinates rigidly connected with the bottom of the vessel as follows:

$$\begin{aligned}\delta W = -\rho \int_{t_1}^{t_2} \int_D \Big\{ \Big[ \frac{\partial \Phi}{\partial t} + \frac{1}{2}(\nabla\Phi)^2 - \nabla\Phi \cdot (\boldsymbol{v}_0 + \boldsymbol{\omega} \times \boldsymbol{r}) \\ - \mathbf{g} \cdot \boldsymbol{r} \Big] \sqrt{q} \Big|_{x^1 = x_0^1 + f} \cdot \delta f + \int_0^{x_0^1+f} \Big[ \nabla\Phi \cdot \nabla(\delta\Phi) + \delta\Phi_t \\ - \nabla(\delta\Phi) \cdot (\boldsymbol{v}_0 + \boldsymbol{\omega} \times \boldsymbol{r}) \Big] \sqrt{q}\, dx^1 \Big\} dx^2 dx^3 dt = 0, \end{aligned} \tag{2.5.25}$$

where $D$ is the domain of variation of the parameters $x^2$ and $x^3$. By using the Green identity (2.5.14), we represent relation (2.5.25) in the form

$$\begin{aligned}\delta W = -\rho \int_{t_1}^{t_2} \Big\{ \int_D \Big[ \frac{\partial \Phi}{\partial t} + \frac{1}{2}(\nabla\Phi)^2 - \nabla\Phi \cdot (\boldsymbol{v}_0 + \boldsymbol{\omega} \times \boldsymbol{r}) - \mathbf{g} \cdot \boldsymbol{r} \Big] \sqrt{q} \Big|_{x^1 = x_0^1 + f} \delta f\, dS \\ + \int_S \Big[ \frac{\partial \Phi}{\partial \nu} - \boldsymbol{v}_0 \cdot \boldsymbol{\nu} - (\boldsymbol{\omega} \times \boldsymbol{r}) \cdot \boldsymbol{\nu} \Big] \delta\Phi\, dS \\ + \int_D \Big[ \frac{\partial \Phi}{\partial \nu} - \boldsymbol{v}_0 \cdot \boldsymbol{\nu} - (\boldsymbol{\omega} \times \boldsymbol{r}) \cdot \boldsymbol{\nu} - u_\nu \Big] \delta\Phi\, dS - \int_Q \nabla^2\Phi \delta\Phi\, dQ \Big\} dt = 0, \end{aligned} \tag{2.5.26}$$

where is an elementary area of the free surface

$$dS = \sqrt{\left[ 1 + q^{22} \left( \frac{\partial f}{\partial x^2} \right)^2 + q^{33} \left( \frac{\partial f}{\partial x^3} \right)^2 - 2q^{12} \frac{\partial f}{\partial x^2} \right] q\, dx^2 dx^3}. \tag{2.5.27}$$

In view of the independence of the variations $\delta f$ and $\delta\Phi$ in (2.5.26), we arrive at the nonlinear boundary problem (2.5.4) in the general case.

## 2.6 Multimodal method for a tank performing prescribed motions

The solution of the free-surface problem (2.5.4) with an aim to find the velocity potential $\mathbf{\Phi}(x, y, z, t)$ and the motions of the free surface is connected with serious mathematical difficulties. The perturbation theory makes it possible to develop constructive algorithms for the solution of this problem. However, their practical realization is complicated [173]. Simpler algorithms can be obtained by using the variational statement of (2.5.4). It is customary to say that approximate methods of this kind belong to direct methods of mathematical physics. As already indicated, these methods possess, in addition to the simplicity of their realization, numerous important advantages typical of the energy methods.

We now present a version of the direct method for the solution of problem (2.5.4) based on the assumption that the translational velocity $\boldsymbol{v}_0(t)$ and the instantaneous angular velocity of the rigid vessel $\boldsymbol{\omega}(t)$ are known functions of time.

First, we consider an upright cylindrical vessel in the vicinity of the free surface such that the perturbed free surface takes the form (2.5.1) and the normal component of the relative velocity of this surface is given in the form (2.5.5). The kinematic boundary conditions (2.5.4) make it possible to reduce the velocity potential $\mathbf{\Phi}(x, y, z, t)$ to the form

$$\mathbf{\Phi}(x, y, z, t) = \boldsymbol{v}_0 \cdot \boldsymbol{V} + \boldsymbol{\omega} \cdot \boldsymbol{\Omega} + \varphi, \tag{2.6.1}$$

where $\boldsymbol{V}(x, y, z)$ and $\boldsymbol{\Omega}(x, y, z)$ are harmonic vectors, i.e., vectors, whose projections $V_1, V_2, V_3$ and $\Omega_1, \Omega_2, \Omega_3$ onto the axes of the system $Oxyz$ are harmonic functions satisfying the boundary conditions

$$\left.\frac{\partial V_1}{\partial \nu}\right|_{S+\Sigma} = \nu_1, \quad \left.\frac{\partial V_2}{\partial \nu}\right|_{S+\Sigma} = \nu_2, \quad \left.\frac{\partial V_3}{\partial \nu}\right|_{S+\Sigma} = \nu_3; \tag{2.6.2}$$

$$\left.\frac{\partial \Omega_1}{\partial \nu}\right|_{S+\Sigma} = y\nu_3 - z\nu_2, \quad \left.\frac{\partial \Omega_2}{\partial \nu}\right|_{S+\Sigma} = z\nu_1 - x\nu_3, \quad \left.\frac{\partial \Omega_3}{\partial \nu}\right|_{S+\Sigma} = x\nu_2 - y\nu_1. \tag{2.6.3}$$

Here, $\nu_1, \nu_2, \nu_3$ are the projections of the unit vector $\boldsymbol{\nu}$ onto the axes of the system $Oxyz$. The vector function $\boldsymbol{\Omega}(x, y, z)$ is called the Stokes–Zhukovsky potential.

The harmonic function $\varphi$ in (2.6.1) satisfies the boundary conditions

$$\left.\frac{\partial \varphi}{\partial \nu}\right|_{S} = 0, \quad \left.\frac{\partial \varphi}{\partial \nu}\right|_{\Sigma} = \frac{f_t}{N}, \quad N = \sqrt{1 + f_y^2 + f_z^2} \tag{2.6.4}$$

and describes the liquid sloshing in an immobile vessel. The solution of (2.6.2) can be readily found (to within an arbitrary function of time), as follows:

$$V_1 = x, \quad V_2 = y, \quad V_3 = z. \tag{2.6.5}$$

Hence, $\boldsymbol{V} = \boldsymbol{r}$. Since $\nu_i$ in the boundary conditions specified on $\Sigma$ depend on time, $\boldsymbol{V}$ is also function a of time and time appears in $\boldsymbol{V}$ only via the space variables $x$, $y$, and $z$. This is also true for the definition of the harmonic vector $\boldsymbol{\Omega}$.

It is known that the solution of the Neumann boundary-value problems is unique under the conditions

$$\int\limits_{S+\Sigma} \frac{\partial \Omega_i}{\partial \nu} dS = 0 \quad \text{and} \quad \int\limits_{\Sigma} \frac{\partial \varphi}{\partial \nu} dS = 0. \tag{2.6.6}$$

In the boundary-value problem for $\Omega_i$, this condition is satisfied. This can easily be verified by applying the Gauss theorem for the transformation of the surface integral into the integral over the volume:

$$\int\limits_{S+\Sigma} F\nu_1 dS = \int\limits_{Q} \frac{\partial F}{\partial x} dQ, \quad \int\limits_{S+\Sigma} F\nu_2 dS = \int\limits_{Q} \frac{\partial F}{\partial y} dQ, \quad \int\limits_{S+\Sigma} F\nu_3 dS = \int\limits_{Q} \frac{\partial F}{\partial z} dQ, \tag{2.6.7}$$

where $F$ is a function continuous together with its partial derivatives in the domain $Q$ and on its boundary. In the case under consideration, we have

$$\int\limits_{S+\Sigma} \frac{\partial \Omega_1}{\partial \nu} dS = \int\limits_{S+\Sigma} (y\nu_3 - z\nu_2) dS = \int\limits_{Q} \left( \frac{\partial y}{\partial z} - \frac{\partial z}{\partial y} \right) dQ \equiv 0. \tag{2.6.8}$$

In order to solve the boundary-value problem for the function $\varphi$, it is necessary to demand that the unknown function $f(y, z, t)$, which describes the shape of the free surface, must satisfy the condition

$$\int\limits_{\Sigma} \frac{\partial \varphi}{\partial \nu} dS = \int\limits_{\Sigma} f_t dS = 0.$$

We denote the radius vector of a liquid particle relative to the vessel (i.e., to the coordinate system $Oxyz$) by $\boldsymbol{r}\{x(t), y(t), z(t)\}$ and its relative velocity by $\overset{*}{\boldsymbol{r}}\{\dot{x}(t), \dot{y}(t), \dot{z}(t)\}$. This yields

$$\boldsymbol{v}_a = \boldsymbol{v}_0 + \boldsymbol{\omega} \times \boldsymbol{r} + \overset{*}{\boldsymbol{r}}. \tag{2.6.9}$$

Here and in what follows, the asterisks mark the vectors whose projections onto the axes of the coordinate system $Oxyz$ rigidly connected with the body are equal to the derivatives of the projections of the corresponding vectors onto these axes. At the same time, in view of (2.6.1), we obtain

$$\begin{aligned} \boldsymbol{v}_a = \nabla\Phi &= \nabla(\boldsymbol{v}_0 \cdot \boldsymbol{r}) + \nabla(\boldsymbol{\omega} \cdot \boldsymbol{\Omega}) + \nabla\varphi \\ &= \boldsymbol{v}_0 + \nabla(\boldsymbol{\omega} \cdot \boldsymbol{\Omega}) + \nabla\varphi. \end{aligned} \tag{2.6.10}$$

The comparison with (2.6.9) gives

$$\overset{*}{\boldsymbol{r}} = \nabla(\boldsymbol{\omega} \cdot \boldsymbol{\Omega}) + \nabla\varphi - (\boldsymbol{\omega} \times \boldsymbol{r}). \tag{2.6.11}$$

Thus, in the case of prescribed motions of the vessel, the flows of liquid can be regarded as known if we know the solutions of the Neumann boundary-value problems for the functions $\boldsymbol{\Omega}_i$ and $\varphi$.

We seek the approximate solution of the nonlinear boundary problem (2.6.5) in the form (2.6.1) by using the equivalent variational problem, i.e., the problem of finding the extreme values of action (2.5.7). In this case, we assume that the function $f(y, z, t)$ can be represented as follows:

$$f = \sum_{i=1}^{\infty} \beta_i(t) f_i(y, z), \tag{2.6.12}$$

where $f_i(y, z)$ is a complete system of functions given on the unperturbed free surface $\Sigma_0$ and orthogonal together with a constant and $\beta_i(t)$ are the generalized Fourier coefficients parametrically depending on time. These coefficients can be regarded as generalized coordinates characterizing the deviations of the free surface:

$$\beta_i(t) = \int_{\Sigma_0} f(y, z, t) f_i(y, z) dS. \tag{2.6.13}$$

The velocity potential is constructively specified in the following way: As above, we seek the solution of the boundary-value problem for the function $\varphi$ in the form

$$\varphi(x, y, z, t) = \sum_{n=1}^{\infty} R_n(t) \varphi_n(x, y, z), \tag{2.6.14}$$

where $\varphi_n(x, y, z)$ is a system of harmonic functions satisfying only the condition of impermeability ($\partial\varphi_n/\partial\nu = 0$) on the wetted surface of the vessel $S$ and $R_n(t)$ are the parameters characterizing the time dependence of $\varphi$. This system of harmonic functions is obtained as a result of the solution of linear spectral boundary problems with parameters in the boundary conditions. The functions $\boldsymbol{\Omega}_i$ depending on $\beta_i(t)$ and the space variables are regarded as known. (The problem of their construction is discussed in what follows.) Hence, it is necessary to find the unknown functions of time $\beta_i(t)$ and $R_n(t)$ which, together with given functions $\boldsymbol{v}_0(t)$ and $\boldsymbol{\omega}(t)$, determine the relative liquid motions [see (2.6.11)].

To find $\beta_i(t)$ and $R_n(t)$, we use the variational equation (2.5.23), where the integrals over the domain $Q$ and the surface $S$ are equal to zero by the conditions imposed on the velocity potential $\mathbf{\Phi}$ (2.6.1). Thus, only the surface integrals over the unperturbed free surface $\Sigma_0$ remain in (2.5.23), i.e., we get

$$\delta W = -\rho \int_{t_1}^{t_0} \left\{ \int_{\Sigma_0} \left[ \frac{\partial \Phi}{\partial t} + \frac{1}{2}(\nabla\Phi)^2 - \nabla\Phi \cdot (\boldsymbol{v}_0 + \boldsymbol{\omega} \times \boldsymbol{r}) - \mathbf{g} \cdot \boldsymbol{r} \right]_{x=f} \delta f dS \right.$$
$$\left. + \int_{\Sigma_0} ([\nabla\Phi \cdot \nabla\zeta - (\boldsymbol{v}_0 \cdot \nabla\zeta) - (\boldsymbol{\omega} \times \boldsymbol{r}) \cdot \nabla\zeta - f_t]\delta\Phi)_{x=f} dS \right\} dt = 0, \tag{2.6.15}$$

where $\zeta(x, y, z, t) = 0$ is the equation of perturbed free surface.

Equation (2.6.15) yields the infinite system of nonlinear differential equations for $\beta_i(t)$ and $R_n(t)$. This system can also be directly derived from (2.5.7) by substituting (2.6.1) in (2.5.7) with subsequent variation over $\beta_i$ and $R_n$. We now substitute relation (2.6.1) for the velocity potential in (2.5.6) and specify the Lagrangian in the following form:

$$L = -\rho \int_Q \left[ \overset{*}{\boldsymbol{v}}_0 \cdot \boldsymbol{r} + \frac{\partial}{\partial t}(\boldsymbol{\omega} \cdot \boldsymbol{\Omega}) + \frac{1}{2}\nabla(\boldsymbol{\omega} \cdot \boldsymbol{\Omega}) \cdot \nabla(\boldsymbol{\omega} \cdot \boldsymbol{\Omega}) \right.$$
$$+ \nabla(\boldsymbol{\omega} \cdot \boldsymbol{\Omega}) \cdot \nabla\varphi - \frac{1}{2}v_0^2 - \boldsymbol{\omega} \cdot (\boldsymbol{r} \times \boldsymbol{v}_0)$$
$$\left. - \boldsymbol{\omega} \cdot (\boldsymbol{r} \times \nabla(\boldsymbol{\omega} \cdot \boldsymbol{\Omega})) - \boldsymbol{\omega} \cdot (\boldsymbol{r} \times \nabla\varphi) \right] dQ + L_r, \tag{2.6.16}$$

where

$$L_r = -\rho \int_Q \left[ \frac{\partial \varphi}{\partial t} + \frac{1}{2}(\nabla\varphi)^2 + U \right] dQ \tag{2.6.17}$$

is the Lagrangian of the liquid motion of an immobile vessel.

In view of (2.6.13) and (2.6.14), relation (2.6.17) implies that

$$L_r = -\int_Q \left[ \sum_{n=1} \dot{R}_n \varphi_n + \frac{1}{2} \sum_n \sum_k R_n R_k (\nabla\varphi_n, \nabla\varphi_k) + U \right] dQ$$
$$= -\left[ \sum_n A_n \dot{R}_n + \frac{1}{2} \sum_n \sum_k A_{nk} R_n R_k - \mathrm{g}_1 l_1 - \mathrm{g}_2 l_2 - \mathrm{g}_3 l_3 - m_1 \mathbf{g} \cdot \boldsymbol{r}'_0 \right], \tag{2.6.18}$$

where the quantities

$$A_n = \rho \int_Q \varphi_n dQ, \; l_1 = \rho \int_Q x dQ, \quad l_2 = \rho \int_Q y dQ, \quad l_3 = \rho \int_Q z dQ,$$

$$A_{nk} = A_{kn} = \rho \int_Q (\nabla\varphi_n, \nabla\varphi_k) dQ = \rho \int_{S+\Sigma} \varphi_n \frac{\partial \varphi_k}{\partial \nu} dS, \tag{2.6.19}$$

are, by virtue of (2.6.12), polynomials in the parameters $\beta_i(t)$.

For our subsequent presentation, it is convenient to rewrite relation (2.6.16) in the form

$$L = -\Big[\dot{v}_{01} l_1 + \dot{v}_{02} l_2 + \dot{v}_{03} l_3 + \dot{\omega}_1 l_{1\omega} + \dot{\omega}_2 l_{2\omega} + \dot{\omega}_3 l_{3\omega} + \omega_1 l_{1\omega t}$$
$$+ \omega_2 l_{2\omega t} + \omega_3 l_{3\omega t} - \tfrac{1}{2}\omega_1^2 J_{11}^1 - \tfrac{1}{2}\omega_2^2 J_{22}^1 - \tfrac{1}{2}\omega_3^2 J_{33}^1 - \omega_1\omega_2 J_{12}^1$$
$$- \omega_1\omega_3 J_{13}^1 - \omega_2\omega_3 J_{23}^1 - \tfrac{1}{2} m_1 (v_{01}^2 + v_{02}^2 + v_{03}^2) + (\omega_2 v_{03} - \omega_3 v_{02}) l_1$$
$$+ (\omega_3 v_{01} - \omega_1 v_{03}) l_2 + (\omega_1 v_{02} - \omega_2 v_{01}) l_3\Big] + L_r. \tag{2.6.20}$$

Here, we have introduced the total mass of liquid $m_1$, the quantities $l_{k\omega}$ and $l_{k\omega t}$, and the inertia tensor of the liquid $J_{ij}^1$ :

$$m_1 = \rho \int_Q dQ, \; l_{k\omega} = \rho \int_Q \Omega_k dQ, \; l_{k\omega t} = \rho \int_Q \frac{\partial \Omega_k}{\partial t} dQ,$$

$$J_{11}^1 = \rho \int_Q \left( y \frac{\partial \Omega_1}{\partial z} - z \frac{\partial \Omega_1}{\partial y} \right) dQ = \rho \int_{S+\Sigma} \Omega_1 \frac{\partial \Omega_1}{\partial \nu} dS,$$

$$J_{22}^1 = \rho \int_Q \left( z \frac{\partial \Omega_2}{\partial x} - x \frac{\partial \Omega_2}{\partial z} \right) dQ = \rho \int_{S+\Sigma} \Omega_2 \frac{\partial \Omega_2}{\partial \nu} dS,$$

$$J_{33}^1 = \rho \int_Q \left( x \frac{\partial \Omega_3}{\partial y} - y \frac{\partial \Omega_3}{\partial x} \right) dQ = \rho \int_{S+\Sigma} \Omega_3 \frac{\partial \Omega_3}{\partial \nu} dS, \tag{2.6.21}$$

$$J_{12}^1 = J_{21}^1 = \rho \int_Q \left( z \frac{\partial \Omega_1}{\partial x} - x \frac{\partial \Omega_1}{\partial z} \right) dQ = \rho \int_Q \left( y \frac{\partial \Omega_2}{\partial z} - z \frac{\partial \Omega_2}{\partial y} \right) dQ$$
$$= \rho \int_{S+\Sigma} \Omega_1 \frac{\partial \Omega_2}{\partial \nu} dS = \rho \int_{S+\Sigma} \Omega_2 \frac{\partial \Omega_1}{\partial \nu} dS,$$

$$J_{13}^1 = J_{31}^1 = \rho \int_Q \left( x \frac{\partial \Omega_1}{\partial y} - y \frac{\partial \Omega_1}{\partial x} \right) dQ = \rho \int_Q \left( y \frac{\partial \Omega_3}{\partial z} - z \frac{\partial \Omega_3}{\partial y} \right) dQ$$

$$= \rho \int\limits_{S+\Sigma} \Omega_1 \frac{\partial \Omega_3}{\partial \nu} dS = \rho \int\limits_{S+\Sigma} \Omega_3 \frac{\partial \Omega_1}{\partial \nu} dS,$$

$$J_{23}^1 = J_{32}^1 = \rho \int\limits_Q \left( x\frac{\partial \Omega_2}{\partial y} - y\frac{\partial \Omega_2}{\partial x} \right) dQ = \rho \int\limits_Q \left( z\frac{\partial \Omega_3}{\partial x} - x\frac{\partial \Omega_3}{\partial z} \right) dQ$$

$$= \rho \int\limits_{S+\Sigma} \Omega_2 \frac{\partial \Omega_3}{\partial \nu} dS = \rho \int\limits_{S+\Sigma} \Omega_3 \frac{\partial \Omega_2}{\partial \nu} dS.$$

Relations (2.6.21) are obtained from the following formulas:

$$\int\limits_Q \left( y\frac{\partial \Omega_k}{\partial z} - z\frac{\partial \Omega_k}{\partial y} \right) dQ = \int\limits_{S+\Sigma} \Omega_k \frac{\partial \Omega_1}{\partial \nu} dS,$$

$$\int\limits_Q \left( z\frac{\partial \Omega_k}{\partial x} - x\frac{\partial \Omega_k}{\partial z} \right) dQ = \int\limits_{S+\Sigma} \Omega_k \frac{\partial \Omega_2}{\partial \nu} dS, \tag{2.6.22}$$

$$\int\limits_Q \left( x\frac{\partial \Omega_k}{\partial y} - y\frac{\partial \Omega_k}{\partial x} \right) dQ = \int\limits_{S+\Sigma} \Omega_k \frac{\partial \Omega_3}{\partial \nu} dS$$

derived from (2.6.7) and the boundary conditions (2.6.3). Indeed, the quantity $J_{11}^1$ is found, according to (2.6.16), as follows:

$$J_{11}^1 = -\rho \int\limits_Q (\nabla \Omega_1, \nabla \Omega_1) dQ + 2\rho \int\limits_Q \left( y\frac{\partial \Omega_1}{\partial z} - z\frac{\partial \Omega_1}{\partial y} \right) dQ. \tag{2.6.23}$$

Applying, first, the Green formula

$$\int\limits_Q (\nabla \Omega_1, \nabla \Omega_1) dQ = \int\limits_{S+\Sigma} \Omega_1 \frac{\partial \Omega_1}{\partial \nu} dS$$

to the first integral in (2.6.23) and using the formulas of transformation (2.6.22), we arrive at the final formula for $J_{11}^1$ presented above.

In a similar way, we show that the corresponding terms in (2.6.20) generated by the fourth and last terms in the integrand of (2.6.16) are mutually annihilated. The first variation of the action (2.5.7) can now be readily found. Assume that $\boldsymbol{v}_0$ and $\boldsymbol{\omega}$ are given. Then, in view of relations (2.6.18) and (2.6.20), we obtain

$$\delta W = \int\limits_{t_1}^{t_2} \Bigg[ \sum_n A_n \delta \dot{R}_n + \frac{1}{2} \sum_n \sum_k A_{nk} R_k \delta R_n + \sum_i \Bigg( \sum_n \dot{R}_n \frac{\partial A_n}{\partial \beta_i}$$

$$+\omega_1 \frac{\partial l_{1\omega t}}{\partial \beta_i} + \omega_2 \frac{\partial l_{2\omega t}}{\partial \beta_i} + \omega_3 \frac{\partial l_{3\omega t}}{\partial \beta_i} + \frac{1}{2} \sum_n \sum_k R_n R_k \frac{\partial A_{nk}}{\partial \beta_i}$$

$$+\dot{\omega}_1 \frac{\partial l_{1\omega}}{\partial \beta_i} + \dot{\omega}_2 \frac{\partial l_{2\omega}}{\partial \beta_i} + \dot{\omega}_3 \frac{\partial l_{3\omega}}{\partial \beta_i} + (\dot{v}_{01} - g_1 + \omega_2 v_{03} - \omega_3 v_{02}) \frac{\partial l_1}{\partial \beta_i}$$

$$+ (\dot{v}_{02} - g_2 + \omega_3 v_{01} - \omega_1 v_{03}) \frac{\partial l_2}{\partial \beta_i} + (\dot{v}_{03} - g_3 + \omega_1 v_{02} - \omega_2 v_{01}) \frac{\partial l_3}{\partial \beta_i}$$

$$- \frac{1}{2} \omega_1^2 \frac{\partial J_{11}^1}{\partial \beta_i} - \frac{1}{2} \omega_2^2 \frac{\partial J_{22}^1}{\partial \beta_i} - \frac{1}{2} \omega_3^2 \frac{\partial J_{33}^1}{\partial \beta_i} - \omega_1 \omega_2 \frac{\partial J_{12}^1}{\partial \beta_i} - \omega_1 \omega_3 \frac{\partial J_{13}^1}{\partial \beta_i}$$

$$- \omega_2 \omega_3 \frac{\partial J_{23}^1}{\partial \beta_i} \Bigg) \delta \beta_i + \left( \omega_1 \frac{\partial l_{1\omega t}}{\partial \dot{\beta}_i} + \omega_2 \frac{\partial l_{2\omega t}}{\partial \dot{\beta}_i} + \omega_3 \frac{\partial l_{3\omega t}}{\partial \dot{\beta}_i} \right) \delta \dot{\beta}_i \Bigg] dt = 0. \qquad (2.6.24)$$

Integrating (2.6.24) by parts and using relations (2.5.9), we get the following infinite-dimensional systems of nonlinear ordinary differential equations for $R_n(t)$ and $\beta_i(t)$:

$$\frac{d}{dt} A_n - \sum_k R_k A_{nk} = 0, \quad n = 1, 2, \dots ; \qquad (2.6.25)$$

$$\sum_n \dot{R}_n \frac{\partial A_n}{\partial \beta_i} + \frac{1}{2} \sum_n \sum_k \frac{\partial A_{nk}}{\partial \beta_i} R_n R_k + \dot{\omega}_1 \frac{\partial l_{1\omega}}{\partial \beta_i} + \dot{\omega}_2 \frac{\partial l_{2\omega}}{\partial \beta_i}$$

$$+ \dot{\omega}_3 \frac{\partial l_{3\omega}}{\partial \beta_i} + \omega_1 \frac{\partial l_{1\omega t}}{\partial \beta_i} + \omega_2 \frac{\partial l_{2\omega t}}{\partial \beta_i} + \omega_3 \frac{\partial l_{3\omega t}}{\partial \beta_i}$$

$$- \frac{d}{dt} \left( \omega_1 \frac{\partial l_{1\omega t}}{\partial \dot{\beta}_i} + \omega_2 \frac{\partial l_{2\omega t}}{\partial \dot{\beta}_i} + \omega_3 \frac{\partial l_{3\omega t}}{\partial \dot{\beta}_i} \right)$$

$$+ (\dot{v}_{01} + \omega_2 v_{03} - \omega_3 v_{02} - g_1) \frac{\partial l_1}{\partial \beta_i} + (\dot{v}_{02} + \omega_3 v_{01} - \omega_1 v_{03} - g_2) \frac{\partial l_2}{\partial \beta_i}$$

$$+ (\dot{v}_{03} + \omega_1 v_{02} - \omega_2 v_{01} - g_3) \frac{\partial l_3}{\partial \beta_i} - \frac{1}{2} \omega_1^2 \frac{\partial J_{11}^1}{\partial \beta_i} - \frac{1}{2} \omega_2^2 \frac{\partial J_{22}^1}{\partial \beta_i}$$

$$- \frac{1}{2} \omega_3^2 \frac{\partial J_{33}^1}{\partial \beta_i} - \omega_1 \omega_2 \frac{\partial J_{12}^1}{\partial \beta_i} - \omega_1 \omega_3 \frac{\partial J_{13}^1}{\partial \beta_i} - \omega_2 \omega_3 \frac{\partial J_{23}^1}{\partial \beta_i} = 0. \qquad (2.6.26)$$

The system of equations (2.6.25) is linear with respect to the parameters $R_n(t)$ characterizing the behavior of the velocity potential. Therefore, this system can be resolved with respect to these parameters. Moreover, system (2.6.26) yields a system of nonlinear differential equations of the second order for the parameters $\beta_i$ characterizing the elevations of the free surface. Note that the quantities $\partial l_k / \partial \beta_i$ can be found relatively easily:

$$\frac{\partial l_1}{\partial \beta_i} = \rho N_i^2 \beta_i = \lambda_{i1} \beta_i, \quad N_i^2 = \int_{\Sigma_0} f_i^2 dS,$$

$$\frac{\partial l_2}{\partial \beta_i} = \rho \int\limits_{\Sigma_0} y f_i dS = \lambda_{i2}, \quad \frac{\partial l_3}{\partial \beta_i} = \rho \int\limits_{\Sigma_0} z f_i dS = \lambda_{i3}. \tag{2.6.27}$$

The nonlinear equations of motion in an immobile vessel derived in the previous sections follow from system (2.6.25), (2.6.26) as a special case.

The case of translational motion of the vessel proves to be quite simple for the analysis (in a sense of finding the actual values of coefficients in the equations of motion). The corresponding nonlinear equations of motion take the form

$$\frac{d}{dt} A_n - \sum_k A_{nk} R_k = 0, \quad n = 1, 2, \dots,$$

$$\sum_n \dot{R}_n \frac{\partial A_n}{\partial \beta_i} + \frac{1}{2} \sum_n \sum_k \frac{\partial A_{nk}}{\partial \beta_i} R_n R_k + (\dot{v}_{01} - g_1)\lambda_{i1}\beta_i \tag{2.6.28}$$

$$+(\dot{v}_{02} - g_2)\lambda_{i2} + (\dot{v}_{03} - g_3)\lambda_{i3} = 0, \quad i = 1, 2, \dots .$$

The outlined method can be generalized for the vessels of more complicated geometric shapes. By using the results presented in [112], this generalization can be obtained for the class of vessels in the form bodies of of revolution. In the introduced curvilinear coordinate system $Ox^1x^2x^3$ rigidly connected with the container, the equation of the free surface and the velocity potential are given in the form

$$x^1 = x_0^1 + \beta_0(t) + \sum_{i=1}^{\infty} \beta_i(t) f_i(x^2, x^3); \tag{2.6.29}$$

$$\varphi(x^1, x^2, x^3, t) = \sum_{n=1}^{\infty} R_n(t) \varphi_n(x^1, x^2, x^3). \tag{2.6.30}$$

For the subsequent construction of the algorithm, it is necessary to take into account the fact that the operations encountered in relations (2.6.15), (2.6.16)–(2.6.19) have the tensor nature.

Note that the problem of liquid sloshing is described by the system of nonlinear first-order ordinary differential equations (2.6.25), (2.6.26) and that system (2.6.25) is linear in $R_k(t)$. The quantities $R_k$ can be represented as homogeneous linear forms with respect to $\dot{\beta}_s$:

$$R_k = \sum_s a_{ks} \dot{\beta}_s, \quad a_{ks} = a_{ks}(\beta_1, \beta_2, \dots, \beta_s, \dots). \tag{2.6.31}$$

These quantities are structurally similar to the generalized momenta studied in the analytic mechanics of holonomic systems [143]. If we choose the parameters $\beta_i(t)$ as generalized coordinates of the system, then, according to (2.6.31), the generalized velocities $\dot{\beta}_i(t)$ are linear forms with respect to the parameters $R_k$. Therefore, for some problems, we can introduce the generalized momenta

$$p_n = \sum_m c_{nm} R_m \tag{2.6.32}$$

and represent the equations of motion, such as (2.6.25) and (2.6.26), in the canonical form

$$\dot{\beta}_n = \frac{\partial H}{\partial p_n}, \quad \dot{p}_n = -\frac{\partial H}{\partial \beta_n}, \tag{2.6.33}$$

where $H$ is the Hamiltonian function. This fact was first mentioned by J. Miles [151].

The investigation of the general system of nonlinear ordinary differential equations (2.6.25), (2.6.26) used to describe the phenomenon of liquid sloshing in mobile vessels encounters serious difficulties. Thus, practical results can be obtained only for a finite number of the parameters $R_n(t)$ and $\beta_i(t)$.

Numerous applied problems of the theory of motion of rigid bodies with cavities containing liquids can be studied in the linear statement, i.e., under the assumption that all parameters characterizing the motion of the mechanical system are small in a sense that their products can be neglected. The corresponding system of linear differential equations can be derived from (2.6.25) and (2.6.26) by neglecting nonlinear terms. To this end, we specify the components of the vectors $\boldsymbol{l}$, $\boldsymbol{l}_\omega$, and $\boldsymbol{l}_{\omega t}$, the tensor $\boldsymbol{J}^1$, and the quantities $A_n$ and $A_{nk}$ appearing in system (2.6.25), (2.6.26) to within terms of the first order of smallness. In this case, it is assumed that the perturbed liquid domain can be split, in the general case, into the domain occupied by the liquid in the unperturbed state and the layer of perturbed liquid with the shape of a cylinder with base $\Sigma_0$ and height $f$ to within the terms of the first order of smallness. We represent the integrand in the vicinity of $\Sigma_0$ in the form of the following Taylor series:

$$F(x, y, z, \beta_i) = F|_{\Sigma_0} + \left.\frac{\partial F}{\partial x}\right|_{\Sigma_0} x + \frac{1}{2}\left.\frac{\partial^2 F}{\partial x^2}\right|_{\Sigma_0} x^2 + \dots. \tag{2.6.34}$$

As above, the sets of functions $f_i(y, z)$ and $\varphi_n(x, y, z)$ are associated with the eigensolutions of the spectral boundary problem (2.2.4) by setting

$$f_i(y, z) = \varkappa_i \varphi_i(0, y, z).$$

As a result, we obtain

$$\begin{aligned} l_1 = \rho \int_Q x dQ &= \rho \int_{Q_0} x dQ + \rho \int_{\Sigma_0} \Big( \int_0^f x dx \Big) dS \\ &= l_{10} + \frac{1}{2} \int_{\Sigma_0} f^2 dS = l_{10} + \frac{1}{2} \sum_{i=1}^{\infty} \lambda_{i1} \beta_i^2; \end{aligned} \tag{2.6.35}$$

$$l_2 = \rho \int\limits_Q y dQ = \rho \int\limits_{Q_0} y dQ + \rho \int\limits_{Q_1} y dQ = l_{20} + \rho \int\limits_{\Sigma_0} f y dS = l_{20} + \sum_{i=1}^{\infty} \lambda_{i2} \beta_i; \tag{2.6.36}$$

$$l_3 = \rho \int\limits_Q z dQ = \rho \int\limits_{Q_0} z dQ + \rho \int\limits_{Q_1} z dQ = l_{30} + \rho \int\limits_{\Sigma_0} f z dS = l_{30} + \sum_{i=1}^{\infty} \lambda_{i3} \beta_i; \tag{2.6.37}$$

$$A_i = \rho \int\limits_Q \varphi_i dQ = \rho \int\limits_{\Sigma_0} \left( \int\limits_0^f \left[ \varphi_i |_{\Sigma_0} + \left. \frac{\partial \varphi_i}{\partial x} \right|_{\Sigma_0} x + \ldots \right] dx \right) dS$$

$$= \rho \int\limits_{\Sigma_0} \varphi_i |_{\Sigma_0} f dS + O(\beta_i^2) = \mu_i \beta_i + O(\beta_i^2); \tag{2.6.38}$$

$$A_{ik} = \rho \int\limits_{\Sigma_0} \left( \varphi_i \frac{\partial \varphi_k}{\partial x} \right)_{\Sigma_0} dS = \begin{cases} \mu_i & \text{for } i = k, \\ 0 & \text{for } i \neq k, \end{cases} \tag{2.6.39}$$

$$l_{k\omega} = \rho \int\limits_Q \Omega_k dS = \rho \int\limits_{Q_0} (\Omega_{k0} + \Omega_{k1}) dQ + \rho \int\limits_{Q_1} (\Omega_{k0} + \Omega_{k1}) dQ$$

$$= \rho \int\limits_{Q_1} (\Omega_{k0} + \Omega_{k1}) dQ = \rho \int\limits_{\Sigma_0} \left[ \int\limits_0^f \left( \Omega_{k0} + \frac{\partial \Omega_{k0}}{\partial x} x + \ldots \right) dx \right] dS$$

$$= \rho \int\limits_{\Sigma_0} \Omega_{k0} f dS + O(\beta_i^2) = \sum_{i=1}^{\infty} \lambda_{0ik} \beta_i. \tag{2.6.40}$$

Similarly, we can show that components of the tensor $\boldsymbol{J}^1$ can be found in the linear approximation by the formulas

$$J_{ij}^1 = \int\limits_{S+\Sigma_0} \Omega_{i0} \frac{\partial \Omega_{j0}}{\partial \nu} dS, \tag{2.6.41}$$

and the components of the vector $\boldsymbol{l}_{\omega t}$ start from the terms of the second order of smallness in $\beta_i(t)$.

In relations (2.6.35)–(2.6.41), we have used the following notation:

$$l_{10} = \rho \int\limits_{Q_0} x dQ, \quad l_{20} = \rho \int\limits_{Q_0} y dQ, \quad l_{30} = \rho \int\limits_{Q_0} z dQ,$$

$$\lambda_{i1} = \rho N_i^2, \quad \lambda_{i2} = \rho \int\limits_{\Sigma_0} y f_i dS, \quad \lambda_{i3} = \rho \int\limits_{\Sigma_0} z f_i dS, \tag{2.6.42}$$

$$\mu_i = \frac{\lambda_{i1}}{\varkappa_i}, \quad \lambda_{0ik} = \rho \int\limits_{\Sigma_0} \Omega_{k0} f_i dS, \quad N_i^2 = \int\limits_{\Sigma_0} f_i^2 dS.$$

Under the simplifying assumptions made above, the system of equations (2.6.25), (2.6.26) takes the form

$$\mu_i(\ddot{\beta}_i + \sigma_i^2\beta_i) + \boldsymbol{\lambda}_i \cdot \boldsymbol{w}_* + \boldsymbol{\lambda}_{0i} \cdot \boldsymbol{\omega} = 0, \tag{2.6.43}$$

where $\boldsymbol{\lambda}_i$ is the vector with components $\{0, \lambda_{i2}, \lambda_{i3}\}$, $\boldsymbol{\lambda}_{0i}$ is the vector with components $\lambda_{0ik}$ (2.6.42), $\boldsymbol{w}_*$ is the vector of apparent acceleration,

$$\boldsymbol{w}_* = \overset{*}{\boldsymbol{v}}_0 + \boldsymbol{\omega} \times \boldsymbol{v}_0 - \mathbf{g}, \tag{2.6.44}$$

whose projections onto the axes of the mobile system are $w_1 = j, w_2$, and $w_3$, and $\sigma_i^2 = \varkappa_i j$.

The equations of liquid motions in an immobile vessel follow from (2.6.43) with $\boldsymbol{v}_0 = 0$ and $\boldsymbol{\omega} = 0$.

Thus, for the prescribed motions of the vessel, the phenomenon of liquid sloshing can be described by integrating the system of differential equations under the corresponding initial conditions (or the periodicity conditions). The hydrodynamic coefficients of these equations are given by relations (2.6.42) based on the eigensolutions of the spectral boundary problem (2.2.4) of natural liquid sloshing in the vessel and the boundary-value problems for the Stokes–Zhukovsky potentials (2.6.3) posed in the unperturbed liquid domains.

## 2.7 Evaluation of the Stokes–Zhukovsky potentials by the variational method

In the previous section, in constructing the equations of motion of liquid in a mobile vessel, it was assumed that the Stokes–Zhukovsky potentials $\boldsymbol{\Omega}_i(x, y, z, t)$ are known. Since the domain of their definition is variable and defined, in fact, by the generalized coordinates $\beta_i(t)$, the relations obtained by using the potentials $\boldsymbol{\Omega}_i$ are also functions of the parameters $\beta_i$. There are two possible ways of finding the potentials $\boldsymbol{\Omega}_i$. First, by analogy with the definition of $\varphi(x, y, z, t)$, they can be represented in the form

$$\Omega_k(x, y, z, t) = \sum_{j=1}^{q} R_j^{(k)}(t)\chi_j^{(k)}(x, y, z). \tag{2.7.1}$$

As a result, we get systems of algebraic equations of the form (2.6.25) for finding $R_j^{(k)}(t)$ in the general scheme of direct method. Second, the boundary-value problems (2.6.3) can be solved in advance by the variational method under the assumption that the shape of the perturbed free surface is known and its location at any time is specified by $\beta_i(t)$. In this case, the solution of the boundary-value problems (2.6.3) for the functions $\boldsymbol{\Omega}_k$ is equivalent to the determination of minimum of the following quadratic functionals:

$$L(\Omega_k) = \int\limits_{Q} (\nabla\Omega_k)^2 dQ - 2 \int\limits_{S+\Sigma} \Omega_k (\boldsymbol{r} \times \boldsymbol{\nu})_k \, dS, \quad k = 1, 2, 3. \tag{2.7.2}$$

As a specific feature of finding approximate solutions of the corresponding variational problems for functionals (2.7.2), we can mention the fact that the functions $\Omega_k$ are nonstationary because the domain $Q$ is a function of time. By analogy with (2.6.14), we determine these functions in the form (2.7.1), where $\chi_j(x, y, z)$ is the system of coordinate functions obeying the well-known requirements, which should be satisfied for the realization of the Ritz method in the variational problems with natural boundary conditions. In what follows, we choose the system of functions $\chi_j(x, y, z)$ from the class of functions harmonic in the domain $Q$. The condition of minimum of functional (2.7.2) yields the following system of algebraic equations for the parameters $R_j^{(k)}(t)$:

$$A\boldsymbol{R}^{(k)} = \boldsymbol{\gamma}^{(k)}, \tag{2.7.3}$$

where $A$ is a symmetric $q \times q$ square matrix with elements

$$A_{jn}^{(k)} = \int\limits_{Q} (\nabla\chi_j^{(k)}, \nabla\chi_n^{(k)}) dQ = \int\limits_{S+\Sigma} \chi_j^{(k)} \frac{\partial\chi_n^{(k)}}{\partial\nu} dS \tag{2.7.4}$$

and $\boldsymbol{\gamma}^{(k)}$ is a $q$-dimensional vector with the components

$$\gamma^{(k)} = \int\limits_{S+\Sigma} \chi_n^{(k)} (\boldsymbol{r} \times \boldsymbol{\nu})_k \, dS. \tag{2.7.5}$$

Note that the components of the matrix $A$, the vector $\boldsymbol{\gamma}^{(k)}$, and the required vector $\boldsymbol{R}^{(k)}(t)$ depend on the parameters $\beta_i(t)$ specifying the perturbed surface $\Sigma$.

Both procedures turn out to be completely equivalent. Therefore, in the investigation of specific cases, the problem of construction of the Stokes–Zhukovsky potentials is studied separately with the maximum possible use of the data obtained in the previous stages of investigation.

## 2.8 Resulting hydrodynamic forces and moments

The problem of finding the forces of interaction of a moving body with a liquid partially filling its cavities is of significant interest for the solution of numerous practically important problems. The case of space motions of the body with liquid sloshing in its cavities proves to be especially difficult for analysis. In what follows, we present the formulas for the pressure of liquid upon the rigid body under sufficiently general assumptions introduced in the previous section of this chapter.

By definition, the resulting force $\boldsymbol{P}$ and moment $\boldsymbol{N}$ caused by the hydrodynamic pressure are as follows:

$$\boldsymbol{P} = \int_S p\boldsymbol{\nu}\, dS \quad \text{and} \quad \boldsymbol{N} = \int_S \boldsymbol{r} \times p\boldsymbol{\nu}\, dS. \tag{2.8.1}$$

If the velocity potential $\Phi(x, y, z, t)$ is a solution of the boundary-value problem (2.5.4), then the pressure $p$ is given by the Bernoulli equation (1.5.2). For this pressure, $\boldsymbol{P}$ and $\boldsymbol{N}$ are determined by quadratures (2.8.1). However, the direct evaluation of $\boldsymbol{P}$ and $\boldsymbol{N}$ leads to cumbersome expressions inconvenient for the practical purposes.

More compact formulas can be obtained by transforming the preliminarily relations (2.8.1) with the help of the integral theorems. We now substitute $p$ from relation (1.5.2) in the formula for $\boldsymbol{P}$ in (2.8.1). This yields

$$\boldsymbol{P} = -\rho \int_S \left[\frac{\partial \Phi}{\partial t} + \frac{1}{2}(\nabla\Phi)^2 - \nabla\Phi \cdot (\boldsymbol{v}_0 + \boldsymbol{\omega} \times \boldsymbol{r}) + U\right] \boldsymbol{\nu}\, dS. \tag{2.8.2}$$

By the Gauss theorem, we obtain

$$\rho \int_{S+\Sigma} \frac{\partial \Phi}{\partial t}\boldsymbol{\nu}\, dS = \rho \int_Q \nabla\left(\frac{\partial \Phi}{\partial t}\right) dQ = \rho \int_Q \frac{\partial}{\partial t}(\nabla\Phi) dQ. \tag{2.8.3}$$

We now transform the integral on the right-hand side of (2.8.3) with the help of the relation

$$\frac{d}{dt}\int_Q f dQ = \int_Q f_t\, dQ + \int_S f u_\nu\, dS$$

valid for any function $f(x, y, z, t)$ in the case where the domain $Q$ may depend on time $t$. In this relation, $u_\nu$ denotes the normal velocity of the boundary $S$ of the domain $Q$ regarded as positive in the direction of the outer normal to $S$. As a result, we get

$$\rho \int_Q \frac{\partial}{\partial t}(\nabla\Phi) dQ = \overset{*}{\boldsymbol{K}} - \rho \int_\Sigma \nabla\Phi u_\nu\, dS, \tag{2.8.4}$$

where

$$\boldsymbol{K} = \rho \int_Q \nabla\Phi dQ$$

is the momentum of the liquid $Q$. By virtue of (2.8.3) and (2.8.4), we find

$$\rho \int_S \frac{\partial \Phi}{\partial t}\boldsymbol{\nu}\, dS = \overset{*}{\boldsymbol{K}} - \rho \int_\Sigma \nabla\Phi u_\nu\, dS - \rho \int_\Sigma \frac{\partial \Phi}{\partial t}\boldsymbol{\nu}\, dS. \tag{2.8.5}$$

To transform the second integral in (2.8.2), we use the formula [159]

$$\frac{1}{2}\int\limits_{S+\Sigma}\left[\boldsymbol{\nu} a^2 - 2(\boldsymbol{\nu}\cdot\boldsymbol{a})\boldsymbol{a}\right]dS = \int\limits_{Q}\left[\boldsymbol{a}(\nabla\cdot\boldsymbol{a}) - \boldsymbol{a}\times(\nabla\times\boldsymbol{a})\right]dQ \tag{2.8.6}$$

obtained as a consequence of the Gauss theorem and the following relations of vector algebra:

$$\nabla(\boldsymbol{a}\cdot\boldsymbol{b}) = \boldsymbol{a}\times(\nabla\times\boldsymbol{b}) + (\boldsymbol{a}\cdot\nabla)\boldsymbol{b} + \boldsymbol{b}\times(\nabla\times\boldsymbol{a}) + (\boldsymbol{b}\cdot\nabla)\boldsymbol{a},$$

$$(\nabla\cdot\boldsymbol{b})\boldsymbol{a} = \boldsymbol{a}(\nabla\cdot\boldsymbol{b}) + (\boldsymbol{b}\cdot\nabla)\boldsymbol{a}.$$

Setting $\boldsymbol{a} = \nabla\Phi$ in (2.8.6), we get

$$\frac{1}{2}\rho\int\limits_{S}(\nabla\Phi)^2\boldsymbol{\nu}dS = \rho\int\limits_{S+\Sigma}\nabla\Phi\frac{\partial\Phi}{\partial\nu}dS - \frac{1}{2}\rho\int\limits_{\Sigma}(\nabla\Phi)^2\boldsymbol{\nu}dS. \tag{2.8.7}$$

The Gauss theorem enables us to deduce the relation

$$\int\limits_{S+\Sigma}[\boldsymbol{\nu}(\boldsymbol{a}\cdot\boldsymbol{b}) - (\boldsymbol{\nu}\cdot\boldsymbol{b})\boldsymbol{a}]dS = \int\limits_{Q}[\boldsymbol{a}\times(\nabla\times\boldsymbol{b}) + (\boldsymbol{a}\cdot\nabla)\boldsymbol{b}$$

$$+\boldsymbol{b}\times(\nabla\times\boldsymbol{a}) - \boldsymbol{a}(\nabla\cdot\boldsymbol{b})]dQ. \tag{2.8.8}$$

For $\boldsymbol{a} = \nabla\Phi$ and $\boldsymbol{b} = \boldsymbol{v}_0 + \boldsymbol{\omega}\times\boldsymbol{r}$, this yields

$$\rho\int\limits_{S}[\nabla\Phi\cdot(\boldsymbol{v}_0 + \boldsymbol{\omega}\times\boldsymbol{r})]\boldsymbol{\nu}dS = \rho\int\limits_{S+\Sigma}\frac{\partial\Phi}{\partial\nu}\nabla\Phi dS - \rho\int\limits_{Q}(\boldsymbol{\omega}\times\nabla\Phi)dQ$$

$$-\rho\int\limits_{\Sigma}[\nabla\Phi\cdot(\boldsymbol{v}_0 + \boldsymbol{\omega}\times\boldsymbol{r})]\boldsymbol{\nu}dS - \rho\int\limits_{\Sigma}u_\nu\nabla\Phi dS. \tag{2.8.9}$$

To get (2.8.9), we have used the fact that

$$\nabla\Phi\times(\nabla\times(\boldsymbol{\omega}\times\boldsymbol{r})) = -2\boldsymbol{\omega}\times\nabla\Phi,\quad (\nabla\Phi\cdot\nabla)(\boldsymbol{\omega}\times\boldsymbol{r}) = \boldsymbol{\omega}\times\nabla\Phi.$$

Since the motion of the rigid body is regarded as known and, hence, $\boldsymbol{r}_0'$ is a known function of time, we obtain

$$\rho\int\limits_{S}U\boldsymbol{\nu}dS = -\rho\int\limits_{S}[\mathbf{g}\cdot(\boldsymbol{r}_0' + \boldsymbol{r})]\boldsymbol{\nu}dS = -\rho\int\limits_{Q}\nabla(\mathbf{g}\cdot\boldsymbol{r})dQ + \rho\int\limits_{\Sigma}[\mathbf{g}\cdot(\boldsymbol{r}_0' + \boldsymbol{r})]\boldsymbol{\nu}dS$$

$$= -m_1\mathbf{g} + \rho\int\limits_{\Sigma}[\mathbf{g}\cdot(\boldsymbol{r}_0' + \boldsymbol{r})]\boldsymbol{\nu}dS, \tag{2.8.10}$$

where $m_1$ is the mass of liquid.

In view of relations (2.8.5), (2.8.7), (2.8.9), and (2.8.10), we replace $\frac{1}{2}(\nabla\Phi)^2$ on $\Sigma$ in relation (2.8.2) by its value from the dynamic boundary condition. As a result, relation (2.8.2) implies that

$$\boldsymbol{P} = m_1\mathbf{g} - \overset{*}{\boldsymbol{K}} - \boldsymbol{\omega} \times \boldsymbol{K}. \tag{2.8.11}$$

We now substitute relation (2.8.7) in the formula for $\boldsymbol{K}$. Then

$$\boldsymbol{K} = \rho \int\limits_Q \nabla\Phi dQ = m_1\boldsymbol{v}_0 + \rho \int\limits_Q \nabla(\boldsymbol{\omega} \cdot \boldsymbol{\Omega})dQ + \rho \int\limits_Q \nabla\varphi dQ. \tag{2.8.12}$$

Further, we use the integral theorems to transform the last two terms on the right-hand side of equality (2.8.12) with regard for the boundary conditions (2.5.4):

$$\begin{aligned}\rho \int\limits_Q \nabla(\boldsymbol{\omega} \cdot \boldsymbol{\Omega})dQ &= \rho \int\limits_{S+\Sigma} (\boldsymbol{\omega} \cdot \boldsymbol{\Omega})\boldsymbol{\nu} dS = \rho \int\limits_{S+\Sigma} \left(\boldsymbol{\omega} \cdot \frac{\partial \boldsymbol{\Omega}}{\partial \boldsymbol{\nu}}\right)\boldsymbol{r} dS \\ &= \rho \int\limits_{S+\Sigma} [\boldsymbol{\omega} \cdot (\boldsymbol{r} \times \boldsymbol{\nu})]\boldsymbol{r}\, dS = \rho \int\limits_Q [\boldsymbol{\omega} \cdot (\boldsymbol{r} \times \nabla)]\boldsymbol{r}\, dQ \\ &= \rho \int\limits_Q (\boldsymbol{\omega} \times \boldsymbol{r})dQ = m_1\boldsymbol{\omega} \times \boldsymbol{r}_{1C},\end{aligned} \tag{2.8.13}$$

where $\boldsymbol{r}_{1C}$ is the radius vector of the center-of-mass of the liquid relative to $O$.

We now transform the last term in (2.8.12) as follows:

$$\rho \int\limits_Q \nabla\varphi dQ = \rho \int\limits_{S+\Sigma} \varphi\boldsymbol{\nu} dS = \rho \int\limits_{S+\Sigma} \boldsymbol{r}\frac{\partial\varphi}{\partial\nu} dS = \rho \int\limits_\Sigma \boldsymbol{r} u_\nu dS = \rho \int\limits_\Sigma \boldsymbol{r}\frac{f_t}{N} dS. \tag{2.8.14}$$

By analogy with (2.8.4), we obtain

$$\rho \int\limits_\Sigma \boldsymbol{r} u_\nu dS = m_1(\boldsymbol{i}_1\dot{x}_{1C} + \boldsymbol{i}_2\dot{y}_{1C} + \boldsymbol{i}_3\dot{z}_{1C}) = m_1\overset{*}{\boldsymbol{r}}_{1C}. \tag{2.8.15}$$

Thus, the momentum of liquid is given by the formula

$$\boldsymbol{K} = m_1(\boldsymbol{v}_0 + \boldsymbol{\omega} \times \boldsymbol{r}_{1C} + \overset{*}{\boldsymbol{r}}_{1C}), \tag{2.8.16}$$

and the resulting hydrodynamic force finally takes the form

$$\begin{aligned}\boldsymbol{P} = m_1\mathbf{g} - m_1[&\overset{*}{\boldsymbol{v}}_0 + \boldsymbol{\omega} \times \boldsymbol{v}_0 + \boldsymbol{\omega} \times (\boldsymbol{\omega} \times \boldsymbol{r}_{1C}) \\ &+ \dot{\boldsymbol{\omega}} \times \boldsymbol{r}_{1C} + 2\boldsymbol{\omega} \times \overset{*}{\boldsymbol{r}}_{1C} + \overset{**}{\boldsymbol{r}}_{1C}].\end{aligned} \tag{2.8.17}$$

The quantity in the square brackets is the absolute acceleration of the center of inertia of liquid in agreement with the general theory of the dynamics of relative motions. In this case, the group of terms

$$\boldsymbol{w}_e = \overset{*}{\boldsymbol{v}}_0 + \boldsymbol{\omega} \times \boldsymbol{v}_0 + \dot{\boldsymbol{\omega}} \times \boldsymbol{r}_{1C} + \boldsymbol{\omega} \times (\boldsymbol{\omega} \times \boldsymbol{r}_{1C}) \tag{2.8.18}$$

specifies the translational acceleration formed by the acceleration of the pole $\boldsymbol{w}_0$, the centripetal acceleration $\boldsymbol{w}^c$, and the rotational acceleration $\boldsymbol{w}^{\text{rot}}$:

$$\boldsymbol{w}_0 = \overset{*}{\boldsymbol{v}}_0 + \boldsymbol{\omega} \times \boldsymbol{v}_0, \quad \boldsymbol{w}^{\text{c}} = \boldsymbol{\omega} \times (\boldsymbol{\omega} \times \boldsymbol{r}_{1C}), \quad \boldsymbol{w}^{\text{rot}} = \dot{\boldsymbol{\omega}} \times \boldsymbol{r}_{1C}. \tag{2.8.19}$$

The terms

$$\boldsymbol{w}_C^{\text{Cor}} = 2\boldsymbol{\omega} \times \overset{*}{\boldsymbol{r}}_{1C} \quad \text{and} \quad \boldsymbol{w}_r = \overset{**}{\boldsymbol{r}}_{1C} \tag{2.8.20}$$

are, respectively, the Coriolis acceleration and the relative acceleration.

For the translational motion of the vessel filled with a liquid in the gravitational field, the projections of the principal vector $\boldsymbol{P}$ onto the axes of the mobile system of coordinates take the following form:

$$\begin{aligned} P_x &= m_1 g_1 - m_1 \Big[ \dot{v}_{01} + \sum_i \lambda_{i1} (\ddot{\beta}_i \beta_i + \dot{\beta}_i^2) \Big], \\ P_y &= m_1 g_2 - m_1 \Big( \dot{v}_{02} + \sum_i \lambda_{i2} \ddot{\beta}_i \Big), \\ P_z &= m_1 g_3 - m_1 \Big( \dot{v}_{03} + \sum_i \lambda_{i3} \ddot{\beta}_i \Big), \end{aligned} \tag{2.8.21}$$

where

$$\lambda_{i1} = \rho N_i^2 = \rho \int_{\Sigma_0} f_i^2 dS, \quad \lambda_{i2} = \rho \int_{\Sigma_0} y f_i dS, \quad \lambda_{i3} = \rho \int_{\Sigma_0} z f_i dS. \tag{2.8.22}$$

Relations (2.8.21) have been obtained by using the equalities (2.5.20) and the representation of the shape of the free surface in the form (2.6.12).

Thus, as already indicated, the resulting hydrodynamic force $\boldsymbol{P}$ is completely determined by the data on the perturbed free surface. Moreover, its projections $P_y$ and $P_z$ are linear functions of the parameters characterizing the evolution of the free surface. However, the projection $P_x$ of the hydrodynamic forces onto the $Ox$-axis is a quadratic function of these parameters. This follows from relation (2.8.21).

We now find the resulting hydrodynamic moment $\boldsymbol{N}$ relative to the point $O$. Substituting the expression for pressure from (1.5.2) in (2.8.1), we get

$$\boldsymbol{N} = -\rho \int_S \boldsymbol{r} \times \left[ \frac{\partial \Phi}{\partial t} + \frac{1}{2} (\nabla \Phi)^2 - \nabla \Phi \cdot (\boldsymbol{v}_0 + \boldsymbol{\omega} \times \boldsymbol{r}) + U \right] \boldsymbol{\nu} \, dS. \tag{2.8.23}$$

Further, we transform relation (2.8.23) by using the integral theorems. For the first term, we obtain

$$\rho \int\limits_{S+\Sigma} \left( \boldsymbol{r} \times \frac{\partial \Phi}{\partial t} \boldsymbol{\nu} \right) dS = \rho \int\limits_{Q} \left( \boldsymbol{r} \times \nabla \frac{\partial \Phi}{\partial t} \right) dQ = \rho \int\limits_{Q} \left[ \boldsymbol{r} \times \frac{\partial}{\partial t} (\nabla \Phi) \right] dQ$$

$$= \rho \int\limits_{Q} \frac{\partial}{\partial t} (\boldsymbol{r} \times \nabla \Phi) dQ - \rho \int\limits_{Q} (\boldsymbol{u} \times \nabla \Phi) dQ$$

$$= \rho \frac{d}{dt} \int\limits_{Q} (\boldsymbol{r} \times \nabla \Phi) dQ - \rho \int\limits_{\Sigma} (\boldsymbol{r} \times \nabla \Phi) u_{\nu} dS - \rho \int\limits_{Q} (\boldsymbol{u} \times \nabla \Phi) dQ, \quad (2.8.24)$$

where $\boldsymbol{u}$ is the relative velocity of motion of the liquid. Therefore,

$$\rho \int\limits_{S} \left( \boldsymbol{r} \times \frac{\partial \Phi}{\partial t} \boldsymbol{\nu} \right) dS = \overset{*}{\boldsymbol{G}} - \rho \int\limits_{\Sigma} (\boldsymbol{r} \times \nabla \Phi) u_{\nu} dS - \rho \int\limits_{Q} (\boldsymbol{u} \times \nabla \Phi) dQ$$

$$- \rho \int\limits_{\Sigma} \left( \boldsymbol{r} \times \frac{\partial \Phi}{\partial t} \boldsymbol{\nu} \right) dS, \quad (2.8.25)$$

where

$$\boldsymbol{G} = \rho \int\limits_{Q} (\boldsymbol{r} \times \nabla \Phi) dQ$$

is the angular momentum of the contained liquid relative to the point $O$.

By analogy with (2.8.6), we obtain the relation

$$\frac{1}{2} \int\limits_{S+\Sigma} \boldsymbol{r} \times [\boldsymbol{\nu} a^2 - 2(\boldsymbol{\nu} \cdot \boldsymbol{a}) \boldsymbol{a}] dS = \int\limits_{Q} \boldsymbol{r} \times [\boldsymbol{a} (\nabla \cdot \boldsymbol{a}) - \boldsymbol{a} \times (\nabla \times \boldsymbol{a})] dQ. \quad (2.8.26)$$

For $\boldsymbol{a} = \nabla \Phi$, this relation implies that

$$\frac{1}{2} \int\limits_{S} [\boldsymbol{r} \times (\nabla \Phi)^2 \boldsymbol{\nu}] dS = \rho \int\limits_{S+\Sigma} \left( \boldsymbol{r} \times \nabla \Phi \frac{\partial \Phi}{\partial \nu} \right) dS - \frac{1}{2} \rho \int\limits_{\Sigma} [\boldsymbol{r} \times (\nabla \Phi)^2 \boldsymbol{\nu}] dS$$

$$= \rho \int\limits_{S+\Sigma} \left( \boldsymbol{r} \times \nabla \Phi \frac{\partial \Phi}{\partial \nu} \right) dS + \rho \int\limits_{\Sigma} \boldsymbol{r} \times \left[ \frac{\partial \Phi}{\partial t} - \nabla \Phi \cdot (\boldsymbol{v}_0 + \boldsymbol{\omega} \times \boldsymbol{r}) + U \right] \boldsymbol{\nu} dS. \quad (2.8.27)$$

In view of (2.8.8), we arrive at the following expression for the third term in (2.8.23):

$$\rho \int\limits_{S} \boldsymbol{r} \times [\nabla \Phi \cdot (\boldsymbol{v}_0 + \boldsymbol{\omega} \times \boldsymbol{r})] \boldsymbol{\nu} dS = \rho \int\limits_{S+\Sigma} \boldsymbol{r} \times \nabla \Phi \frac{\partial \Phi}{\partial \nu} dS$$

$$-\rho \int_S \boldsymbol{r} \times (\boldsymbol{\omega} \times \nabla\Phi) dQ - \rho \int_\Sigma \boldsymbol{r} \times [\nabla\Phi \cdot (\boldsymbol{v}_0 + \boldsymbol{\omega} \times \boldsymbol{r})]\boldsymbol{\nu} dS$$
$$-\rho \int_\Sigma (\boldsymbol{r} \times \nabla\Phi) u_\nu dS. \tag{2.8.28}$$

Thus. we get

$$\rho \int_S \boldsymbol{r} \times U\boldsymbol{\nu} dS = -\rho \int_Q \boldsymbol{r} \times [\mathbf{g} \cdot (\boldsymbol{r}_0' + \boldsymbol{r}_0)]\boldsymbol{\nu} dS$$
$$= -\rho \int_S \boldsymbol{r} \times \nabla(\mathbf{g} \cdot \boldsymbol{r}) dQ + \rho \int_\Sigma \boldsymbol{r} \times [\mathbf{g} \cdot (\boldsymbol{r}_0' + \boldsymbol{r}_0)]\boldsymbol{\nu} dS$$
$$= m_1 \boldsymbol{r}_{1C} \times \mathbf{g} + \rho \int_\Sigma \boldsymbol{r} \times [\mathbf{g} \cdot (\boldsymbol{r}_0' + \boldsymbol{r}_0)]\boldsymbol{\nu} dS. \tag{2.8.29}$$

We now substitute (2.8.25), (2.8.27), (2.8.28), and (2.8.29) in (2.8.23). Hence,

$$\boldsymbol{N} = m_1 \boldsymbol{r}_{1C} \times \mathbf{g} - \overset{*}{\boldsymbol{G}} + \rho \int_Q \boldsymbol{u} \times \nabla\Phi dQ - \rho \int_Q \boldsymbol{r} \times (\boldsymbol{\omega} \times \nabla\Phi) dQ. \tag{2.8.30}$$

Since

$$\boldsymbol{u} \times \nabla\Phi = -\boldsymbol{v}_0 \times \nabla\Phi - (\boldsymbol{\omega} \times \boldsymbol{r}) \times \nabla\Phi,$$
$$\boldsymbol{\omega} \times (\boldsymbol{r} \times \nabla\Phi) = \boldsymbol{r} \times (\boldsymbol{\omega} \times \nabla\Phi) - \nabla\Phi \times (\boldsymbol{\omega} \times \boldsymbol{r}),$$

we find

$$\boldsymbol{N} = m_1 \boldsymbol{r}_{1C} \times \mathbf{g} - \overset{*}{\boldsymbol{G}} - \boldsymbol{\omega} \times \boldsymbol{G} - \boldsymbol{v}_0 \times \boldsymbol{K}. \tag{2.8.31}$$

The angular momentum $\boldsymbol{G}$ of the liquid volume $Q$ relative to the point $O$ is defined with regard for representation (2.6.1):

$$\boldsymbol{G} = \rho \int_Q \boldsymbol{r} \times \nabla\Phi\, dQ = \rho \int_Q \boldsymbol{r} \times \boldsymbol{v}_0\, dQ + \rho \int_Q \boldsymbol{r} \times \nabla(\boldsymbol{\omega} \cdot \boldsymbol{\Omega}) dQ$$
$$+\rho \int_Q \boldsymbol{r} \times \nabla\varphi\, dQ = m_1 \boldsymbol{r}_{1C} \times \boldsymbol{v}_0$$
$$+\rho \int_{S+\Sigma} (\boldsymbol{\omega} \cdot \boldsymbol{\Omega})(\boldsymbol{r} \times \boldsymbol{\nu}) dS + \rho \int_{S+\Sigma} (\boldsymbol{r} \times \boldsymbol{\nu})\varphi\, dS. \tag{2.8.32}$$

By using the boundary conditions (2.5.4), we represent relation (2.8.32) in the form

$$\boldsymbol{G} = m_1 \boldsymbol{r}_{1C} \times \boldsymbol{v}_0 + \rho \int_{S+\Sigma} (\boldsymbol{\omega} \cdot \boldsymbol{\Omega}) \frac{\partial \Omega}{\partial \nu} dS + \rho \int_{S+\Sigma} \boldsymbol{\Omega} \frac{\partial \varphi}{\partial \nu} dS$$
$$= m_1 \boldsymbol{r}_{1C} \times \boldsymbol{v}_0 + \boldsymbol{\omega} \cdot \boldsymbol{J}^1 + \rho \int_\Sigma \boldsymbol{\Omega} \frac{f_t}{N} dS$$

$$= m_1 \boldsymbol{r}_{1C} \times \boldsymbol{v}_0 + \boldsymbol{\omega} \cdot \boldsymbol{J}^1 + \overset{*}{\boldsymbol{l}}_\omega - \boldsymbol{l}_{\omega t}, \tag{2.8.33}$$

where

$$J^1_{ij} = \rho \int\limits_{S+\Sigma} \frac{\partial \Omega_i}{\partial \nu} \Omega_j dS$$

are the components of the tensor of inertia of liquid $\boldsymbol{J}^1$ relative to the point $O$,

$$\boldsymbol{l}_{\omega t} = \rho \int\limits_Q \frac{\partial \boldsymbol{\Omega}}{\partial t} dQ, \quad \boldsymbol{l}_\omega = \rho \int\limits_Q \boldsymbol{\Omega} dQ,$$

and the vector $\overset{*}{\boldsymbol{l}}_\omega$ is such that its projections onto the axes of the rigidly fixed coordinate system are equal to the derivatives of the projections of the vector $\boldsymbol{l}_\omega$ onto these axes. The components of the tensor $\boldsymbol{J}^1$ and the projections of the vectors $\boldsymbol{l}_\omega$ and $\boldsymbol{l}_{\omega t}$ depend on the parameters $\beta_i(t)$ because the free surface $\Sigma$ is determined by these parameters.

Substituting relations (2.8.16) and (2.8.33) in (2.8.31), we finally obtain

$$\boldsymbol{N} = m_1 \boldsymbol{r}_{1C} \times (\mathbf{g} - \boldsymbol{\omega} \times \boldsymbol{v}_0 - \overset{*}{\boldsymbol{v}}_0) - \boldsymbol{J}^1 \cdot \dot{\boldsymbol{\omega}} - \overset{*}{\boldsymbol{J}^1} \cdot \boldsymbol{\omega}$$

$$- \boldsymbol{\omega} \times (\boldsymbol{J}^1 \cdot \boldsymbol{\omega}) - \overset{**}{\boldsymbol{l}}_\omega + \overset{*}{\boldsymbol{l}}_{\omega t} - \boldsymbol{\omega} \times (\overset{*}{\boldsymbol{l}}_\omega - \boldsymbol{l}_{\omega t}). \tag{2.8.34}$$

It is worth noting that

$$\dot{\boldsymbol{\omega}} = \overset{*}{\boldsymbol{\omega}} + \boldsymbol{\omega} \times \boldsymbol{\omega} = \overset{*}{\boldsymbol{\omega}}$$

because the coordinate system $Oxyz$ is rigidly connected with the body.

In deducing relations (2.8.17) and (2.8.34), it is assumed that the equation of the free surface has the form (2.6.12). The generalized coordinates $\beta_i(t)$ specifying the shape of the free surface are usually found as the solutions of a system of nonlinear ordinary differential equations. After this, the forces of interaction between the body and the liquid are determined from relations (2.8.17) and (2.8.34). However, it is worth noting that relations (2.8.11) and (2.8.31) are also true in a more general case of vortex motions of an ideal incompressible liquid [164]. The principal vector and the principal moment of pressure forces in the forms (2.8.17) and (2.8.34) can also be found from the equations of motion of the analyzed mechanical system. These equations are obtained in the next section by the variational method. In the corresponding equations for forces and moments, we separate the terms related to the volume of liquid $Q$ and deduce the equations of motion solely for the rigid body. Then the right-hand sides of the obtained equations contain not only the external forces and moments but also the principal vector and the principal moment of pressure forces.

## 2.9 Nonlinear dynamic equations of the body–liquid system

Earlier, we considered the problem of sloshing in the cavity of a rigid body participating in prescribed motions. As important problems in the theory of motion of the bodies with cavities containing liquids, we can mention the problem of determination both of the motion of the body itself and the motion of liquids inside the cavities in this body, as well as the problem of finding the forces of interaction between the body and the liquid according to the given external forces applied to the body.

The present section is devoted to the construction of the nonlinear equations of motion for the body with liquid and to finding the hydrodynamic coefficients of these equations on the basis of the variational approach to the analyzed problem [116].

The mechanical system under consideration consists of a rigid body and a liquid. These components of the system differ from each other by the tools used to describe their motion and by the way of application of forces responsible for this motion. It is known that the dynamics of rigid bodies deals mainly with the concentrated forces, whereas the principal role in the mechanics of continua is played by the distributed forces acting everywhere in the domain $Q$ or upon every element of the surface of continuum.

In analyzing the problems of dynamics of the mechanical body–liquid system, we denote the forces applied to points of the rigid body by $\boldsymbol{F}_1$, $\boldsymbol{F}_2$, ..., $\boldsymbol{F}_N$. As the parameters characterizing the motion of the rigid body, we use the quasivelocities $v_{0i}$ and $\omega_i$. The motions of liquid are characterized by the parameters $R_n(t)$ and $\beta_i(t)$.

Virtual displacements of a point $M_i$ of the rigid body are described by the equation

$$\delta \boldsymbol{r}_i' = \delta \boldsymbol{r}_0' + \delta \boldsymbol{\theta} \times \boldsymbol{r}_i, \tag{2.9.1}$$

where $\delta \boldsymbol{r}_0'$ is a virtual displacement of the origin $O$ of the coordinate system $Oxyz$ rigidly connected with the body, $\delta\boldsymbol{\theta}$ is the vector of an infinitely small rotation of the body, and $\boldsymbol{r}_i = \boldsymbol{OM}_i$ is the radius vector of the analyzed point. The elementary work of forces acting upon the rigid body is given by the formula

$$\delta' A = \boldsymbol{P} \cdot \delta \boldsymbol{r}_0' + \boldsymbol{M}^0 \cdot \delta \boldsymbol{\theta}. \tag{2.9.2}$$

In terms of the projections onto the axes of the rigidly fixed coordinate system $Oxyz$, we obtain

$$\delta' A = P_1 \delta r_{01} + P_2 \delta r_{02} + P_3 \delta r_{03} + M_1^0 \delta\theta_1 + M_2^0 \delta\theta_2 + M_3^0 \delta\theta_3, \tag{2.9.3}$$

where $\delta r_{01}$, $\delta r_{02}$, ..., $\delta\theta_3$ are variations of quasicoordinates corresponding to the quasivelocities $v_{01}$, $v_{02}$, $v_{03}$, $\omega_1$, $\omega_2$, and $\omega_3$. The quasivelocities $\omega_1$, $\omega_2$, and $\omega_3$ are connected with the generalized velocities by the formulas

$$\omega_1 = \dot{\psi}\cos\vartheta + \dot{\varphi},\ \omega_2 = \dot{\psi}\sin\vartheta\sin\varphi + \dot{\vartheta}\cos\varphi,\ \omega_3 = \dot{\psi}\sin\vartheta\cos\varphi - \dot{\vartheta}\sin\varphi, \quad (2.9.4)$$

where $\psi$, $\vartheta$, and $\varphi$ are the Euler angles. The inverse relations for this group take the form

$$\begin{aligned} \dot{\psi} &= (\sin\vartheta)^{-1}(\omega_2\sin\varphi + \omega_3\cos\varphi), \\ \dot{\vartheta} &= \omega_2\cos\varphi - \omega_3\sin\varphi, \\ \dot{\varphi} &= -(\omega_2\sin\varphi + \omega_3\cos\varphi)\mathrm{ctg}\vartheta + \omega_1. \end{aligned} \quad (2.9.5)$$

The variations of the quasicoordinates $\delta\theta_1$, $\delta\theta_2$, and $\delta\theta_3$ are found via the generalized coordinates $\delta\psi$, $\delta\vartheta$, and $\delta\varphi$ as follows:

$$\begin{aligned} \delta\theta_1 &= \cos\vartheta\delta\psi + \delta\varphi, \\ \delta\theta_2 &= \sin\vartheta\sin\varphi\delta\psi + \cos\varphi\delta\vartheta, \\ \delta\theta_3 &= \sin\vartheta\cos\varphi\delta\psi - \sin\varphi\delta\vartheta. \end{aligned} \quad (2.9.6)$$

The inverse relations are obtained from (2.9.5) if $\dot{\psi}$, $\dot{\vartheta}$, $\dot{\varphi}$, and $\omega_i$ are replaced with $\delta\psi$, $\delta\vartheta$, $\delta\varphi$, and $\delta\theta_i$, respectively.

These relations immediately yield the formulas [143]

$$\begin{aligned} (\delta\theta_1)^{\cdot} - \delta\omega_1 &= \omega_3\delta\theta_2 - \omega_2\delta\theta_3, \\ (\delta\theta_2)^{\cdot} - \delta\omega_2 &= \omega_1\delta\theta_3 - \omega_3\delta\theta_1, \\ (\delta\theta_3)^{\cdot} - \delta\omega_3 &= \omega_2\delta\theta_1 - \omega_1\delta\theta_2, \end{aligned} \quad (2.9.7)$$

which are frequently used in what follows. Similar relations can also be obtained for the quasicoordinates $\delta r_{01}$, $\delta r_{02}$, and $\delta r_{03}$ by using the formulas

$$v_{0k} = \boldsymbol{v}_0 \cdot \boldsymbol{i}_k, \quad \delta r_{0k} = \boldsymbol{\delta r}_0 \cdot \boldsymbol{i}_k, \quad \delta\boldsymbol{i}_k = \boldsymbol{\delta\theta} \times \boldsymbol{i}_k \quad (2.9.8)$$

and the formula of differentiation of unit vectors:

$$(\delta r_{0k})^{\cdot} - \delta v_{0k} = (\boldsymbol{\delta r}_0 \times \boldsymbol{\omega} - \boldsymbol{v}_0 \times \boldsymbol{\delta\theta}) \cdot \boldsymbol{i}_k \quad (2.9.9)$$

or

$$\begin{aligned} (\delta r_{01})^{\cdot} - \delta v_{01} &= \omega_3\delta r_{02} - \omega_2\delta r_{03} + v_{03}\delta\theta_2 - v_{02}\delta\theta_3, \\ (\delta r_{02})^{\cdot} - \delta v_{02} &= \omega_1\delta r_{03} - \omega_3\delta r_{01} + v_{01}\delta\theta_3 - v_{03}\delta\theta_1, \\ (\delta r_{03})^{\cdot} - \delta v_{03} &= \omega_2\delta r_{01} - \omega_1\delta r_{02} + v_{02}\delta\theta_1 - v_{01}\delta\theta_2. \end{aligned} \quad (2.9.10)$$

The introduction of quasivelocities instead of generalized velocities and the use of the Euler–Lagrange equations instead of the Lagrange equations turn out very fruitful for the analysis of the dynamics of systems of rigid bodies.

The Euler–Lagrange equations have a simpler form because the kinetic energy is specified in terms of the quasivelocities in a much simpler way than in terms of the generalized velocities.

The equations of motion of a free rigid body can be obtained from the Lagrange principle. It is known that the action for a time interval $(t_1, t_2)$ is defined by the formula

$$S = \int_{t_1}^{t_2} L_1 dt, \tag{2.9.11}$$

where $L_1 = T - \Pi$ is the kinetic potential. Therefore, the variational equation corresponding to (2.9.11) takes the form

$$\delta S = \int_{t_1}^{t_2} \delta L_1 dt = 0. \tag{2.9.12}$$

In terms of quasivelocities, the kinetic energy is defined as follows:

$$\begin{aligned} T = \tfrac{1}{2} \{ & m_0(v_{01}^2 + v_{02}^2 + v_{03}^2) + 2m_0 \left[ (v_{02}\omega_3 - v_{03}\omega_2)x_C^0 \right. \\ & + (v_{03}\omega_1 - v_{01}\omega_3)y_C^0 + \left. (v_{01}\omega_2 - v_{02}\omega_1)z_C^0 \right] + J_{11}^0\omega_1^2 + J_{22}^0\omega_2^2 \\ & + J_{33}^0\omega_3^2 + 2J_{12}^0\omega_1\omega_2 + 2J_{23}^0\omega_2\omega_3 + 2J_{31}^0\omega_3\omega_1 \} , \end{aligned} \tag{2.9.13}$$

and the potential energy in the gravitational field takes the form

$$\Pi = -m_0 \mathbf{g} \cdot \boldsymbol{r}'_C, \tag{2.9.14}$$

where

$$\boldsymbol{r}'_C = \boldsymbol{r}_0 + \boldsymbol{r}_C^0 = \boldsymbol{r}_0 + \frac{1}{m_0} \int_{(m)} \boldsymbol{r} dm.$$

The quantities $J_{ij}^0$ in (2.9.13) stand for the components of the tensor of inertia $\boldsymbol{J}^0$ of the rigid body at the point $O$. They are connected with the axial and centrifugal moments of inertia of the body by the formulas

$$\begin{gathered} J_{11}^0 = J_x, \quad J_{22}^0 = J_y, \quad J_{33}^0 = J_z, \\ J_{12}^0 = J_{21}^0 = -J_{xy}, \quad J_{23}^0 = J_{32}^0 = -J_{yz}, \quad J_{31}^0 = J_{13}^0 = -J_{zx}. \end{gathered} \tag{2.9.15}$$

Thus, in view of (2.9.1), we get

$$\delta\Pi = -m_0 \mathbf{g} \cdot \delta \boldsymbol{r}'_0 - (m_0 \boldsymbol{r}_C^0 \times \mathbf{g}) \cdot \delta\boldsymbol{\theta} = -\boldsymbol{P}^C \cdot \delta\boldsymbol{r}'_0 - \boldsymbol{M}^C \cdot \delta\boldsymbol{\theta}. \tag{2.9.16}$$

Hence, the variation of the kinetic potential $L_1$ can be represented in the form

$$\delta L_1 = \delta T - \delta\Pi = \frac{\partial T}{\partial v_{01}}\delta v_{01} + \frac{\partial T}{\partial v_{02}}\delta v_{02} + \frac{\partial T}{\partial v_{03}}\delta v_{03} + \frac{\partial T}{\partial \omega_1}\delta\omega_1$$

$$+\frac{\partial T}{\partial \omega_2}\delta\omega_2+\frac{\partial T}{\partial \omega_3}\delta\omega_3+P_1^C\delta r_{01}+P_2^C\delta r_{02}+P_3^C\delta r_{03}+M_1^C\delta\theta_1$$
$$+M_2^C\delta\theta_2+M_3^C\delta\theta_3. \tag{2.9.17}$$

Replacing, in this expression, the variations of the quasivelocities $\delta v_{01}$, $\delta v_{02}$, …, $\delta\omega_3$ by $(\delta r_{01})^{\cdot}$, $(\delta r_{02})^{\cdot}$, …, $(\delta\omega_3)^{\cdot}$ according to relations (2.9.7) and (2.9.10), we substitute the relation obtained as a result in the variational equation (2.9.11). We now perform integration by parts in relations of the form

$$\int_{t_1}^{t_2}\frac{\partial T}{\partial v_{0i}}(\delta r_{0i})^{\cdot}dt=\frac{\partial T}{\partial v_{0i}}\delta r_{0i}\bigg|_{t_1}^{t_2}-\int_{t_1}^{t_2}\delta r_{0i}\frac{d}{dt}\frac{\partial T}{\partial v_{0i}}dt. \tag{2.9.18}$$

Since $\delta r_{0i}(t_1)=0$, $\delta\theta_i(t_1)=0$, $\delta r_{0i}(t_2)=0$, and $\delta\theta_i(t_2)=0$, we arrive at the following equations of motion of the free rigid body in the Euler–Lagrange form:

$$\begin{aligned}&\frac{d}{dt}\frac{\partial T}{\partial v_{01}}+\omega_2\frac{\partial T}{\partial v_{03}}-\omega_3\frac{\partial T}{\partial v_{02}}=P_1^C,\\&\frac{d}{dt}\frac{\partial T}{\partial \omega_1}+\omega_2\frac{\partial T}{\partial \omega_3}-\omega_3\frac{\partial T}{\partial \omega_2}+v_{02}\frac{\partial T}{\partial v_{03}}-v_{03}\frac{\partial T}{\partial v_{02}}=M_1^C.\end{aligned} \tag{2.9.19}$$

The remaining equations follow from (2.9.19) by the cyclic permutation of subscripts.

In view of (2.9.13), relation (2.9.19) finally yields

$$\dot v_{01}+\omega_2 v_{03}-\omega_3 v_{02}+\dot\omega_2 z_C^0-\dot\omega_3 y_C^0-(\omega_2^2+\omega_3^2)x_C^0$$
$$+\omega_1(\omega_2 y_C^0+\omega_3 z_C^0)=P_1^C/m_0; \tag{2.9.20}$$

$$J_{11}^0\dot\omega_1+J_{12}^0\dot\omega_2+J_{31}^0\dot\omega_3+(J_{33}^0-J_{22}^0)\omega_2\omega_3+J_{23}^0(\omega_2^2-\omega_3^2)$$
$$+\omega_1(J_{13}^0\omega_2-J_{12}^0\omega_3)+m_0\left[y_C^0(\dot v_{03}+\omega_1 v_{02}-\omega_2 v_{01})\right.$$
$$\left.-z_C^0(\dot v_{02}+\omega_3 v_{01}-\omega_1 v_{03})\right]=M_1^C. \tag{2.9.21}$$

By the cyclic permutation of subscripts, we obtain the other two equations of system (2.9.20), (2.9.21). In the presence of nonpotential forces acting upon the absolutely rigid body, it is impossible to use the Lagrange principle in the variational statement (2.9.12). In this case, the Euler–Lagrange equations can be obtained from the relation

$$\delta S+\int_{t_1}^{t_2}\delta' A\,dt=\int_{t_1}^{t_2}(\delta L_1+\delta' A)dt=0, \tag{2.9.22}$$

where the elementary work $\delta' A$ is defined as

$$\delta' A=\sum_{i=1}^{n}P_i\delta\theta_i$$

and $\delta\theta_i$ are variations of the quasicoordinates. The generalized forces $P_i$ appear on the right-hand sides of (2.9.20) and (2.9.21).

Relation (2.9.22) is just the variational principle. In this form, it is applied in the problems of mechanics for nonpotential forces and nonholonomic systems. Relation (2.9.22) differs from the similar relation (2.9.12) by the fact that, in this case, there is no functional whose variation leads to (2.9.22). Hence, the problem of variational calculus is not considered in this case.

In the investigation of the problems of dynamics of mechanical systems formed by rigid bodies and liquid masses, we use the principle in the following form:

$$\delta S + \delta W + \int_{t_1}^{t_2} \delta' A dt = \int_{t_1}^{t_2} (\delta L_1 + \delta L + \delta' A) dt = 0, \tag{2.9.23}$$

where $L_1$ is the kinetic potential of the rigid body, $L$ is the Lagrangian of the problem of dynamics of a bounded volume of liquid given by relation (2.5.6), and $\delta' A$ is an elementary work of nonpotential forces applied to the rigid body.

The equations of motion in the quasivelocities $v_{0i}$ and $\omega_i$ characterizing the motion of the rigid body and in the parameters $R_n(t)$ and $\beta_n(t)$ characterizing the motion of liquid can be obtained from (2.9.23) by using the results presented above. To find $\delta W$ by formula (2.6.20), it is necessary to take into account the fact that the quasivelocities are also varied in this case:

$$\begin{aligned}
\delta W = -\int_{t_1}^{t_2} \Big[ & l_1 \delta \dot v_{01} + l_2 \delta \dot v_{02} + l_3 \delta \dot v_{03} + l_{1\omega} \delta \dot\omega_1 + l_{2\omega} \delta \dot\omega_2 + l_{3\omega} \delta \dot\omega_3 + l_{1\omega t} \delta \omega_1 \\
& + l_{2\omega t} \delta \omega_2 + l_{3\omega t} \delta \omega_3 - J_{11}^1 \omega_1 \delta \omega_1 - J_{22}^1 \omega_2 \delta \omega_2 - J_{33}^1 \omega_3 \delta \omega_3 \\
& - J_{12}^1 \omega_2 \delta \omega_1 - J_{12}^1 \omega_1 \delta \omega_2 - J_{13}^1 \omega_1 \delta \omega_3 - J_{13}^1 \omega_3 \delta \omega_1 - J_{23}^1 \omega_2 \delta \omega_3 \\
& - J_{23}^1 \omega_3 \delta \omega_2 - m_1 (v_{01} \delta v_{01} + v_{02} \delta v_{02} + v_{03} \delta v_{03}) \\
& + l_1 (\omega_2 \delta v_{03} + v_{03} \delta \omega_2 - \omega_3 \delta v_{02} - v_{02} \delta \omega_3) + l_2 (\omega_3 \delta v_{01} + v_{01} \delta \omega_3 \\
& - \omega_1 \delta v_{03} - v_{03} \delta \omega_1) + l_3 (\omega_1 \delta v_{02} + v_{02} \delta \omega_1 - \omega_2 \delta v_{01} - v_{01} \delta \omega_2) \\
& - m_1 \mathbf{g} \cdot \delta \mathbf{r}_0' - (m_1 \mathbf{r}_{1C} \times \mathbf{g}) \cdot \delta \boldsymbol{\theta} \Big] dt - \delta W_1,
\end{aligned} \tag{2.9.24}$$

where the quantity $\delta W_1$ corresponds to relation (2.6.24).

We now integrate the expressions

$$\int_{t_1}^{t_2} l_i \delta \dot v_{0i} dt, \quad \int_{t_1}^{t_2} l_{i\omega} \delta \dot\omega_i dt, \quad \int_{t_1}^{t_2} A_n \delta \dot R_n dt$$

by parts. In view of relations (2.5.9), it follows from (2.9.24) that

$$\delta W = \int_{t_1}^{t_2} \Big[ \Big( \frac{dl_1}{dt} + m_1 v_{01} + l_3 \omega_2 - l_2 \omega_3 \Big) \delta v_{01}$$

$$+\left(\frac{dl_2}{dt}+m_1v_{02}+l_1\omega_3-l_3\omega_1\right)\delta v_{02}+\left(\frac{dl_3}{dt}+m_1v_{03}+l_2\omega_1-l_1\omega_2\right)\delta v_{03}$$

$$+\left(\frac{dl_{1\omega}}{dt}+J^1_{11}\omega_1+J^1_{12}\omega_2+J^1_{13}\omega_3-l_{1\omega t}+l_2v_{03}-l_3v_{02}\right)\delta\omega_1$$

$$+\left(\frac{dl_{2\omega}}{dt}+J^1_{22}\omega_2+J^1_{23}\omega_3+J^1_{21}\omega_1-l_{2\omega t}+l_3v_{01}-l_1v_{03}\right)\delta\omega_2$$

$$+\left(\frac{dl_{3\omega}}{dt}+J^1_{33}\omega_3+J^1_{31}\omega_1+J^1_{32}\omega_2-l_{3\omega t}+l_1v_{02}-l_2v_{01}\right)\delta\omega_3$$

$$+\sum_n\left(\frac{d}{dt}A_n-\sum_k R_kA_{nk}\right)\delta R_n-\sum_i\left(\sum_n \dot R_n\frac{\partial A_n}{\partial\beta_i}+\frac{1}{2}\sum_n\sum_k R_nR_k\frac{\partial A_{nk}}{\partial\beta_i}\right.$$

$$+\dot\omega_1\frac{\partial l_{1\omega}}{\partial\beta_i}+\dot\omega_2\frac{\partial l_{2\omega}}{\partial\beta_i}+\dot\omega_3\frac{\partial l_{3\omega}}{\partial\beta_i}-\frac{d}{dt}\left(\omega_1\frac{\partial l_{1\omega t}}{\partial\dot\beta_i}+\omega_2\frac{\partial l_{2\omega t}}{\partial\dot\beta_i}+\omega_3\frac{\partial l_{3\omega t}}{\partial\dot\beta_i}\right)$$

$$+\omega_1\frac{\partial l_{1\omega t}}{\partial\beta_i}+\omega_2\frac{\partial l_{2\omega t}}{\partial\beta_i}+\omega_3\frac{\partial l_{3\omega t}}{\partial\beta_i}+(\dot v_{01}+\omega_2v_{03}-\omega_3v_{02}-\mathrm{g}_1)\frac{\partial l_1}{\partial\beta_i}$$

$$+(\dot v_{02}+\omega_3v_{01}-\omega_1v_{03}-\mathrm{g}_2)\frac{\partial l_2}{\partial\beta_i}+(\dot v_{03}+\omega_1v_{02}-\omega_2v_{01}-\mathrm{g}_3)\frac{\partial l_3}{\partial\beta_i}-\frac{1}{2}\omega_1^2\frac{\partial J^1_{11}}{\partial\beta_i}$$

$$\left.-\frac{1}{2}\omega_2^2\frac{\partial J^1_{22}}{\partial\beta_i}-\frac{1}{2}\omega_3^2\frac{\partial J^1_{33}}{\partial\beta_i}-\omega_1\omega_2\frac{\partial J^1_{12}}{\partial\beta_i}-\omega_1\omega_3\frac{\partial J^1_{13}}{\partial\beta_i}-\omega_2\omega_3\frac{\partial J^1_{23}}{\partial\beta_i}\right)\delta\beta_i$$

$$+m_1\mathbf{g}\cdot\delta\boldsymbol{r}_0+(m_1\boldsymbol{r}_{1C}\times\mathbf{g})\cdot\delta\boldsymbol{\theta}]\,dt=0. \qquad (2.9.25)$$

We now substitute relation (2.9.25), together with (2.9.17), in (2.9.23). In this case, we replace the variations of quasivelocities by the variations of the quasicoordinates by using relations (2.9.7) and (2.9.10). In order that the integrand contain solely the variations of quasicoordinates $\delta r_{0i}$ and $\delta\theta_i$ and generalized coordinates $\delta R_n$ and $\delta\beta_i$, it is necessary to perform integration by parts with an aim to remove the terms of the form (2.9.18). In this case, It is necessary to take into account the fact that the variations of quasicoordinates at the initial and final times are equal to zero. In the obtained integrand, we equate to zero the terms appearing as the coefficients of the variations of quasicoordinates and generalized coordinates $R_n$ and $\beta_i$. Thus, in addition to (2.6.25), (2.6.26), we get the following equations of motion:

$$M\left[\dot v_{01}+\omega_2v_{03}-\omega_3v_{02}-\mathrm{g}_1+\dot\omega_2z_C-\dot\omega_3y_C-x_C(\omega_2^2+\omega_3^2)+\omega_1(\omega_2y_C+\omega_3z_C)\right]$$

$$+2m_1(\omega_2\dot z_{1C}-\omega_3\dot y_{1C})+m_1\ddot x_{1C}=P_1; \qquad (2.9.26)$$

$$J_{11}\dot\omega_1+J_{12}\dot\omega_2+J_{13}\dot\omega_3+\dot J^1_{11}\omega_1+\dot J^1_{12}\omega_2+\dot J^1_{13}\omega_3+J_{32}(\omega_2^2-\omega_3^2)$$

$$+(J_{33}-J_{22})\omega_2\omega_3+\omega_1(J_{31}\omega_2-J_{21}\omega_3)$$

$$+M[y_C(\dot{v}_{03}+\omega_1 v_{02}-\omega_2 v_{01}-g_3)-z_C(\dot{v}_{02}+\omega_3 v_{01}-\omega_1 v_{03}-g_2)]$$

$$+\omega_2\frac{dl_{3\omega}}{dt}-\omega_3\frac{dl_{2\omega}}{dt}+\frac{d^2 l_{1\omega}}{dt^2}-\frac{dl_{1\omega t}}{dt}-\omega_2 l_{3\omega t}+\omega_3 l_{2\omega t}=M_1^0, \qquad (2.9.27)$$

where $M=m_0+m_1$ is the mass of the mechanical system under consideration and $\boldsymbol{r}_C^0$, $\boldsymbol{r}_{1C}$, and $\boldsymbol{r}_C$ are, respectively, the radius vectors of the centers of masses of the rigid body, liquid, and the entire system relative to the point $O$; these vectors are given by the formulas

$$\boldsymbol{r}_C^0=\frac{1}{m_0}\int\limits_{(m_0)}\boldsymbol{r}dm,\quad \boldsymbol{r}_{1C}=\frac{1}{m_1}\int\limits_{(m_1)}\boldsymbol{r}dm,\quad \boldsymbol{r}_C=\frac{m_0\boldsymbol{r}_C^0+m_1\boldsymbol{r}_{1C}}{M}; \qquad (2.9.28)$$

$J_{ij}$ are the components of the tensor of inertia of the entire system formed by the tensor of inertia of the rigid body,

$$\boldsymbol{J}^0=\begin{pmatrix} J_{11}^0 & J_{12}^0 & J_{13}^0\\ J_{21}^0 & J_{22}^0 & J_{23}^0\\ J_{31}^0 & J_{32}^0 & J_{33}^0 \end{pmatrix}=\begin{pmatrix} J_x & -J_{xy} & -J_{zx}\\ -J_{yx} & J_y & -J_{yz}\\ -J_{zx} & -J_{zy} & J_z, \end{pmatrix} \qquad (2.9.29)$$

and a symmetric tensor of the second rank

$$\boldsymbol{J}^1=\begin{pmatrix} J_{11}^1 & J_{12}^1 & J_{13}^1\\ J_{21}^1 & J_{22}^1 & J_{23}^1\\ J_{31}^1 & J_{32}^1 & J_{33}^1. \end{pmatrix}, \qquad (2.9.30)$$

By analogy, tensor (2.9.30) can be called the tensor of inertia of liquid partially filling the cavity in the body at the point $O$.

By the cyclic permutations of indices in (2.9.26) and (2.9.27), we obtain two more pairs of similar equations. Three equations in each group admit the following natural representation in the vector form

$$M[\overset{*}{\boldsymbol{v}}_0+\boldsymbol{\omega}\times\boldsymbol{v}_0-\mathbf{g}+\dot{\boldsymbol{\omega}}\times\boldsymbol{r}_C+\boldsymbol{\omega}\times(\boldsymbol{\omega}\times\boldsymbol{r}_C)]$$

$$+m_1\overset{**}{\boldsymbol{r}}_{1C}+2m_1(\boldsymbol{\omega}\times\overset{*}{\boldsymbol{r}}_{1C})=\boldsymbol{P}; \qquad (2.9.31)$$

$$\boldsymbol{J}\cdot\dot{\boldsymbol{\omega}}+\overset{*}{\boldsymbol{J}^1}\cdot\boldsymbol{\omega}+\boldsymbol{\omega}\times(\boldsymbol{J}\cdot\boldsymbol{\omega})+M\boldsymbol{r}_C\times(\overset{*}{\boldsymbol{v}}_0+\boldsymbol{\omega}\times\boldsymbol{v}_0-\mathbf{g})$$

$$+\boldsymbol{\omega}\times\overset{*}{\boldsymbol{l}}_\omega+\overset{**}{\boldsymbol{l}}_\omega-\overset{*}{\boldsymbol{l}}_{\omega t}-\boldsymbol{\omega}\times\boldsymbol{l}_{\omega t}=\boldsymbol{M}^0, \qquad (2.9.32)$$

where $\boldsymbol{P}$ is the principal vector and $\boldsymbol{M}^0$ is the principal moment of all active forces relative to the pole $O$. In Eqs. (2.9.31) and (2.9.32), the asterisks denote the vectors whose projections onto the axes of the coordinate system $Oxyz$ rigidly connected with the body are equal to the derivatives of the projections of the corresponding vectors onto these axes.

The quantities of the form $\boldsymbol{J} \cdot \boldsymbol{\omega}$ in Eqs. (2.9.31) and (2.9.32) should be understood as the multiplication of the tensor $\boldsymbol{J}$ by the vector $\boldsymbol{\omega}$ from the right. The equations of motion (2.9.31) and (2.9.32) should be considered together with the system of differential equations (2.6.25), (2.6.26) for the relative motion of liquid.

Equations (2.6.25) and (2.6.26) can be represented in a more compact form:

$$\frac{d}{dt}A_n - \sum_k R_k A_{nk} = 0, \quad n = 1, 2, \ldots; \tag{2.9.33}$$

$$\begin{aligned}\sum_n \dot{R}_n \frac{\partial A_n}{\partial \beta_i} + \frac{1}{2}\sum_n \sum_k \frac{\partial A_{nk}}{\partial \beta_i} R_n R_k + \dot{\boldsymbol{\omega}} \cdot \left(\frac{\partial \boldsymbol{l}_\omega}{\partial \beta_i} - \frac{\partial \boldsymbol{l}_{\omega t}}{\partial \dot{\beta}_i}\right) \\ + \boldsymbol{\omega} \cdot \left(\frac{\partial \boldsymbol{l}_{\omega t}}{\partial \beta_i} - \frac{d}{dt}\frac{\partial \boldsymbol{l}_{\omega t}}{\partial \dot{\beta}_i}\right) + (\overset{*}{\boldsymbol{v}}_0 + \boldsymbol{\omega} \times \boldsymbol{v}_0 - \mathbf{g}) \cdot \frac{\partial \boldsymbol{l}}{\partial \beta_i} \\ - \frac{1}{2}\boldsymbol{\omega} \cdot \frac{\partial \boldsymbol{J}^1}{\partial \beta_i} \cdot \boldsymbol{\omega} = 0.\end{aligned} \tag{2.9.34}$$

The system of equations (2.9.31)–(2.9.34) is an infinite-dimensional system of nonlinear differential equations of the first order for the six quasivelocities $v_{0i}$ and $\omega_i$ and the liquid-related generalized coordinates and velocities $\beta_k$ and $R_n$. They should be supplemented with the relations connecting the generalized velocities corresponding to the generalized coordinates of the rigid body with the quasivelocities $v_{0i}$ and $\omega_i$. Since

$$M\overset{**}{\boldsymbol{r}}_C = m_1 \overset{**}{\boldsymbol{r}}_{C_1} \quad \text{and} \quad M\overset{*}{\boldsymbol{r}}_C = m_1 \overset{*}{\boldsymbol{r}}_{C_1},$$

Eq. (2.9.31) can be represented in the form

$$M[\overset{*}{\boldsymbol{v}}_0 + \boldsymbol{\omega} \times \boldsymbol{v}_0 - \mathbf{g} + \dot{\boldsymbol{\omega}} \times \boldsymbol{r}_C + \boldsymbol{\omega} \times (\boldsymbol{\omega} \times \boldsymbol{r}_C) + 2(\boldsymbol{\omega} \times \overset{*}{\boldsymbol{r}}_C) + \overset{**}{\boldsymbol{r}}_C] = \boldsymbol{P}. \tag{2.9.35}$$

In this form, Eq. (2.9.35) expresses the theorem on motion of the center of inertia of the system and completely coincides with the corresponding equation of the dynamics of relative motion of mechanical systems. The expression in the square brackets is the absolute acceleration $\boldsymbol{w}_C$ of the center of inertia of the system.

According to the generally accepted terminology, the group of terms

$$\boldsymbol{w}_e = \overset{*}{\boldsymbol{v}}_0 + \boldsymbol{\omega} \times \boldsymbol{v}_0 + \dot{\boldsymbol{\omega}} \times \boldsymbol{r}_C + \boldsymbol{\omega} \times (\boldsymbol{\omega} \times \boldsymbol{r}_C) \tag{2.9.36}$$

is called the translational acceleration and consists of the acceleration of the pole

$$\boldsymbol{w}_0 = \overset{*}{\boldsymbol{v}}_0 + \boldsymbol{\omega} \times \boldsymbol{v}_0 \tag{2.9.37}$$

and the centripetal and rotational accelerations

$$\boldsymbol{w}^{\mathrm{c}} = \boldsymbol{\omega} \times (\boldsymbol{\omega} \times \boldsymbol{r}_C) \quad \text{and} \quad \boldsymbol{w}^{\mathrm{rot}} = \dot{\boldsymbol{\omega}} \times \boldsymbol{r}_C, \tag{2.9.38}$$

respectively. The term

$$\boldsymbol{w}_C^{\mathrm{Cor}} = 2\boldsymbol{\omega} \times \overset{*}{\boldsymbol{r}}_C$$

is the Coriolis acceleration and $w_r = \overset{**}{\boldsymbol{r}}_C$ is the relative acceleration of the center of inertia of the system.

The second equation of motion of the rigid body with regard for the liquid motions in the cavity can be represented in a somewhat different form. Recall that, by the definition of the principal moment of the relative momenta with respect to the pole $O$, we have

$$\boldsymbol{G}_r^0 = \rho \int\limits_Q \boldsymbol{r} \times \boldsymbol{v}_r \, dQ = \rho \int\limits_Q \boldsymbol{r} \times \nabla\varphi \, dQ. \tag{2.9.39}$$

After simple transformations performed with the help of the formulas of vector analysis and relation (1.1.4), we obtain

$$\begin{aligned} \boldsymbol{G}_r^0 &= \rho \int\limits_Q \boldsymbol{r} \times \nabla\varphi \, dQ = \rho \int\limits_{S+\Sigma} (\boldsymbol{r} \times \boldsymbol{\nu})\varphi \, dS \\ &= \rho \int\limits_{S+\Sigma} \frac{\partial \boldsymbol{\Omega}}{\partial \nu}\varphi \, dS = \rho \int\limits_{S+\Sigma} \boldsymbol{\Omega} \frac{\partial \varphi}{\partial \nu} \, dS = \rho \int\limits_{S+\Sigma} \boldsymbol{\Omega} \, u_\nu dS \\ &= \frac{d}{dt}\rho \int\limits_Q \boldsymbol{\Omega} dQ - \rho \int\limits_Q \frac{\partial \boldsymbol{\Omega}}{\partial t} dQ = \overset{*}{\boldsymbol{l}}_\omega - \boldsymbol{l}_{\omega t}. \end{aligned} \tag{2.9.40}$$

Equation (2.9.32) now takes the form

$$\begin{aligned} &\boldsymbol{J} \cdot \dot{\boldsymbol{\omega}} + \overset{*}{\boldsymbol{J}}{}^1 \cdot \boldsymbol{\omega} + \boldsymbol{\omega} \times (\boldsymbol{J} \cdot \boldsymbol{\omega}) + \overset{*}{\boldsymbol{G}}{}_r^0 + \boldsymbol{\omega} \times \boldsymbol{G}_r^0 \\ &\quad + M\boldsymbol{r}_C \times (\overset{*}{\boldsymbol{v}}_0 + \boldsymbol{\omega} \times \boldsymbol{v}_0 - \mathbf{g}) = \boldsymbol{M}^0, \end{aligned} \tag{2.9.41}$$

which coincides with the corresponding vector Euler– Lagrange equation presented, e.g., in [143]. The group of terms $\overset{*}{\boldsymbol{J}}{}^1 \cdot \boldsymbol{\omega} + \boldsymbol{\omega} \times \boldsymbol{G}_r^0$ transferred into the right-hand side of (2.9.41) is usually interpreted as the moment of the Coriolis inertial forces

$$m_{\mathrm{Cor}}^0 = -\overset{*}{\boldsymbol{J}}{}^1 \cdot \boldsymbol{\omega} - \boldsymbol{\omega} \times \boldsymbol{G}_r^0. \tag{2.9.42}$$

This moment is applied to the rigid body and caused by the liquid motions in the cavity.

In the case where the vessel is filled completely, the flow of liquid in the vessel does not change the position of the center of inertia of the system, i.e., $\overset{*}{\boldsymbol{r}}_C = 0$ and $\overset{**}{\boldsymbol{r}}_C = 0$. In this case, Eq. (2.9.35) takes the form

$$M[\overset{*}{\boldsymbol{v}}_0 + \boldsymbol{\omega} \times \boldsymbol{v}_0 - \mathbf{g} + \dot{\boldsymbol{\omega}} \times \boldsymbol{r}_C + \boldsymbol{\omega} \times (\boldsymbol{\omega} \times \boldsymbol{r}_C)] = \boldsymbol{P}. \tag{2.9.43}$$

For the same reason, the second, fourth, and fifth terms vanish in the equation of moments (2.9.41). After simplifications, this equation can be rewritten in the form

$$\boldsymbol{J} \cdot \dot{\boldsymbol{\omega}} + \boldsymbol{\omega} \times (\boldsymbol{J} \cdot \boldsymbol{\omega}) + M\boldsymbol{r}_C \times (\overset{*}{\boldsymbol{v}}_0 + \boldsymbol{\omega} \times \boldsymbol{v}_0 - \mathbf{g}) = \boldsymbol{M}^0, \tag{2.9.44}$$

which coincides with the equation of motion of the (equivalent) rigid body. Equations (2.9.43) and (2.9.44) were obtained by N.E. Zhukovsky in 1885 [83].

In the case where the motion of the rigid body is prescribed, it is necessary to consider only Eqs. (2.9.33) and (2.9.34). As shown above, the form of these equations can be different in different specific cases. The physical meaning of Eqs. (2.9.31) and (2.9.34) coincides with the physical meaning of similar equations from [173]. They have somewhat different forms, solely due to the methods used for their construction. In deducing equations of this kind in [171], Narimanov solved some nonlinear boundary-value problems of hydrodynamics. His results made it possible to determine the dynamic quantities corresponding to a liquid, such as the momentum, kinetic moment, and kinetic energy necessary for the construction of the equations of motion of the entire system in the Euler–Lagrange form either by applying the theorems on changes in the momentum and angular momentum or by the Lagrange method with the use of the formula for the kinetic energy of the system.

To deduce the equations of motion of the rigid body containing a cavity partially filled with liquid by the method discussed in the present work, we use the Bateman–Luke variational principle. In other words, we use the characteristics specifying a certain property of the actual motions, which differ these motions from all other possible motions not contradicting the constraints.

In the variational method, the Lagrangian is expressed via the parameters characterizing the motion of the system in a quite clear form which does not contradict the necessary conditions of extremum of the functional and the physical conditions of applicability of the variational principles. To satisfy two nonlinear boundary conditions on the free surface, we use the idea based on the direct methods applied in the problems of mathematical physics. In our case, this leads to the investigation of nonlinear systems of ordinary differential equations with time-dependent parameters instead of the system of algebraic equations typical of stationary problems.

The coefficients of the obtained equations of motion are obtained in a fairly simple way. This is realized, as shown above, by a suitable choice of the system of "coordinate" functions. Since we deal with the variational problems under natural boundary conditions, the constructive choice, e.g., of the velocity potential can be different.

The equations of motion (2.9.31)–(2.9.34) are deduced in the most general form. For their practical application, it is necessary to choose, among the parameters characterizing the motion of the system, the parameters predominant in a certain sense and significant for the investigated physical process.

The Euler–Lagrange equations of motion of the rigid body, the equations of motion of the rigid body containing a cavity completely filled with a liquid [83], and the linear equations of motion of the rigid body with cavity partially filled with liquid can be obtained from (2.9.31)–(2.9.34) as special cases [149, 164, 59].

In conclusion, we note that the idea to use variational principles for the construction of the equations of motion of rigid bodies with cavities containing liquids was earlier realized by Rumyantsev [164]. The classical version of the Lagrange variational method based on the introduction of Lagrange multipliers makes it possible to obtain the equations of motion of the mechanical "body–liquid" system with dynamic conditions on the free surface. However, in this approach, it is necessary to assume that the nonlinear kinematic condition is satisfied on the free surface. The indicated difficulty can be overcome, as already indicated, by the construction of a Lagrangian leading to the variational problems with natural boundary conditions in the variational principle. This opens wide possibilities for the construction of approximate solutions of the nonlinear problems of the theory of motion of rigid bodies with cavities containing liquids by the direct methods.

# Chapter 3

# Hydrodynamic theory of motions of the ships transporting liquids

On the basis the Bateman–Luke variational principle, we present a mathematical theory of motions of the ships with tanks filled with liquids. The ships are regarded as floating rigid bodies subjected to given external forces. The outboard water participates in complex motions affected by the incoming waves, diffraction, scattering, and the wave motions caused by the hull of the ship. The inner (contained) liquid fills (either partially or completely) the tanks of the ship and also participates in complex motions caused by its interaction with the walls. Thus, in the presented theory, it is reasonable to speak about a model of the hybrid mechanical system formed by the outer water, a perfectly rigid body, and the inner (contained) liquid. It is also worth noting that the most general statement of the problems of dynamics of the "outer water–body–contained liquid" system is based on the use of a viscous hydrodynamic model. In view of the contemporary results on the dynamics of ships regarded as floating bodies and the dynamics of liquid-fuel rockets, we can propose several physical assumptions under which the mathematical models of the analyzed mechanical system give nontrivial practical results.

First, as in the fundamental works devoted to the dynamics of ships [76, 183, 106, 9, 93], we consider a rigid body with six degrees of freedom. We also assume that the body (ship) can perform small oscillations about the translational motion with constant velocity and the rotational motion with oblique circulation. It is not necessary to write the equations of motion for the outer and inner liquids in the form of the Navier–Stokes equations with the corresponding boundary and initial conditions in order to understand that they make the general statement of the problem more complicated. Due to the presence of the free surfaces of the outer and inner liquids depending on the characteristics of rolling and pitching of the ship, we get complicated free-surface problems. At the same time, the accumulated experience shows that most problems of the liquid–ship interaction can be studied within the framework of potential inviscid incompressible flows. As for the outer water, the friction forces slightly affect the characteristics of surface waves and, in particular, of progressive waves. According to the spectral method, the wind waves can be represented in the form of the superposition of simple harmonic waves. If necessary, the influence of viscosity can be taken into account in dedicated experiments in the form of empirical corrections to the expressions for hydrodynamic forces and moments. Thus, in what follows, we use the mathematical models based on the equations of kinematics and dynamics of rigid bodies and the theory of wave motions of ideal incompressible liquids with irrotational flows.

## 3.1 Statement of the problem

We consider the general case of motion of a rigid body floating in seawater and containing cavities filled with a homogeneous incompressible ideal liquid. We introduce three coordinate systems: Namely, a fixed system and two moving coordinate systems. The absolute motion of the body is described in the body-fixed coordinate system $O'x'y'z'$. The plane $O'x'y'$ coincides with the unperturbed free surface of the outer water and the $O'z'$-axis is directed vertically upward. The first moving coordinate system $O_1\xi\eta\zeta$ is chosen so that the plane $O_1\xi\eta$ coincides with the plane $O'x'y'$ and the $O_1\zeta$-axis is directed vertically upward and passes through the center of gravity of the body. The $O_1\xi$-axis is directed along the mean velocity of the rectilinear motion of the body and makes a constant angle $\beta$ with the fixed $O'x'$-axis. The second mobile coordinate system $Oxyz$ is rigidly connected with the body in such a way that the $Oz$-axis also passes through the center of gravity of the body.

The presented mobile coordinate systems are used to introduce the Euler–Krylov angles $\theta, \chi$, and $\psi$ characterizing the roll, yaw, and pitch of the ship, respectively. In the equilibrium state of the rigid body and the outer and inner liquids, the directions of the axes of both moving systems coincide. In our studies, the first moving system is auxiliary. We characterize the motion of the coordinate system $Oxyz$ relative to the body-fixed coordinate system $O'x'y'z'$ by the vector of translational velocity $\boldsymbol{v}_0$ of the point $O$ and the vector of instantaneous angular velocity $\boldsymbol{\omega}$ with respect to $O$.

It is convenient to split the problem of dynamics of the water–body–liquid system into two parts. First, we focus our attention on the hydrodynamic problems of motion of the outer and inner liquids under the assumption that the kinematic parameters of motion of the rigid body ($\boldsymbol{v}_0$ and $\boldsymbol{\omega}$) are known. This enables us to find the fields of hydrodynamic velocities and pressures as well as the resulting hydrodynamic forces and moments. Thus, we can deduce the equations of motion of the body by using the main theorems of dynamics.

Under the assumption that the motion of liquid in the body-fixed coordinate system is irrotational, we can reduce the problem of finding the hydrodynamic fields to the solution of the Laplace equations in the domains $Q_i$ $(i = 1, 2)$ for the velocity potentials $\Phi_i(x', y', z', t)$ satisfying the corresponding nonlinear kinematic and dynamic conditions on the unknown free surfaces, the slip conditions on the wetted surfaces of the body, the conditions of conservation of volumes, the conditions imposed at infinity, and the initial conditions at $t = 0$. Furthermore, the pressure $p_i$ is given by the Bernoulli equation. The set of the above-mentioned relations can be represented in the form

$$\nabla^2\Phi_i = 0; \quad \boldsymbol{v}_i = \nabla\Phi_i \text{ in } Q_i, \tag{3.1.1}$$

$$\frac{\partial\Phi_i}{\partial t} + \frac{1}{2}(\nabla\Phi_i)^2 + gz' = 0 \text{ on } \Sigma_i, \tag{3.1.2}$$

$$\frac{d\zeta_i(x', y', z', t)}{dt} = 0 \text{ on } \Sigma_i, \tag{3.1.3}$$

$$\frac{dF_i(x', y', z', t)}{dt} = 0 \text{ on } S_i, \tag{3.1.4}$$

where $Q_i$ are the domains occupied by the outer and inner liquids, $\Sigma_i$ are the free surfaces, $S_i$ are the wetted surfaces, $\zeta_i(x', y', z', t) = 0$ are the equations of the free surfaces, and $F_i(x', y', z', t) = 0$ are the equations of the wetted surfaces. Here, the subscripts $i = 1$ and $i = 2$ correspond to the outer and inner liquids, respectively. Conditions (3.1.4) are called the slip conditions. They are often represented in the form

$$\frac{\partial\Phi_i}{\partial n} = V_{in} \text{ on } S_i, \tag{3.1.5}$$

where $V_{in}$ is the projection of the velocity of points of the surface $S_i$ onto the normal to the surface.

Moreover, $\Phi_1(x', y', z', t)$ must decay far from the free surface and far from the body. The potential wave motion must degenerate into a given system of incoming waves with bounded velocities. Thus, $\nabla\Phi_1 \to 0$ as $z' \to \infty$ and $\nabla\Phi_1$ is bounded as $(x'^2 + y'^2)^{\frac{1}{2}} \to \infty$. The velocity potential $\Phi_2(x', y', z', t)$ satisfies the condition of conservation of volume.

As initial conditions, it is customary to specify the initial positions of the free surfaces, the initial values of the velocity potentials $\Phi_i(x', y', z', t)$, and the initial values of the position and velocity of the rigid body for $t = 0$. After finding the velocity potentials $\Phi_i(x', y', z', t)$, the pressure fields in the outer and inner volumes of liquids are given by the Bernoulli equation

$$\frac{p_i}{\rho_i} + \frac{\partial\Phi_i}{\partial t} + \frac{1}{2}(\nabla\Phi_i)^2 + gz' = C(t), \tag{3.1.6}$$

where $p_i(x', y', z', t)$ is pressure and $\rho_i$ is the mass density of liquid which is assumed to be constant.

It is worth noting that the attempts to construct exact solutions of the considered free-surface problems and even their qualitative analysis encounter serious mathematical difficulties. Significant advances in the dynamics of ships achieved for about 100 years are explained solely by the use of linearization. At the same time, numerous important nonlinear effects in the theory of rolling and pitching of ships were established with the use of approximate methods for the solution of nonlinear boundary-value problems in bounded volumes of liquid [106, 9]. In recent years, a considerable progress in the construction of approximate methods for the solution of boundary-value problems of internal hydrodynamics has been achieved in the nonlinear dynamics of bodies filled with liquids in connection with the solution of a series of problems of rocket and space engineering. [124].

The experience accumulated in these two fields shows that the mathematical model of the dynamics of ships with liquid tanks should be constructed with regard for the principal nonlinear factors caused by the wave motions in the outer and inner liquid domains. In the construction of the nonlinear models of rolling and pitching of ships, it is customary to use the first moving coordinate system [183]. At the same time, the moving system rigidly connected with the body is also used in some problems of the dynamics of mechanical "body–liquid" systems. The results of our investigations are also, for the most part, obtained in this coordinate system.

We now present the most frequently used relations of the mathematical theory of absolute motion of liquid in the coordinate system $Oxyz$. These results can be obtained by using the laws of kinematics of the relative motions and the tensor properties of differential operators.

By $\boldsymbol{\Phi}_i(x', y', z', t)$, we denote the velocity potentials of the outer and inner liquids. According to the formulas of transformation of coordinates, in the fixed coordinate system, they take the form $\boldsymbol{\Phi}_i(x, y, z, t)$. The vector of absolute velocity in the coordinate system $(x', y', z')$ is defined via the quantities $\frac{\partial \Phi_i}{\partial x'}$, $\frac{\partial \Phi_i}{\partial y'}$, and $\frac{\partial \Phi_i}{\partial z'}$ by the formula $\boldsymbol{v}_{a,i} = \mathrm{grad}\boldsymbol{\Phi}_i$. In passing to the moving coordinate system $(x, y, z)$, the components of the covariant vector $\mathrm{grad}\,\boldsymbol{\Phi}_i$ are transformed according to the rule

$$\frac{\partial \Phi_i}{\partial x} = \frac{\partial \Phi_i}{\partial x'}\frac{\partial x'}{\partial x} + \frac{\partial \Phi_i}{\partial y'}\frac{\partial y'}{\partial x} + \frac{\partial \Phi_i}{\partial z'}\frac{\partial z'}{\partial x} \tag{3.1.7}$$

(together with two similar expressions).

It is worth noting that the scalar product of two vectors $\boldsymbol{a}$ and $\boldsymbol{b}$

$$\boldsymbol{a}\cdot\boldsymbol{b} = \sum_{k=1}^{3} a_k' b_k' = \sum_{i=1}^{3} a_i b_i$$

is an invariant of the orthogonal transformation of the coordinate system, i.e., an *absolute scalar*. In an arbitrary curvilinear system, this invariant is the quantity

$$\boldsymbol{a}\cdot\boldsymbol{b} = a^i b_i,$$

where $a^i$ and $b_i$ are, respectively, the contravariant and covariant components of the vectors $\boldsymbol{a}$ and $\boldsymbol{b}$. This implies that the investigated potential fields have the invariants $\mathrm{grad}^2\boldsymbol{\Phi}_i$ and

$$\frac{\partial \Phi_i}{\partial n} = \mathrm{grad}\boldsymbol{\Phi} \cdot \boldsymbol{n}.$$

It is also clear that $\boldsymbol{\Phi}_i$ are harmonic functions of the variables $x, y$, and $z$ because they are harmonic functions of $x^{'}, y^{'}$, and $z^{'}$.

To write the equations of absolute liquid motion in the mobile coordinate system, we use the equations of motion in the Euler form and the principal relations of the dynamics of relative motions. If $\boldsymbol{v}$ is the absolute velocity, then the equations of motion take the following vector form:

$$\frac{\partial \boldsymbol{v}}{\partial t} + \tfrac{1}{2}\nabla(v^2) - \boldsymbol{v} \times \mathrm{rot}\boldsymbol{v} = \boldsymbol{F} - \frac{1}{\rho}\nabla p. \tag{3.1.8}$$

Assume that the fixed $(O^{'}x^{'}y^{'}z^{'})$ and moving $(Oxyz)$ coordinate systems coincide at the initial time. By $\boldsymbol{r}$, we denote the radius vector of a liquid particle. Moreover, by $\boldsymbol{v}_r$, we denote the vector of its relative velocity, i.e., the velocity relative to the moving coordinate system. It is known that the absolute velocity of liquid particles is given by the formula $\boldsymbol{v}_a = \boldsymbol{v}_e + \boldsymbol{v}_r$, where $\boldsymbol{v}_e = \boldsymbol{v}_0 + \boldsymbol{\omega} \times \boldsymbol{r}$ is the translational velocity. By the Coriolis theorem, for the absolute acceleration of liquid particles, we get

$$\boldsymbol{w}_a = \boldsymbol{w}_e + \boldsymbol{w}_r + \boldsymbol{w}_c,$$

where

$$\boldsymbol{w}_e = \frac{d\boldsymbol{v}_0}{dt} + \frac{d\boldsymbol{\omega}}{dt} \times \boldsymbol{r} + \boldsymbol{\omega} \times (\boldsymbol{\omega} \times \boldsymbol{r})$$

is the translation acceleration, $\boldsymbol{w}_c = 2(\boldsymbol{\omega} \times \boldsymbol{v}_r)$ is the Coriolis acceleration, and $\boldsymbol{w}_r$ is the relative acceleration.

In the moving coordinate system, the relative acceleration is defined as follows:

$$\boldsymbol{w}_r = \frac{d^{'}\boldsymbol{v}_r}{dt} = \frac{\partial^{'}\boldsymbol{v}_r}{dt} + (\boldsymbol{v}_r \cdot \nabla)\boldsymbol{v}_r = \frac{\partial^{'}\boldsymbol{v}_r}{\partial t} + \mathrm{grad}(\tfrac{1}{2}v_r^2) - (\boldsymbol{v}_r \times \mathrm{rot}\boldsymbol{v}_r), \tag{3.1.9}$$

where the prime in the derivatives means that differentiation is carried out in the moving coordinate system. The equation of relative motions of liquid particles takes the form

$$\frac{\partial^{'}\boldsymbol{v}_r}{dt} + \mathrm{grad}(\tfrac{1}{2}v_r^2) - \boldsymbol{v}_r \times \mathrm{rot}\boldsymbol{v}_r + 2(\boldsymbol{\omega} \times \boldsymbol{v}_r) = \boldsymbol{F} - \frac{1}{\rho}\mathrm{grad}\, p - \boldsymbol{w}_e, \tag{3.1.10}$$

where $\boldsymbol{F}$ is the vector of mass forces and $p$ is pressure.

To write the expression for the absolute motion of liquid particles in the moving coordinate system, we use the equation of relative motion (3.1.10) in the case of complete rest ($\boldsymbol{v}_a = 0$, $\boldsymbol{v}_r = -\boldsymbol{v}_e$, $F = 0$, and $p = \mathrm{const}$) and arrive at the relation

$$\frac{\partial' \boldsymbol{v}_e}{dt} - \operatorname{grad}(\tfrac{1}{2}v_e^2) + \boldsymbol{v}_e \times \operatorname{rot}\boldsymbol{v}_e + 2(\boldsymbol{\omega} \times \boldsymbol{v}_e) = \boldsymbol{w}_e, \tag{3.1.11}$$

which, together with (3.1.10), yields the required result (after simple transformations):

$$\frac{\partial' \boldsymbol{v}_a}{dt} + \operatorname{grad}(\tfrac{1}{2}v_a^2 - \boldsymbol{v}_a \cdot \boldsymbol{v}_e) - (\boldsymbol{v}_a - \boldsymbol{v}_e) \times \operatorname{rot}\boldsymbol{v}_a = \boldsymbol{F} - \frac{1}{\rho}\operatorname{grad} p. \tag{3.1.12}$$

This is the equation of absolute motions in the moving coordinate system $Oxyz$. In the case where the mass forces have the potential $U$, i.e., for the vortex-free absolute liquid motions ($\boldsymbol{v}_a = \operatorname{grad}\Phi$), Eq. (3.1.12) yields the following form of the Bernoulli equation in the moving coordinate system $Oxyz$:

$$\frac{\partial' \Phi}{dt} + \frac{1}{2}(\nabla\Phi)^2 - \nabla\Phi \cdot (\boldsymbol{v}_0 + \boldsymbol{\omega} \times \boldsymbol{r}) + U + \frac{p}{\rho} = C(t), \tag{3.1.13}$$

where the vectors $\boldsymbol{v}_0$ and $\boldsymbol{\omega}$ are given by their projections onto the axes of this coordinate system. Recall that the derivative $\frac{\partial' \Phi}{dt}$ is given at a point of the moving coordinate system corresponding to the location of the liquid particle is present at the corresponding time (in the fixed coordinate system, this point has the coordinates $x, y, z$).

Equation (3.1.13) can be directly derived from (3.1.6) containing the time derivative of $\Phi$ in the fixed coordinate system $O'x'y'z'$:

$$\frac{\partial \Phi}{dt} = \frac{\partial \Phi(x', y', z', t)}{\partial t}. \tag{3.1.14}$$

Since the time derivatives of $x'$, $y'$, and $z'$ specify the projections of the translational velocity $\boldsymbol{v}_e$ of the analyzed liquid particle

$$\frac{dx'}{dt} = v_{ex} = v_{0x} + \omega_y z - \omega_z y, \quad \frac{dy'}{dt} = v_{ey} = v_{0y} + \omega_z x - \omega_x e,$$

$$\frac{dz'}{dt} = v_{ex} = v_{0z} + \omega_x y - \omega_y x, \tag{3.1.15}$$

according to the rules differentiation (for fixed $x$, $y$, and $z$), we get

$$\frac{\partial' \Phi}{dt} = \frac{\partial \Phi}{\partial t} + \frac{\partial \Phi}{\partial x'}\frac{dx'}{dt} + \frac{\partial \Phi}{\partial y}\frac{dy'}{dt} + \frac{\partial \Phi}{\partial z}\frac{dz'}{dt} = \frac{\partial \Phi}{\partial t} + \nabla\Phi \cdot \boldsymbol{v}_e. \tag{3.1.16}$$

Substituting (3.1.16) in (3.1.6), we again arrive at Eq. (3.1.13). By using relation (3.1.16), we can write the kinematic and dynamic boundary conditions on the free surfaces of the outer and inner liquids in the moving coordinate system. The dynamic conditions directly follow from the Bernoulli equation (3.1.13) for $p$ equal to the atmospheric pressure $p_0$:

$$\frac{\partial \Phi_i}{\partial t} + \frac{1}{2}(\nabla \Phi_i)^2 - \nabla \Phi_i \cdot (\boldsymbol{v}_0 + \boldsymbol{\omega} \cdot \boldsymbol{r}) + U_i = 0 \text{ on } \Sigma_i. \tag{3.1.17}$$

To deduce the kinematic conditions on the free surfaces defined, in the fixed coordinate system $O'x'y'z'$, by the equations $\zeta_i(x', y', z', t) = 0$, we also use relation (3.1.16). Assume that, in the moving coordinate system $Oxyz$, the equations of free surfaces take the form $\zeta_i(x', y', z', t) = 0$. We can directly get the following equations from (3.1.3):

$$\frac{d\zeta_i(x', y', z', t)}{dt} = \frac{\partial \zeta_i}{\partial t} + \nabla \zeta_i \cdot \nabla \Phi_i = 0 \text{ on } \Sigma_i. \tag{3.1.18}$$

By using (3.1.16), we reduce (3.1.18) to the form

$$\frac{\partial \zeta_i}{\partial t} - \nabla \zeta_i \cdot \boldsymbol{v}_e + \nabla \zeta_i \cdot \nabla \Phi_i = 0 \text{ on } \Sigma_i. \tag{3.1.19}$$

In view of the fact that the unit vector of normals to the surfaces $\zeta_i(x, y, z, t) = 0$ is defined as $\nabla \zeta_i / \sqrt{|\nabla \zeta_i|^2}$, it follows from (3.1.19) that

$$\frac{\partial \Phi_i}{\partial n} = \boldsymbol{v}_e \cdot \boldsymbol{n} - \frac{1}{\sqrt{|\nabla \zeta_i|^2}} \frac{\partial' \zeta_i}{\partial t} = \boldsymbol{v}_0 \cdot \boldsymbol{n} + \boldsymbol{\omega} \cdot (\boldsymbol{r} \times \boldsymbol{n}) - \frac{1}{\sqrt{|\nabla \zeta_i|^2}} \frac{\partial' \zeta_i}{\partial t}. \tag{3.1.20}$$

We can now write the boundary-value problems that describe the absolute motion of liquids outside and inside the floating ship. These problems take the following form:

$$\nabla^2 \Phi_i = 0 \text{ in } Q_i, \tag{3.1.21}$$

$$\frac{\partial \Phi_i}{\partial n} = \boldsymbol{v}_0 \cdot \boldsymbol{n} + \boldsymbol{\omega} \cdot (\boldsymbol{r} \times \boldsymbol{n}) \text{ on } S_i; \tag{3.1.22}$$

$$\frac{\partial \Phi_i}{\partial n} = \boldsymbol{v}_0 \cdot \boldsymbol{n} + \boldsymbol{\omega} \cdot (\boldsymbol{r} \times \boldsymbol{n}) + v_{ni} \text{ on } \Sigma_i, \tag{3.1.23}$$

$$\frac{\partial \Phi_i}{\partial t} + \frac{1}{2}(\nabla \Phi_i)^2 - \nabla \Phi_i \cdot (\boldsymbol{v}_0 + \boldsymbol{\omega} \times \boldsymbol{r}) + U_i = 0 \text{ on } \Sigma_i. \tag{3.1.24}$$

Furthermore, the normal velocities of the free surfaces $\Sigma_i$ are given by the formulas

$$v_{ni} = -\frac{\zeta_{it}}{\sqrt{|\nabla \zeta_i|^2}}. \tag{3.1.25}$$

The pressure in $Q_i$ is given by the Bernoulli equation (3.1.13). Here and in what follows, if this does not lead to misunderstanding, we omit the primes ($'$) in the time derivatives in relations (3.1.21)–(3.1.25) and (3.1.13) written in the moving coordinate system.

In conclusion, we present one more well-known relation used, e.g., in the proof of integral theorems for the domains with boundaries varying as functions of time. If a function $\Phi(x, y, z, t)$ is given in this domain, then, according to the Reynolds transport theorem (1.1.4),

$$\frac{d}{dt} \int\limits_{Q(t)} \Phi(x, y, z, t) dQ = \int\limits_{Q(t)} \frac{\partial \Phi}{\partial t} dQ + \int\limits_{S(t)} \Phi v_n dS, \tag{3.1.26}$$

where $v_n$ is the velocity of motion of the boundary toward the outer normal, for an observer connected with the moving coordinate system $Oxyz$.

The coordinate system $Oxyz$ rigidly connected with the body has the following advantage: In the case of completely filled cavities, the velocity potential for the inner liquid can be represented in the form (Kirchhoff representation) for which its components are independent of time. The rigidly fixed system also has some advantages in the solution of both outer and inner hydrodynamic problems in the case of partially filled cavities.

## 3.2 Bateman–Luke variational principle for floating bodies containing tanks filled with liquids

The interaction of the outer and inner hydrodynamic fields with rigid bodies specified by relations (3.1.13) and (3.1.21)–(3.1.25) can be described by using a Lagrange-type principle in which the action

$$W = \int\limits_{t_1}^{t_2} L dt$$

uses the Lagrangian

$$L = \sum_{i=1}^{2} \int\limits_{Q_i} p_i dQ_i = -\sum_{i=1}^{2} \rho_i \int\limits_{Q_i} \left[ \frac{\partial \Phi_i}{\partial t} + \frac{1}{2} (\nabla \Phi_i)^2 - \nabla \Phi_i \cdot (\boldsymbol{v}_0 + \boldsymbol{\omega} \times \boldsymbol{r}) + U_i \right] dQ_i. \tag{3.2.1}$$

The Bateman–Luke variational principle can be formulated as follows: The actual motions of ideal incompressible liquids interacting with floating rigid bodies are attained at the stationary points of action under the conditions of conservation of volume and the conditions of decay at infinity provided that the motions of comparison coincide at the initial and finite times.

The variational problem corresponding to this principle is formulated as follows: Among all functions $\Phi_i(x, y, z, t)$ and $\zeta_i(x, y, z, t)$ continuous together with their first derivatives with respect to the space variables and time $t$, it is necessary to find the functions delivering the stationary values of the action.

Here, the "actual motions" are specified by the required functions $\Phi_i(x, y, z, t)$ and $\zeta_i(x, y, z, t)$ and the motions of comparison are specified by the functions $\Phi_i + \delta\Phi_i$ and $\zeta_i + \delta\zeta_i$. Furthermore, the variations $\delta\Phi_i$ and $\delta\zeta_i$ are arbitrary functions from their domain of definition satisfying the conditions

$$\delta\Phi_i(x, y, z, t_1) = 0, \quad \delta\Phi_i(x, y, z, t_2) = 0,$$
$$\delta\zeta_i(x, y, z, t_1) = 0, \quad \delta\zeta_i(x, y, z, t_2) = 0. \tag{3.2.2}$$

Since the parts of the boundaries of the domains $Q_i$ given by the equations $\zeta_i(x, y, z, t) = 0$ are unknown in advance, it is necessary to determine the required functions $\Phi_i(x, y, z, t)$ in the domains $Q_i(t)$ varying as functions of time. Hence, the stationary points of action are pairs formed by the required functions $\Phi_i(x, y, z, t)$ and their domains of definition $Q_i(t)$. In the proof of the variational principle, we select a one-parameter family of pairs $\Phi_1(x, y, z, t, \alpha_1)$ and $\zeta_1(x, y, z, t, \alpha_1)$, $\Phi_2(x, y, z, t, \alpha_2)$ and $\zeta_2(x, y, z, t, \alpha_2)$. These pairs are motions of comparison and arbitrarily depend on the parameters $\alpha_i$ ($\alpha_i$ are sufficiently close to zero). Furthermore, the actual motions $\Phi_i(x, y, z, t) = \Phi_i(x, y, z, t, 0)$ and $\zeta_i(x, y, z, t) = \zeta_i(x, y, z, t, 0)$ occur for $\alpha_i = 0$ and the first variation of action is defined for these motions. As already indicated, this variational principle has some specific features as compared with the existing principles. One of these features is connected with the fact that Lagrangian (3.2.1) consists of two components defined, generally speaking, on different sets of admissible functions. The characteristic property of admissible functions on which the component

$$L_1 = -\rho_1 \int\limits_{Q_1} \left[ \frac{\partial \Phi_1}{\partial t} + \frac{1}{2}(\nabla\Phi_1)^2 - \nabla\Phi_1 \cdot (\boldsymbol{v}_0 + \boldsymbol{\omega} \times \boldsymbol{r}) + U_1 \right] dQ_1, \tag{3.2.3}$$

where $Q_1(t)$ is the unbounded domain of the outer liquid, is defined is their behavior at infinity (together with the first derivatives). At the same time, for the component of Lagrangian

$$L_2 = -\rho_2 \int\limits_{Q_2} \left[ \frac{\partial \Phi_2}{\partial t} + \frac{1}{2}(\nabla\Phi_2)^2 - \nabla\Phi_2 \cdot (\boldsymbol{v}_0 + \boldsymbol{\omega} \times \boldsymbol{r}) + U_2 \right] dQ_2, \tag{3.2.4}$$

the corresponding functional spaces must have all features required for the statement of variational principles in bounded domains. We introduce two parameters $\alpha_1$ and $\alpha_2$. This enables us to distinguish the variations of the velocity potentials $\delta\Phi_1$ and $\delta\Phi_2$ and the variations of the free surfaces $\delta\zeta_1$ and $\delta\zeta_2$ (the domains $Q_1$ and $Q_2$) under the assumption that they are independent.

This point of view is correct because the hydrodynamic fields interact with each other via the boundaries $S_1$ and $S_2$ of the rigid body and these boundaries are specified by the same parameters of motion of the body ($\boldsymbol{v}_0$ and $\boldsymbol{\omega}$) regarded as known in the analyzed problem ($\delta\boldsymbol{v}_0 = 0$ and $\delta\boldsymbol{\omega} = 0$). From the physical point of view, this means that the prescribed motion of the body is governed by given external forces. This means that, for the considered physical problem, the variational principles can be formulated separately for the outer and inner hydrodynamic fields. In the case where the law of motion of the rigid body is unknown and must be found with regard for the motions of the outer and inner liquids, this type of splitting of Lagrangian (3.2.1) is impossible in principle.

In the proof of the variational principle, we first assume that the actual motions of the rigid body containing liquid occur from time $t_1$ to $t_2$ in a certain mass of liquid (domain $Q_1$) bounded by the surface $S_0$ of any geometric shape fixed in the space. Finally, we assume that the surrounding basin is infinitely large and infinitely remote from the moving body in each direction. In view of the fact that the integration domains $Q_i(t)$ vary with time and, under our assumption, must be varied according to the definition of the first variation, as an intermediate result, we get

$$\delta I = -\int_{t_1}^{t_2}\sum_{i=1}^{2}\rho_1\left\{-\int_{\Sigma_i}\left[\frac{\partial\Phi_i}{\partial t}+\frac{1}{2}(\nabla\Phi_i)^2\right.\right.$$

$$\left.-\nabla\Phi_i\cdot(\boldsymbol{v}_0+\boldsymbol{\omega}\times\boldsymbol{r})+\mathbf{g}\cdot\boldsymbol{r}\right]\frac{\delta\zeta_i}{\sqrt{(\nabla\zeta_i)^2}}dS$$

$$\left.+\int_{Q_i}[\nabla\Phi_i\nabla(\delta\Phi_i)+\delta\Phi_{it}-\nabla(\delta\Phi_i)\cdot(\boldsymbol{v}_0+\boldsymbol{\omega}\times\boldsymbol{r})]dQ_i\right\}dt, \tag{3.2.5}$$

where

$$\delta\Phi_i=\left.\frac{\partial\Phi_i}{\partial\alpha_i}\right|_{\alpha_i=0}\alpha_i \quad\text{and}\quad \delta\zeta_i=\left.\frac{\partial\zeta_i}{\partial\alpha_i}\right|_{\alpha_i=0}\alpha_i. \tag{3.2.6}$$

We transform the group of integrals over the volume $Q_i$ in relation (3.2.6) by using the Green formula

$$\int_{Q}\nabla\varphi\cdot\nabla\psi\,dQ=\int_{\partial Q}\psi\frac{\partial\psi}{\partial n}\,dS-\int_{Q}\psi\nabla^2\varphi\,dQ. \tag{3.2.7}$$

Setting

$$\nabla\varphi=\boldsymbol{\omega}\cdot\boldsymbol{r}=\boldsymbol{i}_1(\omega_2 z-\omega_3 y)+\boldsymbol{i}_2(\omega_3 x-\omega_1 z)+\boldsymbol{i}_3(\omega_1 y-\omega_2 x)$$

in this formula and noting that $\boldsymbol{v}_0=\nabla(\boldsymbol{v}_0\cdot\boldsymbol{r})$, we obtain

$$\int_{Q_i}\nabla\Phi_i\cdot\nabla(\delta\Phi_i)\,dQ_i=\int_{\Gamma_i}\delta\Phi_i\frac{\partial\Phi_i}{\partial n}\,dS-\int_{Q_i}\delta\Phi_i\nabla^2\Phi_i\,dQ_i, \tag{3.2.8}$$

$$\int_{Q_i}\boldsymbol{v}_0\cdot\nabla(\delta\Phi_i)\,dQ=\int_{Q_i}\nabla(\boldsymbol{v}_0\cdot\boldsymbol{r})\cdot\nabla(\delta\Phi_i)\,dQ=\int_{\Gamma_i}\delta\Phi_i\boldsymbol{v}_0\cdot\boldsymbol{n}\,dS, \tag{3.2.9}$$

$$\int\limits_{Q_i} (\boldsymbol{\omega} \times \boldsymbol{r}) \cdot \nabla(\delta\Phi_i)\, dQ = \int\limits_{\Gamma_i} \delta\Phi_i (\boldsymbol{\omega} \times \boldsymbol{r}) \cdot \boldsymbol{n}\, dS, \tag{3.2.10}$$

where $\Gamma_i$ are the boundary surfaces of $Q_i$: $\Gamma_1 = S_0 \cup S_1 \cup \Sigma_1$ and $\Gamma_2 = S_2 \cup \Sigma_2$.

Since time is not varied, we find

$$\int\limits_{Q_i} \delta\dot{\Phi}_i\, dQ = \int\limits_{Q_i} \frac{\partial}{\partial t}(\delta\Phi_i)\, dQ \tag{3.2.11}$$

(here, the differentiation with respect to time is performed in the moving coordinate system). By applying the Reynolds transport theorem to (3.2.11), we get

$$\int\limits_{Q_i} \frac{\partial}{\partial t}(\delta\Phi_i)\, dQ = \frac{d}{dt}\int\limits_{Q_i} \delta\Phi_i\, dQ - \int\limits_{\Gamma_i} \delta\Phi_i v_{ni}\, dS, \tag{3.2.12}$$

where $v_{ni}$ is the velocity of motion of the boundary surfaces of the volumes $Q_i$ toward the outer normals relative to the coordinate system connected with the observer.

Prior to summarizing the results of application of the Green theorem, we note that the observer placed in the moving coordinate system sees points fixed in the absolute coordinate system as moving with the velocity $\boldsymbol{v} = -(\boldsymbol{v}_0 + \boldsymbol{\omega} \times \boldsymbol{r})$. This means that the following conditions must be used in integral (3.2.12):

$$v_{ni} = 0 \text{ on } S_i, \quad v_{ni} = -\frac{\zeta_{it}}{\sqrt{|\nabla\zeta_i|^2}} \text{ on } \Sigma_i, \; v_n = -(\boldsymbol{v}_0 + \boldsymbol{\omega} \times \boldsymbol{r}) \cdot \boldsymbol{n} \text{ on } S_0. \tag{3.2.13}$$

Inserting (3.2.8)–(3.2.10) and (3.2.12) in (3.2.5) and taking into account (3.2.13), we find

$$\begin{aligned}
\delta I = -\int\limits_{t_1}^{t_2} \sum_{i=1}^{2} \rho_i \Bigg\{ \int\limits_{\Sigma_i} \Bigg[ & \frac{\partial\Phi_i}{\partial t} + \frac{1}{2}(\nabla\Phi_i)^2 \\
& - \nabla\Phi_i \cdot (\boldsymbol{v}_0 + \boldsymbol{\omega} \times \boldsymbol{r}) + \mathbf{g} \cdot \boldsymbol{r} \Bigg] \frac{\delta\zeta_i}{\sqrt{(\nabla\zeta_i)^2}}\, dS \\
& + \int\limits_{\Sigma_i} \left[ \frac{\partial\Phi_i}{\partial n} - \boldsymbol{v}_0 \cdot \boldsymbol{n} - \boldsymbol{\omega} \cdot (\boldsymbol{r} \times \boldsymbol{n}) + \frac{\zeta_{it}}{\sqrt{(\nabla\zeta_i)^2}} \right] \delta\Phi_i\, dS \\
& + \int\limits_{S_0} \frac{\partial\Phi_1}{\partial n} \delta\Phi_1 dS + \int\limits_{S_i} \left[ \frac{\partial\Phi_i}{\partial n} - \boldsymbol{v}_0 \cdot \boldsymbol{n} - \boldsymbol{\omega} \cdot (\boldsymbol{r} \times \boldsymbol{n}) \right] \delta\Phi_i\, dS \\
& - \int\limits_{Q_i(t)} \nabla^2\Phi_i \delta\Phi_i dQ_i + \frac{d}{dt} \int\limits_{Q_i(t)} \delta\Phi_i dQ_i \Bigg\}\, dt.
\end{aligned} \tag{3.2.14}$$

For this variational problem, the natural condition imposed on the surface $S_0$ is as follows:

$$\nabla\Phi_1 \cdot \boldsymbol{n} = 0 \text{ on } S_0. \tag{3.2.15}$$

As accepted above, the scalar product (3.2.15) is an absolute scalar with respect to the orthogonal transformation of coordinates. If the boundary $S_0$ moves from a floating body to infinity, then condition (3.2.15) is reduced to the condition of decay at infinity, i.e., $\nabla\Phi_1 \to 0$ as $r \to \infty$, where $r$ is the distance from the floating body.

These results enable us to conclude that the stationary values of the functional $W$ are attained on the solutions of the boundary problems (3.1.21)–(3.1.24). This is confirmed by the variational principle formulated above. As a special case of this principle, we get the variational principle formulated in the previous chapters for the nonlinear theory of space motions of a rigid body with cavities partially filled with liquid. In the case where the motion of the rigid body is unknown in advance and there are nonpotential forces among the forces acting upon the rigid body, the equations of motion of the considered mechanical system can be obtained by using the variational principle

$$\int_{t_1}^{t_2} (\delta L_1 + \delta L + \delta' A)\, dt = 0, \tag{3.2.16}$$

where $L_1$ is the kinetic potential of the rigid body, $L$ is Lagrangian (3.2.1), and $\delta' A$ is the elementary work of nonpotential forces applied to the rigid body.

In deducing the variational relation (3.2.16), the kinematic parameters of the rigid body are also varied. Unlike the first problem of dynamics considered above, expression (3.2.16) does not have the action whose variation leads to the variational relation (3.2.16).

## 3.3 Hydrodynamic forces and moments acting upon the floating rigid body

We consider the case in which the velocity potentials $\Phi_i$ are solutions of the boundary-value problems (3.1.21)–(3.1.24) with the corresponding conditions of decay at infinity for $\Phi_1$ and the condition of conservation of volume for $\Phi_2$. In the water-wave theory, the mathematical statement requires initial conditions imposed for $t = 0$. As above, we assume that the motion of the rigid body is known, i.e., the translational velocity $\boldsymbol{v}_0$ and the instantaneous rotational velocity $\boldsymbol{\omega}$ are known functions of time. We represent the velocity potentials $\Phi_1$ and $\Phi_2$ specifying the absolute motion in the unbounded outer and bounded inner domains in the moving coordinate system in the form

$$\Phi_i(x, y, z, t) = \boldsymbol{v}_0 \cdot \boldsymbol{V}_i + \boldsymbol{\omega} \cdot \boldsymbol{\Omega}_i + \varphi_i, \tag{3.3.1}$$

where $\boldsymbol{V}_i$ and $\boldsymbol{\Omega}_i$ are harmonic vector functions whose projections onto the axes of the body-fixed coordinate system satisfy the boundary conditions

$$\left.\frac{\partial V_{ij}}{\partial n}\right|_{\Sigma_i+S_i} = n_j \quad (j = 1, 2, 3); \quad \left.\frac{\partial \Omega_{ik}}{\partial n}\right|_{\Sigma_i+S_i} = [\boldsymbol{r} \times \boldsymbol{n}]_k, \quad k = 1, 2, 3, \tag{3.3.2}$$

and $\varphi_i$ are harmonic functions satisfying the relations

$$\left.\frac{\partial \varphi_i}{\partial n}\right|_{S_i} = 0 \quad \text{and} \quad \left.\frac{\partial \varphi_i}{\partial n}\right|_{\Sigma_i} = -\frac{\partial \zeta_i / \partial t}{\sqrt{|\nabla \zeta_i|^2}}. \tag{3.3.3}$$

Thus, the solutions of harmonic problems with respect to $\varphi_i$ give a superposition of the free surface waves for the outer and inner liquid domains in the case of fixed boundaries of the rigid body.

By using representation (3.3.1) introduced, for the first time, by Kirchhoff in the theory of motion of rigid bodies in infinite volumes of liquid ($\varphi_i \equiv 0$), we can separate the components of the velocity potentials $\boldsymbol{\Phi}_i$ (3.3.1) from the parameters of motion of the rigid body $\boldsymbol{v}_0$ and $\boldsymbol{\omega}$.

In the linear theory, the vector functions $\boldsymbol{V}_i$ and $\boldsymbol{\Omega}_i$ are functions of the space coordinates, whereas the functions $\varphi_i$ also depend on the parameters characterizing the deformation of the free surface $\Sigma_i$.

In the nonlinear case, $\boldsymbol{V}_i$ and $\boldsymbol{\Omega}_i$ also depend on these parameters. It is clear that, under conditions (3.3.2), (3.3.3) and the conditions of decay and conservation of volume, $\boldsymbol{\Phi}_i$ in the form (3.3.1) satisfy all conditions of the boundary-value problems (3.1.21)–(3.1.24).

The integral hydrodynamic loads acting on the surface $S_i$ of the rigid body from the outer and inner liquids are connected with the hydrodynamic force and moment as follows:

$$\boldsymbol{P}_i = \int_{S_i} p_i \boldsymbol{n}\, dS \quad \text{and} \quad \boldsymbol{N}_i = \int_{S_i} \boldsymbol{r} \times p_i \boldsymbol{n}\, dS, \tag{3.3.4}$$

where $\boldsymbol{n}$ is the unit vector of the outer normal to the surfaces bounding the domains $Q_i$. The direct use of relations (3.3.4), especially in the nonlinear case, may lead to cumbersome computations. More convenient expressions for $\boldsymbol{P}_i$ and $\boldsymbol{N}_i$ can be obtained by using the main theorems of dynamics [formulated in what follows on the basis of relations (3.3.4)].

Under the assumptions concerning the potential flows in the outer domain $Q_1$ (relative to the body), we first find the hydrodynamic force $\boldsymbol{P}_1$ and moment $\boldsymbol{N}_1$ relative to the moving origin $O$. To this end, we consider the motion of a rigid body partially immersed in a liquid of finite mass in a fixed closed basin of any shape. In the results thus obtained, we then pass to the limit of infinite domain $Q_1$ and infinite distance from the floating body to the surface $S_0$ of the domain in the required direction. Substituting the expression for the pressure $p_1$ from the Bernoulli equation (3.1.13) with $C_i(t) \equiv 0$ in relation (3.3.4), for $\boldsymbol{P}_1$, we find

$$\boldsymbol{P}_1 = -\rho_1 \int\limits_{S_1} \left[ \frac{\partial \Phi_1}{\partial t} + \frac{1}{2}(\nabla \Phi_1)^2 - \nabla \Phi_1 \cdot (\boldsymbol{v}_0 + \boldsymbol{\omega} \times \boldsymbol{r}) - \mathbf{g} \cdot \boldsymbol{r} \right] \boldsymbol{n}\, dS. \quad (3.3.5)$$

We apply the Gauss theorem to relation (3.3.5) and take into account the fact that the velocity potential $\Phi_1$ is a solution of the boundary problem (3.1.21)–(3.1.24). First, we obtain

$$\rho_1 \int\limits_{\Gamma_1} \frac{\partial \Phi_1}{\partial t} \boldsymbol{n}\, dS = \rho_1 \int\limits_{Q_1} \nabla \left( \frac{\partial \Phi_1}{\partial t} \right) dQ = \rho_1 \int\limits_{Q_1} \frac{\partial}{\partial t} (\nabla \Phi_1)\, dQ, \quad (3.3.6)$$

where $\Gamma_1 = \Sigma_1 \cup S_1 \cup S_0$ and $S_0$ is a fixed surface bounding the basin. Finally, we get

$$\rho_1 \int\limits_{Q_1} \frac{\partial}{\partial t} (\nabla \Phi_1)\, dQ = \overset{*}{\boldsymbol{K}}_1 - \rho_1 \int\limits_{S_0 \cup \Sigma_1} \nabla \Phi_1 v_n\, dS, \quad (3.3.7)$$

where

$$\boldsymbol{K}_1 = \rho_1 \int\limits_{Q_1} \nabla \Phi_1\, dQ$$

is the momentum of the liquid $Q_1$. Relations (3.3.6) and (3.3.7) imply that

$$\rho_1 \int\limits_{S_1} \frac{\partial \Phi}{\partial t} \boldsymbol{n}\, dS = \overset{*}{\boldsymbol{K}}_1 - \rho_1 \int\limits_{S_0 \cup \Sigma_1} \nabla \Phi_1 v_n\, dS$$
$$- \rho_1 \int\limits_{\Sigma_1} \frac{\partial \Phi_1}{\partial t} \boldsymbol{n}\, dS - \rho_1 \int\limits_{S_0} \frac{\partial \Phi_1}{\partial t} \boldsymbol{n}\, dS. \quad (3.3.8)$$

In the course of transformations of the second integral in (3.3.5), we obtain the relation

$$\frac{1}{2} \int\limits_{\Gamma_1} [\boldsymbol{n} a^2 - 2(\boldsymbol{n} \cdot \boldsymbol{a})\boldsymbol{a}]\, dS = \int\limits_{Q_1} [\boldsymbol{a}(\nabla \cdot \boldsymbol{a}) - \boldsymbol{a} \times (\nabla \times \boldsymbol{a})]\, dQ, \quad (3.3.9)$$

which follows (as a corollary) from the Gauss theorem and the following relations of the vector algebra:

$$\nabla(\boldsymbol{a} \cdot \boldsymbol{b}) = \boldsymbol{a} \times (\nabla \times \boldsymbol{b}) + (\boldsymbol{a} \cdot \nabla)\boldsymbol{b} + \boldsymbol{b} \times (\nabla \times \boldsymbol{a}) + (\boldsymbol{b} \cdot \nabla)\boldsymbol{a}$$

and

$$(\nabla \cdot \boldsymbol{b})\boldsymbol{a} = \boldsymbol{a}(\nabla \cdot \boldsymbol{b}) + (\boldsymbol{b} \cdot \nabla)\boldsymbol{a}.$$

Setting $\boldsymbol{a} = \nabla\Phi$ in (3.3.9), we get

$$\frac{1}{2}\rho_1 \int\limits_{S_1} (\nabla\Phi)^2 \boldsymbol{n}\, dS = \rho_1 \int\limits_{S_1} \nabla\Phi_1[(\boldsymbol{v}_0 + \boldsymbol{\omega}\times\boldsymbol{r})\cdot\boldsymbol{n}]\, dS$$
$$+\rho_1 \int\limits_{\Sigma_1} \nabla\Phi_1[(\boldsymbol{v}_0 + \boldsymbol{\omega}\times\boldsymbol{r})\cdot\boldsymbol{n} - \frac{\zeta_{1t}}{\sqrt{|\nabla\zeta_1|^2}}]\, dS$$
$$-\frac{1}{2}\rho_1 \int\limits_{\Sigma_1} (\nabla\Phi_1)^2 \boldsymbol{n}\, dS - \frac{1}{2}\rho_1 \int\limits_{S_0} (\nabla\Phi_1)^2 \boldsymbol{n}\, dS. \qquad (3.3.10)$$

By using the relation

$$\int\limits_{\Gamma_1} [\boldsymbol{n}(\boldsymbol{a}\cdot\boldsymbol{b}) - (\boldsymbol{n}\cdot\boldsymbol{b})\boldsymbol{a}]\, dS = \int\limits_{Q_1} [\boldsymbol{a}\times(\nabla\times\boldsymbol{b}) + (\boldsymbol{a}\cdot\nabla)\boldsymbol{b}$$
$$+\boldsymbol{b}\times(\nabla\times\boldsymbol{a}) - \boldsymbol{a}(\nabla\cdot\boldsymbol{b})]\, dQ, \qquad (3.3.11)$$

which is also a corollary of the Gauss theorem, we now transform the third integral in (3.3.5). Setting $\boldsymbol{a} = \nabla\Phi_1$ and $\boldsymbol{b} = \boldsymbol{v}_0 + \boldsymbol{\omega}\times\boldsymbol{r}$ in (3.3.11) and taking into account the fact that

$$\nabla\Phi_1 \times (\nabla\times(\boldsymbol{\omega}\times\boldsymbol{r})) = -2\boldsymbol{\omega}\times\nabla\Phi_1,\ (\nabla\Phi_1\cdot\nabla)(\boldsymbol{\omega}\times\boldsymbol{r}) = \boldsymbol{\omega}\times\nabla\Phi_1,$$

we get

$$\rho_1 \int\limits_{S_1} \nabla\Phi_1[(\boldsymbol{v}_0 + \boldsymbol{\omega}\times\boldsymbol{r})\cdot\boldsymbol{n}]\, dS = \rho_1 \int\limits_{\Gamma_1} [\boldsymbol{n}\cdot(\boldsymbol{v}_0 + \boldsymbol{\omega}\times\boldsymbol{r})]\nabla\Phi_1\, dS$$
$$-\rho_1 \int\limits_{Q_1} (\boldsymbol{\omega}\times\nabla\Phi_1)\, dQ - \rho_1 \int\limits_{\Sigma_1} \nabla\Phi_1[(\boldsymbol{v}_0 + \boldsymbol{\omega}\times\boldsymbol{r})\cdot\boldsymbol{n}]\, dS$$
$$-\rho_1 \int\limits_{S_0} \nabla\Phi_1[(\boldsymbol{v}_0 + \boldsymbol{\omega}\times\boldsymbol{r})]\boldsymbol{n}\, dS. \qquad (3.3.12)$$

We now summarize the results of transformations by using relations (3.3.8), (3.3.10), and (3.3.12) and substitute them in (3.3.5) with regard for the value of $\frac{1}{2}(\nabla\Phi_1)^2$ on $\Sigma_1$ according to (3.1.24). The hydrodynamic force acting upon the rigid body takes the form

$$\boldsymbol{P}_1 = -\overset{*}{\boldsymbol{K}}_1 - \boldsymbol{\omega}\times\boldsymbol{K}_1 + \boldsymbol{P}_1^{\mathrm{st}}, \qquad (3.3.13)$$

where $\boldsymbol{P}_1^{\mathrm{st}}$ is the principal vector of buoyancy forces caused by the gravitational acceleration

$$\boldsymbol{P}_1^{\mathrm{st}} = \rho_1 \int\limits_{S_1\cup\Sigma_1} (\mathbf{g}\cdot\boldsymbol{r})\boldsymbol{n}\, dS. \qquad (3.3.14)$$

Relation (3.3.13) was obtained with regard for the condition that the surface integral

$$-\rho_1 \int\limits_{S_0} \left[ \frac{\partial \Phi_1}{\partial t} + \frac{1}{2}(\nabla \Phi_1)^2 - \nabla \Phi_1 \cdot (\boldsymbol{v}_0 + \boldsymbol{\omega} \times \boldsymbol{r}) - \mathbf{g} \cdot \boldsymbol{r} \right] \boldsymbol{n}\, dS, \tag{3.3.15}$$

where the surface $S_0$ is infinitely remote from the floating body, approaches zero together with the first derivatives of $\mathbf{\Phi}_1$ with respect to $x, y, z$ and time $t$.

The resulting gravity force acting upon the rigid body from the inner liquid is given by the following formula obtained with the use of transformations similar to those presented in Sec. 2.8, Chap. 2:

$$\boldsymbol{P}_2 = m_2 \mathbf{g} - \overset{*}{\boldsymbol{K}}_2 - \boldsymbol{\omega} \times \boldsymbol{K}_2, \tag{3.3.16}$$

where $m_2$ is the mass of liquid filling the cavity in the rigid body.

Further, we transform the expressions for the momenta $\boldsymbol{K}_1$ and $\boldsymbol{K}_2$ by using the representations for the velocity potentials (3.3.1). This yields

$$\begin{aligned} \boldsymbol{K}_1 = \rho_1 \int\limits_{Q_1} \nabla \Phi_1 \, dQ = \rho_1 \int\limits_{Q_1} \nabla(\boldsymbol{v}_0 \cdot \boldsymbol{V}_1) dQ \\ + \rho_1 \int\limits_{Q_1} \nabla(\boldsymbol{\omega} \cdot \boldsymbol{\Omega}) dQ + \rho_1 \int\limits_{Q_1} \nabla \varphi_1 dQ. \end{aligned} \tag{3.3.17}$$

By using the integral theorems, we transform the integrals over the domain in (3.3.17) as follows:

$$\begin{aligned} \rho_1 \int\limits_{Q_1} \nabla(\boldsymbol{v}_0 \cdot \boldsymbol{V}_1)\, dQ &= \rho_1 \int\limits_{\Gamma_1} (\boldsymbol{v}_0 \cdot \boldsymbol{V}_1)\, \boldsymbol{n}\, dS = \rho_1 \int\limits_{\Gamma_1} (\boldsymbol{v}_0 \cdot \boldsymbol{V}_1) \frac{\partial \boldsymbol{r}}{\partial n}\, dS \\ &= \rho_1 \int\limits_{\Gamma_1} (\boldsymbol{v}_0 \cdot \frac{\partial \boldsymbol{V}_1}{\partial n})\, \boldsymbol{r}\, dS = \rho_1 \int\limits_{S_1 \cup \Sigma_1} (\boldsymbol{v}_0 \cdot \boldsymbol{n})\, \boldsymbol{r}\, dS; \end{aligned} \tag{3.3.18}$$

$$\begin{aligned} \rho_1 \int\limits_{Q_1} \nabla(\boldsymbol{\omega} \cdot \boldsymbol{\Omega}_1)\, dQ &= \rho_1 \int\limits_{\Gamma_1} (\boldsymbol{\omega} \cdot \boldsymbol{\Omega})\, \boldsymbol{n}\, dS = \rho_1 \int\limits_{\Gamma_1} (\boldsymbol{\omega} \cdot \frac{\partial \boldsymbol{\Omega}_1}{\partial n})\, \boldsymbol{r}\, dS \\ &= \rho_1 \int\limits_{S_1 \cup \Sigma_1} [\boldsymbol{\omega} \cdot (\boldsymbol{r} \times \boldsymbol{n})]\, \boldsymbol{r}\, dS; \end{aligned} \tag{3.3.19}$$

$$\begin{aligned} \rho_1 \int\limits_{Q_1} \nabla \varphi_1 \, dQ &= \rho_1 \int\limits_{\Gamma_1} \varphi \boldsymbol{n}\, dS = \rho_1 \int\limits_{\Gamma_1} \varphi_1 \frac{\partial \boldsymbol{r}}{\partial n}\, dS = \rho_1 \int\limits_{\Gamma_1} \boldsymbol{r} \frac{\partial \varphi_1}{\partial n}\, dS \\ &= -\rho_1 \int\limits_{\Sigma_1} \boldsymbol{r} \frac{\zeta_{1t}}{\sqrt{|\nabla \zeta_1|^2}}\, dS. \end{aligned} \tag{3.3.20}$$

In deducing relations (3.3.18)–(3.3.20), we have used the fact that $\nabla V_1 \to 0$, $\nabla \Omega_1 \to 0$, and $\nabla \varphi_1 \to 0$ as the distance between $S_0$ and the floating body infinitely increases. For the velocity potential $\boldsymbol{\Phi}_2$, these conditions are absent. As a result, relations (3.3.18)–(3.3.20) take a more compact form

$$\rho_2 \int\limits_{Q_2} \nabla(\boldsymbol{v}_0 \cdot \boldsymbol{V}_2)\, dQ = \rho_2 \int\limits_{\Gamma_2} (\boldsymbol{v}_0 \cdot \boldsymbol{n}_2)\, \boldsymbol{r}\, dS = \rho_2 \int\limits_{\Gamma_2} (\boldsymbol{v}_0 \cdot \frac{\partial \boldsymbol{V}_2}{\partial n})\, \boldsymbol{r}\, dQ$$

$$= \rho_2 \int\limits_{Q_2} (\boldsymbol{v}_0 \cdot \nabla)\, \boldsymbol{r}\, dQ = m_2 \boldsymbol{v}_0;$$

$$\rho_2 \int\limits_{Q_2} \nabla(\boldsymbol{\omega} \cdot \boldsymbol{\Omega}_2)\, dQ = \rho_2 \int\limits_{\Gamma_2} (\boldsymbol{\omega} \cdot \frac{\partial \boldsymbol{\Omega}_2}{\partial n})\, \boldsymbol{r}\, dS = \rho_2 \int\limits_{\Gamma_2} [\boldsymbol{\omega} \cdot (\boldsymbol{r} \times \boldsymbol{n})]\, \boldsymbol{r}\, dS$$

$$= \rho_2 \int\limits_{Q_2} [\boldsymbol{\omega}(\boldsymbol{r} \times \nabla)]\, \boldsymbol{r}\, dQ = \rho_2 \int\limits_{Q_2} (\boldsymbol{\omega} \times \boldsymbol{r})\, dQ = m_2 \boldsymbol{\omega} \times \boldsymbol{r}_{2c}; \qquad (3.3.21)$$

$$\rho_2 \int\limits_{Q_2} \nabla \varphi_2\, dQ_2 = \rho_2 \int\limits_{\Gamma_2} \boldsymbol{r} \frac{\partial \varphi_2}{\partial n}\, dQ = \rho_2 \int\limits_{\Gamma_2} \boldsymbol{r} v_{n2}\, dS = \frac{d}{dt} \int\limits_{Q_2} \boldsymbol{r}\, dQ$$

$$= m_2(\boldsymbol{i}_1 \dot{x}_{20} + \boldsymbol{i}_2 \dot{y}_{20} + \boldsymbol{i}_3 \dot{z}_{20}) = m_2 \overset{*}{\boldsymbol{r}}_{2c},$$

where $\boldsymbol{r}_{2c}$ is the radius vector of the center of gravity of the inner liquid.

By using the results presented above, we get

$$\boldsymbol{K}_1 = \rho_1 \int\limits_{S_1 \cup \Sigma_1} [(\boldsymbol{v}_0 \cdot \boldsymbol{n}) + \boldsymbol{\omega} \cdot (\boldsymbol{r} \times \boldsymbol{n})]\, \boldsymbol{r}\, dS - \rho_1 \int\limits_{\Sigma_1} \boldsymbol{r} \frac{\zeta_{1t}}{\sqrt{|\nabla \zeta_1|^2}}\, dS, \qquad (3.3.22)$$

$$\boldsymbol{K}_2 = m_2(\boldsymbol{v}_0 + \boldsymbol{\omega} \times \boldsymbol{r}_{2c} + \overset{*}{\boldsymbol{r}}_{2c}). \qquad (3.3.23)$$

Relation (3.3.23) for the momentum $\boldsymbol{K_2}$ enables us to represent $\boldsymbol{P}_2$ in the ordinary form

$$\boldsymbol{P}_2 = m_2 \mathbf{g} - m_2[\overset{*}{\boldsymbol{v}}_0 + \boldsymbol{\omega} \times \boldsymbol{v}_0 + \boldsymbol{\omega} \times (\boldsymbol{\omega} \times \boldsymbol{r}_{2c})$$

$$+ \boldsymbol{\omega} \times \boldsymbol{r}_{2c} + 2\dot{\boldsymbol{\omega}} \times \overset{*}{\boldsymbol{r}}_{2c} + \overset{**}{\boldsymbol{r}}_{2c}], \qquad (3.3.24)$$

where the group of terms

$$\boldsymbol{w}_e = \overset{*}{\boldsymbol{v}}_0 + \boldsymbol{\omega} \times \boldsymbol{v}_0 + \boldsymbol{\omega} \times \boldsymbol{v}_0 + \dot{\boldsymbol{\omega}} \times \boldsymbol{r}_{2c} + \boldsymbol{\omega} \times (\boldsymbol{\omega} \times \boldsymbol{r}_{2c}) \qquad (3.3.25)$$

represents the translational acceleration formed by the acceleration of the pole $\boldsymbol{w}_0$, the centrifugal acceleration $\boldsymbol{w}_c$, and the rotational acceleration $\boldsymbol{w}_r$:

$$\boldsymbol{w}_0 = \overset{*}{\boldsymbol{v}}_0 + \boldsymbol{\omega} \times \boldsymbol{v}_0; \;\; \boldsymbol{w}_c = \boldsymbol{\omega} \times (\boldsymbol{\omega} \times \boldsymbol{r}_{2c}); \;\; \boldsymbol{w}_r = \boldsymbol{\omega} \times \boldsymbol{r}_{2c}. \qquad (3.3.26)$$

The components

$$\boldsymbol{w}_0^{\mathrm{Cor}} = 2\boldsymbol{\omega} \times \overset{*}{\boldsymbol{r}}_{2c} \text{ and } \boldsymbol{w}_r = \overset{**}{\boldsymbol{r}}_{2c}, \tag{3.3.27}$$

are the Coriolis acceleration and the relative acceleration, respectively.

We now define the hydrodynamic moment caused by the outer liquid $\boldsymbol{N}_1$ relative to the moving point $O$. Substituting the pressure $p$ from (3.1.13) in the corresponding relation in (3.3.4), we get

$$\boldsymbol{N}_1 = -\rho_1 \int_{S_1} \boldsymbol{r} \times \left[ \frac{\partial \Phi_1}{\partial t} + \frac{1}{2}(\nabla \Phi_1)^2 - \nabla \Phi_1 \cdot (\boldsymbol{v}_0 + \boldsymbol{\omega} \times \boldsymbol{r}) - \mathbf{g} \cdot \boldsymbol{r} \right] \boldsymbol{n}\, dS. \tag{3.3.28}$$

We now transform relation (3.3.28) with the help of the integral theorems of field theory. For the first term, we obtain

$$\begin{aligned} \rho_1 \int_{\Gamma_1} (\boldsymbol{r} \times \frac{\partial \Phi_1}{\partial n} \boldsymbol{n})\, dS &= \rho_1 \int_{Q_1} (\boldsymbol{r} \times \nabla) \frac{\partial \Phi_1}{\partial t}\, dQ = \rho_1 \int_{Q_1} [\boldsymbol{r} \times \frac{\partial}{\partial t}(\nabla \Phi_1)]\, dQ \\ &= \rho_1 \int_{Q_1} \frac{\partial}{\partial t}(\boldsymbol{r} \times \nabla \Phi_1)\, dQ - \rho_1 \int_{Q_1} (\boldsymbol{u} \times \nabla \Phi_1)\, dQ = \rho_1 \frac{d}{dt} \int_{Q_1} (\boldsymbol{r} \times \nabla \Phi_1)\, dQ \\ &\quad - \rho_1 \int_{\Gamma_1} (\boldsymbol{r} \times \nabla \Phi_1)\, u_{n1}\, dQ - \rho_1 \int_{Q_1} (\boldsymbol{u} \times \nabla \Phi_1)\, dQ, \end{aligned} \tag{3.3.29}$$

where $\boldsymbol{u}$ is the relative velocity of liquid particles. By using (3.3.29), we reduce the first integral in (3.3.28) to the form

$$\begin{aligned} \rho_1 \int_{S_1} \left( \boldsymbol{r} \times \frac{\partial \Phi}{\partial t} \boldsymbol{n} \right) dS &= \overset{*}{\boldsymbol{G}}_1 - \rho_1 \int_{Q_1} (\boldsymbol{u} \times \nabla \Phi_1)\, dQ - \rho_1 \int_{\Gamma_1} (\boldsymbol{r} \times \nabla \Phi_1)\, v_{n1} dQ \\ &\quad - \rho_1 \int_{S_0 \cup \Sigma_1} \left( \boldsymbol{r} \times \frac{\partial \Phi_1}{\partial t} \right) \boldsymbol{n}\, dS, \end{aligned} \tag{3.3.30}$$

where

$$G_1 = \rho_1 \int_{Q_1} (\boldsymbol{r} \times \nabla \Phi_1) dQ$$

is the angular momentum of the outer liquid relative to the point $O$. Further, by using the integral relation

$$\frac{1}{2} \int_{\Gamma_1} \boldsymbol{r} \times [\boldsymbol{n} a^2 - 2(\boldsymbol{n} \cdot \boldsymbol{a})\boldsymbol{a}]\, dS = \int_{Q_1} \boldsymbol{r} \times [\boldsymbol{a}(\nabla \cdot \boldsymbol{a}) - \boldsymbol{a} \times (\nabla \times \boldsymbol{a})]\, dQ, \tag{3.3.31}$$

for $\boldsymbol{a} = \nabla \Phi_1$, we get the following expression for the second surface integral in (3.3.28):

$$\frac{1}{2}\rho_1 \int\limits_{S_1} [\boldsymbol{r} \times (\nabla\Phi_1)^2 \boldsymbol{n}]\, dS$$

$$= \rho_1 \int\limits_{\Gamma_1} \left(\boldsymbol{r} \times \nabla\Phi_1 \frac{\partial \Phi_1}{\partial n}\right) dS - \frac{1}{2}\rho_1 \int\limits_{S_0 \cup \Sigma_1} [\boldsymbol{r} \times (\nabla\Phi_1)^2 \boldsymbol{n}]\, dS$$

$$= \rho_1 \int\limits_{\Gamma_1} \left(\boldsymbol{r} \times \nabla\Phi_1 \frac{\partial \Phi_1}{\partial n}\right) dS - \frac{1}{2}\rho_1 \int\limits_{S_0} [\boldsymbol{r} \times (\nabla\Phi_1)^2 \boldsymbol{n}]\, dS$$

$$+\rho_1 \int\limits_{\Sigma_1} \boldsymbol{r} \times \left[\frac{\partial \Phi_1}{\partial t} - \nabla\Phi_1 \cdot (\boldsymbol{v}_0 + \boldsymbol{\omega} \times \boldsymbol{r}) - \mathbf{g} \cdot \boldsymbol{r}\right] \boldsymbol{n}\, dS. \qquad (3.3.32)$$

By virtue of relations (3.3.11) and (3.3.12), we obtain

$$\rho_1 \int\limits_{S_1} \boldsymbol{r} \times [\nabla\Phi_1 \cdot (\boldsymbol{v}_0 + \boldsymbol{\omega} \times \boldsymbol{r})]\, \boldsymbol{n}\, dS = \rho_1 \int\limits_{\Gamma_1} \boldsymbol{r} \times \frac{\partial \Phi}{\partial n} \nabla\Phi_1\, dS$$

$$-\rho_1 \int\limits_{Q_1} \boldsymbol{r} \times (\boldsymbol{\omega} \times \nabla\Phi_1)\, dQ - \rho_1 \int\limits_{S_0 \cup \Sigma} \boldsymbol{r} \times [\nabla\Phi_1(\boldsymbol{v}_0 + \boldsymbol{\omega} \times \boldsymbol{r})]\, \boldsymbol{n}\, dS$$

$$-\rho_1 \int\limits_{\Gamma_1} (\boldsymbol{r} \times \nabla\Phi_1)\, \boldsymbol{v}_{n1}\, dS. \qquad (3.3.33)$$

We substitute (3.3.30), (3.3.32), and (3.3.33) in relation (3.3.28) for the hydrodynamic moment caused by the outer liquid relative to the point $O$. Since the integral

$$\rho_1 \int\limits_{S_0} \boldsymbol{r} \times \left[\frac{\partial \Phi_1}{\partial t} + \frac{1}{2}(\nabla\Phi_1)^2 - \nabla\Phi_1 \cdot (\boldsymbol{v}_0 + \boldsymbol{\omega} \times \boldsymbol{r})\right] \boldsymbol{n}\, dS \qquad (3.3.34)$$

tends to zero according to the condition of decay, we get the following relation for the moment $\boldsymbol{N}_1$:

$$\boldsymbol{N}_1 = \rho_1 \int\limits_{S_1 \cup \Sigma} \boldsymbol{r} \times (\mathbf{g} \cdot \boldsymbol{r})\, \boldsymbol{n}\, dS - \overset{*}{G} + \rho_1 \int\limits_{Q_1} (\boldsymbol{u} \times \nabla\Phi_1)\, dQ$$

$$-\rho_1 \int\limits_{Q_1} \boldsymbol{r} \times (\boldsymbol{\omega} \times \nabla\Phi_1)\, dQ. \qquad (3.3.35)$$

By using

$$\boldsymbol{u} \times \nabla\Phi_1 = -\boldsymbol{v}_0 \nabla\Phi_1 - (\boldsymbol{\omega} \times \boldsymbol{r}) \times \nabla\Phi_1$$

and

$$\boldsymbol{\omega} \times (\boldsymbol{r} \times \nabla\Phi_1) = \boldsymbol{r} \times (\boldsymbol{\omega} \times \nabla\Phi_1) - \nabla\Phi_1 \times (\boldsymbol{\omega} \times \boldsymbol{r}),$$

we can represent relation (3.3.35) in the form

$$\boldsymbol{N}_1 = \rho_1 \int\limits_{S_1 \cup \Sigma} \boldsymbol{r} \times (\mathbf{g} \cdot \boldsymbol{r})\, \boldsymbol{n}\, dS - \overset{*}{\boldsymbol{G}}_1 - \boldsymbol{\omega} \times \boldsymbol{G}_1 - \boldsymbol{v}_0 \times \boldsymbol{K}_1. \qquad (3.3.36)$$

Similarly, we obtain the following relation for the principal moment of pressure forces of the inner liquid relative to the point $O$:

$$\boldsymbol{N}_2 = m_2 \boldsymbol{r}_{2c} \times \mathbf{g} - \overset{*}{\boldsymbol{G}}_2 - \boldsymbol{\omega} \times \boldsymbol{G}_2 - \boldsymbol{v}_0 \times \boldsymbol{K}_2, \qquad (3.3.37)$$

where $\boldsymbol{K}_2$ and $\boldsymbol{G}_2$ are, respectively, the vectors of momentum and angular momentum of the liquid $m_2$, which partially fills the cavity in the rigid body, relative to the point $O$.

Relations (3.3.13), (3.3.16), (3.3.36), and (3.3.37) represent the mathematical statement of the theorem on variations of the momentum and angular momentum of the outer and inner liquids relative to the origin of the moving coordinate system connected with the body.

The subsequent analysis of the expressions for the vectors of moments $\boldsymbol{N}_1$ (3.3.36) and $\boldsymbol{N}_2$ (3.3.37) can be performed with regard for the separation of the velocity potential $\boldsymbol{\Phi}_i$ into components by using proper structural representations of the shapes of free surfaces $\Sigma_i$. Representations (3.3.1) and the boundary conditions (3.3.2) enable us to reduce the relations for the angular momentum of liquid $\boldsymbol{G}_i$ to the form

$$\boldsymbol{G}_1 = \rho_1 \int\limits_{S_1 \cup \Sigma} \boldsymbol{\Omega} \cdot (\boldsymbol{v}_0 \cdot \boldsymbol{n})\, dS + \rho_1 \int\limits_{S_1 \cup \Sigma_1} (\boldsymbol{\omega} \cdot \boldsymbol{\Omega}_1) \frac{\partial \boldsymbol{\Omega}_1}{\partial n}\, dS - \rho_1 \int\limits_{\Sigma_1} \boldsymbol{\Omega}_1 \frac{\zeta_{1t}}{\sqrt{|\nabla \zeta_1|^2}}\, dS, \qquad (3.3.38)$$

$$\boldsymbol{G}_2 = m_2 \boldsymbol{r}_{2c} \times \boldsymbol{v}_0 + \boldsymbol{\omega} \cdot \boldsymbol{J}^{\,2} - \rho_2 \int\limits_{\Sigma_2} \boldsymbol{\Omega}_2 \frac{\zeta_{2t}}{\sqrt{|\nabla \zeta_2|^2}}\, dS, \qquad (3.3.39)$$

where $\boldsymbol{J}^{\,2}$ is the tensor of inertia of the inner liquid relative to the point O with the components

$$J^2_{ijk} = \rho_2 \int\limits_{S_2 \cup \Sigma_2} \frac{\partial \Omega_{ij}}{\partial n}\, \Omega_{ik}\, dS.$$

The results presented in this section form a basis for the construction of the nonlinear equations of motion of a floating rigid body containing cavities filled with liquids.

Chapter 4

# Nonlinear differential equations of space motions of a rigid body containing an upright cylindrical cavity partially filled with liquid

In the present chapter, we illustrate the applications of the analytic method described in the previous chapters to the case of cavities in the form of upright cylinders. The modal equations of the dynamics of liquid sloshing are proposed for the cases of circular and annular cross sections. These equations are connected with the nonlinear differential equations of space motions of the rigid body. We also propose a new nonlinear mathematical model for the nonlinear sloshing of a "viscous" liquid in the vessel of any shape and construct a finite-dimensional nonlinear model for a vessel in the form of an upright circular cylinder. In addition, we consider the cases of upright cylindrical vessels with bottoms of any geometric shape and cylindrical vessels with elliptic cross sections.

## 4.1 Nonlinear modal equations of liquid sloshing in cylindrical vessels with circular and annular cross sections

We focus our attention on the phenomenon of nonlinear sloshing in an upright cylindrical vessel of circular or annular cross section and the use of the variational method developed in Chapter 2 for getting finite-dimensional approximate nonlinear modal equations and the investigation of resonant sloshing with forcing frequency in the vicinity of the lowest resonance frequency.

We introduce a body-fixed cylindrical coordinate system $Ox\xi\eta$ in which the direction of the $Ox$-axis is opposite to the vector $\mathbf{g}$ and perpendicular to the mean free surface $\Sigma_0$. The meridional cross section of the vessel is shown in Fig. 4.1. We denote the depth of liquid by $h$ and the radii of the inner and outer cylinders by $R_0$ and $R_1$, respectively. According to (2.2.1) and (2.2.3), the free surface and the velocity potential are specified as follows:

$$x = f(\xi, \eta, t) = \sum_{m,n} [r_{mn}(t) \sin m\eta + p_{mn}(t) \cos m\eta] f_{mn}(\xi), \tag{4.1.1}$$

$$\varphi(x, \xi, \eta, t) = \sum_{m,n} [R_{mn}(t) \sin m\eta + P_{mn}(t) \cos m\eta] \psi_{mn}(x, \xi), \tag{4.1.2}$$

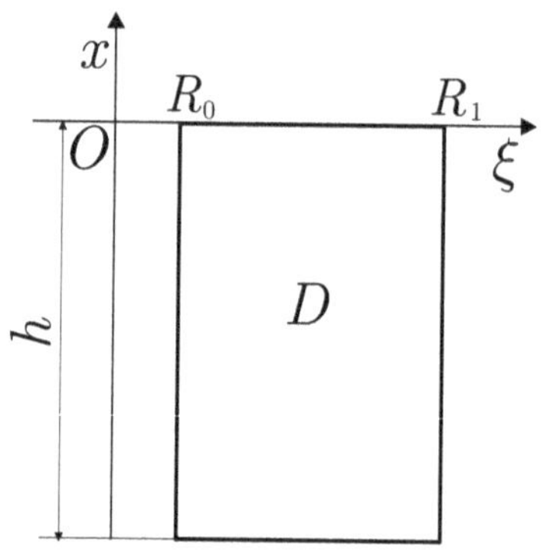


**Figure 4.1**

where

$$Y_m(k_{mn}\xi) = \frac{J_m(k_{mn}\xi)N'_m(\zeta_{mn}) - N_m(k_{mn}\xi)J'_m(\zeta_{mn})}{J_m(\zeta_{mn})N'_m(\zeta_{mn}) - N_m(\zeta_{mn})J'_m(\zeta_{mn})},$$

$$f_{mn}(\xi) = Y_m(k_{mn}\xi),\ \psi_{mn}(x,\xi) = \frac{\cosh k_{mn}(x+h)}{\cosh k_{mn}h} Y_m(k_{mn}\xi); \tag{4.1.3}$$

$J_m(k_{mn}\xi)$ and $N_m(k_{mn}\xi)$ are, respectively, the Bessel and Neumann functions of the $m$th order, and $\zeta_{mn} = k_{mn}R_1$ $(m = 0,\ 1,\ 2,\ \dots;\ n = 1,\ 2,\ \dots)$ are the roots of the equation

$$J'_m(\delta\zeta)N'_m(\zeta) - N'_m(\delta\zeta)J'_m(\zeta) = 0, \quad \delta = R_0/R_1. \tag{4.1.4}$$

We consider the case of resonant sloshing caused by the excitation of the lowest natural sloshing frequency connected with the frequency parameter $\varkappa_{11} = k_{11}\tanh k_{11}h$ $(\zeta_{11} = 1,8412)$ and the so-called Narimanov–Moiseev intermodal asymptotics. In this case, we restrict ourselves to the analysis of quantities of the third order of smallness with respect to the generalized coordinates $r_{11}(t)$ and $p_{11}(t)$ and the quantities $p_{01}(t)$, $r_{21}(t)$, and $p_{21}(t)$ regarded as variables of the same order as $r_{11}^2$ and $p_{11}^2$. The natural sloshing frequencies corresponding to the parameters $p_{01}$, $r_{21}$, and $p_{21}$ $(\zeta_{01} = 3.8317;\ \zeta_{21} = 3.0542)$ do not exceed the second natural frequency of antisymmetric oscillations $(\zeta_{12} = 5.3314)$. In the expansion of the velocity potential, we also keep the fundamental harmonics for $m = 0, 1, 2$.

$$x = f(\xi,\eta,t) = p_0(t)f_{01}(\xi) + [r_1(t)\sin\eta + p_1(t)\cos\eta]f_{11}(\xi)$$
$$+ [r_2(t)\sin 2\eta + p_2(t)\cos 2\eta]f_{21}(\xi); \tag{4.1.5}$$

$$\varphi(x,\xi,\eta,t) = P_0(t)\psi_{01}(x,\xi) + [R_1(t)\sin\eta + P_1(t)\cos\eta]\psi_{11}(x,\xi)$$
$$+ [R_2(t)\sin 2\eta + P_2(t)\cos 2\eta]\psi_{21}(x,\xi). \tag{4.1.6}$$

Here and in what follows, the five generalized coordinates are, for simplicity, marked by the same subscript. As a result, we get a *five-dimensional nonlinear modal equation* in $p_1$, $r_1$, $p_0$, $p_2$, and $r_2$ for the approximation of nonlinear resonant sloshing with five degrees of freedom.

Approximating the free-surface solution by (4.1.5), we deduce a system of nonlinear equations explaining the most pronounced nonlinear effects, such as the formation of swirling waves, mobility of the nodal line, elevation of the height of wave crest over the depth of the trough, dependence of the resonance sloshing frequency on the wave elevations, etc. We substitute relations (4.1.5) and (4.1.6) in (2.2.7). In this case, the collection of parameters $R_n(t)$ and $P_n(t)$ is denoted by $Z_n(t)$. Setting

$$\begin{gathered} P_0(t) = Z_1(t),\ R_1(t) = Z_2(t),\ P_1(t) = Z_3(t),\ R_2(t) = Z_4(t), \\ P_2(t) = Z_5(t),\ \varphi_1 = \psi_{01}(x,\xi),\ \varphi_2 = \psi_{11}(x,\xi)\sin\eta, \\ \varphi_3 = \psi_{11}(x,\xi)\cos\eta,\ \varphi_4 = \psi_{21}(x,\xi)\sin 2\eta,\ \varphi_5 = \psi_{21}(x,\xi)\cos 2\eta, \end{gathered} \tag{4.1.7}$$

for the quantities $A_n$ $(n = 1,2,\dots,5)$ and $A_{nk}$ $(n,k = 1,2,\dots,5)$, we arrive at the following expressions:

$$\begin{gathered} A_1 = a_4(r_1^2 + p_1^2) + a_{17}p_0, \\ A_2 = a_5 r_1 + a_6(r_1^3 + r_1 p_1^2) + a_{18}(p_1 r_2 - r_1 p_2) + a_{14} r_1 p_0, \\ A_3 = a_5 p_1 + a_6(p_1^3 + r_1^2 p_1) + a_{18}(r_1 r_2 + p_1 p_2) + a_{14} p_1 p_0, \\ A_4 = -2a_7 r_1 p_1 + a_{20} r_2,\quad A_5 = a_7(r_1^2 - p_1^2) + a_{20} p_2; \end{gathered} \tag{4.1.8}$$

$$\begin{gathered} A_{11} = 2a_1 + 2a_9(r_1^2 + p_1^2) + 2a_{21}p_0, \\ A_{12} = a_{15} r_1 + a_{26}(p_1 r_2 - r_1 p_2) + a_{23} r_1 p_0, \\ A_{13} = a_{15} p_1 + a_{26}(r_1 r_2 + p_1 p_2) + a_{23} p_1 p_0, \\ A_{14} = -2a_{16} r_1 p_1 + a_{24} r_2,\quad A_{15} = a_{16}(r_1^2 - p_1^2) + a_{24} p_2, \\ A_{22} = 2a_{10} + 2a_{11} r_1^2 + 2a_{12} p_1^2 + 2a_{22} p_0 - 2a_{19} p_2, \\ A_{23} = a_8 r_1 p_1 + 2a_{19} r_2, \\ A_{24} = a_3 p_1 + a_{27} r_1 r_2 + a_{28} p_0 p_1 + a_{29} p_1 p_2 + a_{31} r_1^2 p_1 + a_{32} p_1^3, \\ A_{25} = -a_3 r_1 + a_{27} r_1 p_2 - a_{28} r_1 p_0 - a_{29} p_1 r_2 - a_{33} r_1 p_1^2 - a_{34} r_1^3, \\ A_{33} = 2a_{10} + 2a_{11} p_1^2 + 2a_{12} r_1^2 + 2a_{19} p_2 + 2a_{22} p_0, \\ A_{34} = a_3 r_1 + a_{28} r_1 p_0 - a_{29} r_1 p_2 + a_{27} p_1 r_2 + a_{32} r_1^3 + a_{31} r_1 p_1^2, \\ A_{35} = a_3 p_1 + a_{27}(p_1 p_2 + p_1 r_2) + a_{28} p_0 p_1 + a_{29} r_1 r_2 + a_{34} p_1^3 + a_{33} r_1^2 p_1, \\ A_{44} = 2a_2 + 2a_{13}(r_1^2 + p_1^2) + 2a_{25} p_0,\quad A_{45} = 0, \\ A_{55} = 2a_2 + 2a_{13}(r_1^2 + p_1^2) + 2a_{25} p_0, \end{gathered} \tag{4.1.9}$$

where

$$a_1 = \pi\varkappa_{01}i_6,\quad a_2 = \tfrac{1}{2}\pi\varkappa_{21}i_{12},\quad a_3 = \tfrac{1}{2}\pi(\varkappa_{11}\varkappa_{21}i_5 + i_{22} + 2i_{23}),$$
$$a_4 = \tfrac{1}{2}\pi\varkappa_{01}i_2,\quad a_5 = \pi i_3,\quad a_6 = \tfrac{1}{8}\pi k_{11}^2 i_4,\quad a_7 = -\tfrac{1}{4}\pi\varkappa_{21}i_5,$$
$$a_8 = \tfrac{2}{3}\pi\varkappa_{11}(k_{11}^2 i_4 - i_{24}),\quad a_9 = \tfrac{1}{2}\pi\varkappa_{01}(k_{01}^2 i_8 + i_9),\quad a_{10} = \tfrac{1}{2}\varkappa_{11}a_5,$$
$$a_{11} = 4\varkappa_{11}a_6,\quad a_{12} = a_{11} - \tfrac{1}{2}a_8,\quad a_{13} = \tfrac{1}{4}\pi\varkappa_{21}(k_{21}^2 i_{15} + i_{16} + 4i_{17}),$$
$$a_{14} = \pi\varkappa_{11}i_2,\quad a_{15} = \tfrac{1}{2}\pi(k_{01}^2 + 2\varkappa_{01}\varkappa_{11})i_2,$$
$$a_{16} = -\tfrac{1}{4}\pi(k_{01}^2\varkappa_{21} + k_{12}^2\varkappa_{01})i_{20} - \tfrac{1}{4}\pi(\varkappa_{01} + \varkappa_{21})i_{21},\ a_{17} = 2\pi i_6,$$
$$a_{18} = \tfrac{1}{2}\pi\varkappa_{11}i_5,\quad a_{19} = \tfrac{1}{4}\pi(\varkappa_{11}^2 i_5 - i_{23} + i_{31}),$$
$$a_{20} = \pi i_{12},\quad a_{21} = \tfrac{1}{2}\pi(k_{01}^2 + 2\varkappa_{01}^2)i_{25},\ a_{22} = \tfrac{1}{2}\pi(\varkappa_{11}^2 i_2 + i_{28} + i_{29}),$$
$$a_{23} = \pi[(k_{11}^2\varkappa_{01} + k_{01}^2\varkappa_{11})i_8 + (\varkappa_{11} + \varkappa_{01})i_{34}],\ a_{24} = \pi(\varkappa_{01}\varkappa_{21}i_{30} + i_{36}),$$
$$a_{26} = \tfrac{1}{2}\pi[(k_{11}^2\varkappa_{01} + k_{01}^2\varkappa_{11})i_{20} + (\varkappa_{11} + \varkappa_{01})i_{35}];$$
$$a_{27} = \tfrac{1}{2}\pi[(k_{21}^2\varkappa_{11} + k_{11}^2\varkappa_{21})i_{15} + (\varkappa_{11} + \varkappa_{21})i_{39}];$$
$$a_{28} = \tfrac{1}{2}\pi[(k_{21}^2\varkappa_{11} + k_{11}^2\varkappa_{21})i_{20} + (\varkappa_{11} + \varkappa_{21})(i_{37} + 2i_{38})];$$
$$a_{25} = \tfrac{1}{2}\pi(\varkappa_{21}^2 i_{30} + i_{32} + 4i_{33}),\ a_{29} = \pi(\varkappa_{11} + \varkappa_{21})i_{17},$$
$$\varkappa_{mn} = k_{mn}\tanh k_{mn}h,\quad c_{mn} = \frac{1}{\cosh k_{mn}h},\quad i_2 = \int \xi Y_0(k_{01}\xi)Y_1^2(k_{11}\xi)d\xi,$$
$$i_3 = \int \xi Y_1^2(k_{11}\xi)d\xi,\quad i_4 = \int \xi Y_1^4(k_{11}\xi)d\xi,$$
$$i_5 = \int \xi Y_1^2(k_{11}\xi)Y_2(k_{21}\xi)d\xi,\quad i_6 = \int \xi Y_0^2(k_{01}\xi)d\xi,$$
$$i_7 = \int \xi Y_0'^2(k_{01}\xi)d\xi,\quad i_8 = \int \xi Y_0^2(k_{01}\xi)Y_1^2(k_{11}\xi)d\xi,$$
$$i_9 = \int \xi Y_0'^2(k_{01}\xi)Y_1^2(k_{11}\xi)d\xi,\quad i_{10} = \int \xi Y_1'^2(k_{11}\xi)d\xi, \tag{4.1.10}$$
$$i_{11} = \int \frac{1}{\xi}Y_1^2(k_{11}\xi)d\xi,\quad i_{12} = \int \xi Y_2^2(k_{21}\xi)d\xi,$$
$$i_{13} = \int \xi Y_2'^2(k_{21}\xi)d\xi,\quad i_{14} = \int \frac{1}{\xi}Y_2^2(k_{21}\xi)d\xi,$$
$$i_{15} = \int \xi Y_1^2(k_{11}\xi)Y_2^2(k_{21}\xi)d\xi,\quad i_{16} = \int \xi Y_2'^2(k_{21}\xi)Y_1^2(k_{11}\xi)d\xi,$$
$$i_{17} = \int \frac{1}{\xi}Y_2^2(k_{21}\xi)Y_1^2(k_{11}\xi)d\xi,\quad i_{18} = i_2,$$

$$i_{19} = \int \xi Y_0'(k_{01}\xi)Y_1'(k_{11}\xi)Y_1(k_{11}\xi)d\xi,$$

$$i_{20} = \int \xi Y_0(k_{01}\xi)Y_2(k_{21}\xi)Y_1^2(k_{11}\xi)d\xi,$$

$$i_{21} = \int \xi Y_0'(k_{01}\xi)Y_2'(k_{21}\xi)Y_1^2(k_{11}\xi)d\xi,$$

$$i_{22} = \int \xi Y_2'(k_{21}\xi)Y_1'(k_{11}\xi)Y_1(k_{11}\xi)d\xi,$$

$$i_{23} = \int \frac{1}{\xi}Y_1^2(k_{11}\xi)Y_2(k_{21}\xi)d\xi, \quad i_{24} = \int \frac{1}{\xi}Y_1^4 d\xi;$$

$$i_{25} = \int \xi Y_0^3(k_{01}\xi)d\xi, \quad i_{26} = \int \xi Y_0(k_{01}\xi)Y_0'^2(k_{01}\xi)d\xi,$$

$$i_{28} = \int \xi Y_0(k_{01}\xi)Y_1'^2(k_{11}\xi)d\xi, \quad i_{29} = \int \frac{1}{\xi}Y_0(k_{01}\xi)Y_1^2(k_{11}\xi)d\xi,$$

$$i_{30} = \int \xi Y_0(k_{01}\xi)Y_2^2(k_{21}\xi)d\xi, \quad i_{31} = \int \xi Y_1'^2(k_{11}\xi)Y_2(k_{21}\xi)d\xi,$$

$$i_{32} = \int \xi Y_0(k_{01}\xi)Y_2'^2(k_{21}\xi)d\xi, \quad i_{33} = \int \frac{1}{\xi}Y_0(k_{01}\xi)Y_2^2(k_{21}\xi)d\xi,$$

$$i_{34} = \int \xi Y_0(k_{01}\xi)Y_1(k_{11}\xi)Y_0'(k_{01}\xi)Y_1'(k_{11}\xi)d\xi, \quad i_{35} = \int \xi Y_0'Y_1'Y_1Y_2 d\xi,$$

$$i_{36} = \int \xi Y_0'Y_2'Y_2 d\xi, \quad i_{37} = \int \xi Y_0 Y_1 Y_1' Y_2' d\xi,$$

$$i_{38} = \int \frac{1}{\xi}Y_0 Y_2 Y_1^2 d\xi, \quad i_{39} = \int \xi Y_1 Y_2 Y_1' Y_2' d\xi, \quad i_{40} = \int \xi Y_1^2 Y_1'^2 d\xi.$$

Here and in what follows, the integration over $\xi$ is carried out from $R_0$ to $R_1$.

We now consider the system of equations (2.2.8) in the matrix form

$$AZ = b, \tag{4.1.11}$$

where $A$ is a symmetric square matrix with elements (4.1.9) and $\boldsymbol{b}$ is a vector whose components are the time derivatives of (4.1.8). Solving system (4.1.11), we find $Z_1(t), Z_2(t), \ldots, Z_5(t)$ as functions of $p_0(t), r_1(t), \ldots, p_2(t)$ and their derivatives. The most suitable method for the solution of (4.1.11) is reduced to the representation of the $n$th component of the vector $\boldsymbol{Z}$ in the form

$$\begin{aligned} Z_n &= \alpha_n + \alpha_n^1 p_0 + \alpha_n^2 r_1 + \alpha_n^3 p_1 + \alpha_n^4 r_2 + \alpha_n^5 p_2 + r_1\left(\alpha_{2n}^1 p_0 + \alpha_{2n}^2 r_1\right. \\ &\left. + \alpha_{2n}^3 p_1 + \alpha_{2n}^4 r_2 + \alpha_{2n}^5 p_2\right) + p_1\left(\alpha_{3n}^1 p_0 + \alpha_{3n}^3 p_1 + \alpha_{3n}^4 r_2 + \alpha_{3n}^5 p_2\right) \\ &+ \alpha_{4n} r_1^3 + \alpha_{5n} r_1^2 p_1 + \alpha_{6n} r_1 p_1^2 + \alpha_{7n} p_1^3 \end{aligned} \tag{4.1.12}$$

with subsequent substitution of (4.1.12) in (4.1.11). Gathering the coefficients of the quantities with identical degrees of $p_0$, $r_1$, ..., $p_2$ or their products, we get a system of recurrence relations for the constants $\alpha_n, \alpha_n^1, \ldots, \alpha_{7n}$. This system can easily be solved by the method of successive exclusion of the unknowns. Thus, to within the terms of the third order of smallness, we get

$$P_0(t) = C_0(r_1\dot{r}_1 + p_1\dot{p}_1) + D_0\dot{p}_0,$$

$$R_1(t) = \frac{1}{\varkappa_{11}}\dot{r}_1 + C_2 r_1^2\dot{r}_1 + D_3 p_1^2\dot{r}_1 + C_1 r_1 p_1\dot{p}_1 + D_2(r_2\dot{p}_1 - p_2\dot{r}_1)$$
$$+C_3(p_1\dot{r}_2 - r_1\dot{p}_2) + B_0 p_0\dot{r}_1 + B_3 r_1\dot{p}_0, \tag{4.1.13}$$
$$P_1(t) = \frac{\dot{p}_1}{\varkappa_{11}} + C_2 p_1^2\dot{p}_1 + D_3 r_1^2\dot{p}_1 + D_2(r_2\dot{r}_1 + p_2\dot{p}_1) + C_1 p_1 r_1\dot{r}_1$$
$$+C_3(r_1\dot{r}_2 + p_1\dot{p}_2) + B_0 p_0\dot{p}_1 + B_3 p_1\dot{p}_0,$$
$$R_2(t) = \frac{1}{\varkappa_{21}}\dot{r}_2 - D_1(r_1\dot{p}_1 + p_1\dot{r}_1), \quad P_2(t) = \frac{\dot{p}_2}{\varkappa_{21}} + D_1(r_1\dot{r}_1 - p_1\dot{p}_1),$$

where

$$C_0 = \frac{a_4}{a_1} - \frac{a_5 a_{15}}{4a_1 a_{10}}, \quad C_1 = \frac{1}{a_{10}}\left(a_6 - \frac{a_4 a_{15}}{2a_1} - \frac{a_5 a_8}{4a_{10}} + \frac{a_5 a_{15}^2}{8a_1 a_{10}}\right), \quad D_0 = \frac{a_{17}}{2a_1},$$
$$C_3 = \frac{1}{2a_{10}}\left(a_{18} - \frac{a_3 a_{20}}{2a_2}\right), \quad B_0 = \frac{1}{2a_{10}}\left(a_{14} - \frac{a_5 a_{22}}{a_{10}}\right), \quad D_1 = \frac{a_7}{a_2} + \frac{a_5 a_3}{4a_{10}a_2},$$
$$B_3 = \frac{1}{2a_{10}}\left(a_{14} - \frac{a_{15}a_{17}}{2a_1}\right), \quad D_2 = \frac{1}{2a_{10}}\left(a_{18} - \frac{a_5 a_{19}}{a_{10}}\right), \tag{4.1.14}$$
$$D_3 = \frac{1}{\varkappa_{11}}\left(\frac{a_3^2}{4a_{10}a_2} - \frac{a_{12}}{a_{10}} + \frac{a_3 a_7}{a_2 a_5}\right) + \frac{a_6}{2a_{10}}, \quad C_2 = D_3 + C_1.$$

Substituting (4.1.13) in (2.2.9), we arrive at the following system of nonlinear ordinary differential equations with respect to the generalized coordinates $p_0(t)$, $r_1(t), \ldots, p_2(t)$:

$$\mu_1(\ddot{r}_1 + \sigma_1^2 r_1) + d_1(r_1^2\ddot{r}_1 + r_1\dot{r}_1^2 + r_1p_1\ddot{p}_1 + r_1\dot{p}_1^2) + d_2\left(p_1^2\ddot{r}_1 + 2p_1\dot{r}_1\dot{p}_1\right.$$
$$\left.-r_1p_1\ddot{p}_1 - 2r_1\dot{p}_1^2\right) - d_3(p_2\ddot{r}_1 - r_2\ddot{p}_1 + \dot{r}_1\dot{p}_2 - \dot{p}_1\dot{r}_2)$$
$$+d_4(r_1\ddot{p}_2 - p_1\ddot{r}_2) + d_5(p_0\ddot{r}_1 + \dot{r}_1\dot{p}_0) + d_6 r_1\ddot{p}_0 = 0; \tag{4.1.15}$$
$$\mu_1(\ddot{p}_1 + \sigma_1^2 p_1) + d_1(p_1^2\ddot{p}_1 + r_1p_1\ddot{r}_1 + p_1\dot{r}_1^2 + p_1\dot{p}_1^2)$$
$$+d_2(r_1^2\ddot{p}_1 - r_1p_1\ddot{r}_1 + 2r_1\dot{r}_1\dot{p}_1 - 2p_1\dot{r}_1^2) + d_3(p_2\ddot{p}_1 + r_2\ddot{r}_1 + \dot{r}_1\dot{r}_2 + \dot{p}_1\dot{p}_2)$$
$$-d_4(p_1\ddot{p}_2 + r_1\ddot{r}_2) + d_5(p_0\ddot{p}_1 + \dot{p}_1\dot{p}_0) + d_6 p_1\ddot{p}_0 = 0; \tag{4.1.16}$$
$$\mu_0(\ddot{p}_0 + \sigma_0^2 p_0) + d_6(r_1\ddot{r}_1 + p_1\ddot{p}_1) + d_8(\dot{r}_1^2 + \dot{p}_1^2) = 0; \tag{4.1.17}$$

$$\mu_2(\ddot{r}_2 + \sigma_2^2 r_2) - d_4(\ddot{r}_1 p_1 + \ddot{p}_1 r_1) - 2d_7 \dot{r}_1 \dot{p}_1 = 0; \tag{4.1.18}$$

$$\mu_2(\ddot{p}_2 + \sigma_2^2 p_2) + d_4(\ddot{r}_1 r_1 - \ddot{p}_1 p_1) + d_7(\dot{r}_1^2 - \dot{p}_1^2) = 0, \tag{4.1.19}$$

where

$$\mu_0 = \frac{2\pi\rho i_6}{\varkappa_{01}}, \quad \mu_1 = \frac{\pi\rho i_3}{\varkappa_{11}}, \quad \mu_2 = \frac{\pi\rho i_{12}}{\varkappa_{21}},$$

$$\sigma_0^2 = \varkappa_{01} g, \quad \sigma_1^2 = \varkappa_{11} g, \quad \sigma_2^2 = \varkappa_{21} g,$$

$$d_1 = \rho\left(2a_4 C_0 + 2a_7 D_1 + a_5 C_2 + \frac{3a_6}{\varkappa_{11}}\right), \tag{4.1.20}$$

$$d_2 = \rho\left(a_5 D_3 + \frac{a_6}{\varkappa_{11}} + 2a_7 D_1\right), \quad d_3 = \rho\left(a_5 D_2 + \frac{a_{18}}{\varkappa_{11}}\right),$$

$$d_4 = \rho\left(-a_5 C_3 + \frac{2a_7}{\varkappa_{21}}\right), \quad d_5 = \rho\left(a_5 B_0 + \frac{a_{14}}{\varkappa_{11}}\right),$$

$$d_6 = \rho(2a_4 D_0 + a_5 B_3), \quad d_7 = d_4 + \tfrac{1}{2} d_3, \quad d_8 = d_6 - \tfrac{1}{2} d_5.$$

The structure of the five-dimensional modal equations (4.1.15)–(4.1.19) completely coincides with the structure of the corresponding equations obtained by the method of perturbation theory in Chapter 6. The numerical coefficients of these equations almost coincide. Thus, for a cylinder of unit radius with the depth of liquid $h = 1.0$, by using relations (4.1.20), we obtain $d_1 = 0.6197$ and $d_2 = -0.3819$, which agrees with the values of the coefficients $d_1 = 0.6212$ and $d_2 = -0.3812$ obtained by the Narimanov method. The coefficients of Eqs. (4.1.15)–(4.1.19) were found by Pil'kevich for various geometric parameters of the vessel. Their numerical values obtained for $R_1 = 1.0$ are presented as functions of $\delta = R_0/R_1$ and the dimensionless depth of liquid $h = h/R_0$ in Tables 4.1–4.10. In Tables 4.11–4.13, we show the values of the first roots of the transcendental equation (4.1.4) as functions of the dimensionless radius of the inner cylinder.

The transition from the dimensionless quantities given in the tables to the dimensional quantities is realized by the following formulas:

$$h = \bar{h} R_0, \quad \mu_0 = R_1^3 \bar{\mu}_0, \quad \mu_1 = \bar{\mu}_1 R_1^3, \quad \mu_2 = \bar{\mu}_2 R_1^3,$$

$$\sigma_0^2 = \frac{g\bar{\sigma}_0^2}{R_1}, \quad \sigma_1^2 = \frac{g\bar{\sigma}_1^2}{R_1}, \quad \sigma_2^2 = \frac{g\bar{\sigma}_2^2}{R_1},$$

$$d_1 = \bar{d}_1 R_1, \quad d_2 = \bar{d}_2 R_1, \quad d_3 = \bar{d}_3 R_1^2, \quad d_4 = \bar{d}_4 R_1^2, \tag{4.1.21}$$

$$d_5 = \bar{d}_5 R_1^2, \quad d_6 = \bar{d}_6 R_1^2, \quad d_7 = \bar{d}_7 R_1^2, \quad d_8 = \bar{d}_8 R_1^2.$$

Here, the overbars denote the tabular values of the coefficients of the nonlinear modal equations (4.1.15)–(4.1.19).

**Table 4.1**

| $h$ | $\mu_0$ | $\mu_1$ | $\mu_2$ | $d_1$ | $d_2$ |
|---|---|---|---|---|---|
| | | | $\delta = 0$ | | |
| 0.2 | 1.272 | 1.707 | 0.5393 | 9.287 | 4.992 |
| 0.4 | 0.9001 | 0.9593 | 0.3497 | 1.541 | 0.2769 |
| 0.6 | 0.8366 | 0.7498 | 0.3092 | 0.6918 | −0.1336 |
| 0.8 | 0.8235 | 0.6682 | 0.2982 | 0.4713 | −0.2171 |
| 1.0 | 0.8207 | 0.6325 | 0.2951 | 0.3920 | −0.2416 |
| 1.2 | 0.8201 | 0.6161 | 0.2942 | 0.3590 | −0.2503 |
| 1.4 | 0.8199 | 0.6085 | 0.2939 | 0.3443 | −0.2538 |
| 1.6 | 0.8199 | 0.6048 | 0.2938 | 0.3375 | −0.2554 |
| 1.8 | 0.8199 | 0.6031 | 0.2938 | 0.3343 | −0.2561 |
| 2.0 | 0.8199 | 0.6022 | 0.2938 | 0.3328 | −0.2564 |
| 2.5 | 0.8199 | 0.6016 | 0.2938 | 0.3317 | −0.2566 |
| 3.0 | 0.8199 | 0.6015 | 0.2938 | 0.3315 | −0.2567 |

**Table 4.2**

| $h$ | $d_3$ | $d_4$ | $d_5$ | $d_6$ |
|---|---|---|---|---|
| | | $\delta = 0$ | | |
| 0.2 | 1.821 | 1.502 | 4.714 | −1.625 |
| 0.4 | 0.8847 | 0.2598 | 1.800 | −0.3732 |
| 0.6 | 0.7165 | 0.0399 | 1.276 | −0.1469 |
| 0.8 | 0.6621 | −0.0292 | 1.107 | −0.0731 |
| 1.0 | 0.6403 | −0.0560 | 1.039 | −0.0433 |
| 1.2 | 0.6307 | −0.0675 | 1.009 | −0.0300 |
| 1.4 | 0.6263 | −0.0727 | 0.9955 | −0.0239 |
| 1.6 | 0.6242 | −0.0750 | 0.9890 | −0.0210 |
| 1.8 | 0.6232 | −0.0761 | 0.9860 | −0.0197 |
| 2.0 | 0.6227 | −0.0767 | 0.9845 | −0.0190 |
| 2.5 | 0.6224 | −0.0771 | 0.9834 | −0.0185 |
| 3.0 | 0.6223 | −0.0771 | 0.9832 | −0.0184 |

If we add forcing terms to (4.1.15)–(4.1.19), then it becomes possible to study the resonant liquid sloshing. This includes the visualization of the wave patterns [69] and the classification of the steady-state modes [56]. Despite the fact that the nonlinear modal equations approximate the process of liquid sloshing with five degrees of freedom, their application for the analysis leads to satisfactory qualitative and quantitative agreement with the experiments.

Returning to the modal solution (4.1.1), it is necessary to recall that an infinite set of the generalized coordinates $r_{mn}$ and $p_{mn}$ may be significant.

**Table 4.3**

| $h$ | $\mu_0$ | $\mu_1$ | $\mu_2$ | $d_1$ | $d_2$ |
|---|---|---|---|---|---|
| $\delta = 0.2$ | | | | | |
| 0.2 | 0.9120 | 2.133 | 0.5538 | 11.64 | 7.342 |
| 0.4 | 0.6728 | 1.181 | 0.3582 | 1.920 | 0.4125 |
| 0.6 | 0.6368 | 0.9082 | 0.3163 | 0.8459 | −0.1868 |
| 0.8 | 0.6303 | 0.7983 | 0.3049 | 0.5619 | −0.3072 |
| 1.0 | 0.6292 | 0.7482 | 0.3015 | 0.4569 | −0.3422 |
| 1.2 | 0.6290 | 0.7241 | 0.3006 | 0.4115 | −0.3547 |
| 1.4 | 0.6289 | 0.7123 | 0.3003 | 0.3904 | −0.3599 |
| 1.6 | 0.6289 | 0.7064 | 0.3003 | 0.3802 | −0.3622 |
| 1.8 | 0.6289 | 0.7034 | 0.3003 | 0.3751 | −0.3633 |
| 2.0 | 0.6289 | 0.7019 | 0.3003 | 0.3726 | −0.3638 |
| 2.5 | 0.6289 | 0.7006 | 0.3003 | 0.3705 | −0.3642 |
| 3.0 | 0.6289 | 0.7004 | 0.3003 | 0.3701 | −0.3643 |

**Table 4.4**

| $h$ | $d_3$ | $d_4$ | $d_5$ | $d_6$ |
|---|---|---|---|---|
| $\delta = 0.2$ | | | | |
| 0.2 | 3.061 | 1.909 | 5.780 | −1.274 |
| 0.4 | 1.270 | 0.3778 | 1.970 | −0.3528 |
| 0.6 | 0.9458 | 0.1032 | 1.281 | −0.1793 |
| 0.8 | 0.8392 | 0.0144 | 1.054 | −0.1191 |
| 1.0 | 0.7952 | −0.0214 | 0.9606 | −0.0931 |
| 1.2 | 0.7751 | −0.0375 | 0.9178 | −0.0808 |
| 1.4 | 0.7654 | −0.0451 | 0.8972 | −0.0748 |
| 1.6 | 0.7607 | −0.0489 | 0.8870 | −0.0718 |
| 1.8 | 0.7583 | −0.0507 | 0.8819 | −0.0703 |
| 2.0 | 0.7571 | −0.0517 | 0.8794 | −0.0696 |
| 2.5 | 0.7561 | −0.0524 | 0.8773 | −0.0689 |
| 3.0 | 0.7559 | −0.0525 | 0.8769 | −0.0688 |

In [127], the corresponding infinite-dimensional modal equations were deduced by using the generalized Narimanov–Moiseev asymptotics. The analysis suggested the dimensionless (scaled by the radius of the tank) statement and a low-magnitude horizontal excitations of the tank of order $\epsilon \ll 1$ such that the forcing frequency $\sigma$ is close to the lowest natural frequency $\sigma_{11}$ associated with the two generalized coordinates $p_{11}(t)$ and $r_{11}(t)$. The generalized Narimanov–Moiseev asymptotics require that

$$p_{11} \sim r_{11} = O(\epsilon^{1/3}). \tag{4.1.22}$$

**Table 4.5**

| $h$ | $\mu_0$ | $\mu_1$ | $\mu_2$ | $d_1$ | $d_2$ |
|---|---|---|---|---|---|
| $\delta = 0.4$ | | | | | |
| 0.2 | 0.4571 | 2.927 | 0.6873 | 17.62 | 14.03 |
| 0.4 | 0.3721 | 1.582 | 0.4345 | 2.786 | 1.124 |
| 0.6 | 0.3634 | 1.180 | 0.3776 | 1.155 | −0.0404 |
| 0.8 | 0.3624 | 1.010 | 0.3610 | 0.7209 | −0.2872 |
| 1.0 | 0.3623 | 0.9267 | 0.3558 | 0.5556 | −0.3630 |
| 1.2 | 0.3623 | 0.8836 | 0.3542 | 0.4808 | −0.3918 |
| 1.4 | 0.3623 | 0.8604 | 0.3537 | 0.4436 | −0.4045 |
| 1.6 | 0.3623 | 0.8478 | 0.3535 | 0.4241 | −0.4106 |
| 1.8 | 0.3623 | 0.8408 | 0.3535 | 0.4137 | −0.4137 |
| 2.0 | 0.3623 | 0.8370 | 0.3534 | 0.4080 | −0.4154 |
| 2.5 | 0.3623 | 0.8333 | 0.3534 | 0.4025 | −0.4169 |
| 3.0 | 0.3623 | 0.8324 | 0.3534 | 0.4013 | −0.4173 |

**Table 4.6**

| $h$ | $d_3$ | $d_4$ | $d_5$ | $d_6$ |
|---|---|---|---|---|
| $\delta = 0.4$ | | | | |
| 0.2 | 6.383 | 3.001 | 5.756 | −0.5669 |
| 0.4 | 2.240 | 0.6757 | 1.737 | −0.2052 |
| 0.6 | 1.483 | 0.2514 | 1.002 | −0.1281 |
| 0.8 | 1.228 | 0.1089 | 0.7546 | −0.0977 |
| 1.0 | 1.118 | 0.0478 | 0.6479 | −0.0831 |
| 1.2 | 1.064 | 0.0183 | 0.5962 | −0.0756 |
| 1.4 | 1.037 | 0.0032 | 0.5694 | −0.0715 |
| 1.6 | 1.022 | −0.0049 | 0.5551 | −0.0693 |
| 1.8 | 1.014 | −0.0093 | 0.5474 | −0.0681 |
| 2.0 | 1.010 | −0.0117 | 0.5431 | −0.0675 |
| 2.5 | 1.005 | −0.0141 | 0.5390 | −0.0688 |
| 3.0 | 1.004 | −0.0146 | 0.5380 | −0.0667 |

Postulating (4.1.22) and using the trigonometric algebra with respect to the angular coordinate, we find the second- and third-order generalized coordinates:

$$\begin{aligned} &p_{0n} \sim p_{2n} \sim r_{2n} = O(\epsilon^{2/3}); \quad p_{3n} \sim r_{3n} = O(\epsilon), \quad n = 1, 2, \ldots, \\ &p_{1m} \sim r_{1m} = O(\epsilon), \quad m = 2, 3, \ldots. \end{aligned} \tag{4.1.23}$$

The remaining generalized coordinates have the order $o(\epsilon)$ and can be neglected in our nonlinear multimodal analysis.

**Table 4.7**

| $h$ | $\mu_0$ | $\mu_1$ | $\mu_2$ | $d_1$ | $d_2$ |
|---|---|---|---|---|---|
| $\delta = 0.6$ | | | | | |
| 0.2 | 0.1749 | 3.165 | 0.8018 | 20.26 | 17.65 |
| 0.4 | 0.1641 | 1.679 | 0.4874 | 3.034 | 1.694 |
| 0.6 | 0.1609 | 1.223 | 0.4108 | 1.176 | 0.2068 |
| 0.8 | 0.1608 | 1.022 | 0.3861 | 0.6859 | −0.1208 |
| 1.0 | 0.1608 | 0.9186 | 0.3774 | 0.4986 | −0.2253 |
| 1.2 | 0.1608 | 0.8619 | 0.3743 | 0.4119 | −0.2666 |
| 1.4 | 0.1608 | 0.8294 | 0.3732 | 0.3672 | −0.2854 |
| 1.6 | 0.1608 | 0.8104 | 0.3728 | 0.3426 | −0.2949 |
| 1.8 | 0.1608 | 0.7992 | 0.3726 | 0.3285 | −0.2999 |
| 2.0 | 0.1608 | 0.7924 | 0.3726 | 0.3204 | −0.3028 |
| 2.5 | 0.1608 | 0.7852 | 0.3725 | 0.3116 | −0.3057 |
| 3.0 | 0.1608 | 0.7831 | 0.3725 | 0.3092 | −0.3065 |

By using the same derivations as in the five-dimensional case, the generalized coordinates of the second order, and the generalized coordinates of the third order, we arrive at the nonlinear modal equations including the following two differential equations for the lowest-order generalized coordinates $p_{11}$ and $r_{11}$:

$$\begin{aligned}
\mu_{11}\left[\ddot{p}_{11}+\sigma_{1,1}^2 p_{11}\right] &+ p_{11}\sum_{n=1}^{N} d_{0,n}^{(2)}\ddot{p}_{0n} + \sum_{n=1}^{N} d_{0,n}^{(3)}\left(\ddot{p}_{11}p_{0n}+\dot{p}_{11}\dot{p}_{0n}\right)\\
&+ d_1\left(p_{11}^2\ddot{p}_{11}+p_{11}\dot{p}_{11}^2+r_{11}p_{11}\ddot{r}_{11}+p_{11}\dot{r}_{11}^2\right)\\
&+ d_2\left(r_{11}^2\ddot{p}_{11}+2r_{11}\dot{r}_{11}\dot{p}_{11}-r_{11}p_{11}\ddot{r}_{11}-2p_{11}\dot{r}_{11}^2\right)\\
&+\sum_{n=1}^{N} d_{2,n}^{(3)}\left(\ddot{p}_{11}p_{2n}+\ddot{r}_{11}r_{2n}+\dot{p}_{11}\dot{p}_{2n}+\dot{r}_{11}\dot{r}_{2n}\right)\\
&+\sum_{n=1}^{N} d_{2n}^{(2)}\left(p_{11}\ddot{p}_{2n}+r_{11}\ddot{r}_{2n}\right) = -\frac{\mu_{11}\varkappa_{11}}{k_{11}^2-1}\dot{v}_{O1},
\end{aligned}\tag{4.1.24a}$$

$$\begin{aligned}
\mu_{11}\left[\ddot{r}_{11}+\sigma_{11}^2 r_{11}\right] &+ r_{11}\sum_{n=1}^{N} d_{0,n}^{(2)}\ddot{p}_{0n} + \sum_{n=1}^{N} d_{0,n}^{(3)}\left(\ddot{r}_{11}p_{0n}+\dot{r}_{11}\dot{p}_{0n}\right)\\
&+ d_1\left(r_{11}^2\ddot{r}_{11}+r_{11}\dot{r}_{11}^2+r_{11}p_{11}\ddot{p}_{11}+r_{11}\dot{p}_{11}^2\right)\\
&+ d_2\left(p_{11}^2\ddot{r}_{11}+2p_{11}\dot{r}_{11}\dot{p}_{11}-r_{11}p_{11}\ddot{p}_{11}-2r_{11}\dot{p}_{11}^2\right)\\
&+\sum_{n=1}^{N} d_{2,n}^{(3)}\left(\ddot{p}_{11}r_{2n}-\ddot{r}_{11}p_{2n}+\dot{p}_{11}\dot{r}_{2n}-\dot{r}_{11}\dot{p}_{2n}\right)\\
&+\sum_{n=1}^{N} d_{2,n}^{(2)}\left(p_{11}\ddot{r}_{2n}-r_{11}\ddot{p}_{2n}\right) = -\frac{\mu_{11}\varkappa_{11}}{k_{11}^2-1}\dot{v}_{O2}.
\end{aligned}\tag{4.1.24b}$$

**Table 4.8**

| $h$ | $d_3$ | $d_4$ | $d_5$ | $d_6$ |
|---|---|---|---|---|
| | | $\delta = 0.6$ | | |
| 0.2 | 8.174 | 3.660 | 3.583 | −0.1493 |
| 0.4 | 2.644 | 0.8577 | 1.017 | −0.0671 |
| 0.6 | 1.627 | 0.3427 | 3.5453 | −0.0456 |
| 0.8 | 1.279 | 0.1665 | 0.3839 | −0.0361 |
| 1.0 | 1.125 | 0.0886 | 3.3125 | −0.0313 |
| 1.2 | 1.048 | 0.0493 | 3.2765 | −0.0287 |
| 1.4 | 1.005 | 0.0279 | 3.2569 | −0.0271 |
| 1.6 | 0.9815 | 0.0158 | 3.2458 | −0.0263 |
| 1.8 | 0.9676 | 0.0088 | 3.2393 | −0.0257 |
| 2.0 | 0.9594 | 0.0046 | 3.2355 | −0.0254 |
| 2.5 | 0.9506 | 0.00018 | 3.2314 | −0.0251 |
| 3.0 | 0.9481 | −0.0011 | 3.2303 | −0.0250 |

**Table 4.9**

| $h$ | $\mu_0$ | $\mu_1$ | $\mu_2$ | $d_1$ | $d_2$ |
|---|---|---|---|---|---|
| | | | $\delta = 0.8$ | | |
| 0.2 | 0.0402 | 2.315 | 0.6035 | 14.92 | 13.49 |
| 0.4 | 0.0401 | 1.213 | 0.3545 | 2.148 | 1.396 |
| 0.6 | 0.0401 | 0.8687 | 0.2897 | 0.7896 | 0.2512 |
| 0.8 | 0.0401 | 0.7124 | 0.2670 | 0.4356 | −0.0058 |
| 1.0 | 0.0401 | 0.6297 | 0.2582 | 0.3008 | −0.0892 |
| 1.2 | 0.0401 | 0.5823 | 0.2547 | 0.2379 | −0.1227 |
| 1.4 | 0.0401 | 0.5540 | 0.2533 | 0.2049 | −0.1382 |
| 1.6 | 0.0401 | 0.5366 | 0.2527 | 0.1862 | −0.1461 |
| 1.8 | 0.0401 | 0.5258 | 0.2524 | 0.1752 | −0.1504 |
| 2.0 | 0.0401 | 0.5190 | 0.2523 | 0.1685 | −0.1529 |
| 2.5 | 0.0401 | 0.5109 | 0.2523 | 0.1608 | −0.1556 |
| 3.0 | 0.0401 | 0.5083 | 0.2523 | 0.1583 | −0.1565 |

These equations contain the lowest- and second-order generalized coordinates but the third-order generalized coordinates are absent. The notation used for $k_{mn}$ (the roots of the equation $J'_m(k_{mn}) = 0$), $\varkappa_{mn}$, $\sigma_{mn}$ (natural sloshing frequency), and the components of the translational velocity $\dot{v}_{01}(t)$ and $\dot{v}_{02}(t)$ have been explained earlier. The dimensionless hydrodynamic coefficients of the nonlinear terms are determined at the end of this section.

**Table 4.10**

| $h$ | $d_3$ | $d_4$ | $d_5$ | $d_6$ |
|---|---|---|---|---|
| $\delta = 0.8$ | | | | |
| 0.2 | 6.140 | 2.790 | 1.113 | −0.0170 |
| 0.4 | 1.890 | 0.6640 | 0.3059 | −0.0086 |
| 0.6 | 1.107 | 0.2722 | 0.1572 | −0.0060 |
| 0.8 | 0.8363 | 0.1370 | 0.1059 | −0.0048 |
| 1.0 | 0.7149 | 0.0762 | 0.0828 | −0.0042 |
| 1.2 | 0.6522 | 0.0448 | 0.0709 | −0.0038 |
| 1.4 | 0.6170 | 0.0272 | 0.0643 | −0.0036 |
| 1.6 | 0.5962 | 0.0169 | 0.0603 | −0.0035 |
| 1.8 | 0.5837 | 0.0106 | 0.0579 | −0.0034 |
| 2.0 | 0.5759 | 0.0067 | 0.0565 | −0.0034 |
| 2.5 | 0.5668 | 0.0021 | 0.0547 | −0.0033 |
| 3.0 | 0.5639 | 0.00067 | 0.0542 | −0.0033 |

**Table 4.11**

| $\delta$ | $k_{01}$ | $k_{02}$ | $k_{03}$ | $k_{04}$ | $k_{05}$ |
|---|---|---|---|---|---|
| $m = 0$ | | | | | |
| 0 | 3.831706 | 7.015587 | 10.173468 | 13.323692 | 16.470630 |
| 0.1 | 3.940943 | 7.330568 | 10.748379 | 14.188643 | 17.643300 |
| 0.2 | 4.235748 | 8.055358 | 11.926582 | 15.821042 | 19.727054 |
| 0.4 | 6.391181 | 10.557736 | 15.766458 | 20.988200 | 26.215483 |
| 0.5 | 6.393156 | 12.624702 | 18.888931 | 25.162415 | 31.439708 |
| 0.6 | 7.930086 | 15.747272 | 23.588329 | 31.435762 | 39.285795 |
| 0.8 | 15.737551 | 31.430822 | 47.133810 | 62.839311 | 78.545786 |

**Table 4.12**

| $\delta$ | $k_{11}$ | $k_{12}$ | $k_{13}$ | $k_{14}$ | $k_{15}$ |
|---|---|---|---|---|---|
| $m = 1$ | | | | | |
| 0 | 1.841184 | 5.331443 | 8.536316 | 11.706005 | 14.863589 |
| 0.1 | 1.803469 | 5.137136 | 8.199157 | 11.358797 | 14.634360 |
| 0.2 | 1.705115 | 4.960856 | 8.433066 | 12.165052 | 15.993233 |
| 0.4 | 1.461782 | 5.659099 | 10.683244 | 15.848083 | 21.048785 |
| 0.5 | 1.354673 | 6.564943 | 12.706418 | 18.942662 | 25.202480 |
| 0.6 | 1.262075 | 8.041083 | 15.801057 | 23.623920 | 31.462382 |
| 0.8 | 1.113366 | 15.777725 | 31.450766 | 47.147103 | 62.849279 |

**Table 4.13**

| $\delta$ | $k_{21}$ | $k_{22}$ | $k_{23}$ | $k_{24}$ | $k_{25}$ |
|---|---|---|---|---|---|
| | | | $m = 2$ | | |
| 0 | 3.054237 | 6.706133 | 9.969468 | 13.170371 | 16.347522 |
| 0.1 | 3.052944 | 6.686690 | 9.887514 | 12.969631 | 16.011864 |
| 0.2 | 3.034725 | 6.494953 | 9.549461 | 12.899735 | 16.519252 |
| 0.4 | 2.842401 | 6.415969 | 11.055959 | 16.091693 | 21.229933 |
| 0.5 | 2.681204 | 7.062586 | 12.949411 | 19.103158 | 25.322437 |
| 0.6 | 2.515946 | 8.367073 | 15.961465 | 23.730408 | 31.542123 |
| 0.8 | 2.226458 | 15.897615 | 31.510511 | 43.186895 | 62.879113 |

The differential equations used to find the second-order generalized coordinates $p_{0n}$, $p_{2n}$, and $r_{2n}$ take the form

$$2\mu_{0n}\left[\ddot{p}_{0n} + \sigma_{0n}^2 p_{0n}\right] + d_{0,n}^{(1)}\left(\dot{p}_{11}^2 + \dot{r}_{11}^2\right) + d_{0,n}^{(2)}\left(\ddot{p}_{11}p_{11} + \ddot{r}_{11}r_{11}\right) = 0, \quad (4.1.25a)$$

$$\mu_{2n}\left[\ddot{p}_{2n} + \sigma_{2n}^2 p_{2n}\right] + d_{2,n}^{(1)}\left(\dot{p}_{11}^2 - \dot{r}_{11}^2\right) + d_{2,n}^{(2)}\left(\ddot{p}_{11}p_{11} - \ddot{r}_{11}r_{11}\right) = 0, \quad (4.1.25b)$$

$$\mu_{2n}\left[\ddot{r}_{2n} + \sigma_{2n}^2 r_{2n}\right] + 2d_{2,n}^{(1)}\dot{r}_{11}\dot{p}_{11} + d_{2,n}^{(2)}\left(\ddot{p}_{11}r_{11} + \ddot{r}_{11}p_{11}\right) = 0. \quad (4.1.25c)$$

Here, $n = 1, \ldots, N$, i.e., there are $3N$ ordinary differential equations for these generalized coordinates. Note that Eqs. (4.1.25) contain $p_{11}$ and $r_{11}$ given by (4.1.24) and, therefore, one can say that the first- and second-order generalized coordinates are nonlinearly coupled by our modal equations. However, the third-order generalized coordinates $p_{3n}$ and $r_{3n}$ are absent in (4.1.25). The equations relating these generalized coordinates take the form

$$\begin{aligned}\mu_{3n}\left[\ddot{r}_{3n} + \sigma_{3n}^2 r_{3n}\right] &+ d_3\left(r_{11}\dot{p}_{11}^2 + 2p_{11}\dot{p}_{11}\dot{r}_{11} - r_{11}\dot{r}_{11}^2\right)\\ &+ d_4\left(p_{11}^2\ddot{r}_{11} + 2r_{11}p_{11}\ddot{p}_{11} - r_{11}^2\ddot{r}_{11}\right) + \sum_{n=1}^{N} d_{3,n}^{(1)}\left(\dot{p}_{11}\dot{r}_{2n} + \dot{r}_{11}\dot{p}_{2n}\right)\\ &+ \sum_{n=1}^{N} d_{3,n}^{(2)}\left(p_{11}\ddot{r}_{2n} + r_{11}\ddot{p}_{2n}\right) + \sum_{n=1}^{N} d_{3,n}^{(3)}\left(\ddot{p}_{11}r_{2n} + \ddot{r}_{11}p_{2n}\right) = 0, \quad (4.1.26a)\end{aligned}$$

$$\begin{aligned}\mu_{3n}\left[\ddot{p}_{3n} + \sigma_{3n}^2 p_{3n}\right] &+ d_3\left(p_{11}\dot{p}_{11}^2 - 2r_{11}\dot{p}_{11}\dot{r}_{11} - p_{11}\dot{r}_{11}^2\right)\\ &+ d_4\left(p_{11}^2\ddot{p}_{11} - 2p_{11}r_{11}\ddot{r}_{11} - r_{11}^2\ddot{p}_{11}\right) + \sum_{n=1}^{N} d_{3,n}^{(1)}\left(\dot{p}_{11}\dot{p}_{2n} - \dot{r}_{11}\dot{r}_{2n}\right)\\ &+ \sum_{n=1}^{N} d_{3,n}^{(2)}\left(p_{11}\ddot{p}_{2n} - r_{11}\ddot{r}_{2n}\right) + \sum_{n=1}^{N} d_{3,n}^{(3)}\left(\ddot{p}_{11}p_{2n} - \ddot{r}_{11}r_{2n}\right) = 0, \quad (4.1.26b)\end{aligned}$$

$$\mu_{1n}\left[\ddot{r}_{1n}+\sigma_{1n}^2 r_{1n}\right]+d_5\left(\ddot{r}_{11}r_{11}^2+r_{11}p_{11}\ddot{p}_{11}\right)+d_6\left(r_{11}\dot{r}_{11}^2+r_{11}\dot{p}_{11}^2\right)$$
$$+d_7\left(\ddot{r}_{11}p_{11}^2-r_{11}p_{11}\ddot{p}_{11}\right)+d_8\left(\dot{r}_{11}\dot{p}_{11}p_{11}-r_{11}\dot{p}_{11}^2\right)$$
$$+\sum_{n=1}^N d_{4,n}^{(1)}\left(\dot{p}_{11}\dot{r}_{2n}-\dot{r}_{11}\dot{p}_{2n}\right)+\sum_{n=1}^N d_{4,n}^{(2)}\left(p_{11}\ddot{r}_{2n}-r_{11}\ddot{p}_{2n}\right)$$
$$+\sum_{n=1}^N d_{4,n}^{(3)}\left(\ddot{p}_{11}r_{2n}-\ddot{r}_{11}p_{2n}\right)+\dot{r}_{11}\sum_{n=1}^N d_{5,n}^{(1)}\dot{p}_{0n}$$
$$+r_{11}\sum_{n=1}^N d_{5,n}^{(2)}\ddot{p}_{0n}+\ddot{r}_{11}\sum_{n=1}^N d_{5,n}^{(3)}p_{0n}=-\frac{\mu_{1n}\varkappa_{1n}}{k_{1n}^2-1}\dot{v}_{O2},\quad(4.1.26c)$$

$$\mu_{1n}\left[\ddot{p}_{1n}+\sigma_{1n}^2 p_{1n}\right]+d_5\left(\ddot{p}_{11}p_{11}^2+r_{11}p_{11}\ddot{r}_{11}\right)+d_6\left(p_{11}\dot{p}_{11}^2+p_{11}\dot{r}_{11}^2\right)$$
$$+d_7\left(\ddot{p}_{11}r_{11}^2-r_{11}p_{11}\ddot{r}_{11}\right)+d_8\left(\dot{r}_{11}\dot{p}_{11}p_{11}-p_{11}\dot{r}_{11}^2\right)$$
$$+\sum_{n=1}^N d_{4,n}^{(1)}\left(\dot{p}_{11}\dot{p}_{2n}+\dot{r}_{11}\dot{r}_{2n}\right)+\sum_{n=1}^N d_{4,n}^{(2)}\left(r_{11}\ddot{r}_{2n}+p_{11}\ddot{p}_{2n}\right)$$
$$+\sum_{n=1}^N d_{4,n}^{(3)}\left(\ddot{p}_{11}p_{2n}+\ddot{r}_{11}r_{2n}\right)+\dot{p}_{11}\sum_{n=1}^N d_{5,n}^{(1)}\dot{p}_{0n}$$
$$+p_{11}\sum_{n=1}^N d_{5,n}^{(2)}\ddot{p}_{0n}+\ddot{p}_{11}\sum_{n=1}^N d_{5,n}^{(3)}p_{0n}=-\frac{\mu_{1n}\varkappa_{1n}}{k_{1n}^2-1}\dot{v}_{O1},\quad(4.1.26d)$$

where $n=1,\dots,N$. Equations (4.1.26) are linear in $p_{3n}$ and $r_{3n}$ and contain nonlinear quantities in terms of the first- and second-order generalized coordinates.

The most important result is that the *nonzero* hydrodynamic coefficients in (4.1.24)–(4.1.26) can be efficiently found by using the following simple formulas:

$$d_{0,n}^{(1)}=d_{0,n}^{(2)}-\frac{d_{0,n}^{(3)}}{2};\quad d_{0,n}^{(2)}=\frac{\pi}{2}\left[2-\frac{k_{0,n}^2}{\varkappa_{0,n}\varkappa_{1,1}}\right]j_{(0,n)(1,1)^2},$$

$$d_{0,n}^{(3)}=\pi\left[j_{(0,n)(1,1)^2}-\frac{1}{\varkappa_{1,1}^2}\left(j_{(0,n)}^{(1,1)^2}+i_{(0,n)(1,1)^2}\right)\right];\quad d_{2,n}^{(1)}=d_{2,n}^{(2)}-\frac{d_{2,n}^{(3)}}{2},$$

$$d_{2,n}^{(2)}=\frac{\pi}{2}\left[j_{(2,n)(1,1)^2}-\frac{1}{\varkappa_{2,n}\varkappa_{1,1}}\left(j_{(1,1)}^{(2,n)(1,1)}+2i_{(2,n)(1,1)^2}\right)\right],$$

$$d_{2,n}^{(3)}=\frac{\pi}{2}\left[j_{(2,n)(1,1)^2}-\frac{1}{\varkappa_{1,1}^2}\left(j_{(2,n)}^{(1,1)^2}-i_{(2,n)(1,1)^2}\right)\right],$$

$$d_1=\frac{\pi}{2\varkappa_{1,1}}\left[\frac{k_{0,1}^4\left(j_{(0,1)(1,1)^2}\right)^2}{4\varkappa_{01}\varkappa_{1,1}}\frac{1}{j_{(0,1)^2}}+i_{(1,1)^4}-j_{(1,1)^2}^{(1,1)^2}\right]+d_2,$$

$$d_2 = \frac{\pi}{4\varkappa_{1,1}} \left[ \frac{\left(j_{(1,1)}^{(1,1)(2,1)} + 2i_{(2,1)(1,1)^2}\right)^2}{\varkappa_{1,1}\varkappa_{2,1}} \frac{1}{j_{(2,1)^2}} - 3i_{(1,1)^4} - j_{(1,1)^2}^{(1,1)^2} \right],$$

$$\begin{aligned} d_3 = {} & \frac{\pi}{4\varkappa_{1,1}} \frac{\left(j_{(1,1)}^{(1,1)(2,1)} + 2i_{(2,1)(1,1)^2}\right)}{\varkappa_{1,1}\varkappa_{2,1}} \frac{\left(2i_{(1,1)(2,1)(3,1)} - j_{(3,1)}^{(1,1)(2,1)}\right)}{j_{(2,1)^2}} \\ & + \frac{\pi}{4\varkappa_{1,1}} \left[ j_{(1,1)(3,1)}^{(1,1)^2} - i_{(1,1)^3(3,1)} \right] + 2d_4, \end{aligned}$$

$$\begin{aligned} d_4 = {} & \frac{\pi}{4\varkappa_{1,1}} \frac{\left(j_{(1,1)}^{(1,1)(2,1)} + 2i_{(2,1)(1,1)^2}\right)}{\varkappa_{2,1}\varkappa_{3,1}} \frac{\left(6i_{(1,1)(2,1)(3,1)} + j_{(1,1)}^{(2,1)(3,1)}\right)}{j_{(2,1)^2}} \\ & - \frac{\pi}{4\varkappa_{1,1}} \left[ \frac{(\varkappa_{1,1} + \varkappa_{3,1})}{2\varkappa_{3,1}} \left(3i_{(1,1)^3(3,1)} + j_{(1,1)^2}^{(1,1)(3,1)}\right) \right], \end{aligned}$$

$$d_{3,n}^{(1)} = d_{3,n}^{(2)} + d_{3,n}^{(3)} - \frac{\pi}{2} \left[ j_{(1,1)(2,n)(3,1)} - \frac{j_{(3,1)}^{(1,1)(2,n)} - 2i_{(1,1)(2,n)(3,1)}}{\varkappa_{1,1}\varkappa_{2,n}} \right],$$

$$d_{3,n}^{(2)} = \frac{\pi}{2} \left[ j_{(1,1)(2,n)(3,1)} - \frac{j_{(1,1)}^{(2,n)(3,1)} + 6i_{(1,1)(2,n)(3,1)}}{\varkappa_{2,n}\varkappa_{3,1}} \right],$$

$$d_{3,n}^{(3)} = \frac{\pi}{2} \left[ j_{(1,1)(2,n)(3,1)} - \frac{j_{(2,n)}^{(1,1)(3,1)} + 3i_{(1,1)(2,n)(3,1)}}{\varkappa_{1,1}\varkappa_{3,1}} \right],$$

$$d_{4,n}^{(1)} = d_{4,n}^{(2)} - d_{3,n}^{(3)} - \frac{\pi}{2} \left[ j_{(1,1)(2,n)(1,2)} - \frac{j_{(1,2)}^{(1,1)(2,n)} + 2i_{(1,1)(2,n)(1,2)}}{\varkappa_{1,1}\varkappa_{2,n}} \right],$$

$$d_{4,n}^{(2)} = \frac{\pi}{2} \left[ j_{(1,1)(2,n)(1,2)} - \frac{j_{(1,1)}^{(2,n)(1,1)} + 2i_{(1,1)(2,n)(1,2)}}{\varkappa_{2,n}\varkappa_{1,2}} \right],$$

$$d_{4,n}^{(3)} = \frac{\pi}{2} \left[ j_{(1,1)(2,n)(1,2)} - \frac{j_{(2,n)}^{(1,1)(1,2)} - i_{(1,1)(2,n)(1,2)}}{\varkappa_{1,1}\varkappa_{1,2}} \right],$$

$$d_{5,n}^{(1)} = d_{5,n}^{(2)} + d_{5,n}^{(3)} - \pi \left[ j_{(0,n)(1,1)(1,2)} - \frac{j_{(1,2)}^{(0,n)(1,1)}}{\varkappa_{1,1}\varkappa_{0,n}} \right],$$

$$d_{5,n}^{(2)} = \pi \left[ j_{(0,n)(1,1)(1,2)} - \frac{j_{(1,1)}^{(0,n)(1,2)}}{\varkappa_{0,n}\varkappa_{1,2}} \right],$$

$$d_{5,n}^{(3)} = \pi \left[ j_{(0,n)(1,1)(1,2)} - \frac{j_{(0,n)}^{(1,1)(1,2)} + i_{(0,n)(1,1)(1,2)}}{\varkappa_{1,1}\varkappa_{1,2}} \right].$$

In these formulas, by definition,

$$j_{(a,b)}^{(c,d)} = \int \xi \left(\prod f_{a,b}\left(k_{a,b}\xi\right)\right)\left(\prod \frac{d}{d\xi} f_{c,d}\left(k_{c,d}\xi\right)\right) d\xi,$$

$$i_{(a,b)}^{(c,d)} = \int \frac{1}{\xi} \left(\prod f_{a,b}\left(k_{a,b}\xi\right)\right)\left(\prod \frac{d}{d\xi} f_{c,d}\left(k_{c,d}\xi\right)\right) d\xi.$$

Moreover, there are special indexing rules for $i$ and $j$ exemplified by the formula

$$j_{(0,2)(2,2)(1,1)}^{(1,2)(0,1)(1,2)} = j_{(0,2)(1,1)(2,2)}^{(0,1)(1,2)^2}$$

$$= \int_0^1 \xi \left(f_{0,2}(k_{0,2}\xi)\, f_{1,1}(k_{1,1}\xi)\, f_{2,2}(k_{2,2}\xi)\right)\left(\frac{d}{d\xi} f_{0,1}(k_{0,1}\xi)\left(\frac{d}{d\xi} f_{1,2}(k_{1,2}\xi)\right)^2\right) d\xi.$$

Equations (4.1.26c) and (4.1.26d) contain the coefficients $d_5$, $d_6$, $d_7$, and $d_8$ given by the formulas:

$$d_5 = -\frac{0.51201}{h_{1,1}} - \frac{0.16879}{h_{1,2}} + \frac{0.50224}{h_{1,2}h_{1,1}h_{0,1}} + \frac{0.17969}{h_{1,2}h_{2,1}h_{1,1}},$$

$$d_6 = -\frac{1.34899}{h_{1,1}} - \frac{0.3376}{h_{1,2}} + \frac{1.00448}{h_{1,2}h_{1,1}h_{0,1}} + \frac{0.37908}{h_{1,1}^2 h_{0,1}} + \frac{0.35938}{h_{1,2}h_{2,1}h_{1,1}} + \frac{0.23782}{h_{1,1}^2 h_{2,1}},$$

$$d_7 = -\frac{0.11748}{h_{1,1}} - \frac{0.00307}{h_{1,2}} + \frac{0.17969}{h_{1,2}h_{2,1}h_{1,1}},$$

$$d_8 = -\frac{0.68799}{h_{1,1}} - \frac{0.17186}{h_{1,2}} + \frac{0.37908}{h_{1,2}^2 h_{0,1}} + \frac{0.35938}{h_{1,2}h_{2,1}h_{1,1}} + \frac{0.50224}{h_{1,2}h_{1,1}h_{0,1}},$$

where $h_{m,n} = \tanh(k_{mn}h)$ depends on the dimensionless depth.

In may be important for the practical purposes that the modal equations (4.1.24)–(4.1.26) can be rewritten in the following matrix form:

$$Q(\boldsymbol{q})\ddot{\boldsymbol{q}} + C\boldsymbol{q} + \boldsymbol{\Psi}(\boldsymbol{q}; \dot{\boldsymbol{q}}) = \boldsymbol{V}, \tag{4.1.27}$$

where $\boldsymbol{q} = (q_{1,1}; q_{1,2}; \dots; q_{1,n}; q_{2,1}; q_{2,2}; \dots; q_{2,n}; \dots; q_{7,1}; q_{7,2}; \dots; q_{7,n})^T$. The mass matrix $Q(\boldsymbol{q})$ in (4.1.27) becomes diagonal for $\boldsymbol{q} = 0$ or, more clearly, it becomes diagonal when the first- and second-order generalized coordinates are equal to zero. This means that, as long as the perturbations of the free-surface are relatively small, the system (4.1.27) may serve as a basis for the direct time integrations with appropriate initial conditions.

## 4.2 Nonlinear modal equations for "viscous" liquid

The nonlinear analytic analysis of sloshing of incompressible viscous liquids appears to be in the rudimentary state even in the case of containers of very simple geometric shapes. The main difficulties encountered in this case are connected with the development of the methods aimed at the investigation of singularly perturbed nonlinear initial-boundary-value problems for the Navier–Stokes equations.

As the methods most frequently used within the framework of the Navier–Stokes model, we can mention the methods based on the ideas of the theory of boundary layer [190, 34, 29, 85], the energy-conservation methods [92], and numerical methods similar to the procedures developed in [22]. However, their realization is connected with serious mathematical difficulties, especially in the case of low-viscosity liquids.

The evaluation of the hydrodynamic characteristics, such as the added masses, moments of inertia of liquid, and decrements of sloshing, for liquids with various physical properties seems to be problematic for large Reynolds numbers even if we use contemporary numerical methods.

However, the methods discussed above turn out to be extremely useful for the investigation of the mechanisms of energy dissipation and the appearance of dissipative forces, as well as for a more profound understanding of the specific features of liquid flows in the oscillatory modes. In combination with contemporary experimental methods, these methods make it possible to develop reliable semiempirical theories of sloshing aimed at the adequate description of various actual physical processes.

The analytic and experimental problems of viscous sloshing are most often considered for immobile vessels. For this class of problems, the analytic formulas covering broad ranges of geometric parameters of the vessels and the physical properties of liquids were obtained for the logarithmic decrements of vibrations in [147, 148, 149, 29]. Thus, simple analytic dependences of the logarithmic decrements on the parameters of mechanical systems were established for cylindrical cavities with circular cross sections or with cross sections in the form of an annular sector [147, 29]. If the surface tension of liquid can be neglected (for upright cylindrical tanks in which the diameter of the free surface is greater than 300–400 mm), then the logarithmic decrement for these tanks is given by the formula

$$\delta_{mn} = \frac{\pi}{\sqrt{R_{mn}}}\left[\frac{\zeta_{mn}^2+1}{\zeta_{mn}^2-1} + \frac{2\zeta_{mn}(1-\bar{h})}{\sinh(2\zeta_{mn}\bar{h})} + \frac{8\zeta_{mn}}{\sqrt{R_{mn}}}\right], \tag{4.2.1}$$

which is in good agreement with the experimental data; $R_{mn} = \sigma_{mn}R_1^2/\nu$ is a dimensionless parameter called the sloshing-related Reynolds number (for free oscillations, it is better to use the term “Galileo number”).

Relation (4.2.1) and similar dependences for the logarithmic decrement differ from the formulas given by the laminar boundary-layer theory (without introducing the empirical correction to the hydrodynamical force) by a constant factor of $\sqrt{2}$. The common point of all these formulas is the fact that they are obtained for the irrotational inviscid flows of motion of the ambient liquid. Hence, the use of the notion of matrix of damping rates for the liquid sloshing is in full agreement with the spectral properties of the problem of natural liquid sloshing and can be regarded as well-substantiated. For vessels of different shapes, the data on the damping ratios can be found in [148, 3] and in Chap. 5 of [56].

Unlike the studies based on the Navier–Stokes equations of liquid sloshing, the nonlinear sloshing theory was proposed in [121] on the basis of the Rayleigh hypothesis, which is, in author’s opinion, more natural from the viewpoint of oscillatory hydrodynamic processes. Most likely, Faltinsen [52] was the first who draw attention to this hypothesis in analyzing the problem of steady-state wave motions of liquids in two-dimensional domains. According to this hypothesis, the deviations of liquid particles from their equilibrium positions are hampered by the forces proportional to their relative velocity [96]. Recall that the classical Navier–Stokes equations are based on the Newton hypothesis (1687) characteristic of laminar flows. According to this hypothesis, the tangential stresses $\tau$ in a two-dimensional laminar shear flow can be represented in the form

$$\tau = \pm\mu\frac{\partial v}{\partial n}, \tag{4.2.2}$$

where $\mu$ is the dynamic viscosity, $v$ is the velocity of liquid layer, $n$ is the direction transverse to the flow, and the sign is chosen to guarantee that the stresses are positive. Hypothesis (4.2.2) was established by Newton in analyzing the rotation of coaxial cylindrical containers in infinite liquid media.

As a specific feature of the flows considered in what follows, we can mention the fact that the motion of liquid particles in the problem of free sloshing of low-viscosity liquids in the field of gravity forces occurs along the flow lines. These flow lines also serve as the trajectories of liquid particles along which they perform their oscillatory motions about the equilibrium positions. The oscillations of particles occur in the vertical direction at the wave crests and in the horizontal direction at the nodes. In this case, the amplitude of oscillations of particles sharply decreases with depth. Hence, in the case where the depth of liquid is much greater than the wavelength, the phenomenon of wave formation has a well-pronounced surface character. The property of damping of the amplitude of free oscillations of liquid particles with the distance from the free surface is one of the fundamental properties of the theory of gravity waves.

The oscillatory motions of liquid essentially differ from the flows caused by the rotation of cylindrical surfaces in the infinite liquid. Hence, for the friction forces acting upon liquid particles in the oscillatory mode, it is natural to choose the dependence

$$\boldsymbol{F}_\tau = -\alpha f(\boldsymbol{v}) \approx -\alpha\boldsymbol{v}, \tag{4.2.3}$$

established by Newton in the solution of the problem of slow motions of the balls in air and in water.

In this case, the friction coefficient $\alpha$ can be found experimentally. Most likely, the first experimental procedure of measuring the friction coefficient was proposed by Coulomb. The Coulomb method can be described as follows: The horizontal disk is suspended from a wire both in air and in the investigated liquid and put in the oscillatory motion about its own axis. The drag can be found from the logarithmic decrement of the amplitude of oscillations. For slow oscillations, it was shown that the drag is proportional to the first power of the velocity. Among other important results following from Coulomb's experiments, we can especially mention the conclusion about the character of the influence of the surface operating in contact with liquid and pressure on the logarithmic decrement. First, the disk was covered with a thin layer of fat. This did not affect the logarithmic decrement. Then the disk lubricated with fat was covered with a thin layer of mineral powder. In this case, the effect was also absent. In a special series of experiments, Coulomb discovered that the internal friction (drag) is independent of pressure.

These considerations are of great importance for the statement of initial-boundary-value problems of the theory of viscous liquid sloshing.

First, we consider the problem of free sloshing occurring under the action of friction forces in cavities of arbitrary geometric shapes. We choose the origin of coordinates $Oxyz$ on the unperturbed free surface and assume that the direction of the $Ox$-axis is opposite to the vector of gravitational acceleration **g**. The equations of motion of liquid in the Euler form with regard for the friction forces (4.2.3) are as follows:

$$\frac{\partial \boldsymbol{v}}{\partial t} + (\boldsymbol{v}, \nabla \boldsymbol{v}) = \boldsymbol{F} + \boldsymbol{F}_{\tau} - \frac{\nabla p}{\rho}, \tag{4.2.4}$$

where $\boldsymbol{v}$ is the velocity of motion of liquid particles, $\boldsymbol{F}$ is the vector of potential forces, $\boldsymbol{F}_{\tau} = -\alpha\, \boldsymbol{v}$ is the vector of friction forces, and $\alpha$ is the friction coefficient. The vectors of forces $\boldsymbol{F}$ and $\boldsymbol{F}_{\tau}$ are related to the unit mass. In the formulation of the corresponding boundary and initial conditions, relation (4.2.4), together with the equation of continuity $\operatorname{div}\boldsymbol{v} = 0$, enables one to find the field of velocities $v_x$, $v_y$, $v_z$ and pressure $p$ as functions of the space variables $x$, $y$, $z$ and time $t$. The accepted law of friction does not violate the vortex-free character of liquid motions. We now set

$$\boldsymbol{v} = \operatorname{grad} \boldsymbol{\Phi}.$$

Under the condition $\operatorname{rot} \boldsymbol{v} = 0$, the equation of continuity yields the Laplace equation for the vortex-free velocity potential

$$\nabla^2 \Phi(x, y, z, t) = 0,$$

whereas the generalized Bernoulli equation

$$\frac{\partial \Phi}{\partial t} + \frac{1}{2}(\nabla \Phi)^2 + \alpha \Phi + \mathrm{g}x + \frac{p}{\rho} = F(t) \tag{4.2.5}$$

follows from the Euler equation (4.2.4). For the traditional boundary conditions on the free surface $\Sigma(t)$ and on the rigid wall $S(t)$ of the liquid domain $Q(t)$, we get the following nonlinear boundary-value problem for the velocity potential $\boldsymbol{\Phi}(x, y, z, t)$:

$$\begin{gathered}\nabla^2\Phi = 0, \quad \boldsymbol{r} \in Q(t),\\ \frac{\partial\Phi}{\partial\nu} = -\frac{\zeta_t}{\sqrt{(\nabla\zeta)^2}}, \quad \boldsymbol{r} \in \Sigma(t),\\ \frac{\partial\Phi}{\partial\nu} = 0, \quad \boldsymbol{r} \in S(t),\\ \frac{\partial\Phi}{\partial t} + \frac{1}{2}(\nabla\Phi)^2 + \alpha\Phi + \mathrm{g}x = 0, \quad \boldsymbol{r} \in \Sigma(t),\end{gathered} \tag{4.2.6}$$

where $\zeta(x, y, z, t) = 0$ is the equation of perturbed free surface $\Sigma(t)$ and $\boldsymbol{\nu}$ is the unit vector of the outer normal to the surface of the domain $Q(t)$. The slip condition on the rigid boundary $S$ is taken in the same form as for the ideal incompressible liquid. In this case, we neglect the capillary effects on the free surface and assume that, in this problem, the effect of the properties of the surface $S(t)$ on the logarithmic decrement of liquid sloshing is not strong, as in Coulomb's experiments. As for the nonlinear boundary-value problem (4.2.6), we can formulate either the problem of finding its periodic solutions (the problem of standing waves) or the Cauchy problem with the corresponding initial conditions.

In the analyzed case, the linearized statement of problem (4.2.6) formulated for the domain with fixed boundaries takes the form

$$\begin{gathered}\nabla^2\Phi = 0, \quad \boldsymbol{r} \in Q, \quad \frac{\partial\Phi}{\partial\nu} = 0, \quad \boldsymbol{r} \in S,\\ \frac{\partial^2\Phi}{\partial t^2} + \alpha\frac{\partial\Phi}{\partial t} + \mathrm{g}\frac{\partial\Phi}{\partial x} = 0, \quad \boldsymbol{r} \in \Sigma_0,\end{gathered} \tag{4.2.7}$$

where $\Sigma_0$ is the unperturbed free surface. By the method of separation of variables or with the help of representation of the solutions of problem (4.2.7) in the form

$$\Phi(x, y, z, t) = e^{\omega t}\varphi(x, y, z),$$

the study of the problem of free damped liquid sloshing is reduced to the solution of the already considered problem of natural sloshing modes and frequencies (1.5.34) with respect to $\varphi(x, y, z)$. The resonant natural frequency of sloshing for the viscous liquid is given by the formula

$$\omega_n^2 + \omega_n\alpha_n + \sigma_n^2 = 0, \tag{4.2.8}$$

where $\sigma_n^2 = \varkappa_n \mathrm{g}$ is the natural (circular) sloshing frequency.

The natural sloshing modes correspond to the complex roots of Eq. (4.2.8) and exist under the condition $\frac{1}{2}\alpha_n < \sigma_n$. If this condition is not satisfied, then we get aperiodic wave motions typical of strongly viscous liquid media. The logarithmic decrement of oscillations is given by the formula

$$\delta_n = \frac{\pi \alpha_n}{\sqrt{\sigma_n^2 - (\frac{1}{2}\alpha_n)^2}}. \tag{4.2.9}$$

As already indicated, this important characteristic of the oscillatory processes can be fairly efficiently found from the experiments, at least for the natural sloshing frequencies especially important for the practical purposes. Hence, we can assume that the damping ratios $\alpha_n$ are physical quantities readily found for containers of any shape and any liquid with oscillatory properties in the gravitational and inertial fields.

We now present some results of the nonlinear analysis of the problem of free or forced sloshing of a viscous liquid obtained as a result of generalization of the presented approaches for the case of ideal liquid. We recall that the modal approach in the dynamics of ideal liquid is based on the variational principle of mechanics connected with with the action

$$W = \int_{t_1}^{t_2} L\, dt,$$

where the Lagrangian

$$L = -\rho \int_{Q(t)} \left[ \frac{\partial \Phi}{\partial t} + \frac{1}{2}(\nabla \Phi)^2 + gx \right] dQ. \tag{4.2.10}$$

The stationary values of the action are attained on the solutions of the nonlinear boundary problem (4.2.6) with $\alpha = 0$, which describe the phenomenon of liquid sloshing under the action of gravity forces.

The direct methods using the variational statements of nonlinear problems are based on the constructive representations of the shapes of perturbed free surfaces and the velocity potential in the form

$$\zeta(x, y, z, t) \equiv x - f(y, z, t) = x - \sum_{i=1}^{\infty} \beta_i(t) f_i(y, z) = 0, \tag{4.2.11}$$

$$\Phi(x, y, z, t) = \sum_{n=1}^{\infty} R_n(t)\, \varphi_n(x, y, z), \tag{4.2.12}$$

where $R_n(t)$ are the parameters characterizing the variations of the velocity potential with time, $\beta_i(t)$ are the generalized Fourier coefficients playing the role of generalized coordinates in this problem, $f_i(y, z)$ and $\varphi_n(x, y, z)$ are specially chosen complete systems of functions on the surface $\Sigma_0$ and in the domain $Q$ with the properties of solutions of the spectral boundary problem (1.5.34).

Expansions (4.2.11) and (4.2.12) for the shape of free surface and the velocity potential in terms of the natural sloshing modes allow us, as shown above, to reduce the solutions of boundary-value problems in partial derivatives to the solution of systems of nonlinear ordinary differential equations [modal systems of the form (4.1.15)–(4.1.19)] as a result of the application of the variational method.

For a viscous liquid, similar modal systems can be derived from the Ostrogradskii variational principle

$$\int_{t_1}^{t_2} (\delta L + \delta' A) dt = 0, \qquad (4.2.13)$$

where $L$ is the Lagrangian (4.2.10) and $\delta' A$ is the elementary work of dissipative forces.

Introducing the Rayleigh dissipative function

$$F = \frac{1}{2}\alpha\rho \int_{Q(t)} (\nabla\Phi)^2 dQ \qquad (4.2.14)$$

in one version of the direct methods considered in [120], we arrive at the following system of nonlinear ordinary equations of free sloshing of a viscous liquid [121, 122]:

$$\frac{d^2}{dt^2}\frac{\partial L}{\partial \ddot{\beta}_i} - \frac{d}{dt}\frac{\partial L}{\partial \dot{\beta}_i} + \frac{\partial L}{\partial \beta_i} = \frac{\partial F}{\partial \beta_i}, \quad (i = 1, 2, \ldots) \qquad (4.2.15)$$

provided that the action is transformed as follows:

$$W = \int_{t_1}^{t_2} L(\beta_i, \dot{\beta}_i, \ddot{\beta}_i) dt.$$

This action is called the action in Ostrogradskii's sense [122]. This action differs from the Hamiltonian action by the dependence of the integrand not only on the first time derivatives of the generalized coordinates $\beta_i(t)$ but also on the second derivatives.

We now present the explicit form of this system for the case of forced liquid sloshing in an upright circular cylinder near the lowest resonance. As the parameters $\beta_i(t)$, we choose the generalized coordinates $p_0(t)$, $r_1(t)$, $p_1(t)$, $r_2(t)$, and $p_2(t)$ corresponding to an axisymmetric form of oscillations with one nodal circle and asymmetric forms with one and two nodal diameters (see Sec. 4.1). As a result, we get a five-dimensional modal system for the description of viscous liquid sloshing [121]:

$$L_1(r_1, p_1, p_0, r_2, p_2) \equiv \mu_1(\ddot{r}_1 + \alpha\dot{r}_1 + \sigma_1^2 r_1) + d_1 r_1(r_1\ddot{r}_1 + \dot{r}_1^2 + p_1\ddot{p}_1 + \dot{p}_1^2)$$
$$+d_2(p_1^2\ddot{r}_1 + 2p_1\dot{r}_1\dot{p}_1 - 2r_1\dot{p}_1^2 - r_1 p_1\ddot{p}_1) + d_3(r_2\ddot{p}_1 - p_2\ddot{r}_2 + \dot{p}_1\dot{r}_2 - \dot{r}_1\dot{p}_2)$$
$$+d_4(r_1\ddot{p}_2 - p_1\ddot{r}_2) + d_5(p_0\ddot{r}_1 + \dot{p}_0\dot{r}_1) + d_6 r_1\ddot{p}_0 + \alpha[(d_1 r_1^2 + d_2 p_1^2)\dot{r}_1$$
$$+(d_1 - d_2) r_1 p_1 \dot{p}_1 + d_3(r_2\dot{p}_1 - p_2\dot{r}_1) + d_4(r_1\dot{p}_2 - p_1\dot{r}_2) + d_5 p_0\dot{r}_1 + d_6 r_1\dot{p}_0] = 0,$$

$$L_2(r_1, p_1, p_0, r_2, p_2) \equiv \mu_1(\ddot{p}_1 + \alpha\dot{p}_1 + \sigma_1^2 p_1) + d_1 p_1(r_1\ddot{r}_1 + \dot{r}_1^2 + p_1\ddot{p}_1 + \dot{p}_1^2)$$
$$+d_2(r_1^2\ddot{p}_1 - r_1 p_1\ddot{r}_1 + 2r_1\dot{r}_1\dot{p}_1 - 2p_1\dot{r}_1^2) + d_3(r_2\ddot{r}_1 + p_2\ddot{p}_1 + \dot{r}_1\dot{r}_2 + \dot{p}_1\dot{p}_2)$$
$$-d_4(p_1\ddot{p}_2 + r_1\ddot{r}_2) + d_5(p_0\ddot{p}_1 + \dot{p}_0\dot{p}_1) + d_6 p_1\ddot{p}_0 + \alpha[(d_1 p_1^2 + d_2 r_1^2)\dot{p}_1$$
$$+(d_1 - d_2) r_1 p_1 \dot{r}_1 + d_3(p_2\dot{p}_1 + r_2\dot{r}_1) - d_4(p_1\dot{p}_2 + r_1\dot{r}_2) + d_5 p_0\dot{p}_1 + d_6 p_1\dot{p}_0] = 0,$$

$$L_3(r_1, p_1, p_0) \equiv \mu_0(\ddot{p}_0 + \alpha\dot{p}_0 + \sigma_0^2 p_0) + d_6(r_1\ddot{r}_1 + p_1\ddot{p}_1)$$
$$+d_8(\dot{r}_1^2 + \dot{p}_1^2) + \alpha d_6(r_1\dot{r}_1 + p_1\dot{p}_1) = 0, \tag{4.2.16}$$

$$L_4(r_1, p_1, r_2) \equiv \mu_2(\ddot{r}_2 + \alpha\dot{r}_2 + \sigma_2^2 r_2) - d_4(p_1\ddot{r}_1 + r_1\ddot{p}_1)$$
$$-2d_7\dot{r}_1\dot{p}_1 - \alpha d_4(p_1\dot{r}_1 + r_1\dot{p}_1) = 0,$$

$$L_5(r_1, p_1, p_2) \equiv \mu_2(\ddot{p}_2 + \alpha\dot{p}_2 + \sigma_2^2 p_2) + d_4(r_1\ddot{r}_1 - p_1\ddot{p}_1)$$
$$+d_7(\dot{r}_1^2 - \dot{p}_1^2) + \alpha d_4(r_1\dot{r}_1 - p_1\dot{p}_1) = 0.$$

In view of the close values of the friction coefficients $\alpha_0$, $\alpha_1$, and $\alpha_2$ for the corresponding shapes of natural sloshing modes in the generalized coordinates $p_0(t)$, $r_1(t)$, $p_1(t)$, $r_2(t)$, and $p_2(t)$, the damping ratio $\alpha$ in the equations of system (4.2.16) is represented by the principal resonance forms (in the generalized coordinates $r_1(t)$ and $p_1(t)$).

The presented five-mode nonlinear mathematical model describes the process of free sloshing for a viscous liquid in the vicinity of the main resonance in the form with generalized coordinates $r_1(t)$ and $p_1(t)$. It is obtained in the approximation of the theory of the third order of smallness and can be regarded as a generalization of the system of equations (4.1.15)–(4.1.19) to the case of ideal liquid.

## 4.3 Hydrodynamic coefficients for circular cylindrical tanks with nonflat bottoms

In the aerospace applications, the upright circular cylindrical tanks have, as a rule, nonflat bottoms in the form close to spherical segments. We now consider two procedures used to find the coefficients of the modal equations (4.1.15)–(4.1.19) taking into account the presence nonflat bottom and based on two methods aimed at the solution of the spectral boundary problem (2.2.4).

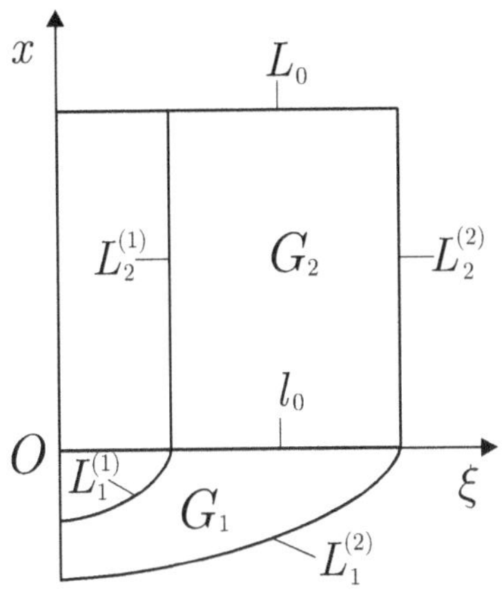


**Figure 4.2**

The *first method* is based on the decomposition of the domain proposed in [59]. Assume that the cylindrical part of the vessel is formed by two coaxial cylindrical surfaces with radii $R_0$ and $R_1$. In the cylindrical coordinate system $Ox\xi\eta$ with origin in the plane of conjugation of the subdomains $Q_1$ and $Q_2$ whose generatrices are shown in Fig. 4.2, problem (2.2.4) takes the form

$$\Delta\varphi^{(1)} = 0, \quad \boldsymbol{r} \in Q_1, \quad \frac{\partial\varphi^{(1)}}{\partial\nu} = 0, \quad \boldsymbol{r} \in S_1, \quad \frac{\partial\varphi^{(1)}}{\partial x} = \nu\varphi^{(1)}, \quad \boldsymbol{r} \in S_0, \tag{4.3.1}$$

$$\begin{gathered}\Delta\varphi^{(2)} = 0, \quad \boldsymbol{r} \in Q_2, \quad \frac{\partial\varphi^{(2)}}{\partial\nu} = \varkappa\varphi^{(2)}, \quad \boldsymbol{r} \in \Sigma_0, \quad \frac{\partial\varphi^{(2)}}{\partial\nu} = 0, \quad \boldsymbol{r} \in S_2, \\ \frac{\partial\varphi^{(2)}}{\partial x} + \nu\varphi^{(2)} = 0, \quad \boldsymbol{r} \in S_0,\end{gathered} \tag{4.3.2}$$

where $Q_1$ is the subdomain formed by the bottoms, $Q_2$ is the subdomain formed solely by the cylindrical surfaces, $\Sigma_0$ is the unperturbed free surface with generatrix $L_0$, $S_1$ is the surface of revolution with the generatrices $L_1^{(1)}$ and $L_1^{(2)}$, $S_2$ is a cylindrical surface with generatrices $L_2^{(1)}$ and $L_2^{(2)}$, and $S_0$ is a conjugating surface with generatrix $l_0$.

Note that, for the accuracy of the proposed method of conjugation, a small part of the cylindrical domain should be regarded as a part of the subdomain $Q_1$ [59].

First, we solve the eigenvalue problem (4.3.1) by using an approximate method (e.g., the variational method). If the parameter $\nu_n$ is known, then the solution of the spectral problem (4.3.2) in the cylindrical subdomain $Q_2$ can be represented in the analytic form by the method of separation of variables:

$$\varphi_1^{(2)} = \psi_{01}^{(2)}(x, \xi), \quad \varphi_2^{(2)} = \psi_{11}^{(2)} \sin\eta, \quad \varphi_3^{(2)} = \psi_{11}^{(2)}(x, \xi)\cos\eta,$$

$$\varphi_4^{(2)} = \psi_{21}^{(2)}(x,\,\xi)\sin 2\eta,\quad \varphi_5^{(2)} = \psi_{21}^{(2)}(x,\,\xi)\cos 2\eta,$$

where

$$\psi_{mn}^{(2)}(x,\xi) = \left[\cosh(k_{mn}x) + \frac{\nu_{mn}}{k_{mn}}\sinh(k_{mn}x)\right]\frac{Y_m(k_{mn}\xi)}{\cosh(k_{mn}h)},$$

$$\varkappa_{mn} = \frac{k_{mn}\tanh(k_{mn}h) + \nu_{mn}}{\frac{\nu_{mn}}{k_{mn}}\tanh(k_{mn}h) + 1}. \tag{4.3.3}$$

For $R_0 = 0$, the function $Y_m(k_{mn}\xi)$ is given by the formula

$$Y_m(k_{mn}) = \frac{J_m(k_{mn}\xi)}{J_m(k_{mn})}.$$

As above, the equation of the free surface can be represented in the form

$$x = h + \sum_{m,n}[r_{mn}\sin m\eta + p_{mn}\cos m\eta]f_{mn}(\xi), \tag{4.3.4}$$

where $h$ is the depth of liquid over the plane $S_0$. In this case, the velocity potential is given by (4.1.2), where the functions $\psi_{mn}(x,\xi)$, as indicated above, are equal to $\psi_{mn}^{(1)}(x,\xi)$ in the subdomain $Q_1$ and to $\psi_{mn}^{(2)}$ in the subdomain $Q_2$.

In the analyzed case, under the restrictions imposed earlier on the parameters $r_{mn}(t)$ and $p_{mn}(t)$, the modal equations remain true as long as the free surface $\Sigma_0$ is located in the cylindrical part of the vessel. In this case, they completely coincide with (4.1.15)–(4.1.19). Their coefficients are computed by using expressions (4.1.14) and (4.1.20) and the parameters appearing in the formulas can be found from the relations

$$a_1 = \pi\left[\bar\varkappa_{01} + \nu_{01}\left(1 + \frac{\bar\varkappa_{01}^2}{k_{01}^2}\right) + \nu_{01}^2\frac{\bar\varkappa_{01}}{k_{01}^2}\right]i_6,$$

$$a_2 = \frac{\pi}{2}\left[\bar\varkappa_{21} + \nu_{21}\left(1 + \frac{\bar\varkappa_{21}^2}{k_{21}^2}\right) + \nu_{21}^2\frac{\bar\varkappa_{21}}{k_{21}^2}\right]i_{12},$$

$$a_3 = \frac{\pi}{2}\left[\left(1 + \nu_{11}\frac{\bar\varkappa_{11}}{k_{11}^2} + \nu_{21}\frac{\bar\varkappa_{21}}{k_{21}^2} + \nu_{11}\nu_{21}\frac{\bar\varkappa_{11}}{k_{11}^2}\frac{\bar\varkappa_{21}}{k_{21}^2}\right)\frac{k_{21}^2}{2}\right.$$
$$\left.+\bar\varkappa_{11}\bar\varkappa_{21} + \nu_{11}\bar\varkappa_{21} + \nu_{21}\bar\varkappa_{11} + \nu_{11}\nu_{21}\right]i_5,$$

$$a_4 = \frac{\pi}{2}(\bar\varkappa_{01} + \nu_{01})i_2,\quad a_5 = \pi\left(1 + \nu_{11}\frac{\bar\varkappa_{11}}{k_{11}^2}\right)i_3,$$

$$a_6 = \frac{\pi}{8}(k_{11}^2 + \nu_{11}\bar\varkappa_{11})i_4,\quad a_7 = -\frac{\pi}{4}(\bar\varkappa_{21} + \nu_{21})i_5,$$

$$a_8 = 2\pi\left[\bar{\varkappa}_{11} + \nu_{11}\left(1 + \frac{\bar{\varkappa}_{11}^2}{k_{11}^2}\right) + \nu_{11}^2\frac{\bar{\varkappa}_{11}}{k_{11}^2}\right]i_{42},$$

$$a_{10} = \frac{a_8 i_3}{4 i_{12}}, \quad a_{11} = \frac{a_{10}k_{11}^2 i_4}{i_3}, \quad a_{12} = a_{11} - \frac{a_8}{2}, \tag{4.3.5}$$

$$a_{14} = \pi(\bar{\varkappa}_{11} + \nu_{11})i_2,$$

$$a_{15} = \pi\left[\frac{k_{01}^2}{2}\left(1 + \nu_{01}\frac{\bar{\varkappa}_{01}}{k_{01}^2} + \nu_{11}\frac{\bar{\varkappa}_{11}}{k_{11}^2} + \nu_{01}\nu_{11}\frac{\bar{\varkappa}_{01}}{k_{01}^2}\frac{\bar{\varkappa}_{11}}{k_{11}^2}\right)\right.$$

$$\left. + \bar{\varkappa}_{01}\bar{\varkappa}_{11} + \nu_{01}\bar{\varkappa}_{11} + \nu_{11}\bar{\varkappa}_{01} + \nu_{01}\nu_{11}\right]i_2,$$

$$a_{17} = 2\pi\left(1 + \nu_{01}\frac{\bar{\varkappa}_{01}}{k_{01}^2}\right)i_6, \quad a_{18} = \frac{\pi}{2}(\bar{\varkappa}_{11} + \nu_{11})i_5,$$

$$a_{19} = \frac{\pi}{4}\left[\left(1 + \nu_{11}\frac{\bar{\varkappa}_{11}}{k_{11}^2}\right)^2(i_{31} - i_{23}) + (\bar{\varkappa}_{11} + \nu_{11})^2 i_5\right],$$

$$a_{20} = \pi\left(1 + \nu_{21}\frac{\bar{\varkappa}_{21}}{k_{22}^2}\right)i_{12}, \quad a_{22} = \frac{\pi}{2}\left[k_{11}^2 + \bar{\varkappa}_{12}^2 + 4\nu_{11}\bar{\varkappa}_{11}\right.$$

$$\left. + \nu_{11}^2\left(1 + \frac{\bar{\varkappa}_{11}^2}{k_{11}^2}\right) - \frac{k_{01}^2}{2}\left(1 + \nu_{11}\frac{\bar{\varkappa}_{11}}{k_{11}^2}\right)^2\right]i_2,$$

$$\bar{\varkappa}_{mn} = k_{mn}\tanh(k_{mn}h) - \frac{(\varkappa_{mn} - \nu_{mn})k_{mn}^2}{k_{mn}^2 - \nu_{mn}\varkappa_{mn}}.$$

In the limiting case (i.e., for $\nu_{mn} = 0$), relations (4.3.5) completely coincide with the corresponding relations (4.1.10) for the cylindrical container with flat bottom.

For the upright circular cylinder with unit radius, the values of quadratures in (4.3.5) are as follows: $i_1 = 0.294989$, $i_2 = 0.144516$, $i_3 = 0.352506$, $i_4 = 0.290301$, $i_5 = 0.287931$, $i_6 = 0.5$, $i_{12} = 0.285600$, $i_{23} = 0.516950$, $i_{31} = 0.150062$, and $i_{42} = 0.153455$.

By using these numbers and the values of the parameters $\nu_{mn}$, one can easily find the hydrodynamic coefficients for any fixed value of $h$ according to relations (4.1.14), (4.1.20), and (4.3.5). In this case, the accuracy of calculations naturally depends on the accuracy of the solution of the eigenvalue problem (4.3.2). In [59], it is shown that, for the above-mentioned method of decomposition of the domain, the eigenvalues are computed with an accuracy to within four or five significant digits.

The *second method* used to find the hydrodynamic coefficients of nonlinear modal equations is based on the approximate solution of the spectral problem (2.2.4) by the variational (Trefftz) method [114]. We consider the case of a vessel with circular unperturbed free surface $\Sigma_0$. In the cylindrical coordinate system $x, \xi, \eta$, we represent the function $\varphi(x, \xi, \eta)$ in the form

$$\varphi(x, \xi, \eta) = \psi_m(x, \xi) \frac{\sin m\eta}{\cos m\eta}. \tag{4.3.6}$$

Then the solution of the boundary-value problem for the functions $\psi_m(x, \xi)$ can be reduced to the solution of the extremal variational problem for the functional

$$J(\psi_m) = \int_G \left[ \xi(\psi_{mx}^2 + \psi_{m\xi}^2) + \frac{m^2}{\xi}\psi_m^2 \right] dx d\xi - \varkappa \int_{L_0} \xi \psi_m^2 dS \tag{4.3.7}$$

in the class of admissible functions continuous in the closed domain $G$ of the meridional cross section of the vessel whose derivatives are piecewise continuous in this domain.

By the Trefftz method, the approximate solution $\psi_{mn}(x, \xi)$ of the variational problem for functional (4.3.7) has the form

$$\psi_{mn}(x, \xi) = \sum_{k,m}^{q} a_k^{(n)} w_k^{(m)}(x, \xi), \tag{4.3.8}$$

where $a_k^{(n)}$ are arbitrary constants obtained from the necessary condition of minimum for functional (4.3.7) and $w_k^{(m)}(x, \xi)$ are linearly independent solutions of the equation $L(\psi) = 0$.

The appropriate system of coordinate functions $w_k^{(m)}$ is given by the recurrence relations

$$\begin{gathered} \frac{\partial w_k^{(m)}}{\partial x} = (k-m) w_{k-1}^{(m)}, \quad \xi \frac{\partial w_k^{(m)}}{\partial \xi} = k w_k^{(m)} - (k-m) x\, w_{k-1}^{(m)}, \\ (k+m+1) w_{k+1}^{(m)} = (2k+1) x\, w_k^{(m)} - (k-m)(x^2+\xi^2) w_{k-1}^{(m)}, \\ (k+m+1)\xi w_k^{(m+1)} = 2(m+1)[(x^2+\xi^2) w_{k-1}^{(m)} - x\, w_k^{(m)}] \end{gathered} \tag{4.3.9}$$

starting from the first two functions of this family: $w_0^{(0)} = 1$ and $w_1^{(0)} = x$. Note that these functions are finite for $\xi = 0$.

To find the weight coefficients $a_k^{(n)}$, the variational principle yields the following spectral algebraic equations:

$$D(\varkappa_m) = |\alpha_{kj}^{(m)} - \varkappa_m \beta_{kj}^{(m)}| = 0, \tag{4.3.10}$$

where

$$\alpha_{kj}^{(m)} = \alpha_{jk}^{(m)} = \int_\Gamma \xi \frac{\partial w_k^{(m)}}{\partial \nu} w_j^{(m)} dS, \quad \beta_{kj}^{(m)} = \beta_{jk}^{(m)} = \int_{L_0} \xi w_k^{(m)} w_j^{(m)} dS. \tag{4.3.11}$$

By analogy with the case considered in the previous section, the velocity potential and the equation of the free surface can be represented in the form

$$\varphi(x,\xi,\eta,t) = \sum_{m,n} [R_{mn}(t)\sin m\eta + P_{mn}(t)\cos m\eta]\psi_{mn}(x,\xi) \tag{4.3.12}$$

and

$$x = f(\xi,\eta,t) = \sum_{m,n} [r_{mn}(t)\sin m\eta + p_{mn}(t)\cos m\eta]\varkappa_{mn}\psi_{mn}(0,\xi), \tag{4.3.13}$$

respectively.

Since our aim is to clarify the principal possibility of construction of the nonlinear modal equations and finding their coefficients in the case where the solution of the boundary-value problem (2.2.4) is found approximately and represented in the form (4.3.8), we restrict ourselves to the *five-dimensional* modal representation

$$\begin{aligned}\varphi(x,\xi,\eta,t) = P_0(t)\,\psi_0(x,\xi) + (R_1(t)\sin\eta + P_1(t)\cos\eta)\psi_1(x,\xi)\\ + (R_2(t)\sin 2\eta + P_2(t)\cos 2\eta)\psi_2(x,\xi),\end{aligned} \tag{4.3.14}$$

$$f(\xi,\eta,t) = r(t)\varkappa_{11}\psi_1(0,\xi)\sin\eta. \tag{4.3.15}$$

This means that we take into account the main harmonics $P_0$, $R_1$, $P_1$, $R_2$, and $P_2$ in the decomposition of the velocity potential for $m = 0, 1, 2$ and the main harmonic with parameter $r(t)$ in the equation of the perturbed free surface for $m = 1$. We perform our subsequent considerations with regard for the results obtained in Section 4.1, where we consider the exact solution of the boundary-value problem for the functions $\psi_{mn}(x,\xi)$. In particular, we take into account the fact that, for free sloshing in the form (4.3.15), the eigenmodes with parameters $P_1(t)$ and $R_2(t)$ are not excited in the velocity potential (4.3.14).

To construct the nonlinear sloshing theory by the variational method, we represent the approximate solution in the form of a power series

$$\psi_m = \sum_{k=m}^{q} a_k w_k^{(m)}(x,\xi) = \sum_{i=0}^{q-m} x^i b_i^{(m)}(\xi) \tag{4.3.16}$$

whose coefficients $b_i^{(m)}(\xi)$ are given by the formula

$$b_i^{(m)}(\xi) = a_{m+i}\xi^m + \sum_{k=1}^{(q-m)/2} (-1)^k a_{2k+m+1}\xi^{2k+m} \times \frac{\prod\limits_{n=1}^{k}(2k-2n+i+2)(2k-2n+i+1)}{\prod\limits_{n=1}^{k}[(2n+m)^2-m^2]}. \tag{4.3.17}$$

As shown in what follows, this form of solution significantly facilitates the construction of the algorithm of solution of the nonlinear problem.

To construct the nonlinear modal equations, we use the variational equation (2.1.22) with regard for the fact that the velocity potential $\varphi$ is a harmonic function and satisfies the boundary condition on the surface $S$ in a sense of convergence in mean. In view of relations (4.3.14) and (4.3.15) and the facts that $P_1(t) \equiv 0$, and $R_2(t) \equiv 0$, we find

$$\frac{\partial \varphi}{\partial t} = \dot{P}_0\psi_0 + \dot{R}_1\psi_1 \sin\eta + \dot{P}_2\psi_2 \cos 2\eta; \tag{4.3.18}$$

$$\begin{aligned}\tfrac{1}{2}(\nabla\varphi)^2 &= \tfrac{1}{2}P_0^2(\psi_{0x}^2 + \psi_{0\xi}^2) + \tfrac{1}{2}R_1^2\left(\sin^2\eta\psi_{1x}^2 + \sin^2\eta\psi_{1\xi}^2 + \xi^{-2}\psi_1^2\cos^2\eta\right)\\ &+\tfrac{1}{2}P_2^2\left(\psi_{2x}^2\cos^2 2\eta + \psi_{2\xi}^2\cos^2 2\eta + \frac{4}{\xi^2}\psi_2^2\sin^2 2\eta\right)\\ &+P_0R_1(\psi_{0x}\psi_{1x} + \psi_{0\xi}\psi_{1\xi})\sin\eta + P_0P_2(\psi_{0x}\psi_{2x} + \psi_{0\xi}\psi_{2\xi})\cos 2\eta\\ &+R_1P_2\Big(\psi_{1x}\psi_{2x}\sin\eta\cos 2\eta + \psi_{1\xi}\psi_{2\xi}\sin\eta\cos 2\eta\\ &\qquad -2\xi^{-2}\psi_1\psi_2\cos\eta\sin 2\eta\Big);\end{aligned} \tag{4.3.19}$$

$$\begin{aligned}&\left(\frac{\partial\varphi}{\partial x} - \frac{\partial\varphi}{\partial\xi}\frac{\partial f}{\partial\xi} - \frac{1}{\xi^2}\frac{\partial\varphi}{\partial\eta}\frac{\partial f}{\partial\eta}\right)\delta\varphi = \Big[P_0(\psi_0\psi_{0x} - f_\xi\psi_0\psi_{0\xi}) + R_1\left(\psi_0\psi_{1x}\sin\eta\right.\\ &\qquad \left.- f_\xi\psi_0\psi_{1\xi}\sin\eta - \xi^{-2}f_\eta\psi_0\psi_1\cos\eta\right)\\ &\qquad +P_2\left(\psi_0\psi_{2x}\cos 2\eta - f_\xi\psi_0\psi_{2\xi}\cos 2\eta + 2\xi^{-2}f_\eta\psi_0\psi_2\sin 2\eta\right)\Big]\delta P_0\\ &+\Big[P_0(\psi_{0x}\psi_1\sin\eta - f_\xi\psi_{0\xi}\psi_1\sin\eta) + R_1\Big(\psi_{1x}\psi_1\sin^2\eta - f_\xi\psi_{1\xi}\psi_1\sin^2\eta\\ &-\frac{1}{\xi^2}f_\eta\psi_1^2\sin\eta\cos\eta\Big) + P_2\Big(\psi_{2x}\psi_1\cos 2\eta\sin\eta - f_\xi\psi_1\psi_{2\xi}\cos 2\eta\sin\eta\\ &+\frac{2}{\xi^2}f_\eta\psi_1\psi_2\sin 2\eta\sin\eta\Big)\Big]\delta R_1 + \Big[P_0(\psi_{0x}\psi_2\cos 2\eta - f_\xi\psi_{0\xi}\psi_2\cos 2\eta)\\ &+R_1\left(\psi_{1x}\psi_2\sin\eta\cos 2\eta - f_\xi\psi_{1\xi}\sin\eta\cos 2\eta - \frac{1}{\xi^2}f_\eta\psi_1\psi_2\cos\eta\cos 2\eta\right)\\ &+P_2\left(\psi_{2x}\psi_2\cos^2 2\eta - f_\xi\psi_{2\xi}\psi_2\cos^2 2\eta + \frac{2}{\xi^2}f_\eta\psi_2^2\sin 2\eta\cos 2\eta\right)\Big]\delta P_2;\end{aligned} \tag{4.3.20}$$

$$f_t\delta\varphi = f_t\psi_0\delta P_0 + f_t\psi_1\sin\eta\delta R_1 + f_t\psi_2\cos 2\eta\delta P_2; \tag{4.3.21}$$

$$\delta f = \varkappa_{11}\psi_1(0,\xi)\sin\eta\delta r. \tag{4.3.22}$$

The functions $\psi_m$, $\psi_{mx}$, and $\psi_{m\xi}$ appearing in relations (4.3.18)–(4.3.22) and their products can be found from relation (4.3.16) as follows:

$$\psi_{mx} = \sum_{i=0}^{q-m} x^i c_i^{(m)}(\xi), \quad \psi_{m\xi} = \sum_{i=0}^{q-m} x^i d_i^{(m)}(\xi),$$

$$\psi_m \psi_n = \sum_{i=0}^{2(q-m)} x^i b_i^{(m,n)}(\xi), \quad \psi_{mx}\psi_{nx} = \sum_{i=0}^{2(q-m)} x^i c_i^{(m,n)}(\xi),$$

$$\psi_{m\xi}\psi_{n\xi} = \sum_{i=0}^{2(q-m)} x^i d_i^{(m,n)}(\xi), \quad \psi_m \psi_{nx} = \sum_{i=0}^{2(q-m)} x^i e_i^{(m,n)}(\xi), \tag{4.3.23}$$

$$\psi_m \psi_{n\xi} = \sum_{i=0}^{2(q-m)} x^i h_i^{(m,n)}(\xi), \quad \psi_{mx}\psi_{n\xi} = \sum_{i=0}^{2(q-m)} x^i l_i^{(m,n)}(\xi),$$

where

$$c_i^{(m)}(\xi) = (i+1) b_{i+1}^{(m)}(\xi), \quad d_i^{(m)}(\xi) = b_{i\xi}^{(m)}(\xi), \tag{4.3.24}$$

$$b_i^{(m,n)} = \sum_{p=0}^{i} b_p^{(m)} b_{i-p}^{(n)} \ (i \leqslant q-m), \quad b_i^{(m,n)} = \sum_{p=i-q+m}^{q-m} b_p^{(m)} b_{i-p}^{(n)}, \quad i > q-m.$$

Substituting relations (4.3.18)–(4.3.22) together with (4.3.23) in relation (2.1.22), integrating over the surface $\Sigma_0$, and restricting ourselves to the terms of the third order of smallness in the parameters of the motion, we arrive at the following system of nonlinear modal equations:

$$a_1 \dot{P}_0 r + a_2 \dot{R}_1 + a_3 \dot{R}_1 r^2 + a_4 \dot{P}_2 r + a_5 P_0^2 r + a_6 R_1^2 r + a_7 P_2^2 r$$
$$+ a_8 P_0 R_1 + a_9 P_0 P_2 r + a_{10} R_1 P_2 + \mathrm{g} a_{11} r = 0; \tag{4.3.25}$$

$$b_1 P_0 + b_2 P_0 r^2 + b_3 R_1 r + b_4 P_2 r^2 - b_5 \dot{r} r = 0; \tag{4.3.26}$$

$$c_1 P_0 r + c_2 R_1 + c_3 R_1 r^2 + c_4 P_2 r - c_5 \dot{r} - c_6 \dot{r} r^2 = 0; \tag{4.3.27}$$

$$d_1 P_2 + d_2 P_2 r^2 + d_3 P_0 r^2 + d_4 R_1 r - d_5 \dot{r} r = 0, \tag{4.3.28}$$

where

$$a_1 = \pi \varkappa_{11}^2 \int_0^{r_0} \xi \psi_1^2 b_1^{(0)} d\xi, \quad a_2 = \pi \varkappa_{11} \int_0^{r_0} \xi \psi_1 b_0^{(1)} d\xi,$$

$$a_3 = \frac{3}{4} \pi \varkappa_{11}^3 \int_0^{r_0} \xi \psi_1^3 b_2^{(1)} d\xi, \quad a_4 = -\frac{\pi}{2} \varkappa_{11}^2 \int_0^{r_0} \xi \psi_1^2 b_1^{(2)} d\xi,$$

$$a_5 = \frac{\pi}{2} \varkappa_{11}^2 \int_0^{r_0} \xi \psi_1^2 [c_1^{(0,0)} + d_1^{(0,0)}] d\xi,$$

$$a_6 = \frac{3}{8}\pi\varkappa_{11}^2 \int_0^{r_0} \xi\psi_1^2[c_1^{(1,1)} + d_1^{(1,1)}]d\xi + \frac{1}{8}\pi\varkappa_{11}^2 \int_0^{r_0} \frac{1}{\xi}\psi_1^2 b_1^{(1,1)} d\xi,$$

$$a_7 = \frac{\pi}{4}\varkappa_{11}^2 \int_0^{r_0} \xi\psi_1^2[c_1^{(2,2)} + d_1^{(2,2)}]d\xi + \pi\varkappa_{11}^2 \int_0^{r_0} \frac{1}{\xi}\psi_1^2 b_1^{(2,2)} d\xi,$$

$$a_8 = \pi\varkappa_{11} \int_0^{r_0} \xi\psi_1[c_0^{(0,1)} + d_0^{(0,1)}]d\xi,$$

$$a_9 = -\frac{\pi}{2}\varkappa_{11}^2 \int_0^{r_0} \xi\psi_1^2[c_1^{(0,2)} + d_1^{(0,2)}]d\xi,$$

$$a_{10} = -\frac{\pi}{2}\varkappa_{11} \int_0^{r_0} \xi\psi_1[c_0^{(1,2)} + d_0^{(1,2)}]d\xi - \pi\varkappa_{11} \int_0^{r_0} \frac{1}{\xi}\psi_1 b_0^{(1,2)} d\xi,$$

$$a_{11} = \pi\varkappa_{11}^2 \int_0^{r_0} \xi\psi_1^2 d\xi, \quad b_1 = 2\pi \int_0^{r_0} \xi e_0^{(0,0)} d\xi,$$

$$b_2 = \pi\varkappa_{11}^2 \int_0^{r_0} \xi\psi_1^2 e_2^{(0,0)} d\xi - \pi\varkappa_{11}^2 \int_0^{r_0} \xi\psi_1\psi_{1\xi} h_1^{(0,0)} d\xi, \tag{4.3.29}$$

$$b_3 = \pi\varkappa_{11} \int_0^{r_0} \xi\psi_1 e_1^{(0,1)} d\xi - \pi\varkappa_{11} \int_0^{r_0} \xi\psi_{1\xi} h_0^{(0,1)} d\xi - \pi\varkappa_{11} \int_0^{r_0} \frac{1}{\xi}\psi_1 b_0^{(0,1)} d\xi,$$

$$b_4 = -\frac{\pi}{2}\varkappa_{11}^2 \int_0^{r_0} \xi\psi_1^2 e_2^{(0,2)} d\xi + \frac{\pi}{2}\varkappa_{11}^2 \int_0^{r_0} \xi\psi_1\psi_{1\xi} h_0^{(0,2)} d\xi + \pi\varkappa_{11}^2 \int_0^{r_0} \frac{1}{\xi}\psi_1^2 b_1^{(0,2)} d\xi,$$

$$b_5 = a_1, \quad c_5 = a_2, \quad c_6 = a_3,$$

$$c_1 = \pi\varkappa_{11} \int_0^{r_0} \xi\psi_1 e_1^{(1,0)} d\xi - \pi\varkappa_{11} \int_0^{r_0} \xi\psi_1 h_0^{(1,0)} d\xi, \quad c_2 = \pi \int_0^{r_0} \xi e_0^{(1,1)} d\xi,$$

$$c_3 = -\frac{3}{4}\pi\varkappa_{11}^2 \int_0^{r_0} \xi\psi_1\psi_{1\xi} h_1^{(1,1)} d\xi - \frac{\pi}{4}\varkappa_{11}^2 \int_0^{r_0} \frac{1}{\xi}\psi_1^2 b_1^{(1,1)} d\xi + \frac{3}{4}\pi\varkappa_{11}^2 \int_0^{r_0} \xi\psi_1^2 e_2^{(1,1)} d\xi,$$

$$c_4 = -\frac{\pi}{2}\varkappa_{11} \int_0^{r_0} \xi\psi_1 e_1^{(1,2)} d\xi + \frac{\pi}{2}\varkappa_{11} \int_0^{r_0} \xi\psi_{1\xi} h_0^{(1,2)} d\xi + \pi\varkappa_{11} \int_0^{r_0} \frac{1}{\xi}\psi_1 b_0^{(1,2)} d\xi,$$

$$d_1 = \pi \int_0^{r_0} \xi e_0^{(2,2)} d\xi,$$

$$d_2 = \frac{\pi}{2}\varkappa_{11}^2 \int_0^{r_0} \xi \psi_1^2 e_2^{(2,2)} d\xi - \frac{\pi}{2}\varkappa_{11}^2 \int_0^{r_0} \xi \psi_{1\xi}\psi_1 h_1^{(2,2)} d\xi,$$

$$d_3 = -\frac{\pi}{2}\varkappa_{11}^2 \int_0^{r_0} \xi \psi_1^2 e_2^{(2,0)} d\xi + \frac{\pi}{2}\varkappa_{11}^2 \int_0^{r_0} \xi \psi_{1\xi}\psi_1 h_1^{(2,0)} d\xi,$$

$$d_4 = -\frac{\pi}{2}\varkappa_{11} \int_0^{r_0} \xi \psi_1 e_1^{(2,1)} d\xi + \frac{\pi}{2}\varkappa_{11} \int_0^{r_0} \xi \psi_{1\xi} h_0^{(2,1)} d\xi - -\frac{\pi}{2}\varkappa_{11} \int_0^{r_0} \frac{1}{\xi}\psi_1 b_0^{(1,2)} d\xi,$$

$$d_5 = a_4.$$

Comparing system (4.3.25)–(4.3.28) with a similar system obtained from (2.2.5), we get the following relations between the coefficients of these equations:

$$a_5 = b_2, \quad a_6 = c_3, \quad a_7 = d_2, \quad a_8 = c_1 = b_3,$$
$$a_9 = 2b_4 = 2d_3, \quad a_{10} = d_4 = c_4. \tag{4.3.30}$$

We now reduce the modal equations (4.3.25)–(4.3.28) to a single nonlinear ordinary differential equation with respect to the parameter $r(t)$. By using (4.3.26)–(4.3.28), we represent the parameters $P_0$, $R_1$, and $P_2$ via the parameter $r$ and substitute the resulting expressions in (4.3.25) with regard for (4.3.30). This yields

$$P_0 = C_0 r\dot{r}, \quad R_1 = B_4\dot{r} - C_2\dot{r}r^2, \quad P_2 = D_1 r\dot{r}; \tag{4.3.31}$$

$$\ddot{r} + \sigma^2 r + \alpha(\ddot{r}r^2 + r\dot{r}^2) = 0, \tag{4.3.32}$$

where

$$\sigma^2 = g\varkappa_{11}, \quad C_0 = \frac{b_5 - b_3 B_4}{b_1}, \quad D_1 = \frac{d_5 - d_4 B_4}{d_1},$$

$$C_2 = \frac{a_8 a_1}{c_2 b_1} + \frac{a_{10} a_4}{c_2 d_1} + \frac{a_2 a_6}{c_2^2} - \frac{a_8^2 a_2}{c_2^2 b_1} - \frac{a_{10}^2 a_2}{c_2^2 d_1} - \frac{a_3}{c_2}, \tag{4.3.33}$$

$$B_4 = \frac{c_5}{c_2}, \quad \alpha = \frac{a_1 C_0 - a_2 C_2 + a_3 B_4 + a_4 D_1}{a_2 B_4}.$$

The numerical realization of the analyzed method was carried out for a cylindrical vessel with plane bottom, the depth of filling $h = 1.0$, and the radius of the unperturbed free surface $R_1 = 1.0$. The eigenfunctions of the boundary problem (2.2.4) corresponding to the eigenvalues $\varkappa_{01}^{(11)} = 3.828117$, $\varkappa_{11}^{(9)} = 1.750798$, and $\varkappa_{21}^{(10)} = 3.040686$ take the following form [112]:

$$\psi_0^{(11)} = 0.261284 w_0^{(0)} + 0.916604 w_1^{(0)} + 2.442378 w_2^{(0)} + 2.334299 w_3^{(0)}$$

$$+1.773754w_4^{(0)} + 1.114694w_5^{(0)} + 0.587500w_6^{(0)} + 0.261036w_7^{(0)}$$

$$+0.0955233w_8^{(0)} + 0.0275600w_9^{(0)} + 0.560905 \cdot 10^{-2}w_{10}^{(0)}$$

$$+0.628392 \cdot 10^{-3}w_{11}^{(0)}; \tag{4.3.34}$$

$$\psi_1^{(9)} = 0.668167w_1^{(1)} + 1.169622w_2^{(1)} + 1.131640w_3^{(1)} + 0.658799w_4^{(1)}$$

$$+0.316405w_5^{(1)} + 0.107734w_6^{(1)} + 0.0322298w_7^{(1)} + 0.636571 \cdot 10^{-2}w_8^{(1)}$$

$$+0.885404 \cdot 10^{-3}w_9^{(1)}; \tag{4.3.35}$$

$$\psi_2^{(10)} = 0.652620w_2^{(2)} + 1.978467w_3^{(2)} + 3.011858w_4^{(2)} + 2.995598w_5^{(2)}$$

$$+2.199189w_6^{(2)} + 1.211926w_7^{(2)} + 0.504293w_8^{(2)}$$

$$+0.142126w_9^{(2)} + 0.0221781w_{10}^{(2)}. \tag{4.3.36}$$

The coefficients of the nonlinear equations (4.3.25)–(4.3.28) obtained by using (4.3.30) take the following values:

$$a_1 = -0.0999626, \quad a_2 = 0.345843, \quad a_3 = 0.197965, \quad a_4 = -0.205088,$$

$$a_5 = 0.551864, \quad a_6 = 0.527906, \quad a_7 = 0.836537, \quad a_8 = -0.209445,$$

$$a_9 = 0.546259, \quad a_{10} = -0.384791, \quad a_{11} = 0.605502, \quad c_2 = 0.345843,$$

$$b_1 = 0.133063, \quad \text{and} \quad d_1 = 0.202979.$$

It follows from relations (4.3.33) that

$$C_0 = 0.822783, \quad C_2 = -0.529295, \quad D_1 = 0.885328,$$

$$B_4 = 1.0, \quad \text{and} \quad \alpha = 0.338883.$$

The invariant quantity $\alpha/a_2B_4 = 0.9799$ is independent of the normalization of solutions of (2.2.4). If the exact solution of the boundary-value problem (2.2.4) is used, then the indicated characteristic value is $\alpha/a_2B_4 = 0.9797$. Thus, the approximate variational method for the solution of the nonlinear sloshing problem is also quite efficient in the cases where the approximate solutions of the spectral boundary problem (2.2.4) are constructed as shown above.

## 4.4 Hydrodynamic coefficients for the upright cylindrical tank of elliptic cross section

As an example of methodical character illustrating the idea of application of the variational method to the solution of the problem of nonlinear sloshing of liquid in vessels with arbitrary cross sections, we consider the solution of this problem in the case of a cylindrical vessel with elliptic cross section [187]. As above, we associate the systems of functions $f_i(y, z)$ and $\varphi_n(x, y, z)$ with the natural sloshing modes (2.2.4) by setting

$$f_i(y, z) = \varkappa_i \varphi_i(h, y, z)$$

(the coordinate system $Oxyz$ is connected with the bottom of the vessel). We now represent the function $\varphi(x, y, z)$ in the form $\varphi(x, y, z) = X(x)\chi(y, z)$ and apply the method of separation of variables to problem (2.2.4). Then the functions $X(x)$ and $\chi(y, z)$ can be found from the equations

$$X'' - k^2 X = 0, \quad X'(0) = 0, \quad X'(h) = \varkappa X(h); \tag{4.4.1}$$

$$\frac{\partial^2 \chi}{\partial y^2} + \frac{\partial^2 \chi}{\partial z^2} + k^2 \chi = 0, \quad y, z \in \Sigma_0, \quad \left.\frac{\partial \chi}{\partial \nu}\right|_C = 0, \tag{4.4.2}$$

where $C$ is the contour of the mean free surface $\Sigma_0$ and $k^2$ is a constant of separation.

There exist countable sets of eigenvalues $k_n$ and nonzero eigenfunctions $\chi_n(y, z)$, orthogonal together with a constant in the domain $\Sigma_0$ and satisfying Eqs. (4.4.2) to within a constant factor. Various types of normalizations used in this problem in connection with the study of water waves are discussed in the works devoted to the linear theory [149, 59]. In the present section, as above, we use the normalization in which the generalized coordinate $\beta_i(t)$ is the $z$-coordinate of the free surface for the $i$th form at a fixed point of the contour $C$.

If $k_n$ is known, then the solution of problem (4.4.1) takes the form

$$X_n(x) = \frac{\cosh(k_n x)}{k_n \sinh(k_n h)}, \quad \varkappa_n = k_n \tanh k_n h. \tag{4.4.3}$$

Hence,

$$\varphi_n(x, y, z) = \frac{\cosh(k_n x)}{k_n \sinh(k_n h)} \chi_n(y, z); \quad f_n(y, z) = \varkappa_n \varphi_n(h, x, z) = \chi_n(y, z). \tag{4.4.4}$$

We now introduce new coordinates $\xi$ and $\eta$ by the formulas

$$y = c \cosh \xi \cos \eta, \quad z = c \sinh \xi \sin \eta. \tag{4.4.5}$$

Then the function $\chi_n(\xi, \eta)$, which is a solution of the Helmholtz equation, can be found in the form of a product of Mathieu functions. However, this procedure is not suitable from the practical viewpoint due to serious difficulties encountered in computations. Therefore, we prefer to use one of the approximate methods used for the solution of problem (4.4.2) and described, e.g., in [124].

In constructing the modal equations, we assume that the natural sloshing modes $\chi_n(y, z)$ and the spectral parameters $k_n$ are known.

In what follows, we introduce two generalized coordinates $\beta_1(t) = r(t)$ and $\beta_2(t) = p(t)$ in the expansion of the free-surface elevations and five functions $R_n(t)$ in the expansion of the velocity potential. All necessary calculations are performed within the framework of the third-order theory under the assumption that $r(t)$ and $p(t)$ have the same order of smallness.

In the systems of equations (2.2.8) and (2.2.9), we find the quantities $A_n$, $A_{nk}$, and $l_k$ depending on the parameters $r(t)$ and $p(t)$ by virtue of the representation

$$x = h + f(y, z, t) = h + r(t) f_1(y, z) + p(t) f_2(y, z). \tag{4.4.6}$$

Thus, we have

$$A_n = \rho \int\limits_Q \varphi_n dQ = \rho \int\limits_{\Sigma_0} f_n(y, z) \int\limits_0^{h+f} \frac{\cosh k_n x}{k_n \sinh k_n h} dx dS$$

$$= \frac{\rho}{k_n^2 \sinh k_n h} \int\limits_{\Sigma_0} f_n(y, z) \sinh k_n(h + f) dS. \tag{4.4.7}$$

To within the terms of the third order of smallness in $f$, relation (4.4.7) takes the form

$$A_n = \rho \int\limits_{\Sigma_0} \left( \frac{1}{k_n^2} + \frac{1}{\varkappa_n} f + \frac{1}{2} f^2 + \frac{k_n^2}{6\varkappa_n} f^3 \right) f_n dS. \tag{4.4.8}$$

In view of representation (4.4.6), we transform relation (4.4.8) as follows:

$$A_n = a_{1n} r + a_{2n} p + a_{3n} r^2 + a_{4n} rp + a_{5n} p^2 + a_{6n} r^3$$
$$+ a_{7n} r^2 p + a_{8n} rp^2 + a_{9n} p^3. \tag{4.4.9}$$

Similarly, we find

$$A_{nk} = a_{0nk} + a_{1nk} r + a_{2nk} p + a_{3nk} r^2 + a_{4nk} rp + a_{5nk} p^2; \tag{4.4.10}$$

$$l_1 = l_{10} + l_{11} r^2 + l_{12} p^2. \tag{4.4.11}$$

In relations (4.4.10)–(4.4.11), we have used the notation

$$a_{1n} = \frac{\rho}{\varkappa_n} \int\limits_{\Sigma_0} f_1 f_n dS, \quad a_{2n} = \frac{\rho}{\varkappa_n} \int\limits_{\Sigma_0} f_2 f_n dS, \quad a_{3n} = \frac{\rho}{2} \int\limits_{\Sigma_0} f_1^2 f_n dS,$$

$$a_{4n} = \rho \int\limits_{\Sigma_0} f_1 f_2 f_n dS, \quad a_{5n} = \frac{\rho}{2} \int\limits_{\Sigma_0} f_2^2 f_n dS, \quad a_{6n} = \frac{\rho k_n^2}{6\varkappa_n} \int\limits_{\Sigma_0} f_1^3 f_n dS,$$

$$a_{7n} = \frac{\rho k_n^2}{2\varkappa_n} \int\limits_{\Sigma_0} f_1^2 f_2 f_n dS, \quad a_{8n} = \frac{\rho k_n^2}{2\varkappa_n} \int\limits_{\Sigma_0} f_1 f_2^2 f_n dS, \quad a_{9n} = \frac{\rho k_n^2}{6\varkappa_n} \int\limits_{\Sigma_0} f_2^3 f_n dS,$$

$$l_{10} = \frac{\rho h^2}{2} \int_{\Sigma_0} f_0^2 dS, \; l_{11} = \frac{\rho}{2} \int_{\Sigma_0} f_1^2 dS, \quad l_{12} = \frac{\rho}{2} \int_{\Sigma_0} f_2^2 dS,$$

$$a_{0nk} = \frac{\rho}{2} \int_{\Sigma_0} \left\{ \left( \frac{1}{\varkappa_n} - \frac{h}{\sinh^2 k_n h} \right) f_n^2 + \frac{1}{k_n^2} \left( \frac{1}{\varkappa_n} + \frac{h}{\sinh^2 k_n h} \right) \right.$$

$$\left. \times \left[ \left( \frac{\partial f_n}{\partial y} \right)^2 + \left( \frac{\partial f_n}{\partial z} \right)^2 \right] \right\} dS = \frac{\rho}{\varkappa_n} \int_{\Sigma_0} f_n^2 dS,$$

$$a_{1nk} = \rho \int_{\Sigma_0} \left[ f_1 f_n f_k + \frac{1}{\varkappa_n \varkappa_k} f_1 \left( \frac{\partial f_n}{\partial y} \frac{\partial f_k}{\partial y} + \frac{\partial f_n}{\partial z} \frac{\partial f_k}{\partial z} \right) \right] dS, \tag{4.4.12}$$

$$a_{2nk} = \rho \int_{\Sigma_0} \left[ f_2 f_n f_k + \frac{1}{\varkappa_n \varkappa_k} f_2 \left( \frac{\partial f_n}{\partial y} \frac{\partial f_k}{\partial y} + \frac{\partial f_n}{\partial z} \frac{\partial f_k}{\partial z} \right) \right] dS,$$

$$a_{3nk} = \frac{\rho}{2} \int_{\Sigma_0} \left[ \left( \frac{k_n^2}{\varkappa_n} + \frac{k_k^2}{\varkappa_k} \right) f_1^2 f_n f_k \right.$$

$$\left. + \left( \frac{1}{\varkappa_n} + \frac{1}{\varkappa_k} \right) f_1^2 \left( \frac{\partial f_n}{\partial y} \frac{\partial f_k}{\partial y} + \frac{\partial f_n}{\partial z} \frac{\partial f_k}{\partial z} \right) \right] dS,$$

$$a_{4nk} = \rho \int_{\Sigma_0} \left[ \left( \frac{k_n^2}{\varkappa_n} + \frac{k_k^2}{\varkappa_k} \right) f_1 f_2 f_n f_k \right.$$

$$\left. + \left( \frac{1}{\varkappa_n} + \frac{1}{\varkappa_k} \right) f_1 f_2 \left( \frac{\partial f_n}{\partial y} \frac{\partial f_k}{\partial y} + \frac{\partial f_n}{\partial z} \frac{\partial f_k}{\partial z} \right) \right] dS,$$

$$a_{5nk} = \frac{\rho}{2} \int_{\Sigma_0} \left[ \left( \frac{k_n^2}{\varkappa_n} + \frac{k_k^2}{\varkappa_k} \right) f_2^2 f_n f_k \right.$$

$$\left. + \left( \frac{1}{\varkappa_n} + \frac{1}{\varkappa_k} \right) f_2^2 \left( \frac{\partial f_n}{\partial y} \frac{\partial f_k}{\partial y} + \frac{\partial f_n}{\partial z} \frac{\partial f_k}{\partial z} \right) \right] dS.$$

We now exclude the parameters $R_k(t)$ from the system of equations (2.2.8). In this case, they take the following form:

$$R_1 = \dot{r} + R_{12} + R_{13}, \quad R_2 = \dot{p} + R_{22} + R_{23}, \quad R_k = R_{k2} + R_{k3}, \tag{4.4.13}$$

$$k = 3, 4, 5,$$

where

$$R_{n2} = (b_{1n} r + b_{2n} p)\dot{r} + (b_{3n} r + b_{4n} p)\dot{p},$$

$$R_{n3} = (c_{1n} r^2 + c_{2n} rp + c_{3n} p^2)\dot{r} + (c_{4n} r^2 + c_{5n} rp + c_{6n} p^2)\dot{p}. \tag{4.4.14}$$

The constants appearing in (4.4.14) are determined via (4.4.12) as follows:

$$b_{1n} = \frac{1}{a_{1nn}}(2a_{3n} - a_{1n1}), \quad b_{2n} = \frac{1}{a_{0nn}}(a_{4n} - a_{2n1}), \quad b_{3n} = \frac{1}{a_{0nn}}(a_{4n} - a_{1n2}),$$

$$b_{4n} = \frac{1}{a_{0nn}}(2a_{5n} - a_{2n2}), \quad c_{1n} = \frac{1}{a_{0nn}}\left(3a_{6n} - a_{3n1} - \sum_{k=1}^{5} a_{1nk}b_{1k}\right),$$

$$c_{2n} = \frac{1}{a_{0nn}}\left[2a_{7n} - a_{4n1} - \sum_{k=1}^{5}(a_{1nk}b_{2k} + a_{2nk}b_{1k})\right],$$

$$c_{3n} = \frac{1}{a_{0nn}}\left(a_{8n} - a_{5n1} - \sum_{k=1}^{5} a_{2nk}b_{2k}\right), \tag{4.4.15}$$

$$c_{4n} = \frac{1}{a_{0nn}}\left(a_{7n} - a_{3n2} - \sum_{k=1}^{5} a_{1nk}b_{3k}\right),$$

$$c_{5n} = \frac{1}{a_{0nn}}\left[2a_{8n} - a_{3n2} - \sum_{k=1}^{5}(a_{1nk}b_{4k} + a_{2nk}b_{3k})\right],$$

$$c_{6n} = \frac{1}{a_{0nn}}\left(3a_{9n} - a_{5n2} - \sum_{k=1}^{5} a_{2nk}b_{4k}\right).$$

These formulas are true for the cylindrical vessels with arbitrary cross sections. By the properties of orthogonality and symmetry of the eigenfunctions of problem (4.4.2) for the domain with elliptic cross section, the general relations (4.4.9) and (4.4.10) are simplified, namely,

$$\begin{gathered}
A_1 = a_{11}r + a_{61}r^3 + a_{81}rp^2, \quad A_2 = a_{22}p + a_{72}r^2p + a_{92}p^3, \\
A_3 = a_{33}r^2 + a_{53}p^2, \quad A_4 = a_{44}rp, \quad A_5 = a_{35}r^2 + a_{55}p^2, \\
A_{11} = a_{011} + a_{311}r^2 + a_{511}p^2, \quad A_{12} = a_{412}rp, \quad A_{13} = a_{113}r, \\
A_{14} = a_{214}p, \quad A_{15} = a_{115}r, \quad A_{22} = a_{022} + a_{322}r^2 + a_{522}p^2, \quad A_{23} = a_{223}p, \\
A_{24} = a_{124}r, \quad A_{25} = a_{225}p, \quad A_{33} = a_{033} + a_{333}r^2 + a_{533}p^2, \quad A_{34} = a_{434}rp, \\
A_{35} = a_{335}r^2 + a_{535}p^2, \quad A_{44} = a_{044} + a_{344}r^2 + a_{544}p^2, \\
A_{45} = 0, \quad A_{55} = a_{055} + a_{355}r^2 + a_{555}p^2.
\end{gathered} \tag{4.4.16}$$

For the same reason, relations (4.4.13) are also simplified:

$$\begin{gathered}
R_1 = \dot{r} + (c_{11}r^2 + c_{31}p^2)\dot{r} + c_{51}rp\dot{p}, \quad R_2 = \dot{p} + c_{22}rp\dot{r} + (c_{42}r^2 + c_{62}p^2)\dot{p}, \\
R_3 = b_{13}r\dot{r} + b_{43}p\dot{p}, \quad R_4 = b_{24}p\dot{r} + b_{34}r\dot{p}, \quad R_5 = b_{15}r\dot{r} + b_{45}p\dot{p}.
\end{gathered} \tag{4.4.17}$$

Excluding the parameters $R_k(t)$ from system (4.4.8) according to (4.4.17), we get the following system of nonlinear ordinary differential (modal) equations in the generalized coordinates $r(t)$ and $p(t)$:

$$\begin{aligned}
&\mu_1(\ddot{r}+\sigma_1^2 r)+d_1(r^2\ddot{r}+r\dot{r}^2)+d_2(p^2\ddot{r}+2p\dot{r}\dot{p})+d_3 rp\ddot{p}+d_4 r\dot{p}^2=0,\\
&\mu_2(\ddot{p}+\sigma_2^2 p)+d_5(p^2\ddot{p}+p\dot{p}^2)+d_6(r^2\ddot{p}+2r\dot{r}\dot{p})+d_3 rp\ddot{r}+d_7 p\dot{r}^2=0,
\end{aligned} \tag{4.4.18}$$

where

$$\begin{gathered}
\mu_1=a_{11},\quad \mu_2=a_{22},\quad \sigma_1^2=\mathrm{g}\varkappa_1,\quad \sigma_2^2=\mathrm{g}\varkappa_2,\\
d_1=3a_{61}+a_{11}c_{11}+2a_{33}b_{13}+2a_{35}b_{15},\quad d_2=a_{81}+a_{11}c_{31}+a_{44}b_{24},\\
d_3=2a_{72}+a_{11}c_{51}+2a_{33}b_{43}+a_{44}b_{34}+2a_{35}b_{45},\\
d_4=a_{322}+a_{124}b_{34}+a_{11}c_{51}+2a_{33}b_{43}+2a_{35}b_{45},\\
d_5=3a_{92}+a_{22}c_{62}+2a_{53}b_{43}+2a_{55}b_{45},\\
d_6=a_{72}+a_{22}c_{42}+a_{44}b_{34},\\
d_7=a_{511}+a_{214}b_{24}+a_{22}c_{22}+2a_{53}b_{13}+2a_{55}b_{15}.
\end{gathered} \tag{4.4.19}$$

This system approximately describes the process of free sloshing of liquid in the vicinity of the basic frequency for these coordinates.

The system of equations (4.4.18) can be easily generalized to the case of the problem of forced liquid sloshing in a vessel participating in translational displacements. The required quantities $\partial l_i/\partial r$ and $\partial l_i/\partial p$, $i=1,2,3$, are computed by using relations (2.6.27).

The numerical realization of the presented algorithm is based on the approximate method for the solution of problem (4.4.2) proposed in [124]. Its main idea is to consider an auxiliary problem for the Helmholtz equation with a parameter in the boundary condition, which can be approximately solved by the method of least squares.

## 4.5 Stokes–Zhukovsky potentials for coaxial cylindrical tanks

We now determine the Stokes–Zhukovsky potentials $\Omega_i(x,y,z,t)$ for the rotational motion of a vessel in the form of two coaxial cylinders with plane bottom partially filled with liquid. We consider the case where the origin of coordinates $Ox\xi\eta$ can be located at any point on the symmetry axis of the vessel. By $x_0$, we denote the coordinate of the center of its bottom . Then the equation of perturbed free surface takes the form

$$\zeta(x,y,z,t)\equiv x-x_0-h-f(y,z,t)=0. \tag{4.5.1}$$

In view of (4.5.1), the unit vector of normal to the free surface $\Sigma$ takes the form

$$\boldsymbol{\nu} = \left\{ \frac{1}{N}, \ -\frac{f_y}{N}, \ -\frac{f_z}{N} \right\}. \tag{4.5.2}$$

We now present the boundary-value problem for the function $\Omega_1(x, \xi, \eta, t)$ in the cylindrical coordinate system $(x, \xi, \eta)$ as follows:

$$\Delta\Omega_1 = 0, \ \ \boldsymbol{r} \in Q; \quad \frac{\partial \Omega_1}{\partial \nu}\bigg|_{\Sigma} = -\frac{f_\eta}{N}; \ \ \frac{\partial \Omega_1}{\partial \nu}\bigg|_{S} = 0, \tag{4.5.3}$$

where $f(\xi, \eta, t)$ is given by (4.3.13). This representation resembles the sloshing problem for an immobile vessel. In this case, it is reasonable to use the system of harmonic functions appearing in expansion (4.3.14) as the system of coordinate functions. Thus, we seek $\Omega_1$ in the form

$$\Omega_1 = \sum_{m,n} [R_{mn}^{(1)}(t) \sin m\eta + P_{mn}^{(1)}(t) \cos m\eta] \frac{\cosh k_{mn}(x - x_0)}{\cosh k_{mn} h} Y_m(k_{mn}\xi). \tag{4.5.4}$$

As above, in the construction of an approximate solution (4.5.4), we restrict ourselves to the terms of the third order of smallness in $r_1(t)$ and $p_1(t)$. In this case, $p_0(t)$, $r_2(t)$, and $p_2(t)$ are regarded as quantities of the orders $r_1^2$ and $p_1^2$. This is connected with the Narimanov–Moiseev asymptotics. In expansion (4.5.4), as in the problem of free sloshing in the immobile tank, we preserve the harmonics with $P_0^{(1)}(t)$, $R_1^{(1)}(t)$, $P_1^{(1)}(t)$, $R_2^{(1)}(t)$, and $P_2^{(1)}(t)$ (for the sake of brevity, we omit the subscripts in what follows). Substituting (4.5.4) in relations (2.7.4) and (2.7.5), we obtain relations for the elements of the matrix $A$, which completely coincide with (2.7.4). In this case, the elements $\gamma_k^{(1)}$ are given by the formulas

$$\begin{gathered}
\gamma_1^{(1)} = 0, \ \ \gamma_2^{(1)} = a_5 p_1 + a_6(r_1^2 p_1 + p_1^3) + a_{18}(r_1 r_2 + p_1 p_2) + a_{14} p_0 p_1, \\
\gamma_3^{(1)} = -a_5 r_1 - a_6(r_1 p_1^2 + r_1^3) + a_{18}(r_1 p_2 - p_1 r_2) - a_{14} r_1 p_0, \\
\gamma_4^{(1)} = 2a_7(r_1^2 - p_1^2) + 2a_{20} p_2, \ \ \gamma_5^{(1)} = 4a_7 r_1 p_1 - 2a_{20} r_2,
\end{gathered} \tag{4.5.5}$$

where the coefficients $a_5$, $a_6$, $a_7$, $\ldots$ are given by (4.1.10). In finding the elements of the matrix $A$ and the right-hand sides of (4.5.5), the chosen system of coordinate functions is arranged in the form of a sequence according to principle (1.2.16). Solving the algebraic system of equations (2.7.3), we get

$$\begin{gathered}
P_0^{(1)} = 0, \ \ R_1^{(1)} = \frac{1}{\varkappa_{11}} p_1 + B_0 p_0 p_1 + B_2(r_1 r_2 + p_1 p_2) + D_3(r_1^2 p_1 + p_1^3), \\
P_1^{(1)} = -\frac{1}{\varkappa_{11}} r_1 - B_0 p_0 r_1 + B_2(r_1 p_2 - p_1 r_2) - D_3(r_1^3 + r_1 p_1^2), \\
R_2^{(1)} = \frac{2}{\varkappa_{21}} p_2 + D_1(r_1^2 - p_1^2), \ \ P_2^{(1)} = -\frac{2}{\varkappa_{21}} r_2 + 2D_1 r_1 p_1,
\end{gathered} \tag{4.5.6}$$

where the coefficients $B_0$, $B_2$, $D_1$, and $D_3$ are given by relations (4.1.14):

$$B_0 = \frac{1}{2a_{10}}\left(a_{14} - \frac{a_{22}a_5}{a_{10}}\right), \quad B_2 = 2C_3 - D_2,$$
$$D_1 = \frac{a_7}{a_2} + \frac{a_3a_5}{4a_2a_{10}}, \quad D_3 = \frac{a_6}{2a_{10}} + \frac{1}{\varkappa_{11}}\left(\frac{a_3^2}{4a_2a_{10}} - \frac{a_{12}}{a_{10}} + \frac{a_3a_7}{a_2a_5}\right). \tag{4.5.7}$$

The function $\Omega_1$ constructed in this way exactly satisfies the equation in the domain $Q$ and the boundary conditions on the lateral wall and at the bottom of the cylinder. As follows from the general theory of variational methods, the boundary condition on the free liquid surface is satisfied approximately in a sense of convergence in mean.

We now find the functions $\Omega_2$ and $\Omega_3$. In what follows, it is convenient to replace them by new functions $F_2$ and $F_3$ as follows:

$$\Omega_2 = -xz + F_{20} + F_2, \quad \Omega_3 = xy - F_{30} + F_3, \tag{4.5.8}$$

where

$$F_{20} = 2\sin\eta\sum_n d_{1n}X_n(x)Y_1(k_{1n}\xi), \quad F_{30} = 2\cos\eta\sum_n d_{1n}X_n(x)Y_1(k_{1n}\xi),$$

$$X_n(x) = \frac{\sinh\left[k_{1n}\left(x - x_0 - \frac{h}{2}\right)\right]}{k_{1n}\cosh\left(k_{1n}\frac{h}{2}\right)}, \tag{4.5.9}$$

$$d_{1n} = \frac{E_{1n}}{N_{1n}^2}, \quad E_{1n} = \int \xi^2 Y_1(k_{1n}\xi)d\xi, \quad N_{1n}^2 = \int \xi Y_1^2(k_{1n}\xi)d\xi.$$

The constants $d_{1n}$ are the coefficients of expansion of the variable $\xi$ in the Fourier series in the system of functions $Y_1(k_{1n}\xi)$. Note that the functions $F_{20}$ and $F_{30}$ were obtained by Zhukovsky in the solution of the problem of motion of a rigid body with cylindrical cavity completely filled with liquid.

By using (2.6.2), (2.6.3), and (4.5.8), the boundary-value problems for the functions $F_2$ and $F_3$ can be represented in the following form:

$$\triangle F_2 = 0, \ \boldsymbol{r} \in Q; \quad \frac{\partial F_2}{\partial \nu} = g_1, \ \boldsymbol{r} \in \Sigma; \quad \frac{\partial F_2}{\partial \nu} = 0, \ \boldsymbol{r} \in S; \tag{4.5.10}$$

$$\triangle F_3 = 0, \ \boldsymbol{r} \in Q; \quad \frac{\partial F_3}{\partial \nu} = g_2, \ \boldsymbol{r} \in \Sigma; \quad \frac{\partial F_3}{\partial \nu} = 0, \ \boldsymbol{r} \in S; \tag{4.5.11}$$

where

$$g_1 = \frac{1}{N}\left(2z - \frac{\partial F_{20}}{\partial x} + \frac{\partial F_{20}}{\partial \xi}\frac{\partial f}{\partial \xi} + \frac{1}{\xi^2}\frac{\partial F_{20}}{\partial \eta}\frac{\partial f}{\partial \eta}\right),$$
$$g_2 = \frac{1}{N}\left(-2y + \frac{\partial F_{30}}{\partial x} - \frac{\partial F_{30}}{\partial \xi}\frac{\partial f}{\partial \xi} - \frac{1}{\xi^2}\frac{\partial F_{30}}{\partial \eta}\frac{\partial f}{\partial \eta}\right). \tag{4.5.12}$$

Hence, the boundary problems for the potentials $\Omega_2$ and $\Omega_3$ can be reduced to the Neumann boundary-value problems (4.5.10) and (4.5.11) with trivial boundary conditions on the rigid wall $S$, i.e., to the problems similar to (4.5.3).

Connecting the solution of these boundary-value problems with variational problems for the functionals

$$J_2(F_2) = \int_Q (\nabla F_2)^2 dQ - 2\int_\Sigma F_2 g_1 dS; \tag{4.5.13}$$

$$J_3(F_3) = \int_Q (\nabla F_3)^2 dQ - 2\int_\Sigma F_3 g_2 dS, \tag{4.5.14}$$

by analogy with the previous case, we set

$$F_2(x,\xi,\eta,t) = \sum_{m,n} [R_{mn}^{(2)}(t)\sin m\eta + P_{mn}^{(2)}(t)\cos m\eta]\psi_{mn}(x,\xi); \tag{4.5.15}$$

$$F_3(x,\xi,\eta,t) = \sum_{m,n} [R_{mn}^{(3)}(t)\sin m\eta + P_{mn}^{(3)}(t)\cos m\eta]\psi_{mn}(x,\xi), \tag{4.5.16}$$

where

$$\psi_{mn}(x,\xi) = \frac{\cosh k_{mn}(x-x_0)}{\cosh k_{mn}h} Y_m(k_{mn}\xi).$$

We now construct the Ritz–Trefftz system for the evaluation of the parameters $R_{mn}^2(t)$, $P_{mn}^{(2)}(t)$, $R_{mn}^{(3)}(t)$, and $P_{mn}^{(3)}(t)$ by keeping $q$ functions in relations (4.5.15) and (4.5.16) for the values of the parameter $m$ equal to $0, 1$, and $2$. Arranging the system of coordinate functions in the form of a sequence

$$\chi_1 = \psi_{01}(x,\xi),\ \chi_2 = \psi_{11}(x,\xi)\sin\eta,\ \chi_3 = \psi_{11}(x,\xi)\cos\eta,$$

$$\chi_4 = \psi_{21}(x,\xi)\sin(2\eta),\ \chi_5 = \psi_{21}(x,\xi)\cos(2\eta),\ \chi_6 = \psi_{02}(x,\xi),$$

$$\chi_7 = \psi_{12}(x,\xi)\sin\eta,\ \ldots,\ \chi_{5q} = \psi_{2q}(x,\xi)\cos(2\eta)$$

in agreement with (1.3.4), we get relations for the elements of the matrix $A$ which are similar to (1.2.20) and completely coincide with these relations for $n = 1$.

We find the components of the vector of right-hand sides of Eqs. (1.3.3) as follows:

$$\gamma_k^{(2)} = \int_{\Sigma_0} \chi_k \left(2z - \frac{\partial F_{20}}{\partial x} + \frac{\partial F_{20}}{\partial \xi}\frac{\partial f}{\partial \xi} + \frac{1}{\xi^2}\frac{\partial F_{20}}{\partial \eta}\frac{\partial f}{\partial \eta}\right) dS; \tag{4.5.17}$$

$$\gamma_k^{(3)} = \int_{\Sigma_0} \chi_k \left(-2y + \frac{\partial F_{30}}{\partial x} - \frac{\partial F_{30}}{\partial \xi}\frac{\partial f}{\partial \xi} - \frac{1}{\xi^2}\frac{\partial F_{30}}{\partial \eta}\frac{\partial f}{\partial \eta}\right) dS. \tag{4.5.18}$$

In quadratures (4.5.17) and (4.5.18), $dS$ is an elementary area of the unperturbed free surface and the integrands are found for $x = x_0 + h + f$. To within terms of the third order of smallness in the parameters characterizing the position of the free surface, relations (4.5.17) imply that

$$\gamma_{1i}^{(2)} = \delta_{1ni}^1 r_1 + \delta_{2ni}^1 r_1 p_0 + \delta_{3ni}^1 p_1 r_2 + \delta_{4ni}^1 r_1 p_2 + \delta_{5ni}^1 r_1^3 + \delta_{6ni}^1 p_1^2 r_1,$$
$$\gamma_{2i}^{(2)} = \delta_{1ni}^2 p_0 + \delta_{2ni}^2 p_2 + \delta_{3ni}^2 r_1^2 + \delta_{4ni}^2 p_1^2, \ \gamma_{3i}^{(2)} = \delta_{1ni}^3 r_2 + \delta_{2ni}^3 r_1 p_1, \quad (4.5.19)$$
$$\gamma_{4i}^{(2)} = \delta_{1ni}^4 p_1 + \delta_{2ni}^4 p_0 p_1 + \delta_{3ni}^4 r_1 r_2 + \delta_{4ni}^4 p_1 p_2 + \delta_{5ni}^4 r_1^2 p_1 + \delta_{6ni}^4 p_1^3,$$
$$\gamma_{5i}^{(2)} = \delta_{1ni}^5 r_1 + \delta_{2ni}^5 r_1 p_0 + \delta_{3ni}^5 r_1 p_2 + \delta_{4ni}^5 p_1 r_2 + \delta_{5ni}^5 r_1^3 + \delta_{6ni}^5 r_1 p_1^2.$$

The elements $\gamma_{ij}^{(2)}$ (4.5.19) should be ordered in correspondence with the coordinate functions $\chi_k$ as follows:

$$\gamma_1^{(2)} = \gamma_{11}^{(2)}, \quad \gamma_2^{(2)} = \gamma_{21}^{(2)}, \quad \gamma_3^{(2)} = \gamma_{31}^{(2)}, \quad \gamma_4^{(2)} = \gamma_{41}^{(2)},$$
$$\gamma_5^{(2)} = \gamma_{51}^{(2)}, \quad \gamma_6^{(2)} = \gamma_{12}^{(2)}, \quad \gamma_7^{(2)} = \gamma_{22}^{(2)}, \ \ldots .$$

Similarly, we get the coefficients

$$\gamma_{1i}^{(3)} = -\delta_{1ni}^1 p_1 - \delta_{2ni}^1 p_0 p_1 - \delta_{3ni}^1 r_1 r_2 + \delta_{4ni}^1 p_1 p_2 - \delta_{5ni}^1 r_1^2 p_1 - \delta_{6ni}^1 p_1^3,$$
$$\gamma_{2i}^{(3)} = -\delta_{1ni}^3 r_2 - \delta_{2ni}^3 p_1 r_1, \quad \gamma_{3i}^{(3)} = -\delta_{1ni}^2 p_0 + \delta_{2ni}^2 p_2 - \delta_{3ni}^2 p_1^2 - \delta_{4ni}^2 r_1^2,$$
$$\gamma_{4i}^{(3)} = -\delta_{1ni}^4 r_1 - \delta_{2ni}^4 r_1 p_0 - \delta_{3ni}^4 p_1 r_2 + \delta_{4ni}^4 r_1 p_2 - \delta_{5ni}^4 r_1 p_1^2 - \delta_{6ni}^4 r_1^3,$$
$$\gamma_{5i}^{(3)} = \delta_{1ni}^5 p_1 + \delta_{2ni}^5 p_0 p_1 - \delta_{3ni}^5 p_1 p_2 + \delta_{4ni}^5 r_1 r_2 + \delta_{5ni}^5 p_1^3 + \delta_{6ni}^5 r_1^2 p_1.$$

Some relations between the coefficients $\delta_{ijl}^k$ prove to be quite useful from the practical viewpoint:

$$\delta_{4ni}^1 = -\delta_{3ni}^1, \quad \delta_{6ni}^2 = \delta_{5ni}^1, \quad \delta_{1ni}^3 = -\delta_{2ni}^2, \quad \delta_{2ni}^3 = \delta_{3ni}^2 - \delta_{4ni}^2, \quad \delta_{1ni}^5 = -\delta_{1ni}^4,$$
$$\delta_{2ni}^5 = -\delta_{2ni}^4, \quad \delta_{3ni}^5 = \delta_{3ni}^4, \quad \delta_{4ni}^5 = -\delta_{4ni}^4, \quad \delta_{5ni}^5 = -\frac{1}{2}(\delta_{5ni}^4 + \delta_{6ni}^4),$$
$$\delta_{6ni}^5 = \frac{1}{2}(\delta_{5ni}^4 - 3\delta_{6ni}^4).$$

The remaining coefficients $\delta_{ijn}^k$ are given by the formulas

$$\delta_{1ni}^1 = \pi \sum_n d_{1n} \left[ -A_{0in}^1 \varepsilon_1^{ni} + B_{0in}^0 (\varepsilon_9^{ni} + \varepsilon_{25}^{ni}) \right],$$

$$\delta_{2ni}^1 = \pi \sum_n d_{1n} \left[ -2A_{0in}^2 \varepsilon_3^{ni} + B_{0in}^1 (\varepsilon_{14}^{ni} + \varepsilon_{13}^{ni} + \varepsilon_{30}^{ni}) \right],$$

$$\delta_{3ni}^1 = \pi \sum_n d_{1n} \left[ -A_{0in}^2 \varepsilon_4^{ni} + \tfrac{1}{2} B_{0in}^1 (\varepsilon_{19}^{ni} + \varepsilon_{16}^{ni} + \varepsilon_{28}^{ni}) \right],$$

$$\delta^1_{5ni} = \pi \sum_n d_{1n} \left[ -\tfrac{3}{4} A^3_{0in} \varepsilon^{ni}_7 + B^2_{0in} \left( \tfrac{3}{4} \varepsilon^{ni}_{23} + \tfrac{1}{4} \varepsilon^{ni}_{31} \right) \right],$$

$$\delta^2_{1ni} = \pi \sum_n d_{1n} \left[ -A^1_{1in} \bar{\varepsilon}^{ni}_1 + B^0_{1in} \varepsilon^{ni}_{10} \right],$$

$$\delta^2_{2ni} = \pi \sum_n d_{1n} \left[ \tfrac{1}{2} A^1_{1in} \varepsilon^{ni}_2 - B^0_{1in} \left( \tfrac{1}{2} \varepsilon^{ni}_{11} + \varepsilon^{ni}_{26} \right) \right],$$

$$\delta^2_{3ni} = \pi \sum_n d_{1n} \left[ -\tfrac{3}{4} A^2_{1in} \varepsilon^{ni}_5 + B^1_{1in} \left( \tfrac{3}{4} \varepsilon^{ni}_{17} + \tfrac{1}{4} \varepsilon^{ni}_{27} \right) \right],$$

$$\delta^2_{4ni} = \pi \sum_n d_{1n} \left[ -\tfrac{1}{4} A^2_{1in} \varepsilon^{ni}_5 + B^1_{1in} \left( \tfrac{1}{4} \varepsilon^{ni}_{17} - \tfrac{1}{4} \varepsilon^{ni}_{27} \right) \right],$$

$$\delta^4_{1ni} = \pi \sum_n d_{1n} \left[ -\tfrac{1}{2} A^1_{2in} \bar{\varepsilon}^{ni}_2 + \tfrac{1}{2} B^0_{2in} \left( \varepsilon^{ni}_{12} - \bar{\varepsilon}^{ni}_{26} \right) \right],$$

$$\delta^4_{2ni} = \pi \sum_n d_{1n} \left[ -A^2_{2in} \bar{\varepsilon}^{ni}_4 + \tfrac{1}{2} B^1_{2in} \left( \varepsilon^{ni}_{18} + \bar{\varepsilon}^{ni}_{19} - \bar{\varepsilon}^{ni}_{28} \right) \right],$$

$$\delta^4_{3ni} = -\pi \sum_n d_{1n} B^1_{2in} \varepsilon^{ni}_{29},$$

$$\delta^4_{4ni} = \pi \sum_n d_{1n} \left[ -A^2_{2in} \varepsilon^{ni}_6 + \tfrac{1}{2} B^1_{2in} \left( \varepsilon^{ni}_{20} + \varepsilon^{ni}_{21} + \varepsilon^{ni}_{29} \right) \right],$$

$$\delta^4_{5ni} = \pi \sum_n d_{1n} \left[ -\tfrac{3}{4} A^3_{2in} \varepsilon^{ni}_8 + \tfrac{1}{4} B^2_{2in} \left( 3\varepsilon^{ni}_{24} + \varepsilon^{ni}_{32} \right) \right],$$

$$\delta^4_{6ni} = \pi \sum_n d_{1n} \left[ -\tfrac{1}{4} A^3_{2in} \varepsilon^{ni}_8 + \tfrac{1}{4} B^2_{2in} \left( \varepsilon^{ni}_{24} - \varepsilon^{ni}_{32} \right) \right],$$

where

$$Y_{mn}(\xi) = Y_m(k_{mn}\xi), \quad F_{mn}(\xi) = Y_{m1}(\xi) Y_{11}(\xi) Y_{1n}(\xi),$$

$$f'_{mn}(\xi) = Y'_{m1}(\xi) Y'_{1n}(\xi),$$

$$\varepsilon^{ni}_1 = \int \xi Y_{11} Y_{0i} Y_{1n} d\xi, \quad \varepsilon^{ni}_2 = \int \xi Y_{1i} Y_{21} Y_{1n} d\xi, \quad \varepsilon^{ni}_3 = \int \xi Y_{0i} Y_{11} Y_{1n} Y_{01} d\xi,$$

$$\bar{\varepsilon}^{ni}_1 = \int \xi Y_{01} Y_{1n} Y_{1i} d\xi, \quad \bar{\varepsilon}^{ni}_2 = \int \xi Y_{11} Y_{2i} Y_{1n} d\xi, \quad \bar{\varepsilon}^{ni}_4 = \int \xi Y_{01} Y_{11} Y_{2i} Y_{1n} d\xi,$$

$$\varepsilon^{ni}_4 = \int \xi Y_{0i} F_{21n} d\xi, \quad \varepsilon^{ni}_5 = \int \xi Y_{1i} F_{11n} d\xi, \quad \varepsilon^{ni}_6 = \int \xi Y_{2i} F_{21n} d\xi,$$

$$\varepsilon^{ni}_7 = \int \xi Y_{0i} Y^3_{11} Y_{1n} d\xi, \quad \varepsilon^{ni}_8 = \int \xi Y^3_{11} Y_{2i} Y_{1n} d\xi, \quad \varepsilon^{ni}_9 = \int \xi Y_{0i} f'_{1n} d\xi,$$

$$\varepsilon^{ni}_{10} = \int \xi Y_{1i} f'_{0n} d\xi, \quad \varepsilon^{ni}_{11} = \int \xi Y_{1i} f'_{2n} d\xi, \quad \varepsilon^{ni}_{12} = \int \xi Y_{2i} f'_{1n} d\xi,$$

$$\varepsilon^{ni}_{13} = \int \xi Y_{01} Y_{0i} f'_{1n} d\xi, \quad \varepsilon^{ni}_{14} = \int \xi Y_{0i} Y_{11} f'_{0n} d\xi, \quad \varepsilon^{ni}_{16} = \int \xi Y_{0i} Y_{11} f'_{2n} d\xi,$$

$$\varepsilon_{17}^{ni} = \int \xi Y_{11} Y_{1i} f_{1n}' d\xi, \ \varepsilon_{18}^{ni} = \int \xi Y_{11} Y_{2i} f_{0n}' d\xi, \ \varepsilon_{19}^{ni} = \int \xi Y_{0i} Y_{21} f_{1n}' d\xi,$$

$$\bar{\varepsilon}_{19}^{ni} = \int \xi Y_{01} Y_{2i} f_{1n}' d\xi, \ \varepsilon_{20}^{ni} = \int \xi Y_{2i} Y_{21} f_{1n}' d\xi, \ \varepsilon_{21}^{ni} = \int \xi Y_{11} Y_{2i} f_{2n}' d\xi,$$

$$\varepsilon_{23}^{ni} = \int \xi Y_{0i} Y_{11}^2 f_{1n}' d\xi, \ \varepsilon_{24}^{ni} = \int \xi Y_{11}^2 Y_{2i} f_{1n}' d\xi, \ \varepsilon_{25}^{ni} = \int \frac{1}{\xi} Y_{0i} Y_{11} Y_{1n} d\xi,$$

$$\bar{\varepsilon}_{26}^{ni} = \int \frac{1}{\xi} Y_{11} Y_{2i} Y_{1n} d\xi, \ \varepsilon_{26}^{ni} = \int \frac{1}{\xi} Y_{1i} Y_{21} Y_{1n} d\xi, \ \varepsilon_{27}^{ni} = \int \frac{1}{\xi} Y_{1i} F_{11n} d\xi,$$

$$\varepsilon_{28}^{ni} = \int \frac{1}{\xi} Y_{0i} F_{21n} d\xi, \ \bar{\varepsilon}_{28}^{ni} = \int \frac{1}{\xi} Y_{01} Y_{11} Y_{2i} Y_{1n} d\xi, \ \varepsilon_{29}^{ni} = \int \frac{1}{\xi} Y_{2i} F_{21n} d\xi,$$

$$\varepsilon_{30}^{ni} = \int \frac{1}{\xi} Y_{0i} F_{01n} d\xi, \ \varepsilon_{31}^{ni} = \int \frac{1}{\xi} Y_{0i} Y_{11}^3 Y_{1n} d\xi, \ \varepsilon_{32}^{ni} = \int \frac{1}{\xi} Y_{11}^3 Y_{2i} Y_{1n} d\xi,$$

$$A_{\min}^1 = 2k_{1n} \tanh(\tfrac{1}{2}k_{1n}h), \ A_{\min}^2 = k_{1n}^2 + 2\varkappa_{mi} k_{1n} \tanh(\tfrac{1}{2}k_{1n}h),$$

$$A_{\min}^3 = \tfrac{1}{3}k_{1n}^3 \tanh(\tfrac{1}{2}k_{1n}h) + \varkappa_{mi} k_{1n}^2 + k_{mi}^2 k_{1n} \tanh(\tfrac{1}{2}k_{1n}h),$$

$$B_{\min}^0 = 2k_{1n}^{-1} \tanh(\tfrac{1}{2}k_{1n}h), \ B_{\min}^1 = 2 + 2\varkappa_{mi} k_{1n}^{-1} \tanh(\tfrac{1}{2}k_{1n}h),$$

$$B_{\min}^2 = (k_{mi}^2 + k_{1n}^2) k_{1n}^{-1} \tanh(\tfrac{1}{2}k_{1n}h) + 2\varkappa_{mi}.$$

We now introduce a vector $\boldsymbol{a}^{(2)}$ with the components

$$a_1^{(2)} = P_{01}^{(2)}, \ a_2^{(2)} = R_{11}^{(2)}, \ a_3^{(2)} = P_{11}^{(2)}, \ a_4^{(2)} = R_{21}^{(2)}, \ a_5^{(2)} = P_{21}^{(2)},$$

$$a_6^{(2)} = P_{02}^{(2)}, \ a_7^{(2)} = R_{12}^{(2)}, \ a_8^{(2)} = P_{12}^{(2)}, \ \ldots .$$

Then the system of algebraic equations

$$A\boldsymbol{a}^{(2)} = \boldsymbol{\gamma}^{(2)}$$

implies that

$$P_{0i}^{(2)} = H_1^i r_1 + H_{01}^i p_0 r_1 + H_{14}^i (r_1 p_2 - p_1 r_2) + H_{11}^{0i} (r_1^3 + r_1 p_1^2),$$

$$R_{1i}^{(2)} = H_0^i p_0 + H_3^i p_2 + H_{11}^i r_1^2 + H_{22}^i p_1^2, \ P_{1i}^{(2)} = -H_3^i r_2 + H_{12}^i r_1 p_1,$$

$$R_{2i}^{(2)} = H_2^i p_1 + H_{13}^i r_1 r_2 + H_{02}^i p_0 p_1 + H_{23}^i p_1 p_2 + H_{12}^{1i} r_1^2 p_1 + H_{22}^{2i} p_1^3,$$

$$P_{2i}^{(2)} = -H_2^i r_1 - H_{02}^i p_0 r_1 + H_{13}^i r_1 p_2 - H_{23}^i p_1 r_2 + H_{11}^{1i} r_1^3 + H_{12}^{2i} r_1 p_1^2,$$

where

$$H_1^i = \frac{\delta_{1ni}^1}{2a_1^i}, \ H_{01}^i = \frac{1}{2a_1^i} \left( \delta_{2ni}^1 - 2\sum_{k=1}^N H_1^k a_{21}^{ik} - \sum_{k=1}^N H_0^k a_{15}^{ik} \right),$$

$$H_{14}^i = \frac{1}{2a_1^i} \left( \delta_{4ni}^1 - \sum_{k=1}^N H_3^k a_{15}^{ik} + \sum_{k=1}^N H_2^k a_{24}^{ik} \right), \ H_0^i = \frac{\delta_{1ni}^2}{2a_{10}^i},$$

$$H_{11}^{0i} = \frac{1}{2a_1^i}\left(\delta_{5ni}^1 - 2\sum_{k=1}^N a_9^{ik} H_1^k - \sum_{k=1}^N a_{15}^{ik} H_{11}^k + \sum_{k=1}^N a_{16}^{ik} H_2^k\right),$$

$$H_3^i = \frac{\delta_{2ni}^2}{2a_{10}^i}, \quad H_{11}^i = \frac{1}{2a_{10}^i}\left(\delta_{3ni}^2 - \sum_{k=1}^N a_{15}^{ik} H_1^k - \sum_{k=1}^N a_3^{ik} H_2^k\right),$$

$$H_{22}^i = \frac{1}{2a_{10}^i}\left(\delta_{4ni}^2 - \sum_{k=1}^N a_3^{ik} H_2^k\right), \quad H_{12}^i = H_{11}^i - H_{22}^i,$$

$$H_2^i = \frac{\delta_{1ni}^4}{2a_2^i}, \quad H_{13}^i = \frac{1}{2a_2^i}\left(\delta_{3ni}^4 - \sum_{k=1}^N a_{24}^{ki} H_1^k - \sum_{k=1}^N a_3^{ki} H_3^k\right), \tag{4.5.20}$$

$$H_{02}^i = \frac{1}{2a_2^i}\left(\delta_{2ni}^4 - \sum_{k=1}^N a_3^{ki} H_0^k - 2\sum_{k=1}^N a_{25}^{ki} H_2^k\right),$$

$$H_{23}^i = \frac{1}{2a_2^i}\left(\delta_{4ni}^4 - \sum_{k=1}^N a_3^{ni} H_3^k\right),$$

$$H_{12}^{1i} = \frac{1}{2a_2^i}\left(\delta_{5ni}^4 + 2\sum_{k=1}^N a_{16}^{ki} H_1^k - \sum_{k=1}^N a_3^{ki} H_{11}^k - 2\sum_{k=1}^N a_{13}^{ki} H_2^k - \sum_{k=1}^N a_3^{ki} H_{12}^k\right),$$

$$H_{22}^{2i} = \frac{1}{2a_2^i}\left(\delta_{6ni}^4 - \sum_{k=1}^N a_3^{ki} H_{22}^k - 2\sum_{k=1}^N a_{13}^{ki} H_2^k\right),$$

$$H_{12}^{1i} + H_{22}^{2i} = -2H_{11}^{1i}, \quad H_{12}^{1i} - 3H_{22}^{2i} = 2H_{12}^{2i}.$$

Similarly, we find the coefficients of expansion of the function $F_3$ (4.5.16). They take the form

$$P_{0i}^{(3)} = -H_1^i p_1 + H_{14}^i(r_1 r_2 + p_1 p_2) - H_{01}^i p_1 p_0 - H_{11}^{0i}(r_1^2 p_1 + p_1^3),$$

$$R_{1i}^{(3)} = H_3^i r_2 - H_{12}^i r_1 p_1, \; P_{1i}^{(3)} = -H_0^i p_0 + H_3^i p_2 - H_{22}^i r_1^2 - H_{11}^i p_1^2,$$

$$R_{2i}^{(3)} = -H_2^i r_1 - H_{02}^i p_0 r_1 + H_{23}^i r_1 p_2 + H_{13}^i p_1 r_2 - H_{22}^{2i} r_1^3 - H_{12}^{1i} r_1 p_1^2,$$

$$P_{2i}^{(3)} = -H_2^i p_1 - H_{02}^i p_0 p_1 - H_{23}^i r_1 r_2 - H_{13}^i p_1 p_2 + H_{12}^{2i} r_1^2 p_1 + H_{11}^{1i} p_1^3.$$

The most efficient method for the solution of the system of equations

$$A\boldsymbol{a} = \boldsymbol{\gamma}. \tag{4.5.21}$$

is based on the representation of the $n$th component of the vector $\boldsymbol{a}$ in the form

$$a_n = \alpha_n + \alpha_n^1 p_0 + \alpha_n^2 r_1 + \alpha_n^3 p_1 + \alpha_n^4 r_2 + \alpha_n^5 p_2 + r_1(\alpha_{2n}^1 p_0 + \alpha_{2n}^2 r_1$$

$$+\alpha_{2n}^3 p_1 + \alpha_{2n}^4 r_2 + \alpha_{2n}^5 p_2) + p_1(\alpha_{3n}^1 p_0 + \alpha_{3n}^3 p_1 + \alpha_{3n}^4 r_2 + \alpha_{3n}^5 p_2)$$
$$+\alpha_{4n} r_1^3 + \alpha_{5n} r_1^2 p + \alpha_{6n} r_1 p_1^2 + \alpha_{7n} p_1^3 \qquad (4.5.22)$$

with subsequent substitution of (4.5.22) in (4.5.21). Gathering the coefficients at the same powers of the parameters $p_0$, $r_1$, ..., $p_2$ and their products, we arrive at the system of recurrence relations for the constants $\alpha_n$, $\alpha_n'$, ..., $\alpha_{7n}$, which is readily solved by the method of successive elimination of the unknown quantities.

In conclusion, we present the formulas for some coefficients appearing in relations (4.5.20):

$$a_1^s = \pi \varkappa_{0s} i_6^s; \quad a_{10}^s = \frac{\pi}{2}\varkappa_{1s} i_3^s; \quad a_2^s = \frac{\pi}{2}\varkappa_{2s} i_{12}^s,$$
$$a_3^{sj} = \pi\left(\tfrac{1}{2}\varkappa_{1s}\varkappa_{2j} i_5^{sj} + \tfrac{1}{2} i_{22}^{sj} + i_{23}^{sj}\right),$$
$$a_{15}^{sj} = \pi(\varkappa_{0s}\varkappa_{1j} i_{18}^{sj} + i_{19}^{sj}); \quad a_{21}^{sj} = \tfrac{1}{2}\pi(\varkappa_{0s}\varkappa_{0j} i_{25}^{sj} + i_{26}^{sj}),$$
$$a_{24}^{sj} = \pi(\varkappa_{0s}\varkappa_{2j} i_{30}^{sj} + i_{36}^{sj}); \quad a_{25}^{sj} = \tfrac{1}{2}\pi(\varkappa_{2s}\varkappa_{2j} i_{40}^{sj} + i_{32}^{sj} + 4 i_{33}^{sj}),$$

where

$$i_3^s = \int \xi Y_{1s}^2 d\xi, \quad i_6^s = \int \xi Y_{0s}^2 d\xi, \quad i_{12}^s = \int \xi Y_{2s}^2 d\xi,$$
$$i_5^{sj} = \int \xi Y_{11} Y_{1s} Y_{2j} d\xi, \quad i_{22}^{sj} = \int \xi Y_{11} Y_{1s}' Y_{2j}' d\xi, \quad i_{23}^{sj} = \int \frac{1}{\xi} Y_{11} Y_{1s} Y_{2j} d\xi,$$
$$i_{18}^{sj} = \int \xi Y_{0s} Y_{1j} Y_{11} d\xi, \quad i_{19}^{sj} = \int \xi Y_{11} Y_{0s}' Y_{1j}' d\xi, \quad i_{25}^{sj} = \int \xi Y_{01} Y_{0s} Y_{0j} d\xi,$$
$$i_{26}^{sj} = \int \xi Y_{01} Y_{0s}' Y_{0j}' d\xi, \quad i_{30}^{sj} = \int \xi Y_{21} Y_{0s} Y_{2j} d\xi, \quad i_{36}^{sj} = \int \xi Y_{21} Y_{0s}' Y_{2j}' d\xi,$$
$$i_{40}^{sj} = \int \xi Y_{01} Y_{2s} Y_{2j} d\xi, \quad i_{32}^{sj} = \int \xi Y_{01} Y_{2s}' Y_{2j}' d\xi, \quad i_{33}^{sj} = \int \frac{1}{\xi} Y_{01} Y_{2s} Y_{2j} d\xi.$$

Finding the functions $\Omega_1$, $\Omega_2$, and $\Omega_3$, we complete the procedure of determination of the velocity potential.

## 4.6 Hydrodynamic parameters connected with motions of the rigid body

We now present the formulas for the added masses and moments of inertia appearing in the general equations of motion (2.9.31)–(2.9.34) for the vessel formed by coaxial cylinders. For their determination, we use the following formulas:

$$J_{11}^1 = \rho \int\limits_Q \left( y\frac{\partial \Omega_1}{\partial z} - z\frac{\partial \Omega_1}{\partial y}\right) dQ, \quad J_{12}^1 = \rho \int\limits_Q \left( y\frac{\partial \Omega_2}{\partial z} - z\frac{\partial \Omega_2}{\partial y}\right) dQ,$$

$$J_{13}^1 = \rho \int\limits_Q \left( y\frac{\partial \Omega_3}{\partial z} - z\frac{\partial \Omega_3}{\partial y} \right) dQ, \quad J_{22}^1 = \rho \int\limits_Q \left( z\frac{\partial \Omega_2}{\partial x} - x\frac{\partial \Omega_2}{\partial z} \right) dQ,$$

$$J_{23}^1 = \rho \int\limits_Q \left( z\frac{\partial \Omega_3}{\partial x} - x\frac{\partial \Omega_3}{\partial z} \right) dQ, \quad J_{33}^1 = \rho \int\limits_Q \left( x\frac{\partial \Omega_3}{\partial y} - y\frac{\partial \Omega_3}{\partial x} \right) dQ, \qquad (4.6.1)$$

$$l_1 = m_1 x_{C_1} = \rho \int\limits_Q x dQ, \quad l_2 = m_1 y_{C_1} = \rho \int\limits_Q y dQ, \quad l_3 = m_1 z_{C_1} = \rho \int\limits_Q z dQ,$$

$$l_{k\omega} = \rho \int\limits_Q \Omega_k dQ, \quad l_{k\omega t} = \rho \int\limits_Q \frac{\partial \Omega_k}{\partial t} dQ.$$

Substituting (4.1.5), (4.5.4), (4.5.15), and (4.5.16) in (4.6.1), we get

$$J_{11}^1 = G_{11}^1 (r_1^2 + p_1^2),$$

$$J_{12}^1 = G_{12}^1 p_1 + G_{12}^2 p_0 p_1 + G_{12}^3 (r_1 r_2 + p_1 p_2) + G_{12}^4 (r_1^2 p_1 + p_1^3),$$

$$J_{13}^1 = G_{12}^1 r_1 + G_{12}^2 p_0 r_1 + G_{12}^3 (p_1 r_2 - r_1 p_2) + G_{12}^4 (r_1 p_1^2 + r_1^3),$$

$$J_{22}^1 = G_{22}^0 + G_{22}^1 p_0 + G_{22}^2 p_2 + G_{22}^3 r_1^2 + G_{22}^4 p_1^2, \qquad (4.6.2)$$

$$J_{23}^1 = G_{23}^1 r_2 + G_{23}^2 r_1 p_1, \quad J_{33}^1 = G_{22}^0 + G_{22}^1 p_0 - G_{22}^2 p_2 + G_{22}^3 p_1^2 + G_{22}^4 r_1^2;$$

$$l_{1\omega} = l_{1\omega}^1 (r_1^2 r_2 - p_1^2 r_2 + 2 r_1 p_1 p_2), \qquad (4.6.3)$$

$$l_{2\omega} = l_{2\omega}^1 r_1 + l_{2\omega}^2 p_0 r_1 + l_{2\omega}^3 (r_1 p_2 - p_1 r_2) + l_{2\omega}^4 (r_1^3 + r_1 p_1^2),$$

$$l_{3\omega} = -l_{2\omega}^1 p_1 - l_{2\omega}^2 p_0 p_1 + l_{2\omega}^3 (r_1 r_2 + p_1 p_2) - l_{2\omega}^4 (r_1^2 p_1 + p_1^3),$$

$$\begin{aligned} l_{1\omega t} = {} & l_{1\omega t}^1 (r_1 \dot p_1 - \dot r_1 p_1) + l_{1\omega t}^2 (r_1 \dot p_1 - \dot r_1 p_1) p_0 + l_{1\omega t}^3 (\dot r_1 r_1 r_2 + \dot r_1 p_1 p_2 \\ & + r_1 \dot p_1 p_2 - \dot p_1 p_1 r_2) + l_{1\omega t}^4 (r_1^2 \dot r_2 + 2 r_1 p_1 \dot p_2 - p_1^2 \dot r_2) \\ & + l_{1\omega t}^5 (r_1^3 \dot p_1 - \dot r_1 p_1^3 - \dot r_1 r_1^2 p_1 + r_1 \dot p_1 p_1^2) + l_{1\omega t}^6 (r_2 \dot p_2 - \dot r_2 p_2), \end{aligned} \qquad (4.6.4)$$

$$\begin{aligned} l_{2\omega t} = {} & l_{2\omega t}^1 \dot r_1 p_0 + l_{2\omega t}^2 \dot p_0 r_1 + l_{2\omega t}^3 (r_1 \dot p_2 - p_1 \dot r_2) + l_{2\omega t}^4 (\dot p_1 r_2 - \dot r_1 p_2) \\ & + l_{2\omega t}^5 \dot r_1 r_1^2 + l_{2\omega t}^6 \dot r_1 p_1^2 + l_{2\omega t}^7 r_1 p_1 \dot p_1, \end{aligned}$$

$$\begin{aligned} l_{3\omega t} = {} & -l_{2\omega t}^1 p_0 \dot p_1 - l_{2\omega t}^2 \dot p_0 p_1 + l_{2\omega t}^3 (r_1 \dot r_2 + p_1 \dot p_2) - l_{2\omega t}^4 (\dot r_1 r_2 + \dot p_1 p_2) \\ & - l_{2\omega t}^5 \dot p_1 p_1^2 - l_{2\omega t}^6 r_1^2 \dot p_1 - l_{2\omega t}^7 \dot r_1 p_1 r_1; \end{aligned}$$

$$l_1 = m_1 x_{C_1} = l_1^0 + \tfrac{1}{2}\lambda_{1x}(r_1^2 + p_1^2) + \tfrac{1}{2}\lambda_{0x} p_0^2 + \tfrac{1}{2}\lambda_{2x}(r_2^2 + p_2^2),$$

$$l_2 = m_1 y_{C_1} = l_2^0 + \lambda_2^1 p_1; \quad l_3 = m_1 z_{C_1} = l_3^0 + \lambda_3^1 r_1. \qquad (4.6.5)$$

In relations (4.6.2)–(4.6.5), we use the notation

$$G_{11}^1 = \frac{\pi \rho N_{11}^2}{\varkappa_{11}}, \quad G_{12}^1 = G_{012}^1 + (x_0 + h) G_{112}^1,$$

$$G_{012}^1 = \frac{2\pi\rho s_n}{k_{11}} E_{11}, \quad G_{112}^1 = -\pi\rho E_{11}, \quad s_n = \tanh\left(\tfrac{1}{2}k_{1n}h\right),$$

$$G_{12}^2 = \pi\rho(\tau_{01} + H_0^1 N_{11}^2), \quad G_{12}^3 = \pi\rho\left(\tfrac{1}{2}\tau_{21} + H_3^1 N_{11}^2 + 2H_2^1 N_{21}^2\right),$$

$$G_{12}^4 = \pi\rho\left[H_{22}^1 N_{11}^2 + \tfrac{1}{4}\sum_n d_{1n}k_{1n}s_n\varepsilon_5^{n1} + \tfrac{1}{2}\sum_n H_2^n \varkappa_{2n} i_{5n}\right],$$

$$G_{22}^0 = \pi\rho\left[\left(x_0^2 h + x_0 h^2 + \tfrac{1}{3}h^3\right)\left(R_1^2 - R_0^2\right) - \tfrac{3}{4}h(R_1^4 - R_0^4)\right] + G_{022}^0,$$

$$G_{022}^0 = 8\pi\rho\sum_n \frac{s_n}{k_{1n}} d_{1n}E_{1n},$$

$$G_{22}^1 = \pi\rho\left[\tau_3 - \sum_i \frac{2d_{1i}s_i}{k_{1i}}\sum_n \frac{2d_{1n}s_n}{k_{1n}}(\tau_{1n}^i + \tau_{2n}^i)\right],$$

$$G_{22}^2 = \pi\rho\left[-\frac{1}{2}\tau_4 + \sum_i \frac{d_{1i}s_i}{k_{1i}}\sum_n \frac{2d_{1n}s_n}{k_{1n}}(\tau_{4n}^i - \tau_{3n}^i)\right],$$

$$G_{22}^3 = \pi\rho\left[\sum_n\left(\varkappa_{0n}H_1^n\tau_{0n} + \tfrac{1}{2}\varkappa_{2n}H_2^n\tau_{2n} + 2\frac{\varkappa_{1n}s_n}{k_{1n}}H_{11}^n E_{1n}\right)\right.$$
$$\left. + \frac{1}{4}\sum_n \frac{d_{1n}}{k_{1n}}s_n(3k_{1n}^2\tau_{3n} - 3\tau_{4n} - \tau_{5n})\right]$$
$$+ \pi\rho(x_0 + h)\left[\tfrac{3}{4}N_{11}^2 - \tfrac{3}{4}\sum_n d_{1n}\tau_{4n} - \sum_n \varkappa_{1n}H_{11}^n E_{1n}\right.$$
$$\left. - \sum_n H_1^n\tau_{6n} - \sum_n H_2^n\left(\tfrac{1}{2}\tau_{7n} + \tau_{8n}\right)\right] = G_{022}^3 + (x_0 + h)G_{122}^3,$$

$$G_{22}^4 = \pi\rho\left[2\sum_n \frac{\varkappa_{1n}s_n}{k_{1n}}H_{22}^n E_{1n} + \tfrac{1}{2}\sum_n \varkappa_{2n}H_2^n\tau_{2n}\right.$$
$$\left. + \tfrac{1}{4}\sum_n \frac{d_{1n}s_n}{k_{1n}}(k_{1n}^2\tau_{3n} - \tau_{4n} - 3\tau_{5n})\right]$$
$$+ \pi\rho(x_0 + h)\left[\tfrac{1}{4}N_{11}^2 - \sum_n H_2^n\left(\tfrac{1}{2}\tau_{7n} + \tau_{8n}\right) - \sum_n \varkappa_{1n}H_{22}^n E_{1n}\right.$$
$$\left. - \tfrac{1}{4}\sum_n d_{1n}\tau_{4n}\right] = G_{022}^4 + (x_0 + h)G_{122}^4,$$

$$G_{23}^1 = G_{22}^2, \quad G_{23}^2 = G_{22}^4 - G_{22}^3,$$

$$l_{1\omega}^1 = \pi\rho[B_2 N_{11}^2 + D_1 N_{21}^2], \quad l_{2\omega}^1 = G_{12}^1, \tag{4.6.6}$$

$$l_{2\omega}^2 = \pi\rho(\tau_{01} + 2H_1^1 N_{01}^2 + H_0^1 N_{11}^2), \; l_{2\omega}^3 = \pi\rho\left(-\tfrac{1}{2}\tau_{21} + H_3^1 N_{11}^2 - H_2^1 N_{21}^2\right),$$

$$l_{2\omega}^{4}=\pi\rho\left[H_{11}^{1}N_{11}^{2}+\sum_{n}\left(\tfrac{1}{2}\varkappa_{0n}H_{1}^{n}i_{2n}+\tfrac{1}{4}\varkappa_{2n}H_{2}^{n}i_{5n}+\tfrac{1}{4}d_{1n}k_{1n}s_{n}\varepsilon_{5}^{n1}\right)\right],$$

$$l_{1\omega t}^{1}=G_{11}^{1},\ l_{1\omega t}^{2}=\pi\rho(B_{0}N_{11}^{2}+i_{2}),\ l_{1\omega t}^{3}=\pi\rho\left(B_{2}N_{11}^{2}-\tfrac{1}{2}i_{5}+2D_{1}N_{21}^{2}\right),$$

$$l_{1\omega t}^{4}=\pi\rho\left(B_{2}N_{11}^{2}+\tfrac{1}{2}i_{5}\right),\ l_{1\omega t}^{5}=\pi\rho\left(B_{1}N_{11}^{2}+\tfrac{1}{8}\frac{i_{4}}{\varkappa_{11}}k_{11}^{2}-\tfrac{1}{2}D_{1}i_{5}\varkappa_{21}\right),$$

$$l_{1\omega t}^{6}=2\pi\rho\frac{N_{21}^{2}}{\varkappa_{21}},\quad l_{2\omega t}^{1}=2\pi\rho N_{1}^{1}N_{01}^{2},\quad l_{2\omega t}^{2}=\pi\rho H_{0}^{1}N_{11}^{2},$$

$$l_{2\omega t}^{3}=\pi\rho H_{3}^{1}N_{11}^{2},\quad l_{2\omega t}^{4}=\pi\rho H_{2}^{1}N_{21}^{2},$$

$$l_{2\omega t}^{5}=\pi\rho\left[\sum_{n}\left(\tfrac{1}{2}H_{1}^{n}\varkappa_{0n}i_{2n}+\tfrac{1}{4}H_{2}^{n}\varkappa_{2n}i_{5n}\right)+2H_{11}^{1}N_{11}^{2}\right],$$

$$l_{2\omega t}^{6}=\pi\rho\left(\tfrac{1}{2}\sum_{n}H_{1}^{n}\varkappa_{0n}i_{2n}+H_{12}^{1}N_{11}^{2}-\tfrac{1}{4}\sum_{n}H_{2}^{n}\varkappa_{2n}i_{5n}\right),$$

$$l_{2\omega t}^{7}=l_{2\omega t}^{5}-l_{2\omega t}^{6},\ l_{1}^{0}=\pi\rho\left(x_{0}h+\tfrac{1}{2}h^{2}\right)\left(R_{1}^{2}-R_{0}^{2}\right),$$

$$\lambda_{0x}=2\pi\rho N_{01}^{2},\quad \lambda_{1x}=\pi\rho N_{11}^{2},\quad \lambda_{2x}=\pi\rho N_{21}^{2},$$

$$\lambda_{2}^{1}=\lambda_{3}^{1}=\lambda=\pi\rho E_{11}=-G_{112}^{1},\ l_{2}^{0}=l_{3}^{0}=0.$$

Note that, in relations (4.6.6), in addition to the already introduced notation, we also denote

$$i_{2n}=\int\xi Y_{0n}Y_{11}^{2}d\xi,\ i_{5n}=\int\xi Y_{2n}Y_{11}^{2}d\xi,\ \tau_{0n}=\int\xi^{2}Y_{0n}Y_{11}d\xi,$$

$$\tau_{2n}=\int\xi^{2}Y_{2n}Y_{11}d\xi,\ \tau_{3n}=\int\xi^{2}Y_{11}^{2}Y_{1n}d\xi,\ \tau_{4n}=\int\xi Y_{11}^{2}Y_{1n}'d\xi,$$

$$\tau_{5n}=\int\xi Y_{11}^{2}Y_{1n}d\xi,\ \tau_{6n}=\int\xi Y_{11}Y_{0n}'d\xi,\ \tau_{7n}=\int\xi Y_{11}Y_{2n}'d\xi,$$

$$\tau_{8n}=\int\xi Y_{11}Y_{2n}d\xi,\ \tau_{1n}^{i}=\int\frac{1}{\xi}Y_{1i}Y_{1n}Y_{01}d\xi,\ \tau_{2n}^{i}=\int\xi Y_{01}Y_{1i}'Y_{1n}'d\xi,$$

$$\tau_{3n}^{i}=\int\frac{1}{\xi}Y_{21}Y_{1i}Y_{1n}d\xi,\ \tau_{4n}^{i}=\int\xi Y_{21}Y_{1i}'Y_{1n}'d\xi,$$

$$\tau_{3}=\int\xi^{3}Y_{01}d\xi,\quad \tau_{4}=\int\xi^{3}Y_{21}d\xi.$$

The presented expressions for the hydrodynamic characteristics enable us to construct fairly general equations of motion for mechanical body–liquid systems capable of description of the most significant nonlinear phenomena in the vicinity of the main resonance.

By using Eqs. (2.9.31)–(2.9.34) and relations (4.6.2)–(4.6.5), we conclude that the translational motions of the system in the vicinity of the main resonance in the parameters $r_1(t)$ and $p_1(t)$ can be described with an accuracy to within the terms of the third order of smallness in these parameters without any restrictions imposed on the velocity of motion of the rigid body $\boldsymbol{v}_0(t)$.

In analyzing the general case of space motions of the system within the framework of the accepted model, we assume that the quantities $\omega_i^3$ are of the fourth order of smallness. Thus, we preserve the quantities $\omega_i r_1$, $\omega_i r_1^2$, $\omega_i^2 r_1$, etc., in the equations of motion.

## 4.7 Scalar equations in some special cases

From the practical viewpoint, it is desirable to reduce the equations of motion to a simpler and more convenient form. For this purpose, it is possible to apply the traditional methods of analytic mechanics (connected with the rational choice of the origin and orientation of the coordinate axes rigidly connected with the body) in the nonlinear theory. We now introduce the vector of geometric sum

$$\boldsymbol{w}_* = \boldsymbol{w}_0 - \mathbf{g}.$$

It is called vector of apparent acceleration. By $w_S$, we denote the projections of $\boldsymbol{w}_*$ onto the axes of the system $Oxyz$. The scalar form of the equations of forces is as follows:

$$\begin{aligned}
&M[w_1 + \dot{\omega}_2 z_C - \dot{\omega}_3 y_C - x_C(\omega_2^2 + \omega_3^2) + \omega_1(\omega_2 y_C + \omega_3 z_C)] \\
&\qquad + 2m_1(\omega_2 \dot{z}_{C_1} - \omega_3 \dot{y}_{C_1}) + m_1 \ddot{x}_{C_1} = P_1, \\
&M[w_2 + \dot{\omega}_3 x_C - \dot{\omega}_1 z_C - y_C(\omega_3^2 + \omega_1^2) + \omega_2(\omega_3 z_C + \omega_1 x_C)] \\
&\qquad + 2m_1(\omega_3 \dot{x}_{C_1} - \omega_1 \dot{z}_{C_1}) + m_1 \ddot{y}_{C_1} = P_2, \\
&M[w_3 + \dot{\omega}_1 y_C - \dot{\omega}_2 x_C - z_C(\omega_1^2 + \omega_2^2) + \omega_3(\omega_1 x_C + \omega_2 y_C)] \\
&\qquad + 2m_1(\omega_1 \dot{y}_{C_1} - \omega_2 \dot{x}_{C_1}) + m_1 \ddot{z}_{C_1} = P_3.
\end{aligned} \tag{4.7.1}$$

The vector equation of moments (2.9.32) yield the second group of the scalar equations of motion for the body–liquid system:

$$\begin{aligned}
&J_{11}\dot{\omega}_1 + J_{12}\dot{\omega}_2 + J_{13}\dot{\omega}_3 + \dot{J}_{11}^1\omega_1 + \dot{J}_{12}^1\omega_2 + \dot{J}_{13}^1\omega_3 + J_{32}(\omega_2^2 - \omega_3^2) \\
&\qquad + (J_{33} - J_{22})\omega_2\omega_3 + \omega_1(J_{31}\omega_2 - J_{21}\omega_3) + M(y_C w_3 - z_C w_2) \\
&\qquad + \omega_2 \dot{l}_{3\omega} - \omega_3 \dot{l}_{2\omega} + \omega_3 l_{2\omega t} - \omega_2 l_{3\omega t} - \dot{l}_{1\omega t} + \ddot{l}_{1\omega} = M_1^0, \\
&J_{22}\dot{\omega}_2 + J_{23}\dot{\omega}_3 + J_{21}\dot{\omega}_1 + \dot{J}_{22}^1\omega_2 + \dot{J}_{23}^1\omega_3 + \dot{J}_{21}^1\omega_1 + J_{13}(\omega_3^2 - \omega_1^2) \\
&\qquad + (J_{11} - J_{33})\omega_3\omega_1 + \omega_2(J_{12}\omega_3 - J_{32}\omega_1) + M(z_C w_1 - x_C w_3)
\end{aligned}$$

$$+\omega_3 \dot{l}_{1\omega} - \omega_1 \dot{l}_{3\omega} + \omega_1 l_{3\omega t} - \omega_3 l_{1\omega t} - \dot{l}_{2\omega t} + \ddot{l}_{2\omega} = M_2^0, \tag{4.7.2}$$

$$J_{33}\dot{\omega}_3 + J_{31}\dot{\omega}_1 + J_{32}\dot{\omega}_2 + \dot{J}_{33}^1\omega_3 + \dot{J}_{31}^1\omega_1 + \dot{J}_{32}^1\omega_2 + J_{21}(\omega_1^2 - \omega_2^2)$$
$$+(J_{22} - J_{11})\omega_1\omega_2 + \omega_3(J_{23}\omega_1 - J_{13}\omega_2) + M(x_C w_2 - y_C w_1)$$
$$+\omega_1 \dot{l}_{2\omega} - \omega_2 \dot{l}_{1\omega} + \omega_2 l_{1\omega t} - \omega_1 l_{2\omega t} - \dot{l}_{3\omega t} + \ddot{l}_{3\omega} = M_3^0.$$

It is known that the maximum simplification of the system of equations (4.7.1), (4.7.2) is attained if the origin $O$ of the moving coordinate system coincides with the center of inertia of the analyzed mechanical system and the coordinate axes of this system are oriented in the directions of the principal axes of inertia at this point. However, in the analyzed case, the radius vector of the center of inertia determined according to (4.6.5) by the components

$$x_C = M^{-1}\left[m_0 x_{C_0} + l_1^0 + \tfrac{1}{2}\lambda_{0x}p_0^2 + \tfrac{1}{2}\lambda_{1x}(r_1^2 + p_1^2) + \tfrac{1}{2}\lambda_{2x}(r_2^2 + p_2^2)\right],$$

$$y_C = M^{-1}[m_0 y_{C_0} + l_2^0 + \lambda p_1], \quad z_C = M^{-1}[m_0 z_{C_0} + l_3^0 + \lambda r_1], \tag{4.7.3}$$

and the moment of inertia of the entire system depend on the parameters $p_0(t)$, $r_1(t)$, $p_1(t)$, … , $p_2(t)$, which should be found in the process of solution of the problem. Therefore, it is difficult to connect the system $Oxyz$ with the principal axes of inertia of the analyzed system. Nevertheless, the system of equations (4.7.1), (4.7.2) can be partially simplified if the point $O$ is placed at the center of inertia $G_0$ of the body–liquid system in which the liquid is frozen in the unperturbed state and the directions of the principal axes of inertia of the rigid body are chosen as the directions of the axes $xyz$ at this point.

Specifying the position of the point $G_0$ in the general case by the formulas

$$M x_{G_0} = m_0 x_{C_0} + l_1^0, \quad M y_{G_0} = m_0 y_{C_0} + l_2^0, \quad M z_{G_0} = m_0 z_{C_0} + l_3^0, \tag{4.7.4}$$

we replace the quantities $x_C, y_C$, and $z_C$ in (4.7.1) and (4.7.2) by their values

$$x_C = x_{G_0} + x_f, \quad y_C = y_{G_0} + y_f, \quad z_C = z_{G_0} + z_f, \tag{4.7.5}$$

for

$$x_{G_0} = y_{G_0} = z_{G_0} = 0, \tag{4.7.6}$$

where

$$x_f = M^{-1}\left[\tfrac{1}{2}\lambda_{0x}p_0^2 + \tfrac{1}{2}\lambda_{1x}(r_1^2 + p_1^2) + \tfrac{1}{2}\lambda_{2x}(r_2^2 + p_2^2)\right],$$

$$y_f = M^{-1}\lambda p_1, \quad z_f = M^{-1}\lambda r_1 \tag{4.7.7}$$

are the displacements of the center of inertia of the system caused by the liquid sloshing in the cavity in the direction of the axes connected with the body.

For the accepted choice of the directions of axes of the coordinate system $Oxyz$, the system of equations (4.7.1) takes the form

$$
\begin{aligned}
&(m_0+m_1)w_1+\lambda_{1x}(\dot r_1 r_1+\dot p_1 p_1)^{\cdot}+\lambda[2(\omega_2\dot r_1-\omega_3\dot p_1)\\
&\qquad+r_1(\dot\omega_2+\omega_1\omega_3)-p_1(\dot\omega_3-\omega_1\omega_2)]=P_1,\\
&(m_0+m_1)w_2+\lambda[\ddot p_1-2\omega_1\dot r_1-r_1(\dot\omega_1-\omega_2\omega_3)-p_1(\omega_1^2+\omega_3^2)]\\
&\qquad+\lambda_{1x}\left[\omega_3(r_1^2+p_1^2)^{\cdot}+\tfrac{1}{2}\dot\omega_1(r_1^2+p_1^2)\right]=P_2, \qquad (4.7.8)\\
&(m_0+m_1)w_3+\lambda[\ddot r_1+2\omega_1\dot p_1+p_1(\dot\omega_1+\omega_2\omega_3)-r_1(\omega_1^2+\omega_2^2)]\\
&\qquad-\lambda_{1x}\left[\omega_2(r_1^2+p_1^2)^{\cdot}+\tfrac{1}{2}\dot\omega_2(r_1^2+p_1^2)\right]=P_3.
\end{aligned}
$$

Thus, the equations of moments relative to the principal axes of inertia at the point $G_0$ take the form

$$
\begin{aligned}
&J_{11}^0\dot\omega_1+(J_{33}^0-J_{22}^0)\omega_2\omega_3+\underline{\mu_1[\dot\omega_1(r_1^2+p_1^2)+\omega_1(r_1^2+p_1^2)^{\cdot}-r_1\ddot p_1+\ddot r_1 p_1]}\\
&\quad\underline{-\lambda_0(\dot\omega_2 p_1+\dot\omega_3 r_1+\omega_1\omega_2 r_1-\omega_1\omega_3 p_1)+\lambda(p_1 w_3-r_1 w_2)}=M_1^{G_0}; \qquad (4.7.9)
\end{aligned}
$$

$$
\begin{aligned}
&(J_{22}^0+G_{22}^0)\dot\omega_2-\lambda_0[\ddot r_1+2\omega_1\dot p_1+p_1(\dot\omega_1+\omega_2\omega_3)+r_1(\omega_3^2-\omega_1^2)]\\
&\quad+c_1(\dot r_1 r_1^2+r_1 p_1\dot p_1)^{\cdot}+c_2(r_1 p_1\dot p_1-\dot r_1 p_1^2)^{\cdot}+c_3(\dot r_1 p_0)^{\cdot}+c_4(\dot p_1 r_2-\dot r_1 p_2)^{\cdot}\\
&\quad+c_5(r_1\dot p_0)^{\cdot}+c_6(p_1\dot r_2-r_1\dot p_2)^{\cdot}+(J_{11}^0-J_{33}^0-G_{22}^0)\omega_1\omega_3\\
&\quad+G_{22}^1(p_0\omega_2)^{\cdot}+G_{22}^2(p_2\omega_2+r_2\omega_3)^{\cdot}+G_{22}^3(r_1^2\omega_2-r_1 p_1\omega_3)^{\cdot}\\
&\quad+G_{22}^4(p_1^2\omega_2+r_1 p_1\omega_3)^{\cdot}-\mu_1(r_1\dot p_1-\dot r_1 p_1)\omega_3+\lambda r_1 w_1\\
&\quad-\tfrac{1}{2}\lambda_{1x}(r_1^2+p_1^2)w_3=M_2^{G_0}; \qquad (4.7.10)
\end{aligned}
$$

$$
\begin{aligned}
&(J_{33}^0+G_{22}^0)\dot\omega_3+\lambda_0[\ddot p_1-2\dot r_1\omega_1-r_1(\dot\omega_1-\omega_2\omega_3)-p_1(\omega_1^2-\omega_2^2)]\\
&\quad-c_1(\dot p_1 p_1^2+r_1 p_1\dot r_1)^{\cdot}+c_2(\dot p_1 r_1^2-r_1 p_1\dot r_1)^{\cdot}-c_3(\dot p_1 p_0)^{\cdot}-c_4(\dot p_1 p_2+\dot r_1 r_2)^{\cdot}\\
&\quad-c_5(p_1\dot p_0)^{\cdot}-c_6(p_1\dot p_2+r_1\dot r_2)^{\cdot}-(J_{11}^0-J_{22}^0-G_{22}^0)\omega_1\omega_2+G_{22}^1(p_0\omega_3)^{\cdot}\\
&\quad-G_{22}^2(p_2\omega_3-r_2\omega_2)^{\cdot}+G_{22}^3(p_1^2\omega_3-r_1 p_1\omega_2)^{\cdot}+G_{22}^4(r_1^2\omega_3+r_1 p_1\omega_2)^{\cdot}\\
&\quad+\mu_1(r_1\dot p_1-\dot r_1 p_1)\omega_2-\lambda p_1 w_1+\tfrac{1}{2}\lambda_{1x}(r_2^2+p_1^2)w_2=M_3^{G_0}. \qquad (4.7.11)
\end{aligned}
$$

The last group of nonlinear differential equations, which completes system (4.7.8)–(4.7.11), can be represented in the expanded form as follows:

$$
\begin{aligned}
&\mu_1(\ddot r_1+w_1\varkappa_{11}r_1)+d_1 r_1(r_1\dot r_1+p_1\dot p_1)^{\cdot}+d_2(p_1^2\ddot r_1+2p_1\dot r_1\dot p_1\\
&\quad-r_1 p_1\ddot p_1-2r_1\dot p_1^2)-d_3(p_2\dot r_1-r_2\dot p_1)^{\cdot}+d_4(r_1\ddot p_2-p_1\ddot r_2)\\
&\quad+d_5(p_0\dot r_1)^{\cdot}+d_6 r_1\ddot p_0+c_1(r_1^2\dot\omega_2-r_1 p_1\dot\omega_3-r_1\dot p_1\omega_3-p_1\dot p_1\omega_2)\\
&\quad-c_2(p_1^2\dot\omega_2+r_1 p_1\dot\omega_3+3r_1\dot p_1\omega_3+3p_1\dot p_1\omega_2)+c_3(p_0\omega_2)^{\cdot}\\
&\quad-c_4(p_2\omega_2+r_2\omega_3)^{\cdot}-c_5\omega_2\dot p_0+c_6(\omega_2\dot p_2+\omega_3\dot r_2)
\end{aligned}
$$

$$
\begin{aligned}
&+\mu_1(p_1\dot\omega_1 + 2\omega_1\dot p_1 - r_1\omega_1^2) - \lambda_0(\dot\omega_2 - \omega_1\omega_3) + G_{22}^3\omega_2(\omega_3 p_1 - \omega_2 r_1)\\
&\qquad -G_{22}^4\omega_3(\omega_2 p_1 + \omega_3 r_1) + \lambda w_3 = 0; \qquad (4.7.12)
\end{aligned}
$$

$$
\begin{aligned}
&\mu_1(\ddot p_1 + w_1\varkappa_{11}p_1) + d_1 p_1(r_1\dot r_1 + p_1\dot p_1)^{\cdot} + d_2(r_1^2\ddot p_1 - r_1 p_1\ddot r_1\\
&\quad +2r_1\dot r_1\dot p_1 - 2p_1\dot r_1^2) + d_3(p_2\dot p_1 + r_2\dot r_1)^{\cdot} - d_4(p_1\ddot p_2 + r_1\ddot r_2)\\
&\quad +d_5(p_0\dot p_1)^{\cdot} + d_6 p_1\ddot p_0 - c_1(p_1^2\dot\omega_3 - r_1 p_1\dot\omega_2 - r_1\omega_3\dot r_1 - p_1\omega_2\dot r_1)\\
&\quad +c_2(r_1^2\dot\omega_3 + r_1 p_1\dot\omega_2 + 3r_1\dot r_1\omega_3 + 3p_1\dot r_1\omega_2) - c_3(p_0\omega_3)^{\cdot}\\
&\quad -c_4(p_2\omega_3 - r_2\omega_2)^{\cdot} + c_5\dot p_0\omega_3 - c_6(\dot r_2\omega_2 - \dot p_2\omega_3)\\
&\quad -\mu_1(r_1\dot\omega_1 + 2\dot r_1\omega_1 + \omega_1^2 p_1) + \lambda_0(\dot\omega_3 + \omega_1\omega_2)\\
&\quad +G_{22}^3(r_1\omega_2\omega_3 - \omega_3^2 p_1) - G_{22}^4(\omega_2\omega_3 r_1 + \omega_2^2 p_1) + \lambda w_2 = 0; \qquad (4.7.13)
\end{aligned}
$$

$$
\begin{aligned}
&\mu_0(\ddot p_0 + w_1\varkappa_{01}p_0) + d_6(r_1\ddot r_1 + p_1\ddot p_1) + d_8(\dot r_1^2 + \dot p_1^2)\\
&\quad +c_3(-\omega_2\dot r_1 + \omega_3\dot p_1 - p_1\omega_1\omega_2 - r_1\omega_1\omega_3) + c_7\omega_1(r_1\dot p_1 - \dot r_1 p_1)\\
&\quad +c_5(\omega_2 r_1 - \omega_3 p_1)^{\cdot} - \frac{1}{2}G_{22}^1(\omega_2^2 + \omega_3^2) = 0; \qquad (4.7.14)
\end{aligned}
$$

$$
\begin{aligned}
&\mu_2(\ddot r_2 + w_1\varkappa_{21}r_2) - d_4(\ddot r_1 p_1 + \ddot p_1 r_1) - 2d_7\dot r_1\dot p_1 - c_4(\omega_2\dot p_1 - \omega_3\dot r_1)\\
&\quad +c_6(\omega_2 p_1 - \omega_3 r_1)^{\cdot} + c_8\dot\omega_1(r_1^2 - p_1^2) + c_{10}\omega_1(\dot r_1 r_1 - p_1\dot p_1)\\
&\quad +c_9(\dot\omega_1 p_2 + 2\omega_1\dot p_2) - G_{22}^2\omega_2\omega_3 - G_{12}^3\omega_1(\omega_3 p_1 + \omega_2 r_1) = 0; \qquad (4.7.15)
\end{aligned}
$$

$$
\begin{aligned}
&\mu_2(\ddot p_2 + w_1\varkappa_{21}p_2) + d_4(\ddot r_1 r_1 - \ddot p_1 p_1) + d_7(\dot r_1^2 - \dot p_1^2) + c_4(\dot r_1\omega_2 + \dot p_1\omega_3)\\
&\quad -c_6(\omega_2 r_1 + \omega_3 p_1)^{\cdot} + 2c_8\dot\omega_1 r_1 p_1 + c_{10}\omega_1(r_1 p_1)^{\cdot} - c_9(\dot\omega_1 r_2 + 2\omega_1\dot r_2)\\
&\quad +G_{12}^3(\omega_3 r_1 - \omega_2 p_1)\omega_1 + \tfrac{1}{2}G_{22}^2(\omega_3^2 - \omega_2^2) = 0. \qquad (4.7.16)
\end{aligned}
$$

The coefficients of the system of equations (4.7.8)–(4.7.16) are given by the formulas

$$
\begin{aligned}
&-\lambda_0 = l_{2\omega}^1 = G_{12}^1,\quad c_1 = 3l_{2\omega}^4 - l_{2\omega t}^5,\quad c_2 = -l_{2\omega}^4 + l_{2\omega t}^6 = -G_{12}^4,\\
&c_3 = l_{2\omega}^2 - l_{2\omega t}^1 = G_{12}^2,\quad c_4 = -(l_{2\omega}^3 + l_{2\omega t}^4),\quad c_5 = l_{2\omega}^2 - l_{2\omega t}^2,\\
&c_6 = -l_{2\omega}^3 + l_{2\omega t}^3,\quad \mu_1 = l_{1\omega t}^1 = G_{11}^1,\quad c_7 = l_{1\omega t}^2,\quad c_8 = l_{1\omega}^1 - l_{1\omega t}^4,\\
&c_9 = l_{1\omega t}^6,\quad c_{10} = l_{1\omega t}^3 - 2l_{1\omega t}^4. \qquad (4.7.17)
\end{aligned}
$$

The system of nonlinear differential equations (4.7.8)–(4.7.16) is a system of 16th order in the quasivelocities and generalized coordinates $r_i(t)$ and $p_i(t)$. If the right-hand sides of this system depend not only on these parameters but also on the angular position of the body, then the investigated system should be supplemented with kinematic equations connecting the angular coordinates of the body with its angular velocities.

System (4.7.8)–(4.7.16) is the most general system among the known nonlinear systems used to describe the motions of absolutely rigid bodies with cavities containing liquids. The method aimed at the construction of these equations presented in our book is, in fact, an approximate procedure of reduction of the original nonlinear problem of continuum mechanics to a nonlinear problem of analytic mechanics with finitely many degrees of freedom. Parallel with various versions of the method of perturbation theory, in the analyzed problem, this method should be regarded as an approximate countermethod, which allows one not only to improve the reliability of the results obtained as a result of the solution of specific problems but also to simplify the algorithm of construction of the equations of motion and remove the errors in the numerical values of the coefficients of these equations available in the literature. The variational nature of the algorithm allows us to use, for this purpose, the solutions of the corresponding linear boundary-value problems constructed with the use of variational methods and characterized, generally speaking, solely by weak convergence.

As compared with similar equations presented in [173], the equations obtained in our book contain additional nonlinear terms, which should be taken into account under the accepted restrictions for the orders of the values of generalized coordinates of the problem.

Thus, it is worth noting that the nonlinear system of equations (4.7.8)–(4.7.16) is a finite-dimensional analog of the infinite system of nonlinear ordinary equations of motion (2.9.31)–(2.9.34). The problem of justifying the possibility of replacement of a system of equations of the form (2.9.31)–(2.9.34) by a "truncated" system was studied in [65].

For the coefficients of the nonlinear equations of motion (4.7.8)–(4.7.16), we performed the required numerical calculations in a broad range of the geometric parameters $\delta = R_0/R_1$ and $\bar{h} = h/R_1$. Their numerical values are presented in Tables 4.14–4.28, which give, together with Tables 4.1–4.13, the complete information about the nonlinear mathematical model of rigid-body–liquid systems.

The transition from the dimensionless coefficients computed for $R_1 = 1.0$ is realized by the formulas

$$\begin{gathered}
G^1_{012} = R_1^4\bar{G}^1_{012}, \quad G^0_{022} = R_1^5\bar{G}^0_{022}, \quad G^1_{22} = R_1^4\bar{G}^1_{22}, \\
G^2_{22} = R_1^4\bar{G}^2_{22}, \quad G^3_{022} = R_1^3\bar{G}^3_{022}, \quad G^3_{122} = R_1^2\bar{G}^3_{122}, \quad G^4_{022} = R_1^3\bar{G}^4_{022}, \\
G^4_{122} = R_1^2\bar{G}^4_{122}, \quad c_1 = R_1^2\bar{c}_1, \quad c_2 = R_1^2\bar{c}_2, \quad c_3 = R_1^3\bar{c}_3, \\
c_4 = R_1^3\bar{c}_4, \quad c_5 = R_1^3\bar{c}_5, \quad c_6 = R_1^3\bar{c}_6, \quad c_7 = R_1^2\bar{c}_7, \\
c_8 = R_1^2\bar{c}_8, \quad c_9 = R_1^3\bar{c}_9, \quad c_{10} = R_1^2\bar{c}_{10}, \quad l_1^0 = R_1^4\bar{l}_1^0, \\
\lambda = R_1^3\bar{\lambda}, \quad \lambda_{0x} = R_1^2\bar{\lambda}_{0x}, \quad \lambda_{1x} = R_1^2\bar{\lambda}_{1x}, \quad \lambda_{2x} = R_1^2\bar{\lambda}_{2x},
\end{gathered} \tag{4.7.18}$$

**Table 4.14**

| $h$ | $c_1$ | $c_2$ | $c_3$ | $c_4$ | $c_5$ | $c_6$ |
|---|---|---|---|---|---|---|
| $\delta = 0$ | | | | | | |
| 0.2 | 0.0678 | 1.074 | 0.745 | 0.503 | 0.126 | 0.169 |
| 0.4 | 0.136 | 0.650 | 0.800 | 0.519 | 0.0295 | 0.105 |
| 0.6 | 0.208 | 0.540 | 0.869 | 0.540 | −0.0793 | 0.0361 |
| 0.8 | 0.278 | 0.495 | 0.939 | 0.562 | −0.177 | −0.0320 |
| 1.0 | 0.341 | 0.479 | 1.003 | 0.582 | −0.255 | −0.0904 |
| 1.2 | 0.394 | 0.458 | 1.057 | 0.599 | −0.316 | −0.137 |
| 1.4 | 0.437 | 0.448 | 1.100 | 0.612 | −0.361 | −0.172 |
| 1.6 | 0.470 | 0.441 | 1.132 | 0.623 | −0.393 | −0.198 |
| 1.8 | 0.495 | 0.436 | 1.157 | 0.631 | −0.417 | −0.216 |
| 2.0 | 0.513 | 0.432 | 1.174 | 0.636 | −0.434 | −0.230 |
| 2.5 | 0.539 | 0.426 | 1.200 | 0.644 | −0.457 | −0.248 |
| 3.0 | 0.549 | 0.424 | 1.210 | 0.648 | −0.466 | −0.255 |

**Table 4.15**

| $h$ | $c_7$ | $c_8$ | $c_9$ | $c_{10}$ | $G^0_{022}$ | $G^1_{22}$ |
|---|---|---|---|---|---|---|
| $\delta = 0$ | | | | | | |
| 0.2 | 4.714 | 1.502 | 1.079 | 7.827 | 0.620 | 0.440 |
| 0.4 | 1.800 | 0.200 | 0.699 | 1.924 | 1.198 | 0.501 |
| 0.6 | 1.276 | 0.0399 | 0.618 | 0.876 | 1.705 | 0.620 |
| 0.8 | 1.107 | −0.0292 | 0.596 | 0.545 | 2.126 | 0.773 |
| 1.0 | 1.039 | −0.0560 | 0.590 | 0.416 | 2.460 | 0.930 |
| 1.2 | 1.009 | −0.0675 | 0.588 | 0.361 | 2.717 | 1.072 |
| 1.4 | 0.996 | −0.0727 | 0.588 | 0.336 | 2.907 | 1.189 |
| 1.6 | 0.989 | −0.0750 | 0.588 | 0.324 | 3.047 | 1.281 |
| 1.8 | 0.986 | −0.0762 | 0.588 | 0.319 | 3.147 | 1.350 |
| 2.0 | 0.985 | −0.0767 | 0.588 | 0.316 | 3.218 | 1.401 |
| 2.5 | 0.983 | −0.0771 | 0.588 | 0.314 | 3.316 | 1.473 |
| 3.0 | 0.983 | −0.0771 | 0.588 | 0.314 | 3.356 | 1.503 |

and

$$\lambda_0 = -G^1_{012} + (x_0 + h)\lambda, \quad G^3_{12} = 2c_6 - c_4,$$

$$G^k_{22} = G^k_{022} + (x_0 + h)G^k_{122}, \quad k = 3,\ 4.$$

We now present some linear systems of scalar equations obtained from the vector equations of motion (2.9.31)–(2.9.34). As follows from the subsequent considerations, the linear equations can be extended onto a broader class of cavities under the main assumption of smallness of the quantities $\beta_i(t)$.

For the components of the radius vector of the center of inertia of the mechanical system, relations (4.7.3) take the form

$$x_C = M^{-1}\left(m_0 x_{C_0} + l_{10} + \tfrac{1}{2}\sum_{i=1}^{\infty}\lambda_{i1}\beta_i^2\right),$$
$$y_C = M^{-1}\left(m_0 y_{C_0} + l_{20} + \tfrac{1}{2}\sum_{i=1}^{\infty}\lambda_{i2}\beta_i\right), \tag{4.7.19}$$
$$z_C = M^{-1}\left(m_0 z_{C_0} + l_{30} + \tfrac{1}{2}\sum_{i=1}^{\infty}\lambda_{i3}\beta_i\right).$$

In the system of coordinates with origin at the point $G_0$, the linear equations of motion take the form

$$(m_0 + m_1)j = P_1,$$
$$(m_0 + m_1)w_2 + \sum_{i=1}^{\infty}\lambda_{i2}\ddot{\beta}_i = P_2, \tag{4.7.20}$$
$$(m_0 + m_1)w_3 + \sum_{i=1}^{\infty}\lambda_{i3}\ddot{\beta}_i = P_3,$$

$$(J_{11}^0 + J_{11}^1)\dot{\omega}_1 + J_{12}^1\dot{\omega}_2 + J_{13}^1\dot{\omega}_3 + w_3\sum_{i=1}^{\infty}\lambda_{i2}\beta_i - w_2\sum_{i=1}^{\infty}\lambda_{i3}\beta_i +$$
$$+\sum_{i=1}^{\infty}\lambda_{0i1}\ddot{\beta}_i = M_1^{G_0},$$
$$(J_{22}^0 + J_{22}^1)\dot{\omega}_2 + J_{23}^1\dot{\omega}_3 + J_{21}^1\dot{\omega}_1 + j\sum_{i=1}^{\infty}\lambda_{i3}\beta_i + \sum_{i=1}^{\infty}\lambda_{0i2}\ddot{\beta}_i = M_2^{G_0}, \tag{4.7.21}$$
$$(J_{33}^0 + J_{33}^1)\dot{\omega}_3 + J_{31}^1\dot{\omega}_1 + J_{32}^1\dot{\omega}_2 - j\sum_{i=1}^{\infty}\lambda_{i2}\beta_i + \sum_{i=1}^{\infty}\lambda_{0i3}\ddot{\beta}_i = M_3^{G_0},$$

$$\mu_i(\ddot{\beta}_i + \sigma_i^2\beta_i) + \lambda_{0i1}\dot{\omega}_1 + \lambda_{0i2}\dot{\omega}_2 + \lambda_{0i3}\dot{\omega}_3 + \lambda_{i2}w_2 + \lambda_{i3}w_3 = 0 \tag{4.7.22}$$

($i = 1, 2, \dots,$), where the quantities $j = w_1, w_2$, and $w_3$ denote the projections of the apparent acceleration onto the axes of the rigidly fixed coordinate system. We now consider the case where the transverse apparent accelerations $w_2$ and $w_3$ are small. Thus, in the first equation of system (4.7.21), we can neglect the second and third terms corresponding to the contributions (to the axial component) of the moments of external forces relative to the center of gravity of the rigid body caused by sloshing in the transverse planes.

We now present some specific cases of the system of equations (4.7.20)–(4.7.22).

**Table 4.16**

| $h$ | $G_{22}^2$ | $G_{022}^3$ | $G_{122}^3$ | $G_{022}^4$ | $G_{122}^4$ | $G_{012}^1$ |
|---|---|---|---|---|---|---|
| $\delta = 0$ | | | | | | |
| 0.2 | −0.342 | −0.157 | 1.113 | −0.306 | 1.110 | 0.183 |
| 0.4 | −0.366 | −0.179 | 1.116 | −0.537 | 1.111 | 0.355 |
| 0.6 | −0.411 | −0.0899 | 1.116 | −0.682 | 1.111 | 0.506 |
| 0.8 | −0.466 | 0.0622 | 1.116 | −0.754 | 1.111 | 0.631 |
| 1.0 | −0.521 | 0.233 | 1.117 | −0.779 | 1.111 | 0.731 |
| 1.2 | −0.570 | 0.394 | 1.117 | −0.779 | 1.111 | 0.808 |
| 1.4 | −0.610 | 0.530 | 1.117 | −0.768 | 1.111 | 0.865 |
| 1.6 | −0.642 | 0.638 | 1.117 | −0.754 | 1.111 | 0.906 |
| 1.8 | −0.665 | 0.720 | 1.117 | −0.741 | 1.111 | 0.986 |
| 2.0 | −0.683 | 0.781 | 1.117 | −0.731 | 1.111 | 0.957 |
| 2.5 | −0.707 | 0.868 | 1.117 | −0.714 | 1.111 | 0.937 |
| 3.0 | −0.717 | 0.904 | 1.117 | −0.707 | 1.111 | 0.999 |

**Table 4.17**

| $h$ | $c_1$ | $c_2$ | $c_3$ | $c_4$ | $c_5$ | $c_6$ |
|---|---|---|---|---|---|---|
| $\delta = 0.2$ | | | | | | |
| 0.2 | 0.0846 | 1.429 | 0.635 | 0.575 | 0.0552 | 0 157 |
| 0.4 | 0.109 | 0.839 | 0.691 | 0.597 | −0.0401 | 0.100 |
| 0.6 | 0.251 | 0.675 | 0.701 | 0.627 | −0.136 | 0.0280 |
| 0.8 | 0.327 | 0.607 | 0.833 | 0.658 | −0 214 | −0.0446 |
| 1.0 | 0.397 | 0.570 | 0.901 | 0.688 | −0.275 | −0.108 |
| 1.2 | 0.457 | 0.548 | 0.959 | 0.715 | −0.321 | −0.161 |
| 1.4 | 0.507 | 0.532 | 1.007 | 0.737 | −0.356 | −0.201 |
| 1.6 | 0.547 | 0.521 | 1.045 | 0.755 | −0.382 | −0.232 |
| 1.8 | 0.578 | 0.513 | 1.074 | 0.768 | −0.401 | −0.255 |
| 2.0 | 0.601 | 0.507 | 1.096 | 0.779 | −0.415 | −0.272 |
| 2.4 | 0.636 | 0.498 | 1.129 | 0.794 | −0.436 | −0.297 |
| 3.0 | 0.652 | 0.494 | 1.144 | 0.801 | −0.445 | −0.307 |

*Body containing a cavity with two planes of symmetry intersecting along the line $Ox$.* Assume that the body and the mean liquid domain have two common planes with geometric and mass symmetries, namely, the coordinate planes $Oxy$ and $Oxz$.

In view of the symmetry of the domain occupied by the liquid, the functions $V_2$ and $\Omega_{30}$ are antisymmetric about the plane $Oxz$ and symmetric about the plane $Oxy$. At the same time, the functions $V_3$ and $\Omega_{20}$ are symmetric about the plane $Oxz$ and antisymmetric about the plane $Oxy$. The function $\Omega_{10}$ is an-

**Table 4.18**

| $h$ | $c_7$ | $c_8$ | $c_9$ | $c_{10}$ | $G^0_{022}$ | $G^1_{22}$ |
|---|---|---|---|---|---|---|
| | | | $\delta = 0.2$ | | | |
| 0.2 | 5.780 | 1.909 | 1.108 | 10.70 | 0.620 | 0.386 |
| 0.4 | 1.970 | 0.378 | 0.716 | 2.781 | 1.200 | 0.431 |
| 0.6 | 1.281 | 0.103 | 0.633 | 1.358 | 1.716 | 0.526 |
| 0.8 | 1.054 | 0.0144 | 0.610 | 0.897 | 2.153 | 0.657 |
| 1.0 | 0.961 | −0.0211 | 0.603 | 0.710 | 2.510 | 0.799 |
| 1.2 | 0.918 | −0.0375 | 0.601 | 0.625 | 2.791 | 0.934 |
| 1.4 | 0.897 | −0.0451 | 0.601 | 0.585 | 3.008 | 0.051 |
| 1.6 | 0.887 | −0.0489 | 0.600 | 0.565 | 3.171 | 1.147 |
| 1.8 | 0.842 | −0.0507 | 0.600 | 0.555 | 3.292 | 1.223 |
| 2.0 | 0.879 | −0.0516 | 0.600 | 0.551 | 3.381 | 1.281 |
| 2.5 | 0.877 | −0.0524 | 0.600 | 0.547 | 3.511 | 1.368 |
| 3.0 | 0.877 | 0.0525 | 0.600 | 0.546 | 3.567 | 1.407 |

**Table 4.19**

| $h$ | $G^2_{22}$ | $G^3_{022}$ | $G^3_{122}$ | $G^4_{022}$ | $G^4_{122}$ | $G^1_{012}$ |
|---|---|---|---|---|---|---|
| | | | $\delta = 0.2$ | | | |
| 0.2 | −0.344 | −0.171 | 1.198 | −0.333 | 1.195 | 0 189 |
| 0.4 | −0.372 | −0.195 | 1.198 | −0.539 | 1.195 | 0 367 |
| 0.6 | −0.427 | −0.102 | 1.197 | −0.757 | 1.195 | 0.527 |
| 0.8 | −0.500 | 0.059 | 1.196 | −0.846 | 1.195 | 0.603 |
| 1.0 | −0.577 | 0.230 | 1.195 | −0.880 | 1.195 | 0.775 |
| 1.2 | −0.648 | 0.400 | 1.195 | −0.883 | 1.195 | 0.863 |
| 1.4 | −0.710 | 0.549 | 1.194 | −0.870 | 1.195 | 0.931 |
| 1.6 | −0.760 | 0.673 | 1.194 | −0.852 | 1.195 | 0.982 |
| 1.8 | −0.799 | 0.770 | 1.194 | −0.834 | 1.195 | 1.020 |
| 2.0 | −0.828 | 0.844 | 1.193 | −0.819 | 1.195 | 1.047 |
| 2.5 | −0.873 | 0.957 | 1.193 | −0.792 | 1.195 | 1.088 |
| 3.0 | −0.893 | 1.008 | 1.193 | −0.779 | 1.195 | 1.106 |

tisymmetric about both planes $Oxy$ and $Oxz$. The system of solutions $\varphi_i$ of the boundary-value problem (2.2.4) splits into three subsystems of functions $\varphi_{ip}$, $\varphi_{is}$, and $\varphi_{iq}$ mutually orthogonal on $\Sigma_0$. Moreover, the functions $\varphi_{ip}$ are antisymmetric about $Oxy$ and symmetric about $Oxz$; $\varphi_{is}$ are symmetric about $Oxy$ and antisymmetric about the plane $Oxz$, and $\varphi_{iq}$ are antisymmetric about both planes $Oxy$ and $Oxz$. We denote the generalized coordinates $\beta_i(t)$ corresponding to the functions $\varphi_{ip}$, $\varphi_{is}$, and $\varphi_{iq}$ by $p_i$, $s_i$, and $q_i$, respectively. By virtue of the indicated symmetry, the principal axes of inertia of the rigid body at the

**Table 4.20**

| $h$ | $c_1$ | $c_2$ | $c_3$ | $c_4$ | $c_5$ | $c_6$ |
|---|---|---|---|---|---|---|
| | | | $\delta = 0.4$ | | | |
| 0.2 | 0.0868 | 2.005 | 0.409 | 0.742 | −0.0168 | 0.173 |
| 0.4 | 0.166 | 1.131 | 0.455 | 0.773 | −0.0948 | 0.119 |
| 0.6 | 0.238 | 0.873 | 0.506 | 0.813 | −0.152 | 0.0454 |
| 0.8 | 0.305 | 0.760 | 0.557 | 0.857 | −0.189 | −0.0326 |
| 1.0 | 0.369 | 0.702 | 0.605 | 0.901 | −0.216 | −0.105 |
| 1.2 | 0.427 | 0.667 | 0.648 | 0.942 | −0.235 | −0.169 |
| 1.4 | 0.478 | 0.643 | 0.685 | 0.979 | −0.251 | −0.221 |
| 1.6 | 0.521 | 0.626 | 0.717 | 1.011 | −0.263 | −0.264 |
| 1.8 | 0.577 | 0.614 | 0.743 | 1.037 | −0.273 | −0.297 |
| 2.0 | 0.585 | 0.604 | 0.763 | 1.058 | −0.280 | −0.323 |
| 2.4 | 0.633 | 0.588 | 0.798 | 1.093 | −0.292 | −0.365 |
| 3.0 | 0.658 | 0.580 | 0.816 | 1.112 | −0.298 | −0.385 |

**Table 4.21**

| $h$ | $c_7$ | $c_8$ | $c_9$ | $c_{10}$ | $G^0_{022}$ | $G^1_{22}$ |
|---|---|---|---|---|---|---|
| | | | $\delta = 0.4$ | | | |
| 0.2 | 5.756 | 3.001 | 1.375 | 18.39 | 0.606 | 0.274 |
| 0.4 | 1.737 | 0.676 | 0.869 | 4.943 | 1.180 | 0.302 |
| 0.6 | 1.002 | 0.251 | 0.755 | 2.489 | 1.703 | 0.362 |
| 0.8 | 0.755 | 0.109 | 0.722 | 1.663 | 2.164 | 0.446 |
| 1.0 | 0.648 | 0.0478 | 0.712 | 1.309 | 2.560 | 0.542 |
| 1.2 | 0.596 | 0.0183 | 0.708 | 1.138 | 2.889 | 0.639 |
| 1.4 | 0.569 | 0.00316 | 0.707 | 1.049 | 3.157 | 0.728 |
| 1.6 | 0.555 | −0.00493 | 0.707 | 1.002 | 3.371 | 0.807 |
| 1.8 | 0.547 | −0.00933 | 0.707 | 0.977 | 3.540 | 0.874 |
| 2.0 | 0.543 | −0.0117 | 0.707 | 0.963 | 3.670 | 0.928 |
| 2.5 | 0.539 | −0.0141 | 0.707 | 0.949 | 3.879 | 1.020 |
| 3.0 | 0.538 | −0.0146 | 0.707 | 0.946 | 3.985 | 1.068 |

point $G_0$ coincide with the directions of the axes $G_0x$, $G_0y$, and $G_0z$. Moreover, the following equalities are true:

$$J^1_{12} = J^1_{21} = 0, \quad J^1_{13} = J^1_{31} = 0, \quad J^1_{23} = J^1_{32} = 0. \tag{4.7.23}$$

In this case, the equations of motion (4.7.20)–(4.7.22) split into the equations of motion of the system as a rigid body in the direction of the $G_0x$-axis, the system of equations used to describe the rotation of the mechanical system about the $G_0x$-axis, and the equations of motion in the principal planes $G_0xy$

**Table 4.22**

| $h$ | $G_{22}^{2}$ | $G_{022}^{3}$ | $G_{122}^{3}$ | $G_{022}^{4}$ | $G_{122}^{4}$ | $G_{012}^{1}$ |
|---|---|---|---|---|---|---|
| | | | $\delta = 0.4$ | | | |
| 0.2 | −0.356 | −0.163 | 1.217 | −0.337 | 1.217 | 0.188 |
| 0.4 | −0.394 | −0.171 | 1.217 | −0.603 | 1.217 | 0.369 |
| 0.6 | −0.474 | −0.0970 | 1.217 | −0.797 | 1.217 | 0.535 |
| 0.8 | −0.581 | 0.0180 | 1.217 | −0.919 | 1.217 | 0.683 |
| 1.0 | −0.098 | 0.156 | 1.216 | −0.983 | 1.217 | 0.809 |
| 1.2 | −0.813 | 0.301 | 1.216 | −1.007 | 1.217 | 0.915 |
| 1.4 | −0.919 | 0.440 | 1.216 | −1.007 | 1.217 | 1.001 |
| 1.6 | −1.011 | 0.563 | 1.216 | −0.994 | 1.217 | 1.069 |
| 1.8 | −1.087 | 0.669 | 1.216 | −0.977 | 1.217 | 1.123 |
| 2.0 | −1.150 | 0.756 | 1.216 | −0.968 | 1.217 | 1.165 |
| 2.5 | −1.254 | 0.904 | 1.216 | −0.920 | 1.217 | 1.232 |
| 3.0 | −1.308 | 0.983 | 1.216 | −0.897 | 1.217 | 1.266 |

and $G_0xz$:

$$(m_0 + m_1)j = P_1; \tag{4.7.24}$$

$$(J_{11}^0 + J_{11}^1)\dot{\omega}_1 + \sum_{i=1}^{\infty} \lambda_{0i1}\ddot{q}_i = M_1^{G_0},$$

$$\mu_{iq}(\ddot{q}_i + \sigma_{iq}^2 q_i) + \lambda_{0i1}\dot{\omega}_1 = 0, \quad i = 1, 2, \dots; \tag{4.7.25}$$

$$(m_0 + m_1)w_2 + \sum_{i=1}^{\infty} \lambda_{i2}\ddot{s}_i = P_2,$$

$$(J_{33}^0 + J_{33}^1)\dot{\omega}_3 - j\sum_{i=1}^{\infty} \lambda_{i2}s_i + \sum_{i=1}^{\infty} \lambda_{0i3}\ddot{s}_i = M_3^{G_0}, \tag{4.7.26}$$

$$\mu_{is}(\ddot{s}_i + \sigma_{is}^2 s_i) + \lambda_{i2}w_2 + \lambda_{i03}\dot{\omega}_3 = 0, \quad i = 1, 2, \dots;$$

$$(m_0 + m_1)w_3 + \sum_{i=1}^{\infty} \lambda_{i3}\ddot{p}_i = P_3,$$

$$(J_{22}^0 + J_{22}^1)\dot{\omega}_2 + j\sum_{i=1}^{\infty} \lambda_{i3}p_i + \sum_{i=1}^{\infty} \lambda_{0i2}\ddot{p}_i = M_2^{G_0}, \tag{4.7.27}$$

$$\mu_{ip}(\ddot{p}_i + \sigma_{ip}^2 p_i) + \lambda_{i3}w_3 + \lambda_{0i2}\dot{\omega}_2 = 0, \quad i = 1, 2, \dots .$$

It is clear that, for the axisymmetric body with axisymmetric liquid domain, the motion of the system is studied with the use not only of Eqs. (4.7.24) but also of the system of equations (4.7.26) [or (4.7.27)] because the process of liquid sloshing is not excited by the motion of the body relative to the $Ox$-axis. This

**Table 4.23**

| $h$ | $c_1$ | $c_2$ | $c_3$ | $c_4$ | $c_5$ | $c_6$ |
|---|---|---|---|---|---|---|
| | | | $\delta = 0.5$ | | | |
| 0.2 | 0.0749 | 2.088 | 0.299 | 0.787 | −0.0312 | 0.180 |
| 0.4 | 0.139 | 1.158 | 0.330 | 0.816 | −0.0925 | 0.132 |
| 0.6 | 0.198 | 0.878 | 0.375 | 0.854 | −0.128 | 0.063 |
| 0.8 | 0.255 | 0.750 | 0.410 | 0.896 | −0.148 | −0.0107 |
| 1.0 | 0.309 | 0.692 | 0.444 | 0.939 | −0.161 | −0 0827 |
| 1.2 | 0.360 | 0.654 | 0.475 | 0.981 | −0.172 | −0 147 |
| 1.4 | 0.405 | 0.630 | 0.502 | 1.020 | −0.180 | −0.203 |
| 1.6 | 0.445 | 0.612 | 0.526 | 1.055 | −0.187 | −0.247 |
| 1.8 | 0.479 | 0.599 | 0.546 | 1.084 | −0.193 | −0.284 |
| 2.0 | 0.507 | 0.589 | 0.563 | 1.109 | −0.197 | −0.314 |
| 2.4 | 0.556 | 0.572 | 0.592 | 1.152 | −0.205 | −0.363 |
| 3.0 | 0.583 | 0.563 | 0.608 | 1.176 | −0.209 | −0.389 |

**Table 4.24**

| $h$ | $c_7$ | $c_8$ | $c_9$ | $c_{10}$ | $G^0_{022}$ | $G^1_{22}$ |
|---|---|---|---|---|---|---|
| | | | $\delta = 0.125$ | | | |
| 0.2 | 4.832 | 3.479 | 1.541 | 21.62 | 0.584 | 0.211 |
| 0.4 | 1.407 | 0.804 | 0.956 | 5.795 | 1.140 | 0 232 |
| 0.6 | 0.778 | 0.313 | 0.818 | 1.654 | 1.654 | 0.276 |
| 0.8 | 0.565 | 0.147 | 0.776 | 1.908 | 2.116 | 0.337 |
| 1.0 | 0.471 | 0.0742 | 0.763 | 1.477 | 2.521 | 0.408 |
| 1.2 | 0.425 | 0.0383 | 0.758 | 1.264 | 2.865 | 0.481 |
| 1.4 | 0.401 | 0.0191 | 0.756 | 1.151 | 3.152 | 0.551 |
| 1.6 | 0.387 | 0.00853 | 0.756 | 1.088 | 8.388 | 0.614 |
| 1.8 | 0.379 | 0.00259 | 0.756 | 1.053 | 3.578 | 0.668 |
| 2.0 | 0.375 | $-0.803 \cdot 10^{-3}$ | 0.755 | 1.033 | 3.729 | 0.714 |
| 2.5 | 0.370 | −0.00426 | 0.755 | 1.012 | 3.981 | 0.795 |
| 3.0 | 0.369 | −0.00514 | 0.755 | 1.007 | 4.114 | 0.841 |

fact follows from the direct analysis of solutions of the corresponding boundary-value problems for cavities in the form of bodies of revolution.

*Body containing a cavity with the plane of symmetry parallel to the $Ox$-axis.* Assume that the body and the unperturbed liquid domain have a single common plane of geometric and mass symmetries. Without loss of generality, we can assume that this is the coordinate plane $G_0xz$. Thus,

$$J^0_{21} + J^1_{21} = J^0_{12} + J^1_{12} = 0. \tag{4.7.28}$$

**Table 4.25**

| $h$ | $G_{22}^2$ | $G_{022}^3$ | $G_{122}^3$ | $G_{022}^4$ | $G_{122}^4$ | $G_{012}^1$ |
|---|---|---|---|---|---|---|
| $\delta = 0.5$ | | | | | | |
| 0.2 | −0.356 | −0.139 | 1.131 | −0.309 | 1.131 | 0.179 |
| 0.4 | −0.399 | −0.136 | 1.131 | −0.555 | 1.131 | 0.352 |
| 0.6 | −0.485 | −0.0829 | 1.131 | −0.744 | 1.131 | 0.512 |
| 0.8 | −0.598 | $-0.254 \cdot 10^{-3}$ | 1.131 | −0.872 | 1.131 | 0.657 |
| 1.0 | −0.723 | 0.107 | 1.131 | −0.947 | 1.131 | 0.784 |
| 1.2 | −0.849 | 0.226 | 1.131 | −0.983 | 1.131 | 0.892 |
| 1.4 | −0.967 | 0.346 | 1.131 | −0.993 | 1.131 | 0.983 |
| 1.6 | −1.072 | 0.458 | 1.131 | −0.988 | 1.131 | 1.056 |
| 1.8 | −1.162 | 0.557 | 1.131 | −0.975 | 1.131 | 1.116 |
| 2.0 | −1.238 | 0.641 | 1.131 | −0.959 | 1.131 | 1.184 |
| 2.5 | −1.371 | 0.793 | 1.131 | −0.923 | 1.131 | 1.243 |
| 3.0 | −1.446 | 0.879 | 1.131 | −0.889 | 1.131 | 1.285 |

**Table 4.26**

| $h$ | $c_1$ | $c_2$ | $c_3$ | $c_4$ | $c_5$ | $c_6$ |
|---|---|---|---|---|---|---|
| $\delta = 0.8$ | | | | | | |
| 0.2 | 0.0273 | 1.298 | 0.0590 | 0.510 | −0.0203 | 0.119 |
| 0.4 | 0.0492 | 0.691 | 0.0668 | 0.520 | −0.0246 | 0.0974 |
| 0.6 | 0.0723 | 0.505 | 0.0718 | 0.534 | −0.0255 | 0.0659 |
| 0.8 | 0.0954 | 0.422 | 0.0763 | 0.552 | −0.0262 | 0.0290 |
| 1.0 | 0.118 | 0.378 | 0.0807 | 0.573 | −0.0268 | −0.00909 |
| 1.2 | 0.140 | 0.352 | 0.0852 | 0.595 | −0.0273 | −0.0457 |
| 1.4 | 0.160 | 0.335 | 0.0895 | 0.616 | −0.0278 | −0.0790 |
| 1.6 | 0.178 | 0.324 | 0.0935 | 0.637 | −0.0282 | −0.109 |
| 1.8 | 0.195 | 0.316 | 0.0972 | 0.656 | −0.0285 | −0.134 |
| 2.0 | 0.210 | 0.309 | 0.100 | 0.673 | −0.0288 | −0.157 |
| 2.4 | 0.238 | 0.298 | 0.107 | 0.707 | −0.0294 | −0.195 |
| 3.0 | 0.257 | 0.291 | 0.111 | 0.729 | −0.0297 | −0.220 |

Due to the symmetry of the cavity, the functions $\Omega_1$, $\Omega_2$, and $\Omega_3$ have the following properties of symmetry and antisymmetry:

$$\Omega_{30}(x,y,z) = -\Omega_{30}(x,-y,z), \quad \Omega_{20}(x,y,z) = \Omega_{20}(x,-y,z),$$

$$\Omega_{10}(x,y,z) = \Omega_{10}(x,-y,z).$$

The system of functions $\varphi_i$ decomposes into two subsystems $\varphi_{ip}$ and $\varphi_{is}$ mutually orthogonal on the surface $\Sigma_0$, where $\varphi_{ip}$ are functions symmetric about

**Table 4.27**

| $h$ | $c_7$ | $c_8$ | $c_9$ | $c_{10}$ | $G^0_{022}$ | $G^1_{22}$ |
|---|---|---|---|---|---|---|
| | | | $\delta = 0.8$ | | | |
| 0.2 | 1.113 | 2.790 | 1.207 | 17.30 | 0.369 | 0.0446 |
| 0.4 | 0.306 | 0.664 | 0.709 | 4.546 | 0.728 | 0.0494 |
| 0.6 | 0.157 | 0.272 | 0.579 | 2.195 | 1.070 | 0.0579 |
| 0.8 | 0.106 | 0.137 | 0.534 | 1.384 | 1.388 | 0.0691 |
| 1.0 | 0.0821 | 0.0762 | 0.516 | 1.020 | 1.679 | 0.0821 |
| 1.2 | 0.0709 | 0.0448 | 0.509 | 0.831 | 1.938 | 0.0960 |
| 1.4 | 0.0643 | 0.0272 | 0.507 | 0.726 | 2.166 | 0.110 |
| 1.6 | 0.0603 | 0.0169 | 0.505 | 0.664 | 2.363 | 0.123 |
| 1.8 | 0.0579 | 0.0106 | 0.505 | 0.626 | 2.531 | 0.136 |
| 2.0 | 0.0565 | 0.00667 | 0.505 | 0.603 | 2.673 | 0.147 |
| 2.5 | 0.0547 | 0.00213 | 0.505 | 0.575 | 2.933 | 0.169 |
| 3.0 | 0.0542 | $0.674 \cdot 10^{-3}$ | 0.505 | 0.567 | 3.092 | 0.184 |

**Table 4.28**

| $h$ | $G^2_{22}$ | $G^3_{022}$ | $G^3_{122}$ | $G^4_{022}$ | $G^4_{122}$ | $G^1_{012}$ |
|---|---|---|---|---|---|---|
| | | | $\delta = 0.8$ | | | |
| 0.2 | −0.239 | −0.0386 | 0.587 | −0.147 | 0.587 | 0.102 |
| 0.4 | −0.271 | −0.0489 | 0.586 | −0.273 | 0.586 | 0.201 |
| 0.6 | −0.322 | −0.0478 | 0.584 | −0.378 | 0.584 | 0.295 |
| 0.8 | −0.387 | −0.0284 | 0.582 | −0.457 | 0.582 | 0.383 |
| 1.0 | −0.462 | 0.00681 | 0.581 | −0.511 | 0.581 | 0.464 |
| 1.2 | −0.540 | 0.0529 | 0.580 | −0.546 | 0.580 | 0.535 |
| 1.4 | −0.618 | 0.105 | 0.578 | −0.564 | 0.578 | 0.598 |
| 1.6 | −0.692 | 0.158 | 0.578 | −0.572 | 0.578 | 0.653 |
| 1.8 | −0.761 | 0.210 | 0.577 | −0.573 | 0.577 | 0.699 |
| 2.0 | −0.822 | 0.258 | 0.576 | −0.569 | 0.576 | 0.739 |
| 2.5 | −0.944 | 0.355 | 0.575 | −0.551 | 0.575 | 0.810 |
| 3.0 | −1.024 | 0.422 | 0.574 | −0.534 | 0.574 | 0.854 |

the plane $Oxz$ and $\varphi_{is}$ are the functions antisymmetric about the same plane. By $p_i$ and $s_i$, we denote the generalized coordinates corresponding to these functions. Then the equations of motion of the mechanical body–liquid system take the form

$$(m_0 + m_1)j = P_1; \quad (m_0 + m_1)w_2 + \sum_{i=1}^{\infty} \lambda_{i2}\ddot{s}_i = P_2, \tag{4.7.29}$$

$$(J_{11}^0 + J_{11}^1)\dot{\omega}_1 + (J_{13}^0 + J_{13}^1)\dot{\omega}_3 + \sum_{i=1}^{\infty} \lambda_{0i1}\ddot{s}_i = M_1^{G_0},$$

$$(J_{33}^0 + J_{33}^1)\dot{\omega}_3 + (J_{13}^0 + J_{13}^1)\dot{\omega}_1 - j\sum_{i=1}^{\infty} \lambda_{i2}s_i + \sum_{i=1}^{\infty} \lambda_{0i3}\ddot{s}_i = M_3^{G_0}, \qquad (4.7.30)$$

$$\mu_{is}(\ddot{s}_i + \sigma_{is}^2 s_i) + \lambda_{i2}w_2 + \lambda_{0i1}\dot{\omega}_1 + \lambda_{i03}\dot{\omega}_3 = 0, \;\; i = 1,\, 2,\, \ldots\,;$$

$$(m_0 + m_1)w_3 + \sum_{i=1}^{\infty} \lambda_{i3}\ddot{p}_i = P_3,$$

$$(J_{22}^0 + J_{22}^1)\dot{\omega}_2 + j\sum_{i=1}^{\infty} \lambda_{i3}p_i + \sum_{i=1}^{\infty} \lambda_{0i2}\ddot{p}_i = M_2^{G_0}, \qquad (4.7.31)$$

$$\mu_{ip}(\ddot{p}_i + \sigma_{ip}^2 p_i) + \lambda_{i3}w_3 + \lambda_{0i2}\dot{\omega}_2 = 0, \;\; i = 1,\, 2,\, \ldots\,.$$

Hence, in the analyzed case, the system of equations (4.7.20)–(4.7.22) splits into two independent subsystems (4.7.30) and (4.7.31). The first subsystem describes the motion in the plane $Oxy$ and the rotation about the axis $Ox$. The second subsystem describes the motion in the plane $Oxz$. Equation (4.7.29) describes the motions of the entire system as a rigid body in the direction of the $Ox$-axis. The results presented in the monograph [149] enable us to construct the equations of motion for the other cases of geometric and mass symmetries.

The presented equations of motion can easily be generalized for the case where the body has several cavities. In this case, the hydrodynamic forces and moments should be found as the sums of the forces and moments for the corresponding cavities.

Note that the hydrodynamic coefficients $\lambda_i$, $\lambda_{0i}$, and $\mu_i$ depend on arbitrary normalizing factors $a_i$ connected with the normalizing factor for the natural sloshing modes $\varphi_i$. There are several procedures of choosing the factors $a_i$. It is also worth noting that there exist mechanical invariants independent of the normalization. These invariants are discussed in [149].

For simplicity, we consider the motion of a mechanical system in the case where the rigid body and liquid domain have the shapes of bodies of revolution with the common longitudinal axis. In this case, we introduce normalizing factors $a_i$ with the dimension of length in the generalized coordinates $p_i$ of system (4.7.27) and represent the analyzed system in the form

$$(m_0 + m_1)w_3 + \sum_{i=1}^{\infty} \Lambda_i\ddot{\zeta}_i = P_3,$$

$$(J_{22}^0 + J_{22}^1)\dot{\omega}_2 + j\sum_{i=1}^{\infty} \Lambda_i\zeta_i + \sum_{i=1}^{\infty} \Lambda_{0i}\ddot{\zeta}_i = M_2^{G_0}, \qquad (4.7.32)$$

$$M_i(\ddot{\zeta}_i + \sigma_i^2\zeta_i) + \Lambda_i w_3 + \Lambda_{0i}\dot{\omega}_2 = 0, \;\; i = 1,\, 2,\, 3,\, \ldots\,,$$

where

$$\zeta_i = \frac{p_i}{a_i}, \quad M_i = a_i^2 \mu_i, \quad \Lambda_i = a_i \lambda_i, \quad \Lambda_{0i} = a_i \lambda_{0i}. \tag{4.7.33}$$

Further, we choose the normalizing factors $a_i$ so that the generalized coordinate $\zeta_i$ becomes equal to the angular deviation of the plane approximating the unperturbed free liquid surface for the $i$th form of oscillations in a sense of minimizing of the mean-square deviation. This problem is reduced to finding the extremum of the functional

$$J = \int_{\Sigma_0} (p_i f_i - \zeta_i z)^2 dS$$

in the parameters $\zeta_i$. The necessary condition

$$\frac{\partial J}{\partial \zeta_i} = 2 \int_{\Sigma_0} z(p_i f_i - \zeta_i z) dS = 0, \quad i = 1, 2, \ldots,$$

implies that

$$\begin{gathered} \zeta_i = \frac{p_i}{a_i}, \quad a_i = \frac{G}{\lambda_i}, \quad \Lambda_i = G, \quad \Lambda_{0i} = -c_i G, \\ G = \rho J_y, \quad M_i = \frac{G^2}{m_i}, \quad m_i = \frac{\lambda_i^2}{\mu_i}, \quad c_i = -\frac{\lambda_{0i}}{\lambda_i}, \end{gathered} \tag{4.7.34}$$

where $J_y = \int_{\Sigma_0} z^2 dS$.

Thus, the coefficients of Eqs. (4.7.32) are represented via the following parameters invariant under the choice of normalization of the functions $\varphi_i$:

$$\sigma_i^2, \quad J_{22}^1, \quad m_i = \frac{\lambda_i^2}{\mu_i}, \quad c_i = -\frac{\lambda_{0i}}{\lambda_i}. \tag{4.7.35}$$

The role of the coefficients $J$ and $m_i$ in the equations of rotation (4.7.25) about the longitudinal axis (transformed in a similar way) is played by

$$J_{11}^1 \quad \text{and} \quad m_{iq} = \frac{\lambda_{0i1}^2}{\mu_{iq}}, \tag{4.7.36}$$

respectively.

Finally, we present the formulas of transformation of the hydrodynamic coefficients $\sigma_i^2, \mu_i, \lambda_i, \lambda_{0i}$ and the components of the tensor $\boldsymbol{J}^1$ under a parallel transfer of the connected system of coordinates from the position $O'x'y'z'$ to the position $Oxyz$. To deduce these formulas, we represent the radius vector of any point of the domain $Q$ relative to the point $O$ in the form

$$\boldsymbol{r} = \boldsymbol{r}_{O'} + \boldsymbol{r}', \tag{4.7.37}$$

where $\boldsymbol{r}_{O'}$ is the radius vector of $O'$ relative to the point $O$ and $\boldsymbol{r}'$ is the radius vector of any point of the liquid domain relative to $O'$.

Starting from the boundary-value problems (2.6.2) and (2.6.3) for the functions $\boldsymbol{V}$ and $\boldsymbol{\Omega}$, we can easily establish the following relations:

$$\boldsymbol{\Omega} = \boldsymbol{\Omega}' + \boldsymbol{r}_{O'} \times \boldsymbol{r}', \quad \boldsymbol{V} = \boldsymbol{V}' + \boldsymbol{r}_{O'}. \tag{4.7.38}$$

Substituting (4.7.38) in relations (2.6.42), we get

$$\mu_i = \mu_i', \ \sigma_i^2 = \sigma'^2_i, \ \lambda_{i2} = \lambda_{i2}', \ \lambda_{i3} = \lambda_{i3}', \ \lambda_{0i1} = \lambda_{0i1}' + y_{O'}\lambda_{i3} - z_{O'}\lambda_{i2},$$
$$\lambda_{0i2} = \lambda_{0i2}' - x_{O'}\lambda_{i3}, \quad \lambda_{0i3} = \lambda_{0i3}' + x_{O'}\lambda_{i2} \tag{4.7.39}$$

or, in the vector form,

$$\boldsymbol{\lambda}_i = \boldsymbol{\lambda}_i', \quad \boldsymbol{\lambda}_{0i} = \boldsymbol{\lambda}_{0i}' + \boldsymbol{r}_{O'} \times \boldsymbol{\lambda}_i. \tag{4.7.40}$$

For the components $J_{ij}^1$ of the tensor of inertia of liquid, the corresponding formulas of transition can be obtained from relation (2.6.41):

$$\begin{aligned} J_{ij}^1 &= \rho \int\limits_{S+\Sigma_0} [\Omega_{i0}' + (\boldsymbol{r}_{O'} + \boldsymbol{r}')_i][(\boldsymbol{r}_{O'} + \boldsymbol{r}') \times \boldsymbol{\nu}]_j dS = \rho \int\limits_{S+\Sigma_0} \Omega_{i0}'(\boldsymbol{r}' \times \boldsymbol{\nu})_j dS \\ &\quad + \rho \int\limits_{S+\Sigma_0} \Omega_{i0}'(\boldsymbol{r}_{O'} \times \boldsymbol{\nu})_j dS + \rho \int\limits_{S+\Sigma_0} (\boldsymbol{r}_{O'} \times \boldsymbol{\nu})_j (\boldsymbol{r}_{O'} \\ &\quad + \boldsymbol{r}')_i dS + \rho \int\limits_{S+\Sigma_0} (\boldsymbol{r}' \times \boldsymbol{\nu})_j (\boldsymbol{r}_{O'} \times \boldsymbol{r}')_i dS \\ &= J_{ij}^{1'} + \rho \int\limits_{S+\Sigma_0} \Omega_{i0}'(\boldsymbol{r}_{O'} \times \boldsymbol{\nu})_j dS + \rho \int\limits_{S+\Sigma_0} (\boldsymbol{r}_{O'} \times \boldsymbol{\nu})_j (\boldsymbol{r}_{O'} + \boldsymbol{r}')_i dS \\ &\quad + \rho \int\limits_{S+\Sigma_0} (\boldsymbol{r}' \times \boldsymbol{\nu})_j (\boldsymbol{r}_{O'} \times \boldsymbol{r}')_i dS. \end{aligned} \tag{4.7.41}$$

By using the Gauss theorem, we get

$$\begin{aligned} J_{11}^1 &= J_{11}^{1'} + 2m_1(y_{O'}y_{C_1}' + z_{O'}z_{C_1}') + m_1(y_{O'}^2 + z_{O'}^2), \\ J_{22}^1 &= J_{22}^{1'} + 2m_1(z_{O'}z_{C_1}' + x_{O'}x_{C_1}') + m_1(z_{O'}^2 + x_{O'}^2), \\ J_{33}^1 &= J_{33}^{1'} + 2m_1(x_{O'}x_{C_1}' + y_{O'}y_{C_1}') + m_1(x_{O'}^2 + y_{O'}^2), \\ J_{12}^1 &= J_{21}^1 = J_{12}^{1'} - m_1(x_{O'}y_{C_1}' + y_{O'}x_{C_1}') - m_1x_{O'}y_{O'}, \\ J_{13}^1 &= J_{31}^1 = J_{13}^{1'} - m_1(x_{O'}z_{C_1}' + z_{O'}x_{C_1}') - m_1x_{O'}z_{O'}, \\ J_{23}^1 &= J_{32}^1 = J_{23}^{1'} - m_1(y_{O'}z_{C_1}' + z_{O'}y_{C_1}') - m_1y_{O'}z_{O'}, \end{aligned} \tag{4.7.42}$$

where $x_{O'}, y_{O'}$, and $z_{O'}$ denote the coordinates of the point $O'$ in the coordinate system $Oxyz$ (with origin at $O$) and $x'_{C_1}, y'_{C_1}$, and $z'_{C_1}$ are the coordinates of the center of inertia of the liquid domain with unperturbed boundaries in the coordinate system $O'x'y'z'$ whose axes are parallel to the axis of the first system. Relations (4.7.42) are simplified if $O'$ is the center of inertia of the unperturbed liquid domain $C_1$. Thus,

$$J_{11}^1 = J_{11}^{1'} + m_1(y_{C_1}^2 + z_{C_1}^2), \quad J_{12}^1 = J_{12}^{1'} - m_1 x_{C_1} y_{C_1}. \tag{4.7.43}$$

Note that relations (4.7.42) express the Steiner (parallel axis) theorem well-known in the dynamics of rigid body .

The presented linear systems of equations are close or completely coincide with the systems presented in [149, 191, 59]. It is clear that only finite-dimensional analogs of the equations of motion (4.7.24)–(4.7.27), (4.7.32) are used for practical purposes,. Moreover, most often, it is reasonable to use solely the equations for the predominant natural sloshing modes.

## 4.8 Nonlinear equations of perturbed motion of the body containing a cylindrical cavity partially filled with liquid

For practical purposes, it is often necessary to get the equations of perturbed motion of a mechanical body–liquid system suitable for the analytic investigation of various types of motion of this system. As usual, the unperturbed motion of the system is characterized by a certain specific collection of the parameters (generalized coordinates) obtained as a partial solution of the nonlinear system of equations of motion. These parameters possess numerous characteristic properties connected with the behavior of the analyzed system. Hence, the problem of construction of the equations of perturbed motions is always very closely connected with the problem of existence of physically meaningful partial solutions of the original system of nonlinear equations of motion. Note that it is reasonable to test these solutions for stability.

We now deduce the nonlinear equations of perturbed motion of the body filled with liquid suitable for the investigation of stability of the system in the cases where the hypothesis of smallness of all parameters of motion is not true. In particular, this is observed for the motion of a system characterized by the resonance excitation of some natural sloshing modes. This excitation results in complex space motions of the free surface with finite amplitude even for small linear or angular displacements of the rigid body. In this case, our analysis is based on the system of hypotheses traditionally used in the dynamics of aircrafts. The other special cases of the equations of perturbed motion based on the constructed system of nonlinear equations are considered in Chapter 8.

**Table 4.29**

| Coordinate axes | $O_1X$ | $O_1Y$ | $O_1Z$ |
|---|---|---|---|
| $Ox$ | $\cos\vartheta\cos\psi$ | $\sin\vartheta\cos\psi$ | $-\sin\psi$ |
| $Oy$ | $\cos\vartheta\sin\psi\sin\gamma$ $-\sin\vartheta\cos\gamma$ | $\sin\vartheta\sin\psi\sin\gamma$ $+\cos\vartheta\cos\gamma$ | $\cos\psi\sin\gamma$ |
| $Oz$ | $\cos\vartheta\sin\psi\cos\gamma$ $+\sin\vartheta\sin\gamma$ | $\sin\vartheta\sin\psi\cos\gamma$ $-\cos\vartheta\sin\gamma$ | $\cos\psi\cos\gamma$ |

The rigid body is referred to a coordinate system $Oxyz$ moving together with the body relative to the absolute coordinates system $O_1XYZ$ in which the field of mass forces is specified by a potential function $U$. The absolute coordinate system is regarded as inertial. The axes of this system are connected with the Earth so that the direction of the $O_1Y$-axis is opposite to the direction of the gravity force. The axes $O_1X$ and $O_1Z$ are placed in the horizontal plane. At the initial time, the points $O_1$ and $O$ coincide and the directions of the $Ox$- and $O_1Y$-axes also coincide. Assume that the $Oz$-axis is parallel to the $O_1Z$-axis at the initial time. In this case, the $Oy$-axis is directed along the $O_1X$-axis but in the opposite direction. The directions of the $Ox$-, $Oy$-, and $Oz$-axes relative to the absolute coordinate system are uniquely determined by three angles: the pitch angle $\vartheta$ (the angle between the $O_1X$-axis and the plane $OXZ$), the yaw angle $\psi$ (theangle between the $Ox$-axis and the plane $OXY$), and the roll angle $\gamma$ (the angle between the $Oy$-axis and the line of intersection of the planes $Oyz$ and $OXY$). The cosines of the angles between the axes of the absolute and fixed coordinates systems are presented in Table 4.29.

The motion of the rigid body is characterized by the vector of translational velocity $\boldsymbol{v}_0$ of the point $O$ and the vector of instantaneous angular velocity $\boldsymbol{\omega}$ relative to the point $O$. The kinematic equations used to establish the relationships between the projections of the angular velocity onto the axes of the fixed system with angular parameters characterizing the position of the body relative to the absolute system of coordinates are as follows:

$$\begin{aligned}&\omega_1 = \dot\gamma - \dot\vartheta\sin\psi, \quad \omega_2 = \dot\psi\cos\gamma + \dot\vartheta\cos\psi\sin\gamma,\\ &\omega_3 = -\dot\psi\sin\gamma + \dot\vartheta\cos\psi\cos\gamma.\end{aligned} \tag{4.8.1}$$

We transform the nonlinear system of equations (4.7.8)–(4.7.16) by choosing its program motion described by the simplified equations of motion of a material point located at the center-of-mass of the system as its unperturbed motion [1, 191]. The right-hand sides of the equations of forces (4.7.8) and moments (4.7.9)–(4.7.11) are specified by the character of the problem. Thus, we distinguish the following forces acting upon the aircraft: traction of the engine, aerodynamic forces, controlling forces, Coriolis forces caused by the relative motion of particles inside the rotating hull of the aircraft, and the forces caused by the displacements of the center-of-mass of the system relative to the hull. The free liquid surface in the process of unperturbed motion is close to the plane perpendicular to the longitudinal axis and the parameters $r_i$, $p_i$ are set equal to zero.

In the program motion, the yaw and roll angles are, as a rule, equal to zero. We also assume that

$$\psi = 0 \quad \text{and} \quad \gamma = 0 \tag{4.8.2}$$

for the nonperturbed motion.

In constructing the differential equations of perturbed motion, we agree to consider

(a) the perturbations of the main parameters characterizing the motion of the rigid body;

(b) the perturbations of mass $\Delta m$ and moments of inertia of the rigid body;

(c) the perturbations caused the operation of engine;

(d) the time derivatives of the yaw, roll, and pitch angles, as well as the angle of attack and its time derivative

as small values.

By virtue of the last assumption, the projections of the velocity $\boldsymbol{v}_G$ onto the $G_y$- and $G_z$-axes and their time derivatives are also small quantities. We consider the parameters characterizing the motions of liquid as finite quantities. The perturbed values of the main parameters of motion are marked by the primes. Then the perturbations of the main parameters is found as the difference between their perturbed and unperturbed values:

$$\begin{gathered}\vartheta(t) = \vartheta'(t) - \vartheta^0(t), \quad \psi(t) = \psi'(t) - \psi^0(t) \equiv \psi'(t), \\ \gamma(t) = \gamma'(t) - \gamma^0(t) \equiv \gamma'(t), \quad \boldsymbol{v}_{G_0}(t) = \boldsymbol{v}'_G(t) - \boldsymbol{v}^0_G(t), \\ r_i(t) = r'_i(t) - r^0_i(t) \equiv r'_i(t), \quad p_i(t) = p'_i(t) - p^0_i(t) \equiv p'_i(t).\end{gathered} \tag{4.8.3}$$

The principal relationships between the actual and unperturbed quantities are established with regard for the fact that the positions of the fixed coordinate system at time $t$ for the program and actual motions are different: $Gxyz$ and $Gx'y'z'$, respectively. The transition from the coordinate system in the actual motion to the coordinate system in the unperturbed motion is performed by using the transition matrix [1]:

$$M = \begin{vmatrix} 1 & \vartheta & -\psi \\ -\vartheta & 1 & \gamma \\ \psi & -\gamma & 1 \end{vmatrix}. \tag{4.8.4}$$

To within the quantities of the second order of smallness, the following relations are true:

$$\omega'_{x'} = \dot{\gamma}, \;\; \omega'_{y'} = \dot{\psi}, \;\; \omega'_{z'} = \dot{\vartheta}^0 + \dot{\vartheta}\,; \tag{4.8.5}$$

$$v'_{G_{x'}} = v^0_{G_x}\vartheta + v_x, \;\; v'_{G_{y'}} = -v^0_{G_x}\vartheta + v_y + v^0_{G_y},$$
$$v'_{G_{z'}} = v^0_{G_x}\psi + v^0_{G_z} + v_z\,; \tag{4.8.6}$$

$$g_{x'} = g_x + g_y\vartheta, \;\; g_{y'} = -g_x\vartheta + g_y, \;\; g_{z'} = g_x\psi - g_y\gamma + g_z. \tag{4.8.7}$$

As a result of the transformation of the system of nonlinear equations (4.7.8)–(4.7.16) with regard for the restrictions introduced above, we obtain the equations of perturbed motion of the object in the form [118]

$$\begin{aligned}
&M\ddot{x}_{G'} + a\dot{x}_{G'} - mg_y\gamma + \lambda_{1x}(\dot{r}_1 r_1 + \dot{p}_1 p_1)^{\cdot} \\
&\qquad + \lambda(2\dot{r}_1\dot{\psi} - \dot{\vartheta}\dot{p}_1 + r_1\ddot{\psi} - p_1\ddot{\vartheta} - \ddot{\vartheta}^0\dot{p}_1 - \dot{\vartheta}^0 p_1) = c_{x\delta}\delta_x + \triangle F_x, \\
&M\ddot{y}_{G'} + b\dot{y}_{G'} + v_y\dot{\vartheta} - (P + bv_{G_x})\vartheta + \lambda(\ddot{p}_1 - 2\dot{\gamma}\dot{r}_1 - r_1\ddot{\gamma}) \\
&\qquad = c_{y\delta}\delta_\vartheta + \triangle F_y, \\
&M\ddot{z}_{G'} + b\dot{z}_{G'} + v_z\dot{\psi} + (P + dv_{G_x})\psi + mg_y\gamma \\
&\qquad + \lambda(\ddot{r}_1 + 2\dot{\gamma}\dot{p}_1 + p_1\ddot{\gamma}) = c_{z\delta}\delta_\psi + \triangle F_z;
\end{aligned} \tag{4.8.8}$$

$$\begin{aligned}
&J^0_{11}\ddot{\gamma} + \mu_x\dot{\gamma} - \lambda_0(p_1\ddot{\psi} + r_1\ddot{\vartheta}^0 + r_1\ddot{\vartheta}) - \mu_1(r_1\ddot{p}_1 - \ddot{r}_1 p_1) \\
&\qquad + \lambda p_1(\dot{v}_{G_z} - \dot{v}_{G_x}\psi + \ddot{z}_{G'} - g_x\psi + g_y\gamma - g_z) \\
&\qquad - \lambda r_1(\dot{v}_{G_y} - \dot{v}_{G_x}\vartheta + \ddot{y}_{G'} + \dot{\vartheta}^0 v_{G_x} + g_x\vartheta - g_y) = c_{\gamma\delta}\delta_\gamma + \triangle M_x, \\
&(J^0_{22} + G^0_{22})\ddot{\psi} + \mu_y\dot{\psi} - x_F b\dot{z}_{G'} - x_F b v_{G_x}\psi - \lambda_0(2\dot{\gamma}\dot{p}_1 + p_1\ddot{\gamma}) \\
&\qquad - \lambda r_1(\dot{v}_{G_x} - g_x + \ddot{x}_{G'} - g_y\vartheta) - M_{yf} = c_{\psi\delta}\delta_\psi + \triangle M_y, \\
&(J^0_{33} + G^0_{33})\ddot{\vartheta} + \mu_z\dot{\vartheta} + x_F b\dot{y}_{G'} - x_F b v_{G_x}\vartheta - \lambda_0(2\dot{r}_1\dot{\gamma} + r_1\ddot{\gamma}) \\
&\qquad - \lambda p_1(\dot{v}_{G_x} - g_x + \ddot{x}_{G'} - g_y\vartheta) - M_{zf} = c_{\vartheta\delta}\delta_\vartheta + \triangle M_z;
\end{aligned} \tag{4.8.9}$$

$$\begin{aligned}
&\mu_1(\ddot{r}_1 + \sigma_1^2 r_1) + \mu_1(p_1\ddot{\psi} + 2\dot{\gamma}\dot{p}_1) - \lambda_0\ddot{\psi} + \lambda(\dot{v}_{G_x} - g_x)\psi \\
&\qquad + \lambda\ddot{z}_{G'} + \lambda g_y\gamma + L_1(r_i, p_i) = 0, \\
&\mu_1(\ddot{p}_1 + \sigma_1^2 p_1) - \mu_1(r_1\ddot{\gamma} + 2\dot{r}_1\dot{\gamma}) + \lambda_0\ddot{\vartheta} - \lambda(\dot{v}_{G_x} - g_x)\vartheta \\
&\qquad + \lambda\ddot{y}_{G'} + L_2(r_i, p_i) = 0,
\end{aligned} \tag{4.8.10}$$

$$\begin{aligned}
&\mu_0(\ddot{p}_0 + \sigma_0^2 p_0) + c_3\dot{p}_1(\dot{\vartheta}^0 + \dot{\vartheta}) - c_3\dot{r}_1\dot{\psi} + c_5(\dot{\psi}r_1 - \dot{\vartheta}^0 p_1 - \dot{\vartheta}p_1)^{\cdot} \\
&\qquad + L_3(r_i, p_i) = 0, \\
&\mu_2(\ddot{r}_2 + \sigma_2^2 r_2) - c_4(\dot{\psi}\dot{p}_1 - \dot{\vartheta}^0\dot{r}_1 - \dot{\vartheta}\dot{r}_1) + c_6(\dot{\psi}p_1 - \dot{\vartheta}^0 r_1 - \dot{\vartheta}r_1)^{\cdot} \\
&\qquad + L_4(r_i, p_i) = 0,
\end{aligned}$$

$$\mu_2(\ddot{p}_2+\sigma_2^2 p_2)+c_4(\dot{r}_1\dot{\psi}+\dot{p}_1\dot{\vartheta}^0+\dot{p}_1\dot{\vartheta})$$
$$-c_6(\dot{\psi}r_1+\dot{\vartheta}^0 p_1+\dot{\vartheta}p_1)^{\cdot}+L_5(r_i,\,p_i)=0,$$

where

$$\begin{aligned}
M_{\text{yf}} &= \lambda_0\ddot{r}_1 - c_1(\dot{r}_1 r_1^2 + r_1p_1\dot{p}_1)^{\cdot} - c_2(r_1p_1\dot{p}_1 - \dot{r}_1p_1^2)^{\cdot}\\
&\quad -c_3(\dot{r}_1p_0)^{\cdot} - c_4(\dot{p}_1r_2 - r_1\dot{p}_2)^{\cdot} - c_5(r_1\dot{p}_0)^{\cdot} - c_6(p_1\dot{r}_2 - r_1\dot{p}_2)^{\cdot},\\
M_{\text{zf}} &= -\lambda_0\ddot{p}_1 + c_1(\dot{p}_1p_1^2 + r_1p_1\dot{r}_1)^{\cdot} - c_2(\dot{p}_1r_1^2 - r_1p_1\dot{r}_1)^{\cdot}\\
&\quad +c_3(\dot{p}_1p_0)^{\cdot} + c_4(\dot{p}_1p_2 + \dot{r}_1r_2)^{\cdot} + c_5(p_1\dot{p}_0)^{\cdot} + c_6(p_1\dot{p}_2 + r_1\dot{r}_2)^{\cdot}, \qquad (4.8.11)\\
L_1(r_i,\,p_i) &= d_1r_1(r_1\dot{r}_1 + p_1\dot{p}_1) + d_2(p_1^2\ddot{r}_1 + 2p_1\dot{r}_1\dot{p}_1 - r_1p_1\ddot{p}_1 - 2r_1\dot{p}_1^2)\\
&\quad -d_3(p_2\dot{r}_1 - r_2\dot{p}_1)^{\cdot} + d_4(r_1\ddot{p}_2 - p_1\ddot{r}_2) + d_5(p_0\dot{r}_1)^{\cdot} + d_6r_1\ddot{p}_0,\\
L_2(r_i,\,p_i) &= d_1p_1(r_1\dot{r}_1 + p_1\dot{p}_1)^{\cdot} + d_2(r_1^2\ddot{p}_1 - r_1p_1\ddot{r}_1 + 2r_1\dot{r}_1\dot{p}_1 - 2p_1\dot{r}_1^2)\\
&\quad +d_3(p_2\dot{p}_1 + r_2\dot{r}_1)^{\cdot} - d_4(p_1\ddot{p}_2 + r_1\ddot{r}_2) + d_5(p_0\dot{p}_1)^{\cdot} + d_6p_1\ddot{p}_0,\\
L_3(r_i,\,p_i) &= d_6(r_1\ddot{r}_1 + p_1\ddot{p}_1) + d_8(\dot{r}_1^2 + \dot{p}_1^2),\\
L_4(r_i,\,p_i) &= -d_4(\ddot{r}_1p_1 + \ddot{p}_1r_1) - 2d_7\dot{r}_1\dot{p}_1,\\
L_5(r_i,\,p_i) &= d_4(\ddot{r}_1r_1 - \ddot{p}_1p_1) + d_7(\dot{r}_1^2 - \dot{p}_1^2).
\end{aligned}$$

By $x_{G'}$, $y_{G'}$, and $z_{G'}$, we denote the coordinates of the center of inertia of the object for the perturbed motion in the system of coordinates $Gxyz$. The right-hand sides of Eqs. (4.8.8) and (4.8.9) include the projections of the perturbed forces and moments in the directions of the axes of the coordinate system $Gxyz$. If the positions of the axes of fixed coordinate system in the unperturbed motion are known, then the parameters of motion $x_{G'}$, $y_{G'}$, $z_{G'}$, $\vartheta$, $\psi$, $\gamma$, $r_i$, and $p_i$, completely specify the configuration of the system at time $t$ and can be regarded as its generalized coordinates.

For the construction of the equations of perturbed forces (4.8.8) and moments (4.8.9), we used similar equations of perturbed motion of the objects in the absence of liquid available in the literature [1]. The possibilities of subsequent simplification of the presented system of equations of perturbed motion can be clarified as a result of more comprehensive analyses of the kinematic and dynamic properties of specific mechanical systems.

# Chapter 5

# Nonlinear modal equations for noncylindical axisymmetric tanks

The present chapter is devoted to the analytic approaches [114, 131, 123, 132, 70, 73] aimed at the development of finite-dimensional nonlinear mathematical models (modal equations) approximating the processes of liquid sloshing in axisymmetric tanks of complex shape that are not cylindrical on the free surface.

The application of the variational modal method may become complicated, first of all, due to the absence of exact solutions of the spectral boundary problem (2.2.4), which can be used to specify the velocity potential and the shape of the perturbed free surface as in the case of cylindrical cavities. However, the indicated solutions are, as a rule, given in the Cartesian coordinate system and, hence, for noncylindrical domains with unperturbed boundaries, they do not possess, with rare exceptions, the required properties, including the property of having the domain of definition that coincides with the unperturbed free surface. Furthermore, special approximate analytic approaches are required to guarantee, e.g., the validity of the slip condition $\partial\varphi/\partial\nu = 0$ on the entire rigid wall or, at least, on its wetted part. The validity of this condition on the rigid wall over the mean free surface is required to guarantee the conservation of the mass of liquid in the presence of deviations of the finite free surface.

The problem of choosing the generalized coordinates characterizing the free-surface perturbations relative to the unperturbed state is also of high importance. Since, according to the accumulated results, the time-dependent coefficients of the Fourier expansions for the perturbed free surface play the role of these coordinates, the problem of construction of modal solutions in a neighborhood of the unperturbed free surface is a fundamental problem of the analyzed theory.

In the present chapter, we follow the ideas of the so-called technique of nonconformal mappings and describe its details for conical and, as an illustration, spherical containers. This requires the analysis of the original free-boundary problem in nonorthogonal systems of curvilinear coordinates by using the tensor form of equations and boundary conditions, as described in Sec. 1.5. For truncated circular conical tanks, we use the nonconformal transformation of the original domain into an upright circular cylinder. This makes it possible to construct special solutions of the original spectral boundary problem in partly separated variables. The eigenfunctions (natural sloshing modes) exactly satisfy the slip condition $\partial\varphi/\partial\nu = 0$ on the conical walls.

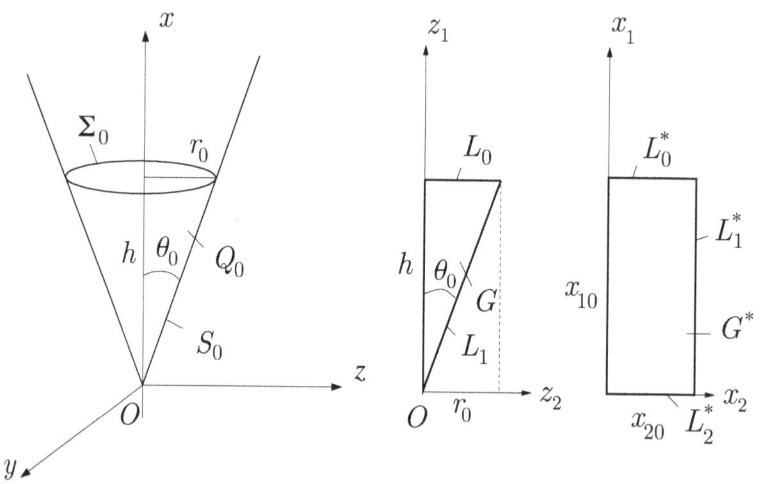


**Figure 5.1.** Schematic of the nontruncated conical tank and its meridional cross section.

## 5.1 Natural sloshing modes for the nontruncated conical tank

For a circular conical tank, the relationship between the Cartesian and curvilinear coordinates is specified as follows [112]:

$$x = x_1, \quad y = x_1 x_2 \cos x_3, \quad z = x_1 x_2 \sin x_3;$$
$$x_2 = \frac{\sqrt{y^2 + z^2}}{x}, \quad x_3 = \operatorname{atan}(z/y), \tag{5.1.1}$$

where the $Ox$-axis is directed along the axis of symmetry. In the coordinate system $Ox_1x_2x_3$, we have $x_2 = \tan\theta_0$, and $x_3 = \eta$ is the polar angle in the plane $Oyz$ measured from the $Oy$-axis in the direction of the $Oz$-axis. In these coordinates, the unperturbed liquid domain $Q_0$ takes the form of a right circular cylinder with meridional cross section $G^*$ (Fig. 5.1) in the plane $Ox_1x_2$ in the form of a rectangle with sides $h$ and $x_{20} = \tan\theta_0$, where $h$ and $\theta_0$ are the depth of liquid and the semiapex angle of the cone, respectively.

The perturbed free surface $\Sigma^*(t)$ can be projected in the transformed domain onto the unperturbed surface $\Sigma_0^*$ of radius $x_{20} = \tan\theta_0$.

By using the results obtained in Sec. 1.5, we reduce the Laplace equation for the function $\varphi(x, y, z)$ to the following differential equation:

$$\frac{\partial^2\varphi}{\partial x_1^2} - \frac{2x_2}{x_1}\frac{\partial^2\varphi}{\partial x_1 \partial x_2} + \frac{(1+x_2^2)}{x_1^2}\frac{\partial^2\varphi}{\partial x_2^2} + \frac{(2x_2^2+1)}{x_1^2 x_2}\frac{\partial\varphi}{\partial x_2} + \frac{1}{x_1^2 x_2^2}\frac{\partial^2\varphi}{\partial x_3^2} = 0 \tag{5.1.2}$$

and, hence, for the natural sloshing modes, we get

$$\varphi(x_1, x_2, x_3) = \psi_m(x_1, x_2) {\sin m\eta \atop \cos m\eta}, \quad x_3 = \eta, \quad m = 0, 1, 2, \dots . \tag{5.1.3}$$

Thus, in the domain $G^*$ of the variables $x_1$ and $x_2$, we arrive at the following spectral boundary problem for $\psi_m(x_1,\, x_2)$:

$$L(\psi_m) = p\frac{\partial^2 \psi_m}{\partial x_1^2} + 2q\frac{\partial^2 \psi_m}{\partial x_1 \partial x_2} + s\frac{\partial^2 \psi_m}{\partial x_2^2} + d\frac{\partial \psi_m}{\partial x_2} - m^2 c\, \psi_m = 0 \text{ in } G^*, \tag{5.1.4}$$

$$p\frac{\partial \psi_m}{\partial x_1} + q\frac{\partial \psi_m}{\partial x_2} = \varkappa p\, \psi_m \text{ on } L_0^*, \tag{5.1.5}$$

$$s\frac{\partial \psi_m}{\partial x_2} + q\frac{\partial \psi_m}{\partial x_1} = 0 \text{ on } L_1^*, \quad p\frac{\partial \psi_m}{\partial x_1} + q\frac{\partial \psi_m}{\partial x_2} = 0 \text{ on } L_2^*, \tag{5.1.6}$$

$$\int_0^{x_{20}} \psi_0 x_2 \, dx_2 = 0, \tag{5.1.7}$$

where

$$p = x_1^2 x_2; \quad q = -x_1 x_2^2; \quad s = x_2(x_2^2 + 1); \quad d = 1 + 2x_2^2; \quad c = \frac{1}{x_2};$$

$$G^* = \{(x_1, x_2) : \, h_0 \leqslant x_1 \leqslant h_1, \; 0 \leqslant x_2 \leqslant x_{20}\},$$

and $L_0^*$, $L_1^*$, and $L_2^*$ are the lines of intersection of the meridional plane with the free boundary, the lateral surface of the cone, and the bottom of the cavity, respectively.

The dependences of the coefficients $p$, $q$, ..., $c$ on $x_1$ and $x_2$ are such that Eq. (5.1.4) admits the separation of variables and, moreover,

$$\psi_m(x_1, x_2) = x_1^\nu V_\nu^{(m)}(x_2), \quad \bar{\psi}_m(x_1, x_2) = x_1^{-(\nu+1)} \bar{V}_\nu(x_2). \tag{5.1.8}$$

To find $V_\nu^{(m)}(x_2)$ from (5.1.4)–(5.1.7), we get the following eigenvalue problem containing a parameter $\nu$ both in the differential equation and in the boundary conditions [123]:

$$x_2(1 + x_2^2)V_\nu''^{(m)} + (1 + 2x_2^2 - 2\nu x_2^2)V_\nu'^{(m)} + \left[\nu(\nu - 1)x_2 - \frac{m^2}{x_2}\right] V_\nu^{(m)} = 0,$$
$$V_\nu'(x_{20}) = \nu\, a(x_{20}) V_\nu(x_{20}), \quad a(x_{20}) = \frac{x_{20}}{1 + x_{20}^2}, \tag{5.1.9}$$

with the condition of boundedness of the solution for $x_2 = 0$. By the change of variables

$$\mu = (1 + x_2^2)^{-\frac{1}{2}} \tag{5.1.10}$$

we reduce Eq. (5.1.9) to the form

$$V_\nu''^{(m)}(\mu)+\left(\frac{2\nu}{\mu}-\frac{2\mu}{1-\mu^2}\right)V_\nu'^{(m)}(\mu)$$
$$+\left[\frac{\nu(\nu-1)(1-\mu^2)-m^2\mu^2}{\mu^2(1-\mu^2)^2}\right]V_\nu^{(m)}(\mu)=0. \quad (5.1.11)$$

By the substitution $y(\mu)=\mu^\nu V(\mu)$, $\mu=\cos\theta$, we obtain the following equation from (5.1.11):

$$(1-\mu^2)y''(\mu)-2\mu y'(\mu)+\left[\nu(\nu+1)-\frac{m^2}{1-\mu^2}\right]y(\mu)=0. \quad (5.1.12)$$

Its solutions are the associated Legendre functions of the first kind $y(\mu)=P_\nu^{(m)}(\cos\theta)$ but the corresponding boundary condition leads to the transcendental equation

$$\left.\frac{\partial P_\nu^{(m)}(\cos\theta)}{\partial\theta}\right|_{\theta=\theta_0}=0 \quad (5.1.13)$$

which has an infinite sequence of roots $\nu_{mi}$ for every fixed $m$.

The associated Legendre functions of the first kind can be represented in the form of a hypergeometric series

$$P_\nu^m(\cos\theta)=\left(-\frac{\sin\theta}{2}\right)^m$$
$$\times\sum_{p=1}^{\infty}\frac{(\nu-m-p+1)(\nu-m-p+2)\,\dots\,(\nu+m+p)}{p!(m+p)!}\left(-\sin^2\frac{\theta}{2}\right)^p. \quad (5.1.14)$$

Thus, we obtain the following partial solutions of Eq. (5.1.9):

$$V_{\nu_{mk}}^{(m)}(x_2)=a_{\nu_{mk}}^{(m)}\left(\sqrt{1+x_2^2}\right)^{\nu_{mk}}P_{\nu_{mk}}^m\left(\frac{1}{\sqrt{1+x_2^2}}\right), \quad (5.1.15)$$

where $a_{\nu_{mk}}^{(m)}$ is a scaling factor, which is set equal to

$$\frac{2^m m!(\nu_{mk}-m)!}{(\nu_{mk}+m)!}.$$

For (5.1.15), we get the following recurrence relations: [123]

$$(\nu+m+1)V_{\nu+1}^{(m)}=(2\nu+1)V_\nu^{(m)}-(\nu-m)V_{\nu-1}^{(m)}(1+x_2^2),$$
$$x_2\frac{dV_\nu^{(m)}}{dx_2}=\nu V_\nu^{(m)}-(\nu-m)V_{\nu-1}^{(m)}, \quad (5.1.16)$$
$$(\nu+m+1)x_2V_\nu^{(m+1)}=2(m+1)[(1+x_2^2)V_{\nu-1}^{(m)}-V_\nu^{(m)}]$$

playing an important role in the numerical analysis.

The second family of partial solutions from (5.1.8) is connected with the eigensolutions of the problem

$$x_2^2(1+x_2^2)\bar{V}''(x_2)+x_2(1+4x_2^2+2\nu x_2^2)\bar{V}'(x_2) \\ +[(\nu+1)(\nu+2)x_2^2-m^2]\bar{V}(x_2)=0, \\ \bar{V}'(x_{20})+(\nu+1)\,a(x_{20})\,\bar{V}(x_{20})=0. \tag{5.1.17}$$

By the change of variables (5.1.10) and the substitution

$$y(\mu)=\mu^{-\nu-1}\,\bar{V}(\mu),\quad \mu=\cos\theta,$$

we find

$$\bar{V}_{\nu_{mk}}^{(m)}(x_2)=a_{\nu_{mk}}^{(m)}\left(\sqrt{1+x_2^2}\right)^{-\nu_{mk}-1}P_{\nu_{mk}}^{m}\left((1+x_2^2)^{-1/2}\right). \tag{5.1.18}$$

The family of partial solutions (5.1.18) obeys the recurrence relations

$$(\nu+m+1)(1+x_2^2)\bar{V}_{\nu+1}^{(m)}=(2\nu+1)\bar{V}_{\nu}^{(m)}-(\nu-m)\bar{V}_{\nu-1}^{(m)},$$

$$x_2(1+x_2^2)\frac{d\bar{V}_{\nu}^{(m)}}{dx_2}=[\nu-(\nu+1)x_2^2]\bar{V}_{\nu}^{(m)}-(\nu-m)\bar{V}_{\nu-1}^{(m)}, \tag{5.1.19}$$

$$(\nu+m+1)\,x_2\,\bar{V}_{\nu}^{(m+1)}=2(m+1)[\bar{V}_{\nu-1}^{(m)}-\bar{V}_{\nu}^{(m)}].$$

Thus, for every integer superscript $m$, we get two infinite-dimensional families of partial solutions characterized by the real numbers $\nu_{mk}$

$$\psi_k^{(m)}=x_1^{\nu_{mk}}V_{\nu_{mk}}^{(m)}(x_2),\quad \bar{\psi}_k^{(m)}=x_1^{-\nu_{mk}-1}\bar{V}_{\nu_{mk}}^{(m)}(x_2) \tag{5.1.20}$$

($m=0,1,2,\ldots,k=1,2,\ldots$). This enables us to exactly satisfy the boundary condition of the original problem on the conical walls.

The first family is regular at the apex of the cone ($x_0=0$). Therefore, it is used for the solution of the spectral problems of sloshing for nontruncated conical tanks. The second family (5.1.20) is singular for $x_1=0$. It is used, together with the first family, for truncated conical tanks.

Note that these two families do not exhaust all practically possible conical domains. Thus, for the conical tanks formed by two coaxial conical surfaces, solutions (5.1.20) should be supplemented by partial solutions with singularities along the axis of symmetry:

$$\psi_k^{*(m)}=x_1^{\nu_{mk}}V_{\nu_{mk}}^{*(m)}(x_2),\quad \bar{\psi}_k^{*(m)}=x_1^{-\nu_{mk}-1}\bar{V}_{\nu_{mk}}^{*(m)}(x_2), \tag{5.1.21}$$

where

$$V_{\nu_{mk}}^{*(m)}(x_2)=a_{\nu_{mk}}^{*(m)}\left(\sqrt{1+x_2^2}\right)^{\nu_{mk}}Q_{\nu_{mk}}^{m}\left((1+x_2^2)^{-1/2}\right),$$

$$\bar{V}^{*(m)}_{\nu_{mk}}(x_2) = \bar{a}^{*(m)}_{\nu_{mk}} \left(\sqrt{1+x_2^2}\right)^{-\nu_{mk}-1} Q^m_{\nu_{mk}}\left((1+x_2^2)^{-1/2}\right), \tag{5.1.22}$$

and $Q^m_{\nu_{mk}}$ is the Legendre function of the second kind.

The solutions $V^{*(m)}_\nu(x_2)$ and $\bar{V}^{*(m)}_\nu(x_2)$ satisfy the recurrence relations (5.1.16) and (5.1.19).

Depending on the considered conical shape, the required solutions of the corresponding spectral boundary problems (5.1.4)–(5.1.7) are represented in the form of linear combinations of solutions (5.1.20) and (5.1.21). To satisfy the slip condition, we choose $\nu_{mk}$ as the $k$th real root of the equation

$$\frac{d}{d\theta}\left[P^m_\nu(\cos\theta_1)\right]\frac{d}{d\theta}\left[Q^m_\nu(\cos\theta_0)\right] - \frac{d}{d\theta}\left[Q^m_\nu(\cos\theta_1)\right]\frac{d}{d\theta}\left[P^m_\nu(\cos\theta_0)\right] = 0, \tag{5.1.23}$$

where $\theta_0$ is the semiapex angle of the outer conical surface and $\theta_1$ is the semiapex angle of the inner conical surface.

Earlier, we have constructed partial solutions of Eq. (5.1.4) satisfying condition (5.1.6) on the conical surface. In the case where this condition is not satisfied, Eq. (5.1.4) admits, for integer $k$, solutions of the polynomial type

$$\psi_m = x_1^k v_k^{(m)}(x_2). \tag{5.1.24}$$

Thus, for $m = 0$, the polynomials $v_k(x_2)$ bounded for $x_2 = 0$ take the form

$$v_0^{(0)} = 1,\ v_1^{(0)} = 1,\ v_2^{(0)} = 1 - \tfrac{1}{2}x_2^2,\ v_3^{(0)} = 1 - \tfrac{3}{2}x_2^2, \ldots. \tag{5.1.25}$$

In this case, the recurrence relations (5.1.16) are also true. By using these relations, one can obtain similar polynomials for any integer numbers $m$ and $k$ according to the first values $v_0^{(0)}$ and $v_1^{(0)}$ of sequence (5.1.25).

In some cases, partial solutions (5.1.24) should be supplemented by the other families of solutions obtained for integer $k$ and possessing the properties of boundedness at a point or on the axis of the mean liquid domain [124].

Note that, for a class of partial solutions (5.1.24), the exact solutions of the sloshing problem for the first natural sloshing frequency in containers of "inverted" conical shape with semiapex angle $\theta_0 = \operatorname{atan}\sqrt{m}$ are obtained in [42].

In the general case, different approximate methods can be used for the solution of the spectral problem (5.1.4)–(5.1.7). Among these methods, the variational method is used especially extensively [124, 132, 67]. We now proceed to the solution of the spectral boundary problem with a parameter $\varkappa$ in the boundary condition on the unperturbed free surface (5.1.4)–(5.1.7). To find the eigensolutions of this problem, we use the partial solutions of Eq. (5.1.4) in the form

$$\psi_{mn}(x_1, x_2) = \sum_{k=0}^{q} \bar{a}_k^{(m)} w_k^{(n)}(x_1, x_2). \tag{5.1.26}$$

We now consider this spectral boundary problem for a conical cavity whose vertex is directed downward. According to the comments made above, we can use the system of functions

$$w_k^{(m)}(x_1, x_2) = \left(\frac{x_1}{x_{10}}\right)^{\nu_{mk}} V_{\nu_{mk}}^{(m)}(x_2). \tag{5.1.27}$$

By virtue of the condition of conservation of the volume of liquid, which is automatically satisfied for $m \neq 0$, the constant $\bar{a}_{n0}^{(m)}$ in relation (5.1.26) should be specified as follows:

$$\bar{a}_{n0}^{(m)} = \begin{cases} 0, \quad m = 0, \\ -\sum_{k=1}^{q} \bar{a}_{nk} c_{\nu_{0k}}, \quad m \neq 0, \end{cases}$$

where

$$c_{\nu_{0k}} = \frac{2}{x_{20}^2} \int_0^{x_{20}} w_{\nu_{0k}}^{(0)}(x_{10}, x_2) x_2 \, dx_2.$$

The spectral problem (5.1.4)–(5.1.7) can be associated with the variational problem of minimization of the quadratic functional

$$J(\psi_m) = \int_{G^*} \left[ p \left(\frac{\partial \psi_m}{\partial x_1}\right)^2 + 2q \frac{\partial \psi_m}{\partial x_1} \frac{\partial \psi_m}{\partial x_2} \right.$$
$$\left. + s \left(\frac{\partial \psi_m}{\partial x_2}\right)^2 + \frac{m^2}{x_2} \psi_m^2 \right] dx_1 \, dx_2 - \varkappa \int_{L_0^*} p \psi_m^2 \, dx_2. \tag{5.1.28}$$

By using the Ritz–Trefftz method, we find the coefficients $\bar{a}_k^{(m)}$ from the condition of minimum of functional (5.1.28). This gives the following matrix spectral problem:

$$\sum_{k=1}^{q} \bar{a}_k^{(m)} (\alpha_{ik}^{(m)} - \varkappa \beta_{ik}) = 0, \quad i = 1, 2, \ldots, n, \tag{5.1.29}$$

where, by the properties of the coordinate functions (5.1.27),

$$\alpha_{ik}^{(m)} = x_{10} \int_0^{x_{20}} x_2 \left[ \left( x_1 \frac{\partial w_{\nu_{mi}}}{\partial x_1} - x_2 \frac{\partial w_{\nu_{mi}}}{\partial x_2} \right) w_{\nu_{mk}}^{(m)} \right]_{x_1 = x_{10}} dx_2,$$

$$\beta_{ik}^{(m)} = x_{10}^2 \int_0^{x_{20}} x_2 \left[ w_{\nu_{mi}} w_{\nu_{mk}} \right]_{x_1 = x_{10}} dx_2, \quad i, k = 1, 2, \ldots, q. \tag{5.1.30}$$

The approximate eigenvalues $\varkappa_{mn}$ are obtained as the roots of the equation

$$\det |\alpha_{ik}^{(m)} - \varkappa \beta_{ik}^{(m)}| = 0, \quad i, k = 1, 2, \ldots, q. \tag{5.1.31}$$

**Table 5.1**

| $\theta_0^\circ$ | $m$ | $\nu_{m1}$ | $\nu_{m2}$ | $\nu_{m3}$ | $\nu_{m4}$ |
|---|---|---|---|---|---|
| 5 | 0 | 43.410 | 79.894 | 116.08 | 152.17 |
| | 1 | 20.615 | 60.598 | 97.321 | 133.64 |
| | 2 | 34.525 | 76.357 | 113.74 | 150.42 |
| 10 | 0 | 21.459 | 39.699 | 57.791 | 75.840 |
| | 1 | 10.083 | 30.056 | 48.415 | 66.574 |
| | 2 | 17.051 | 37.944 | 56.634 | 74.971 |
| 15 | 0 | 14.144 | 26.302 | 38.363 | 50.395 |
| | 1 | 6.5842 | 19.879 | 32.115 | 44.220 |
| | 2 | 11.245 | 25.147 | 37.601 | 49.822 |
| 20 | 0 | 10.488 | 19.604 | 28.649 | 37.672 |
| | 1 | 4.8432 | 14.793 | 23.966 | 33.044 |
| | 2 | 8.3552 | 18.754 | 28.088 | 37.251 |
| 30 | 0 | 6.8353 | 12.908 | 18.936 | 24.951 |
| | 1 | 3.1195 | 9.7120 | 15.821 | 21.870 |
| | 2 | 5.4928 | 12.372 | 18.583 | 24.685 |
| 40 | 0 | 5.0120 | 9.5619 | 14.081 | 18.591 |
| | 1 | 2.2753 | 7.1769 | 11.752 | 16.285 |
| | 2 | 4.0904 | 9.1928 | 13.838 | 18.408 |
| 50 | 0 | 3.9207 | 7.5556 | 11.169 | 14.776 |
| | 1 | 1.7833 | 5.6606 | 9.3135 | 12.937 |
| | 2 | 3.2739 | 7.2955 | 10.997 | 14.647 |
| 60 | 0 | 3.1956 | 6.2195 | 9.2288 | 12.233 |
| | 1 | 1.4679 | 4.6541 | 7.6905 | 10.706 |
| | 2 | 2.7525 | 6.0404 | 9.1109 | 12.145 |
| 70 | 0 | 2.6802 | 5.2665 | 7.8438 | 10.418 |
| | 1 | 1.2542 | 3.9396 | 6.5339 | 9.1155 |
| | 2 | 2.4024 | 5.1534 | 7.7694 | 10.362 |
| 80 | 0 | 2.2961 | 4.5532 | 6.8060 | 9.0575 |
| | 1 | 1.1050 | 3.4082 | 5.6693 | 7.9242 |
| | 2 | 2.1625 | 4.4983 | 6.7699 | 9.0304 |

The eigenvalues $\varkappa_{mn}$ $(m=0,1,\ldots;\ n=1,2,\ldots,q)$ and the eigenvectors $\bar{a}_{nk}^{(m)}$ $(k=1,2,\ldots,q,\ n=1,2,\ldots,q)$ form the approximate eigensolutions in the Ritz–Trefftz sense.

Table 5.1 contains the numerical values of $\nu_{mn}$ specifying the set of functions (5.1.25) for some values of the semiapex angle $\theta_0$. For the values of the roots of the transcendental equations (5.1.13) and (5.1.23), see [149, 59, 14].

The numerical values of the spectral parameter $\varkappa_{mn}$ $(m=0,1,2,\ldots;\ n=1,2,\ldots)$ versus the number $q$ of coordinate functions (5.1.27) are presented in Table 5.2. They reveal rapid convergence to $\varkappa_{mn}$ for a broad range of $\theta_0$.

**Table 5.2**

| $\alpha = \pi/20(rad) = 9^0, x_{10} = 6.31375151$ | | | | | | | |
|---|---|---|---|---|---|---|---|
| $q$ | $\varkappa_{11}$ | $\varkappa_{21}$ | $\varkappa_{01}$ | $\varkappa_{31}$ | $\varkappa_{41}$ | $\varkappa_{12}$ | $\varkappa_{51}$ |
| 2 | 1.691615 | 2.845734 | 3.740941 | 3.940773 | 5.009527 | 5.250674 | 6.063005 |
| 3 | 1.691607 | 2.845702 | 3.740763 | 3.940706 | 5.009419 | 5.240481 | 6.062854 |
| 4 | 1.691606 | 2.845697 | 3.740736 | 3.940695 | 5.009400 | 5.240034 | 6.062825 |
| 5 | 1.691606 | 2.845696 | 3.740729 | 3.940692 | 5.009394 | 5.239957 | 6.062817 |
| 6 | 1.691606 | 2.845695 | 3.740726 | 3.940690 | 5.009392 | 5.239936 | 6.062814 |
| $\alpha = \pi/18(rad) = 10^0, x_{10} = 5.67128182$ | | | | | | | |
| $q$ | $\varkappa_{11}$ | $\varkappa_{21}$ | $\varkappa_{01}$ | $\varkappa_{31}$ | $\varkappa_{41}$ | $\varkappa_{12}$ | $\varkappa_{51}$ |
| 2 | 1.674355 | 2.821056 | 3.729116 | 3.909489 | 4.972118 | 5.241575 | 6.019802 |
| 3 | 1.674346 | 2.821020 | 3.728920 | 3.909416 | 4.972000 | 5.229299 | 6.01964 |
| 4 | 1.674345 | 2.821015 | 3.728890 | 3.909403 | 4.971979 | 5.228803 | 6.019606 |
| 5 | 1.674345 | 2.821014 | 3.728882 | 3.909400 | 4.971973 | 5.228720 | 6.019597 |
| 6 | 1.674345 | 2.821013 | 3.728880 | 3.909399 | 4.971971 | 5.228698 | 6.019594 |
| $\alpha = \pi/12(rad) = 15^0, x_{10} = 3.73205081$ | | | | | | | |
| $q$ | $\varkappa_{11}$ | $\varkappa_{21}$ | $\varkappa_{01}$ | $\varkappa_{31}$ | $\varkappa_{41}$ | $\varkappa_{12}$ | $\varkappa_{51}$ |
| 2 | 1.586196 | 2.693060 | 3.663679 | 3.745779 | 4.775070 | 5.193123 | 5.791041 |
| 3 | 1.586187 | 2.693019 | 3.663438 | 3.745692 | 4.774930 | 5.169025 | 5.790847 |
| 4 | 1.586186 | 2.693014 | 3.663406 | 3.745680 | 4.774909 | 5.168412 | 5.790816 |
| 5 | 1.586186 | 2.693013 | 3.663399 | 3.745677 | 4.774903 | 5.168320 | 5.790807 |
| 6 | 1.586186 | 2.693013 | 3.663397 | 3.745676 | 4.774902 | 5.168297 | 5.790805 |
| $\alpha = 17\pi/180(rad) = 17^0, x_{10} = 3.27085262$ | | | | | | | |
| $q$ | $\varkappa_{11}$ | $\varkappa_{21}$ | $\varkappa_{01}$ | $\varkappa_{31}$ | $\varkappa_{41}$ | $\varkappa_{12}$ | $\varkappa_{51}$ |
| 2 | 1.550088 | 2.639709 | 3.634260 | 3.676856 | 4.691504 | 5.171680 | 5.693465 |
| 3 | 1.550079 | 2.639670 | 3.634023 | 3.676769 | 4.691368 | 5.142563 | 5.693276 |
| 4 | 1.550078 | 2.639665 | 3.633994 | 3.676758 | 4.691349 | 5.141963 | 5.693247 |
| 5 | 1.550078 | 2.639664 | 3.633988 | 3.676755 | 4.691344 | 5.141876 | 5.693240 |
| 6 | 1.550078 | 2.639664 | 3.633986 | 3.676754 | 4.691342 | 5.141856 | 5.693238 |
| $\alpha = \pi/6(rad) = 30^0, x_{10} = 1.73205081$ | | | | | | | |
| $q$ | $\varkappa_{11}$ | $\varkappa_{21}$ | $\varkappa_{31}$ | $\varkappa_{01}$ | $\varkappa_{41}$ | $\varkappa_{12}$ | $\varkappa_{51}$ |
| 2 | 1.304396 | 2.263162 | 3.180280 | 3.385675 | 4.080591 | 4.976944 | 4.971900 |
| 3 | 1.304394835 | 2.263150 | 3.180251 | 3.385606 | 4.080541 | 4.922898 | 4.971824 |
| 4 | 1.304395 | 2.263150 | 3.180249 | 3.385601 | 4.080537 | 4.922767 | 4.971817 |
| 5 | 1.304395 | 2.263150 | 3.180249 | 3.385600 | 4.080536 | 4.922747 | 4.971816 |
| 6 | 1.304395 | 2.263150 | 3.180249 | 3.385600 | 4.080536 | 4.922744 | 4.971816 |
| $\alpha = \pi/4(rad) = 45^0, x_{10} = 1$ | | | | | | | |
| $q$ | $\varkappa_{11}$ | $\varkappa_{21}$ | $\varkappa_{31}$ | $\varkappa_{01}$ | $\varkappa_{41}$ | $\varkappa_{51}$ | $\varkappa_{12}$ |
| 2 | 1.0 | 1.767377 | 2.504929 | 2.926575 | 3.231123 | 3.951541 | 4.525856 |
| 3 | 1.0 | 1.767377 | 2.504928 | 2.926574 | 3.231121 | 3.951538 | 4.483063 |
| 4 | 1.0 | 1.767377 | 2.504928 | 2.926574 | 3.231121 | 3.951538 | 4.483019 |
| 5 | 1.0 | 1.767377 | 2.504928 | 2.926574 | 3.231121 | 3.951538 | 4.483019 |
| 6 | 1.0 | 1.767377 | 2.504928 | 2.926574 | 3.231121 | 3.951538 | 4.483019 |
| $\alpha = \pi/3(rad) = 60^0, x_{10} = 0.577350269$ | | | | | | | |
| $q$ | $\varkappa_{11}$ | $\varkappa_{21}$ | $\varkappa_{31}$ | $\varkappa_{01}$ | $\varkappa_{41}$ | $\varkappa_{51}$ | $\varkappa_{12}$ |
| 2 | 0.677683 | 1.214432 | 1.732051 | 2.206459 | 2.242654 | 2.749810 | 3.254991 |
| 3 | 0.677680 | 1.214432 | 1.732051 | 2.206458 | 2.242654 | 2.749810 | 3.254991 |
| 4 | 0.677680 | 1.214432 | 1.732051 | 2.206458 | 2.242654 | 2.749810 | 3.254991 |
| 5 | 0.677680 | 1.214432 | 1.732051 | 2.206458 | 2.242654 | 2.749810 | 3.254991 |
| 6 | 0.677680 | 1.214432 | 1.732051 | 2.206458 | 2.242654 | 2.749810 | 3.254991 |

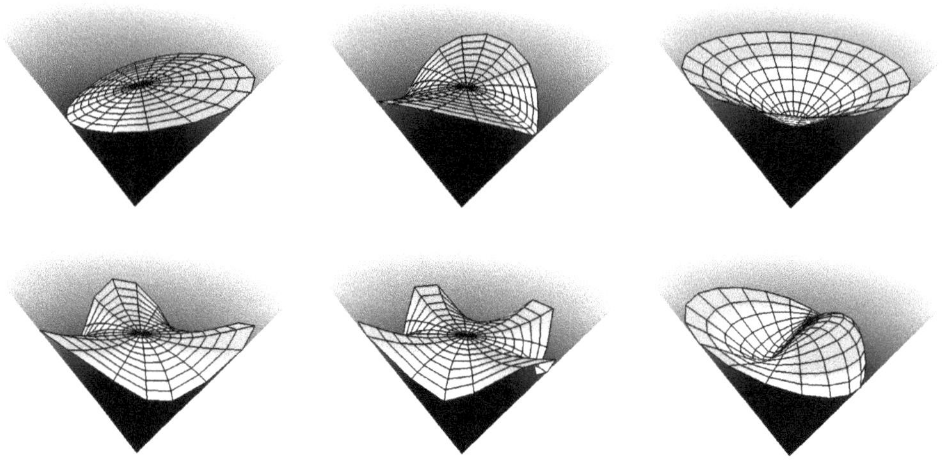

**Figure 5.2.** Natural sloshing modes.

For low values of $\theta_0$, the method is stabilized for lower eigenvalues $\varkappa_{mn}$ with an accuracy of up to 5–6 significant digits for 5–6 coordinate functions preserved in decomposition (5.1.26). Among the presented numerical values $\varkappa_{mn}$, the eigenvalues $\varkappa_{11}$ ($\theta_0 = 45^o$ ) and $\varkappa_{31}$ ($\theta_0 = 60^o$) are exact. We also detect a "migration" of the eigenvalues relative to their absolute values depending on $\theta_0$. In the nonlinear sloshing theory, this may lead to the appearance of physical phenomena of new types in the case of resonance interaction of liquid with the walls in mobile conical tanks [73]. The dimensional circular natural sloshing frequencies are given by the formula

$$\sigma_{mn} = \sqrt{\frac{g\varkappa_{mn}}{r_0}} = \sqrt{\frac{g\varkappa_{mn}}{h\tan\theta_0}} \tag{5.1.32}$$

and the natural free-surface sloshing modes in Fig. 5.2 are described as follows:

$$x_1 = F_{mn}(x_2, x_3) = \varphi_{mn}(x_{10}, x_2, x_3), \tag{5.1.33}$$
$$0 \leqslant x_2 \leqslant x_{20}, \ 0 \leqslant x_3 \leqslant 2\pi,$$

where $\varphi_{mn}(x_1, x_2, x_3)$ was introduced in (5.1.3). The coordinate $x_1$ in these relations is a dimensionless quantity equal to $x/r_0$ (Fig. 5.1).

In conclusion, we also note that partial solutions of the polynomial type (5.1.24) are traditionally used as systems of coordinate systems in the variational method [112, 149, 59]. The detailed comparative analysis of the efficiency of two systems of coordinate systems of the form (5.1.15) and (5.1.24) in the solution of the spectral problems for conical cavities (including the case of truncated cones) is performed in [132, 67, 135].

## 5.2 Natural sloshing modes for truncated conical tanks

In the present section, following [67], we construct, by using *two alternative approximate analytic methods*, the natural sloshing modes and frequencies for the truncated conical tanks depicted in Fig. 5.3.

An ideal incompressible liquid with irrotational flows partly occupies an Earth-fixed rigid conical tank with semiapex angle $\theta_0$. The mean (hydrostatic) volume of liquid coincides with the domain $Q_0$. The gravitational acceleration is directed downward along the axis of symmetry $Ox$. The wetted conical walls are denoted by $S_1$. The circle $S_2$ is the bottom of the tank $S = S_1 \cup S_2$ and $\Sigma_0$ is the unperturbed (hydrostatic) liquid surface. The origin is placed at the artificial apex of the conical surface. In what follows, the problem is considered in the dimensionless statement under the assumption that $r_0$ (the radius of the bottom for $\wedge$-shaped tanks and the radius of the liquid surface for $\vee$-shaped tanks) is chosen as a characteristic geometric dimension. Scaling by $r_0$, we get $h := h/r_0 \to 0$ ($h$ is the depth of liquid) and $r_1 := r_1/r_0$. The dimensionless radius $r_1$ and angle $\theta_0$ completely determine the geometric proportions of $Q_0$. The limit $r_1 \to 1$ implies that $h \to 0$, i.e., the water becomes shallow. For fixed $r_1$, the scaled depth of liquid $h$ tends to zero as $\theta_0 \to \pi/2$.

We use a cylindrical coordinate system $(X, \xi, \eta)$ connected with the original Cartesian coordinates by the formulas $x = X + X_0$, $y = \xi \cos \eta$, $z = \xi \sin \eta$, where the shift $X_0$ along the vertical axis is introduced to place the origin of the cylindrical coordinate system on the liquid surface. The eigensolutions are sought in the form

$$\psi(X, \xi, \eta) = \varphi_m(X, \xi) \begin{pmatrix} \sin m\eta \\ \cos m\eta \end{pmatrix}, \quad m = 0, 1, 2, \ldots, \tag{5.2.1}$$

which enables us to separate the angular coordinate $\eta$ and reduce the three-

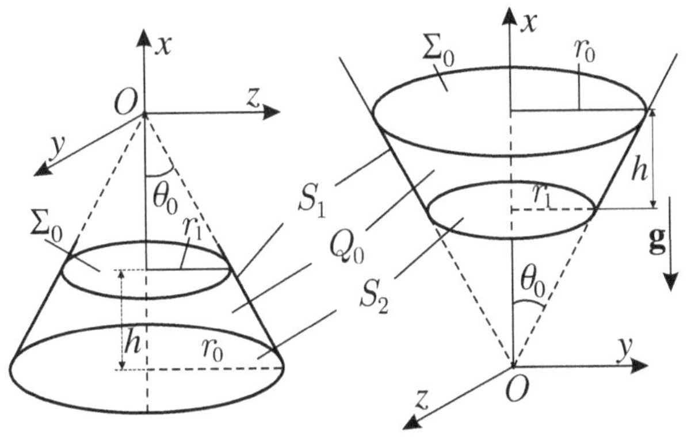


**Figure 5.3.** Hydrostatic liquid domains in $\wedge$- and $\vee$-shaped tanks.

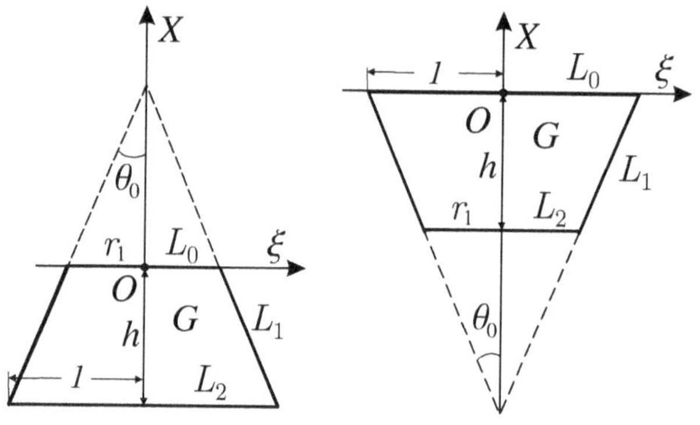


**Figure 5.4.** Meridional planes of ∧- and ∨-shaped tanks.

dimensional spectral boundary problem to the following $m$-parameter family ($m$ is a nonnegative integer) of two-dimensional spectral problems

$$\begin{aligned}
&\frac{\partial}{\partial X}\left(\xi\frac{\partial\varphi_m}{\partial X}\right)+\frac{\partial}{\partial\xi}\left(\xi\frac{\partial\varphi_m}{\partial\xi}\right)-\frac{m^2}{\xi}\varphi_m=0\ \text{ in }\ G;\\
&\frac{\partial\varphi_m}{\partial X}=\varkappa_m\varphi_m\ \text{ on }\ L_0;\quad \frac{\partial\varphi_m}{\partial n}=0\ \text{ on }\ L;\\
&|\varphi_m(X,0)|<\infty;\quad \int\limits_{L_0}\xi\frac{\partial\varphi_0}{\partial X}d\xi=0.
\end{aligned}\tag{5.2.2}$$

Problem (5.2.2) is posed in the meridional plane of $Q_0$ and for $L=L_1+L_2$ (see Fig. 5.4). This means that the eigenvalues of the original three-dimensional problem form a two-parameter set $\varkappa=\varkappa_{mi}$ ($m=0,1,\ldots;\ i=1,2,\ldots$), where $i\geq 1$ enumerates the eigenvalues of (5.2.2) in the ascending order. The corresponding eigenfunctions have the form (5.2.1) with $\varphi_m=\varphi_{mi}(X,\xi)$.

According to [124], the *first method* takes the following form in the meridional plane (after separation of the $\eta$-coordinate):

$$w_k^{(m)}(X,\xi)=\frac{2(k-m)!}{(k+m)!}R^kP_k^{(m)}\left(\frac{X}{R}\right),\ k\geq m,\quad R=\sqrt{X^2+\xi^2},\tag{5.2.3}$$

where $P_k^{(m)}$ are Legendre functions of the first kind. The functions $\left\{w_k^{(m)}\right\}$ are the solutions of the first equation in (5.2.2). It can be shown that $w_k^{(m)}$ are indeed polynomials as functions of $X$ and $\xi$. The first functions of the set (5.2.3) take the form:

$$\begin{aligned}
&w_0^{(0)}=1,\quad w_1^{(0)}=X,\quad w_2^{(0)}=X^2-\tfrac{1}{2}\xi^2,\quad \ldots\quad (m=0),\\
&w_1^{(1)}=\xi,\quad w_2^{(1)}=X\xi,\quad w_3^{(1)}=X^2\xi-\tfrac{1}{4}\xi^3,\quad \ldots\quad (m=1),
\end{aligned}$$

$$w_2^{(2)} = \xi^2, \quad w_3^{(2)} = X\xi^2, \quad w_4^{(2)} = X^2, \xi^2 - \tfrac{1}{6}\xi^4, \quad \ldots \quad (m=2),$$
$$w_3^{(3)} = \xi^3, \quad w_4^{(3)} = X\xi^3, \quad w_5^{(3)} = X^2\xi^3 - \tfrac{1}{8}\xi^5, \quad \ldots \quad (m=3).$$

The functions $\left\{w_k^{(m)}\right\}$ can be found by using the recurrence relations

$$\frac{\partial\, w_k^{(m)}}{\partial X} = (k-m)w_{k-1}^{(m)}; \quad \xi\frac{\partial w_k^{(m)}}{\partial \xi} = k w_k^{(m)} - (k-m)X w_{k-1}^{(m)},$$
$$(k-m+1)w_{k+1}^{(m)} = (2k+1)X w_k^{(m)} - (k-m)(X^2+\xi^2)w_{k-1}^{(m)},$$
$$(k-m+1)\xi w_k^{(m+1)} = 2(m+1)\left((X^2+\xi^2)w_{k-1}^{(m)} - X w_k^{(m)}\right).$$

Separating the $\eta$-coordinate in the variational formulation, as in the previous section, we arrive at the following $m$-parameter families ($m$ is a nonnegative integer) of approximate solutions

$$\varphi_m(X,\xi) = \sum_{k=1}^{q} a_k^{(m)}\, w_{k+m-1}^{(m)}(X,\xi), \tag{5.2.4}$$

and the spectral matrix problems

$$\sum_{k=1}^{q}\left(\left\{\alpha_{ik}^{(m)}\right\} - \varkappa_m\left\{\beta_{ik}^{(m)}\right\}\right)a_k = 0 \quad (i=1,\ldots,q). \tag{5.2.5}$$

The elements $\left\{\alpha_{ik}^{(m)}\right\}$ and $\left\{\beta_{ik}^{(m)}\right\}$ are computed for the $\wedge$-cones by the formulas

$$\begin{aligned}
\alpha_{ij}^{(m)} = {} & \int_0^{r_1}\left(\xi\frac{\partial\, w_{i+m-1}^{(m)}}{\partial X}\, w_{j+m-1}^{(m)}\right)_{X=0} d\xi \\
& + \int_{-h}^{0}\left(\xi\,\frac{\partial\, w_{i+m-1}^{(m)}}{\partial \xi}\, w_{j+m-1}^{(m)}\right)_{\xi=\tan\theta_0 X - r_1} dX \\
& - \tan\theta_0\int_{-h}^{0}\left(\xi\frac{\partial\, w_{i+m-1}^{(m)}}{\partial X}\, w_{j+m-1}^{(m)}\right)_{\xi=\tan\theta_0 X - r_1} dX \\
& - \int_0^{1}\left(\xi\frac{\partial\, w_{i+m-1}^{(m)}}{\partial X}\, w_{j+m-1}^{(m)}\right)_{X=-h} d\xi,
\end{aligned}$$

$$\beta_{ij}^{(m)} = \int_0^{r_1}\left(\xi\, w_{i+m-1}^{(m)}\, w_{j+m-1}^{(m)}\right)_{X=0} d\xi,$$

and for the V cones by the formulas

$$\alpha_{ij}^{(m)} = \int_0^1 \left( \xi \frac{\partial w_{i+m-1}^{(m)}}{\partial X} w_{j+m-1}^{(m)} \right)_{X=0} d\xi$$

$$+ \int_{-h}^0 \left( \xi \frac{\partial w_{i+m-1}^{(m)}}{\partial \xi} w_{j+m-1}^{(m)} \right)_{\xi = \tan\theta_0 X + 1} dX$$

$$- \tan\theta_0 \int_{-h}^0 \left( \xi \frac{\partial w_{i+m-1}^{(m)}}{\partial X} w_{j+m-1}^{(m)} \right)_{\xi = \tan\theta_0 X + 1} dX$$

$$- \int_0^{r_1} \left( \xi \frac{\partial w_{i+m-1}^{(m)}}{\partial X} w_{j+m-1}^{(m)} \right)_{X=-h} d\xi,$$

$$\beta_{ij}^{(m)} = \int_0^1 \left( \xi\, w_{i+m-1}^{(m)} w_{j+m-1}^{(m)} \right)_{X=0} d\xi.$$

For any fixed $m$, the second equation in (5.2.5) has $q$ positive roots $\varkappa_{mn}$ ($n = 1, 2, \ldots, q$). The multiplicities should be taken into account. Since the Ritz–Trefftz method is based on the minimization of the functional, the approximate eigenvalues $\varkappa_{mn}$ converge from above. This enables us to check the convergence by the number of significant digits that do not change as $q$ increases. The method gives the best approximation for the lowest eigenvalue $\varkappa_{m1}$.

In our numerical experiments, we mainly determined the eigenvalues $\varkappa_{m1}$, $m = 0, 1, 2, 3$. These eigenvalues are responsible for the lowest natural modes making a decisive contribution to the hydrodynamic loads. In the case of V-tanks, the method exhibits a rapid convergence to $\varkappa_{m1}$ and guarantees satisfactory accuracy also for $\varkappa_{m2}$ and $\varkappa_{m3}$. A slower convergence is observed for the Λ-tanks. Furthermore, the rate of convergence depends not only on the type of the tank (V or Λ-shaped) but also on the semiapex angle $\theta_0$ and the dimensionless parameter $0 < r_1 < 1$.

The results presented in Table 5.3 (A) reveal a typical (convergence) behavior for the V-tanks with $10° \le \theta_0 \le 75°$ and $0.2 \le r_1 \le 0.9$. The table also shows the stabilization of 5–6 significant digits as $q \ge 14$. The best accuracy is observed for the lowest spectral parameter $\varkappa_{11}$. The accuracy increases with $q$ for $m \ne 0$. However, the evaluation of the value $\varkappa_{01}$ responsible for the axisymmetric natural mode may become unstable for $q > 17$. This explains why we do not present the numerical results for this eigenvalue obtained with $q = 20$. For $m \ne 1$ and $0.2 \le r_1 \le 0.9$, as $\theta_0$ increases ($\theta_0 > 75°$), we get a slower convergence. In this case, even for $q = 17, \ldots, 20$, we can guarantee only 3–4 significant digits (required engineering accuracy). The same number

**Table 5.3.** Convergence to $\varkappa_{m1}$, $m = 0, 1, 2, 3$ for different $r_1$ versus the number of base functions $q$ in (5.2.4). Column (A) corresponds to the $\vee$-shaped tank; column (B) corresponds to the $\wedge$-shaped tank, $\theta_0 = 30°$.

| A | | | | | | B | | | | | |
|---|---|---|---|---|---|---|---|---|---|---|---|
| $q$ | $r_1 = .2$ | $r_1 = .4$ | $r_1 = .6$ | $r_1 = .8$ | $r_1 = .9$ | $q$ | $r_1 = .2$ | $r_1 = .4$ | $r_1 = .6$ | $r_1 = .8$ | $r_1 = .9$ |
| $\varkappa_{01}$ | | | | | | | | | | | |
| 2 | 5.03242 | 4.57783 | 4.15235 | 3.80550 | 2.96140 | 8 | 70.0634 | 15.1409 | 6.89999 | 4.56099 | 2.68341 |
| 5 | 3.39478 | 3.39229 | 3.38487 | 3.14117 | 2.20034 | 10 | 50.4676 | 11.1245 | 6.67227 | 4.55115 | 2.68312 |
| 8 | 3.38560 | 3.38559 | 3.38185 | 3.13886 | 2.19744 | 12 | 39.0611 | 10.1856 | 6.66670 | 4.55088 | 2.68293 |
| 11 | 3.38560 | 3.38559 | 3.38183 | 3.13872 | 2.19724 | 14 | 32.0757 | 10.0175 | 6.66568 | 4.55059 | 2.68280 |
| 14 | 3.38560 | 3.38559 | 3.38182 | 3.13867 | 2.19716 | 16 | 27.6490 | 10.0161 | 6.66506 | 4.55042 | 2.68275 |
| 17 | 3.38560 | 3.38559 | 3.38182 | 3.13864 | 2.19714 | 18 | 25.1530 | 10.0143 | 6.66489 | 4.55035 | 2.68273 |
| $\varkappa_{11}$ | | | | | | | | | | | |
| 2 | 1.34363 | 1.33572 | 1.29998 | 1.00726 | 0.60734 | 8 | 16.5049 | 5.69483 | 3.51678 | 1.66182 | 0.72656 |
| 5 | 1.30441 | 1.30179 | 1.25416 | 0.93441 | 0.54250 | 10 | 13.9478 | 5.63373 | 3.51586 | 1.66168 | 0.72651 |
| 8 | 1.30438 | 1.30169 | 1.25405 | 0.93389 | 0.54228 | 12 | 13.4877 | 5.63190 | 3.51555 | 1.66161 | 0.72649 |
| 11 | 1.30438 | 1.30169 | 1.25398 | 0.93384 | 0.54226 | 14 | 12.2472 | 5.63056 | 3.51537 | 1.66156 | 0.72648 |
| 14 | 1.30438 | 1.30169 | 1.25397 | 0.93382 | 0.54225 | 16 | 11.6635 | 5.62990 | 3.51527 | 1.66155 | 0.72648 |
| 17 | 1.30438 | 1.30169 | 1.25397 | 0.93382 | 0.54225 | 18 | 11.4507 | 5.62978 | 3.51524 | 1.66154 | 0.72648 |
| 20 | 1.30438 | 1.30169 | 1.25397 | 0.93382 | 0.54225 | 20 | 11.3318 | 5.62989 | 3.51521 | 1.66153 | 0.72648 |
| $\varkappa_{21}$ | | | | | | | | | | | |
| 2 | 2.44374 | 2.42439 | 2.38663 | 2.19137 | 1.57205 | 8 | 45.0241 | 9.79780 | 5.95093 | 3.72425 | 1.92323 |
| 5 | 2.26355 | 2.26335 | 2.25510 | 2.01560 | 1.36191 | 10 | 28.6525 | 9.09613 | 5.94191 | 3.72405 | 1.92302 |
| 8 | 2.26315 | 2.26309 | 2.25500 | 2.01492 | 1.36093 | 12 | 25.1839 | 8.97538 | 5.94111 | 3.72371 | 1.92295 |
| 11 | 2.26315 | 2.26309 | 2.25498 | 2.01484 | 1.36084 | 14 | 24.7717 | 8.96803 | 5.94100 | 3.72364 | 1.92293 |
| 14 | 2.26315 | 2.26309 | 2.25497 | 2.01481 | 1.36081 | 16 | 21.5127 | 8.96666 | 5.94065 | 3.72359 | 1.92291 |
| 17 | 2.26315 | 2.26309 | 2.25497 | 2.01480 | 1.36080 | 18 | 19.7386 | 8.96653 | 5.94060 | 3.72355 | 1.92290 |
| 20 | 2.26315 | 2.26309 | 2.25497 | 2.01480 | 1.36080 | 20 | 17.7602 | 8.96652 | 5.94059 | 3.72354 | 1.92290 |
| $\varkappa_{31}$ | | | | | | | | | | | |
| 2 | 3.53317 | 3.52088 | 3.45905 | 3.31620 | 2.69743 | 8 | 139.880 | 15.6706 | 8.14608 | 5.63407 | 3.40404 |
| 5 | 3.18153 | 3.18130 | 3.17954 | 3.04718 | 2.32902 | 10 | 83.8225 | 12.9549 | 8.04675 | 5.63400 | 3.40357 |
| 8 | 3.18025 | 3.18025 | 3.17908 | 3.04674 | 2.32704 | 12 | 64.5885 | 12.2296 | 8.04492 | 5.63357 | 3.40346 |
| 11 | 3.18025 | 3.18025 | 3.17908 | 3.04665 | 2.32688 | 14 | 53.4108 | 12.0886 | 8.04446 | 5.63350 | 3.40341 |
| 14 | 3.18025 | 3.18025 | 3.17907 | 3.04662 | 2.32681 | 16 | 53.1387 | 12.0796 | 8.04423 | 5.63338 | 3.40336 |
| 17 | 3.18025 | 3.18025 | 3.17907 | 3.04661 | 2.32679 | 18 | 38.8552 | 12.0840 | 8.04403 | 5.63334 | 3.40334 |
| 20 | 3.18025 | 3.18025 | 3.17907 | 3.04660 | 2.32678 | 20 | 36.4675 | 12.0849 | 8.04399 | 5.63332 | 3.40333 |

of base functions guarantees the same number of significant digits for the tanks with $r_1 < 0.2$ and $10° \leq \theta_0 \leq 75°$. As $r_1$ tends to zero (truncated tank close to a nontruncated tank), the approximations obtained for $\varkappa_{m1}$ are confirmed by the numerical results reported in [70] as well as by the experimental data presented in [13].

In Table 5.3 (B), we illustrate a typical convergence behavior for the $\wedge$-shaped tanks with $10° \leq \theta_0 \leq 75°$. The comparative analysis of the parts (A) and (B) shows that, in the second case, the method is less efficient. In particular, the numerical analyses of $\varkappa_{01}$ are not so precise. Moreover, 18–20 base functions

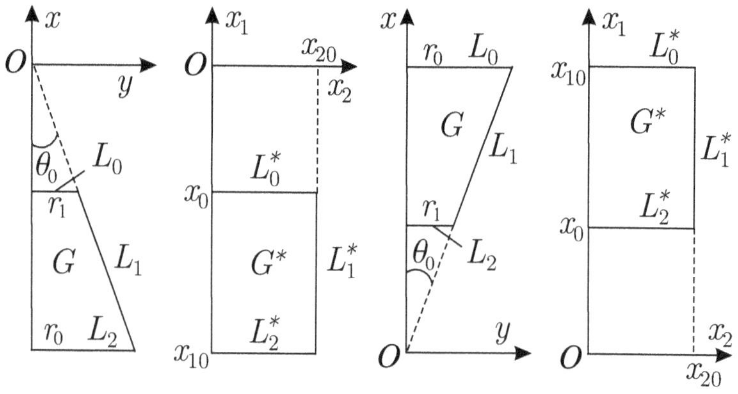


**Figure 5.5.** Meridional cross sections of the original and transformed domains.

give 4–5 significant digits of $\varkappa_{m1}$ only for $r_1 \geq 0.4$. This is not true for lower values of $r_1$. Furthermore, for $r_1 \leq 0.2$, the values $q = 17, \ldots, 20$ can guarantee at most 2–3 significant digits for $\varkappa_{11}$.

The fact of slower convergence observed for $\wedge$-shaped tanks can be partly explained by the presence of singular first derivatives of the eigenfunctions $\psi_m$ at the inner vertex between $L_0$ and $L_1$ [124]. The HPS are smooth in the $(\xi, X)$-plane and, therefore, do not reflect this singular behavior. The singularity disappears if the corner angle is smaller than 90°. This occurs only for the $\wedge$-shaped tanks.

The $\vee$-shaped tanks are characterized by a similar singularity at the vertex formed by $L_1$ and $L_2$. However, since the natural modes (eigenfunctions $\psi_m$) should "decay" (exponentially downward), the method may be sensitive to this singularity only for shallow water. Our numerical experiments confirm this fact for $r_1 < 0.1$.

The *second method* is based on the curvilinear parametrization from Sec. 5.1 connecting the $x, y, z$ coordinates with $x_1, x_2, x_3$ by (5.1.1). The variable $x_3 = \eta$ is the polar angle $\eta$ in the $Oyz$-plane. In Fig. 5.5, we show that, in the $(x_1, x_2, x_3)$-system, the hydrostatic liquid domain $Q_0$ takes the form of an upright rectangular base cylinder $(x_0 \leq x_1 \leq x_{10},\ 0 \leq x_2 \leq x_{20}, 0 \leq x_3 \leq 2\pi)$. The domain $G^*$ represents a rectangle with the sides $h = x_{10} - x_0$ and $x_{20} = \tan\theta_0$ in the $Ox_2x_1$-plane. Here, the radius of the unperturbed liquid surface is $r_t = 1$ for the $\vee$-tanks and $r_t = r_1$ for the $\wedge$-tanks. By using (5.1.3), we conclude that the original three-dimensional spectral problem admits separation of the space variable $x_3$. Furthermore, transformation (5.1.1) generates the following $m$-parameter family of spectral problems with respect to $\psi_m(x_1, x_2)$:

$$p\frac{\partial^2\psi_m}{\partial x_1^2} + 2q\frac{\partial^2\psi_m}{\partial x_1 \partial x_2} + s\frac{\partial^2\psi_m}{\partial x_2^2} + d\frac{\partial\psi_m}{\partial x_2} - m^2 c\psi_m = 0 \quad \text{in} \quad G^*, \qquad (5.2.6)$$

$$p\frac{\partial\psi_m}{\partial x_1}+q\frac{\partial\psi_m}{\partial x_2}=\varkappa_m p\psi_m \quad \text{on} \quad L_0^*, \tag{5.2.7}$$

$$s\frac{\partial\psi_m}{\partial x_2}+q\frac{\partial\psi_m}{\partial x_1}=0 \quad \text{on} \quad L_1^*, \tag{5.2.8}$$

$$p\frac{\partial\psi_m}{\partial x_1}+q\frac{\partial\psi_m}{\partial x_2}=0 \quad \text{on} \quad L_2^*, \tag{5.2.9}$$

$$|\psi_m(x_1,0)|<\infty, \quad m=0,1,2,\dots, \quad \int_0^{x_{20}}\psi_0 x_2 dx_2=0, \tag{5.2.10}$$

where

$$G^*=\{(x_1,x_2):\, x_0\le x_1\le x_{10},\ 0\le x_2\le x_{20}\},$$

$$p=x_1^2x_2, \quad q=-x_1x_2^2, \quad s=x_2(x_2^2+1), \quad d=1+2x_2^2, \quad c=1/x_2,$$

and the boundary of $G^*$ consists of the fragments $L_0^*$, $L_1^*$, and $L_2^*$.

It is possible to show that (5.2.6) and (5.2.8) admit the separation of the space variables $x_1$ and $x_2$. This leads to the following solutions:

$$x_1^{\nu}T_{\nu}^{(m)}(x_2) \ \text{and} \ x_1^{-1-\nu}\bar{T}_{\nu}^{(m)}(x_2), \quad \nu\ge 0. \tag{5.2.11}$$

To find $T_\nu^{(m)}$, it is possible to consider the following homogeneous boundary-value problem, which depends on the real parameter $\nu$:

$$x_2^2(1+x_2^2)T_\nu^{\prime\prime(m)}+x_2(1+2x_2^2-2\nu x_2^2)T_\nu^{\prime(m)}+\left[\nu(\nu-1)x_2^2-m^2\right]T_\nu^{(m)}=0,$$

$$T_\nu^{\prime(m)}(x_{20})=\nu\frac{x_{20}}{1+x_{20}^2}T_\nu^{(m)}(x_{20}), \quad |T_\nu^{(m)}(0)|<\infty. \tag{5.2.12}$$

It can be shown that problem (5.2.12) has only nontrivial solutions for a countable set of values $\nu=\nu_{mn}>0$ ($m=0,1,\dots;\ n=1,2,\dots$).

The second class of functions $\bar{T}_\nu^{(m)}$ appears only for $x_0\ne 0$, i.e., when the conical tank is truncated. The evaluation of $\bar{T}_\nu^{(m)}$ leads to the following $\nu$-parameter problem

$$x_2^2(1+x_2^2)\bar{T}^{\prime\prime(m)}+x_2(1+4x_2^2+2\nu x_2^2)\bar{T}^{\prime(m)}$$
$$+\left[(\nu+1)(\nu+2)x_2^2-m^2\right]\bar{T}^{(m)}=0,$$

$$\bar{T}^{\prime(m)}(x_{20})+(\nu+1)\frac{x_{20}}{1+x_{20}^2}\bar{T}^{(m)}(x_{20})=0. \tag{5.2.13}$$

Clearly, the nontrivial solutions of (5.2.13) exist only for a countable set of nonnegative values $\nu$.

We now show that the solution of (5.2.12) and (5.2.13) can be expressed in terms of the spheroidal harmonics and the set $\{\nu_{mn}\}$ is the same for problems (5.2.12) and (5.2.13). For this purpose, we change the variables in (5.2.12) and (5.2.13) by the formula $\mu = (1 + x_2^2)^{-\frac{1}{2}}$ and substitute $y(\mu) = \mu^\nu T(\mu)$ and $y(\mu) = \mu^{-1-\nu}\bar{T}(\mu)$ in (5.2.12) and (5.2.13), respectively. This reduces the indicated two equations to problem (5.1.12) whose solutions are Legendre functions of the first kind, i.e., $y(\mu) = P_\nu^{(m)}(\mu)$. Treating the boundary conditions in (5.2.12) in the same way and using the substitution $\mu = \cos\theta$, we obtain the transcendental equation (5.1.13) for the evaluation of $\{\nu_{mn}\}$. As a consequence, we get the following nontrivial solutions:

$$T_{\nu_{mk}}^{(m)}(x_2) = (1 + x_2^2)^{\frac{\nu_{mk}}{2}}\, P_{\nu_{mk}}^{(m)}\left((1 + x_2^2)^{-1/2}\right), \tag{5.2.14}$$

$$\bar{T}_{\nu_{mk}}^{(m)}(x_2) = (1 + x_2^2)^{\frac{-1-\nu_{mk}}{2}}\, P_{\nu_{mk}}^{(m)}\left((1 + x_2^2)^{-1/2}\right). \tag{5.2.15}$$

Solutions (5.2.11), (5.2.14) and (5.2.15) can be rewritten as

$$\begin{aligned} W_k^{(m)}(x_1, x_2) &= N_k^{(m)}\, x_1^{\nu_{mk}}\, T_{\nu_{mk}}^{(m)}(x_2), \\ \bar{W}_k^{(m)}(x_1, x_2) &= \bar{N}_k^{(m)}\, x_1^{-1-\nu_{mk}}\, \bar{T}_{\nu_{mk}}^{(m)}(x_2), \end{aligned} \tag{5.2.16}$$

where $N_k^{(m)}$ and $\bar{N}_k^{(m)}$ are factors chosen to satisfy the condition

$$\begin{aligned} 1 = ||W_k^{(m)}||^2_{L_2^*+L_0^*} &= \int_0^{x_{20}} x_2[(W_k^{(m)}|_{x_1=x_{10}})^2 + (W_k^{(m)}|_{x_1=x_0})^2]dx_2 \\ &= \int_0^{x_{20}} x_2[(\bar{W}_k^{(m)}|_{x_1=x_{10}})^2 + (\bar{W}_k^{(m)}|_{x_1=x_0})^2]dx_2 = ||\bar{W}_k^{(m)}||^2_{L_2^*+L_0^*}. \end{aligned} \tag{5.2.17}$$

Equation (5.2.17) shows that $W_k^{(m)}$ and $\bar{W}_k^{(m)}$ have the unit norm (in the mean-square metric) on the boundary $L_2^* + L_0^*$, where Eqs. (5.2.7) and (5.2.9) should be approximatively satisfied. The explicit formulas for these normalizing factors take the form

$$N_k^{(m)} = \frac{1}{\sqrt{x_{10}^{2\nu_{mk}} + x_0^{2\nu_{mk}}}} \frac{1}{\sqrt{\int\limits_0^{x_{20}} (1 + x_2^2)^{\nu_{mk}} \left(P_{\nu_{mk}}^{(m)}\right)^2 dx_2}},$$

$$\bar{N}_k^{(m)} = \frac{1}{\sqrt{x_{10}^{-2-2\nu_{mk}} + x_0^{-2-2\nu_{mk}}}} \frac{1}{\sqrt{\int_0^{x_{20}} (1 + x_2^2)^{-1-\nu_{mk}} \left(P_{\nu_{mk}}^{(m)}\right)^2 dx_2}}.$$

The case $m = 0$ also requires the condition of conservation of volume (5.2.10). This means that the functions $W_k^{(0)}$ and $\bar{W}_k^{(0)}$ should be redefined as follows:

$$W_k^{(0)} := W_k^{(0)} - c_k^{(0)}, \quad \bar{W}_k^{(0)} := \bar{W}_k^{(0)} - \bar{c}_k^{(0)},$$

where

$$c_k^{(0)} = \frac{2}{x_{20}^2} \int_0^{x_{20}} x_2 \, W_k^{(0)}(x_{10}, x_2) \, d\,x_2 \quad \text{and} \quad \bar{c}_k^{(0)} = \frac{2}{x_{20}^2} \int_0^{x_{20}} x_2 \, \bar{W}_k^{(0)}(x_{10}, x_2) \, d\,x_2.$$

According to the Ritz-Trefftz scheme, we represent approximate solutions of (5.2.6)–(5.2.10) in the form

$$\psi_m(x_1, x_2) = \sum_{k=1}^{q_1} a_k^{(m)} W_k^{(m)} + \sum_{l=1}^{q_2} \bar{a}_l^{(m)} \bar{W}_l^{(m)}. \tag{5.2.18}$$

Separating the $x_3$-coordinate in the corresponding variational formulation [after the substitution of (5.1.3)], we see that representation (5.2.18) yields the $m$-parameter family of the corresponding matrix spectral problems:

$$\sum_{k=1}^{Q} \left( \left\{ \alpha_{ik}^{(m)} \right\} - \varkappa_m \left\{ \beta_{ik}^{(m)} \right\} \right) a_k = 0, \quad i = 1, \ldots, Q. \tag{5.2.19}$$

The spectral problem (5.2.19) has $Q = q_1 + q_2$ eigenvalues. Since representation (5.2.18) contains two types of functions, namely, $W_k^{(m)}$ and $\bar{W}_l^{(m)}$, there exist four submatrices of $\left\{ \tilde{\alpha}_{ij}^{(m)} \right\}$ and $\left\{ \tilde{\beta}_{ij}^{(m)} \right\}$ such that

$$\tilde{\alpha}_{ij}^{(m)} = \begin{pmatrix} \alpha_{ij1}^{(m)} & \alpha_{ij2}^{(m)} \\ \alpha_{ij3}^{(m)} & \alpha_{ij4}^{(m)} \end{pmatrix}, \qquad \tilde{\beta}_{ij}^{(m)} = \begin{pmatrix} \beta_{ij1}^{(m)} & \beta_{ij2}^{(m)} \\ \beta_{ij3}^{(m)} & \beta_{ij4}^{(m)} \end{pmatrix}.$$

The elements $\left\{ \alpha_{ijs}^{(m)} \right\}$ and $\left\{ \beta_{ijs}^{(m)} \right\}$, $s = 1, \ldots, 4$, are given by the formulas

$$\alpha_{ij1}^{(m)} = \int_0^{x_{20}} \left( x_1^2 x_2 \frac{\partial W_i^{(m)}}{\partial x_1} - x_1 x_2^2 \frac{\partial W_i^{(m)}}{\partial x_2} \right)_{x_1 = h_t} W_j^{(m)} dx_2$$
$$- \int_0^{x_{20}} \left( x_1^2 x_2 \frac{\partial W_i^{(m)}}{\partial x_1} - x_1 x_2^2 \frac{\partial W_i^{(m)}}{\partial x_2} \right)_{x_1 = h_b} W_j^{(m)} dx_2,$$

$$\alpha_{ij2}^{(m)} = \int_0^{x_{20}} \left( x_1^2 x_2 \frac{\partial W_i^{(m)}}{\partial x_1} - x_1 x_2^2 \frac{\partial W_i^{(m)}}{\partial x_2} \right)_{x_1 = h_t} \bar{W}_j^{(m)} dx_2$$
$$- \int_0^{x_{20}} \left( x_1^2 x_2 \frac{\partial W_i^{(m)}}{\partial x_1} - x_1 x_2^2 \frac{\partial W_i^{(m)}}{\partial x_2} \right)_{x_1 = h_b} \bar{W}_j^{(m)} dx_2,$$

$$\alpha_{ij3}^{(m)}=\int\limits_0^{x_{20}}\left(x_1^2x_2\,\frac{\partial \bar W_i^{(m)}}{\partial x_1}-x_1x_2^2\,\frac{\partial \bar W_i^{(m)}}{\partial x_2}\right)_{x_1=h_t} W_j^{(m)}dx_2 -\int\limits_0^{x_{20}}\left(x_1^2x_2\frac{\partial \bar W_i^{(m)}}{\partial x_1}-x_1x_2^2\,\frac{\partial \bar W_i^{(m)}}{\partial x_2}\right)_{x_1=h_b} W_j^{(m)}dx_2,$$

$$\alpha_{ij4}^{(m)}=\int\limits_0^{x_{20}}\left(x_1^2x_2\,\frac{\partial \bar W_i^{(m)}}{\partial x_1}-x_1x_2^2\frac{\partial \bar W_i^{(m)}}{\partial x_2}\right)_{x_1=h_t} \bar W_j^{(m)}dx_2 -\int\limits_0^{x_{20}}\left(x_1^2x_2\frac{\partial \bar W_i^{(m)}}{\partial x_1}-x_1x_2^2\,\frac{\partial \bar W_i^{(m)}}{\partial x_2}\right)_{x_1=h_b} \bar W_j^{(m)}dx_2,$$

$$\beta_{ij1}^{(m)}=h_t^2\int\limits_0^{x_{20}}x_2\left(W_i^{(m)}W_j^{(m)}\right)_{x_1=h_t}dx_2;\ \beta_{ij2}^{(m)}=h_t^2\int\limits_0^{x_{20}}x_2\left(W_i^{(m)}\bar W_j^{(m)}\right)_{x_1=h_t}dx_2,$$

$$\beta_{ij3}^{(m)}=h_t^2\int\limits_0^{x_{20}}x_2\left(\bar W_i^{(m)}W_j^{(m)}\right)_{x_1=h_t}dx_2;\ \beta_{ij4}^{(m)}=h_t^2\int\limits_0^{x_{20}}x_2\left(\bar W_i^{(m)}\bar W_j^{(m)}\right)_{x_1=h_t}dx_2.$$

For $\wedge$- and $\vee$-tanks, we have $h_t = r_1/\tan\theta_0$, $h_b = 1/\tan\theta_0$ and $h_t = 1/\tan\theta_0$, $h_b = r_1/\tan\theta_0$, respectively.

Column A in Table 5.4 shows a typical convergence behavior in the case of $\vee$-tanks with $10^\circ \le \theta_0 \le 75^\circ$ and $0.2 \le r_1 \le 0.9$. For $0.2 \le r_1 \le 0.55$, the method generates 4–5 significant digits of $\varkappa_{m1}$ for $q = q_1 = q_2 = 7, \dots, 10$ $(14, \dots, 20$ base functions). This agrees with the convergence results presented in Sec. 3. However, the second method is also characterized by a rapid convergence to $\varkappa_{m1}$ for $r_1 < 0.2$. This includes the case of $\varkappa_{01}$, which is not satisfactorily treated by the HPS. Moreover, for $0 < r_1 \le 0.4$, the number of significant digits is larger (for the same number of base functions) for $15^\circ \le \theta_0 < 30^\circ$ but it is only slightly smaller for $15^\circ \le \theta_0$. We explain the indicated slower convergence of a similar method for lower semiapex angles by the asymptotic behavior of the exact solution along the vertical axis. The conclusion is that the eigenfunctions $\psi_m$ should exponentially (downward) decay $Ox$ for a circular cylindrical tank approached by the conical domain as $\theta_0$ decreases. However, $W_k^{(m)}$ and $\bar W_k^{(m)}$ do not reflect this decay. Furthermore, the decrease in the dimensionless depth of liquid $h$ ($r_1 \to 1$ or $\theta_0 \to 90^\circ$) may result in a lower accuracy (3–4 significant digits for 18–24 base functions).

Column B in Table 5.4 describes the convergence for a $\wedge$-tank. We choose the same $r_1$ and $\theta_0$ as in the Column A. It is easy to see that the numerical results may be less precise than the results presented in Sec. 3. Thus, the same

**Table 5.4.** Convergence to $\varkappa_{m1}$, $m = 0, 1, 2, 3$, for different $r_1$ versus the number of base functions $q = q_1 = q_2$ in (5.2.18). Column (A) corresponds to a $\vee$-shaped tank; column (B) corresponds to a $\wedge$-shaped tank; $\theta_0 = 30°$.

| | A | | | | | B | | | | |
|---|---|---|---|---|---|---|---|---|---|---|
| $q$ | $r_1 = .2$ | $r_1 = .4$ | $r_1 = .6$ | $r_1 = .8$ | $r_1 = .9$ | $r_1 = .2$ | $r_1 = .4$ | $r_1 = .6$ | $r_1 = .8$ | $r_1 = .9$ |
| | | | | | $\varkappa_{01}$ | | | | | |
| 4 | 3.38560 | 3.38559 | 3.38188 | 3.14194 | 2.20618 | 20.1518 | 10.0759 | 6.70758 | 4.58886 | 2.71591 |
| 5 | 3.38560 | 3.38559 | 3.38186 | 3.14089 | 2.20328 | 20.0971 | 10.0485 | 6.68925 | 4.57279 | 2.70251 |
| 6 | 3.38560 | 3.38559 | 3.38185 | 3.14027 | 2.20158 | 20.0707 | 10.0353 | 6.68040 | 4.56492 | 2.69582 |
| 7 | 3.38560 | 3.38559 | 3.38184 | 3.13987 | 2.20050 | 20.0561 | 10.0280 | 6.67550 | 4.56052 | 2.69201 |
| 8 | 3.38560 | 3.38559 | 3.38184 | 3.13960 | 2.19977 | 20.0472 | 10.0236 | 6.67253 | 4.55782 | 2.68964 |
| 9 | 3.38560 | 3.38559 | 3.38183 | 3.13941 | 2.19925 | 20.0415 | 10.0207 | 6.67060 | 4.55605 | 2.68807 |
| 10 | 3.38560 | 3.38559 | 3.38183 | 3.13927 | 2.19887 | 20.0375 | 10.0187 | 6.66927 | 4.55484 | 2.68697 |
| 11 | 3.38560 | 3.38559 | 3.38183 | 3.13916 | 2.19859 | 20.0347 | 10.0173 | 6.66833 | 4.55396 | 2.68618 |
| 12 | 3.38560 | 3.38559 | 3.38183 | 3.13908 | 2.19836 | 20.0326 | 10.0163 | 6.66763 | 4.55332 | 2.68559 |
| | | | | | $\varkappa_{11}$ | | | | | |
| 4 | 1.30438 | 1.30169 | 1.25407 | 0.93444 | 0.54298 | 11.3088 | 5.63370 | 3.51859 | 1.66461 | 0.72809 |
| 5 | 1.30438 | 1.30169 | 1.25404 | 0.93423 | 0.54273 | 11.3056 | 5.63208 | 3.51734 | 1.66353 | 0.72754 |
| 6 | 1.30438 | 1.30169 | 1.25402 | 0.93411 | 0.54259 | 11.3038 | 5.63120 | 3.51667 | 1.66293 | 0.72722 |
| 7 | 1.30438 | 1.30169 | 1.25400 | 0.93403 | 0.54250 | 11.3028 | 5.63068 | 3.51626 | 1.66256 | 0.72703 |
| 8 | 1.30438 | 1.30169 | 1.25399 | 0.93398 | 0.54245 | 11.3021 | 5.63034 | 3.51600 | 1.66232 | 0.72691 |
| 9 | 1.30438 | 1.30169 | 1.25399 | 0.93395 | 0.54241 | 11.3017 | 5.63011 | 3.51582 | 1.66215 | 0.72682 |
| 10 | 1.30438 | 1.30169 | 1.25398 | 0.93392 | 0.54238 | 11.3014 | 5.62995 | 3.51569 | 1.66203 | 0.72675 |
| 11 | 1.30438 | 1.30169 | 1.25398 | 0.93391 | 0.54235 | 11.3011 | 5.62983 | 3.51559 | 1.66195 | 0.72671 |
| 12 | 1.30438 | 1.30169 | 1.25398 | 0.93389 | 0.54234 | 11.3009 | 5.62974 | 3.51552 | 1.66188 | 0.72667 |
| | | | | | $\varkappa_{21}$ | | | | | |
| 4 | 2.26315 | 2.26309 | 2.25503 | 2.01634 | 1.36423 | 17.9601 | 8.97965 | 5.95067 | 3.73488 | 1.93222 |
| 5 | 2.26315 | 2.26309 | 2.25501 | 2.01585 | 1.36315 | 17.9495 | 8.97433 | 5.94704 | 3.73101 | 1.92917 |
| 6 | 2.26315 | 2.26309 | 2.25500 | 2.01556 | 1.36251 | 17.9436 | 8.97138 | 5.94502 | 3.72882 | 1.92740 |
| 7 | 2.26315 | 2.26309 | 2.25499 | 2.01538 | 1.36210 | 17.9400 | 8.96959 | 5.94379 | 3.72746 | 1.92629 |
| 8 | 2.26315 | 2.26309 | 2.25499 | 2.01525 | 1.36181 | 17.9377 | 8.96842 | 5.94299 | 3.72656 | 1.92554 |
| 9 | 2.26315 | 2.26309 | 2.25498 | 2.01516 | 1.36162 | 17.9361 | 8.96761 | 5.94244 | 3.72593 | 1.92501 |
| 10 | 2.26315 | 2.26309 | 2.25498 | 2.01510 | 1.36147 | 17.9349 | 8.96704 | 5.94204 | 3.72548 | 1.92463 |
| 11 | 2.26315 | 2.26309 | 2.25498 | 2.01505 | 1.36136 | 17.9341 | 8.96661 | 5.94175 | 3.72514 | 1.92434 |
| 12 | 2.26315 | 2.26309 | 2.25498 | 2.01501 | 1.36128 | 17.9334 | 8.96629 | 5.94153 | 3.72488 | 1.92412 |
| | | | | | $\varkappa_{31}$ | | | | | |
| 4 | 3.18025 | 3.18025 | 3.17909 | 3.04834 | 2.33433 | 24.2103 | 12.1051 | 8.06519 | 5.65429 | 3.42735 |
| 5 | 3.18025 | 3.18025 | 3.17909 | 3.04783 | 2.33212 | 24.1882 | 12.0941 | 8.05781 | 5.64731 | 3.41980 |
| 6 | 3.18025 | 3.18025 | 3.17908 | 3.04751 | 2.33075 | 24.1757 | 12.0878 | 8.05361 | 5.64328 | 3.41532 |
| 7 | 3.18025 | 3.18025 | 3.17908 | 3.04731 | 2.32985 | 24.1679 | 12.0840 | 8.05101 | 5.64075 | 3.41244 |
| 8 | 3.18025 | 3.18025 | 3.17908 | 3.04716 | 2.32923 | 24.1628 | 12.0814 | 8.04929 | 5.63906 | 3.41048 |
| 9 | 3.18025 | 3.18025 | 3.17908 | 3.04706 | 2.32878 | 24.1592 | 12.0796 | 8.04810 | 5.63788 | 3.40909 |
| 10 | 3.18025 | 3.18025 | 3.17908 | 3.04698 | 2.32844 | 24.1567 | 12.0783 | 8.04724 | 5.63703 | 3.40807 |
| 11 | 3.18025 | 3.18025 | 3.17908 | 3.04692 | 2.32818 | 24.1548 | 12.0774 | 8.04660 | 5.63639 | 3.40729 |
| 12 | 3.18025 | 3.18025 | 3.17908 | 3.04687 | 2.32798 | 24.1533 | 12.0766 | 8.04611 | 5.63589 | 3.40669 |

number of base functions gives only 2–3 significant digits for $0.2 \leq r_1 \leq 0.9$. However, unlike the first method, the second method makes it possible to perform reliable computations in the axisymmetric mode $\varkappa_{01}$. In addition, for $0.05 \leq r_1 \leq 0.4$, the lowest eigenvalue $\varkappa_{11}$ is computed with better accuracy. For the same $r_1$, the convergence may become slower as the semiapex angle increases. If $q_1 = q_2 = 12$, then the number of significant digits also decreases as $r_1 \to 1$. This "shallow-water" case is realized with 2–3 significant digits for $q_1 = q_2 = 12, \ldots, 14$.

The presence of two types of base functions in representation (5.2.18) makes it possible to vary $q_1$ and $q_2$ in order to get a better approximation for the same total number of base functions $Q = q_1 + q_2$. The variations of $q_1$ and $q_2$ with fixed $Q \geq 16$ show that a better accuracy of $\varkappa_{m1}$ can be attained for $q_2 > q_1$. In particular, this is true for smaller depths of liquid. Thus, if the $\vee$-shaped tank is characterized by $\theta_0 = 30°$ and $r_1 = 0.9$, then the approximate value $\varkappa_{11} = 0.54233738$ can be obtained either for $q_1 = q_2 = 12$ $(Q = 24)$ or for $q_1 = 7$, $q_2 = 12$ $(Q = 19)$.

The comparison of the numerical experiments performed by using two different functional bases shows that the method based on the HPS is more accurate for smaller depths $(0.6 \leq r_1)$. However, the case of larger depths $(r_1 \leq 0.4)$ is better treated by the second method. This can clearly be seen for $\wedge$-shaped tanks: The calculations performed by the second method preserve their robustness as the number of base functions increases, while the first method fails for larger dimensions. Generally speaking, the accuracy of both methods is similar only for $\vee$-tanks with $0.2 \leq r_1 \leq 0.55$, $\theta_0 > 10^0$.

Despite the fact that the number of base functions is small, the proposed two methods give accurate approximations of the lowest eigenvalue $\varkappa_{11}$. The lowest eigenvalue determines the lowest natural frequency by the formula $\sigma_{11} = \sqrt{g\varkappa_{11}}$. This frequency is of primary interest for modeling tower vibrations in the case of $\vee$-shaped tanks. Therefore, we made a special emphasis on the comparison of the numerical results obtained by the two methods for $\varkappa_{11}$. These results are illustrated in Figs. 5.6a and b. In the figures, the domains in the $(r_1, \theta_0)$-plane are identified for which each method gives the same number of significant digits for 20 base functions. One can see that the accuracy of the first method (HPS) may become low only for small $\theta_0$ and $r_1$, e.g., for large depths of the liquid. In the other cases, the method guarantees rapid convergence and high accuracy. On the other hand, the case of small $\theta_0$ and $r_1$ is satisfactory handled by the second method. However, this method slowly converges as $r_1 \to 1$ and $\theta_0 > 45°$, e.g., for small depths of liquid.

The natural sloshing frequency $\sigma_{11}$ is of high practical importance for the design of water towers. Having in mind this fact, in Table 5.5, we present the values of $\varkappa_{11}$ versus $\theta_0$ and $r_1$. The corresponding computations were performed to get up to five significant digits. The dimensional natural sloshing frequency $\sigma_{11}$

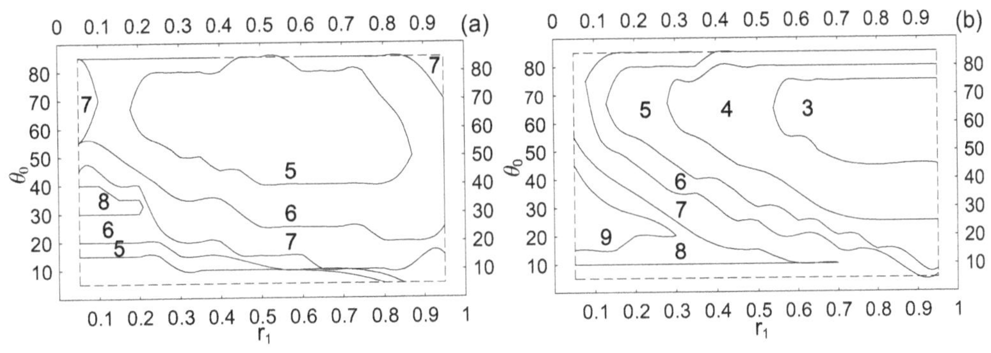


**Figure 5.6.** The number of significant digits of $\varkappa_{11}$ obtained for V-tanks and 20 base functions by the procedures based on the first (Case a) and second (Case b) methods.

**Table 5.5.** $\varkappa_{11}$ versus $\theta_0$ and $r_1$ for V-shaped tanks.

| $r_1$ | 10° | 15° | 20° | 25° | 30° | 35° | 40° | 45° | 50° | 55° | 60° | 65° | 70° |
|---|---|---|---|---|---|---|---|---|---|---|---|---|---|
| .05 | 1.674 | 1.586 | 1.495 | 1.401 | 1.304 | 1.205 | 1.104 | 1.000 | .8943 | .7868 | .6777 | .5671 | .4553 |
| .10 | 1.674 | 1.586 | 1.495 | 1.401 | 1.304 | 1.205 | 1.104 | 1.000 | .8943 | .7868 | .6776 | .5670 | .4551 |
| .15 | 1.674 | 1.586 | 1.495 | 1.401 | 1.304 | 1.205 | 1.104 | 1.000 | .8941 | .7865 | .6772 | .5665 | .4547 |
| .20 | 1.674 | 1.586 | 1.495 | 1.401 | 1.304 | 1.205 | 1.104 | 1.000 | .8935 | .7857 | .6763 | .5655 | .4536 |
| .25 | 1.674 | 1.586 | 1.495 | 1.401 | 1.304 | 1.205 | 1.103 | 0.999 | .8922 | .7840 | .6742 | .5634 | .4517 |
| .30 | 1.674 | 1.586 | 1.495 | 1.401 | 1.304 | 1.204 | 1.102 | 0.999 | .8894 | .7806 | .6706 | .5598 | .4484 |
| .35 | 1.674 | 1.586 | 1.495 | 1.401 | 1.303 | 1.203 | 1.099 | 0.993 | .8845 | .7750 | .6647 | .5541 | .4433 |
| .40 | 1.674 | 1.586 | 1.495 | 1.400 | 1.302 | 1.199 | 1.094 | 0.986 | .8762 | .7660 | .6556 | .5456 | .4359 |
| .45 | 1.674 | 1.586 | 1.495 | 1.399 | 1.298 | 1.193 | 1.085 | 0.974 | .8633 | .7525 | .6425 | .5335 | .4256 |
| .50 | 1.674 | 1.586 | 1.494 | 1.395 | 1.291 | 1.182 | 1.070 | 0.957 | .8442 | .7332 | .6242 | .5171 | .4117 |
| .55 | 1.674 | 1.586 | 1.491 | 1.388 | 1.277 | 1.163 | 1.046 | 0.930 | .8171 | .7067 | .5996 | .4954 | .3936 |
| .60 | 1.674 | 1.584 | 1.484 | 1.373 | 1.254 | 1.132 | 1.011 | 0.893 | .7800 | .6715 | .5676 | .4676 | .3707 |
| .65 | 1.674 | 1.580 | 1.470 | 1.346 | 1.215 | 1.086 | 0.961 | 0.842 | .7309 | .6261 | .5271 | .4330 | .3425 |
| .70 | 1.673 | 1.568 | 1.440 | 1.297 | 1.154 | 1.018 | 0.892 | 0.775 | .6681 | .5693 | .4775 | .3910 | .3086 |
| .75 | 1.667 | 1.539 | 1.381 | 1.217 | 1.063 | 0.924 | 0.800 | 0.690 | .5905 | .5007 | .4183 | .3417 | .2691 |
| .80 | 1.647 | 1.471 | 1.274 | 1.092 | 0.934 | 0.799 | 0.684 | 0.585 | .4978 | .4202 | .3499 | .2851 | .2241 |
| .85 | 1.578 | 1.325 | 1.095 | 0.908 | 0.761 | 0.642 | 0.544 | 0.461 | .3906 | .3284 | .2727 | .2218 | .1741 |
| .90 | 1.370 | 1.047 | 0.820 | 0.660 | 0.542 | 0.452 | 0.380 | 0.321 | .2704 | .2267 | .1879 | .1526 | .1197 |
| .95 | 0.863 | 0.596 | 0.446 | 0.351 | 0.285 | 0.236 | 0.197 | 0.166 | .1394 | .1167 | .0966 | .0784 | .0615 |

is computed from $\varkappa_{11}$ by the formula

$$\sigma_{11} = \sqrt{\frac{g\ \varkappa_{11}(\theta_0, \frac{r_1}{r_0})}{r_0}}, \qquad (5.2.20)$$

where $g$ and $r_1$ are not scaled by $r_0$. Hence, the numerical data from Table 5.5 can be used both for the structural design and for the verification of the other numerical methods.

## 5.3 Nonlinear modal equations for nontruncated conical tanks

In this section, we construct a mathematical model (nonlinear modal equations) for the description of nonlinear sloshing in axially symmetric conical tanks. The method is based on the variational modal approach formulated in the second chapter and the technique of nonconformal mappings. We use the approximate solutions of the spectral problems (5.1.4)–(5.1.7) obtained in Sec. 5.2 and preserve the level of generality of the mathematical model obtained in Sec. 4.1 for the upright circular cylinder. The analytic form of the approximate solution (5.1.26) of (5.1.4)–(5.1.7) enables us to completely perform this procedure.

The ordinary approximate solutions of the linear problem obtained in the Cartesian coordinate system for a domain with unperturbed boundaries do not possess, with rare exceptions, the required properties outside the domain of their definition, which is necessary for their application in the nonlinear theory, i.e., in the domains with perturbed boundaries. As already indicated, one of the fundamental problems of the nonlinear theory for noncylindrical cavities is the problem of effective representation of the perturbed free surface in the vicinity of its equilibrium.

Passing to the construction of the nonlinear mathematical model of free and forced sloshing in conical containers, we introduce a curvilinear coordinate system $Ox_1x_2x_3$ rigidly connected with the tank whose variables $x_1, x_2$, and $x_3$ are expressed via the variables of the Cartesian coordinate system $Oxyz$ by relations (5.1.1). It is convenient to represent the equation of the free surface $\zeta(x_1, x_2, x_3, t) = 0$ in the case of a conical vessel in the form

$$x_1 = x_{10} + f(x_2, x_3, t). \tag{5.3.1}$$

Thus, the domain of definition of $f(x_2, x_3, t)$ with respect to the space variables $x_2$ and $x_3$ is independent of time and coincides with the unperturbed surface $\Sigma_0$. In this case, we represent the Lagrangian $L$ of the action in the form

$$L = -\rho \int_D \left( \int_0^{x_{10}+f} \left[ \varphi_t + \frac{1}{2}(\nabla\varphi)^2 + gx_1 \right] \sqrt{q}\, dx_1 \right) dx_2\, dx_3, \tag{5.3.2}$$

where $\sqrt{q} = x_1^2 x_2$ and $D$ is the range of the parameters $x_2$ and $x_3$. According to (5.3.1), the upper limit of the inner integral depends on $x_2$ and $x_3$.

Further, for the sake of definiteness, we consider a conical container with the shape of inverse circular cone (see Fig. 5.1) for which the equation of the perturbed free surface and the velocity potential $\varphi(x_1, x_2, x_3, t)$ in the coordinate system with origin at the apex of the cone, by analogy with (4.1.5) and (4.1.6), can be reduced to

$$x_1 = x_{10} + \beta_0(t) + p_0(t) f_0(x_2) + [\, r_1(t) \sin x_3 + p_1(t) \cos x_3 \,] f_1(x_2)$$
$$+ [\, r_2(t) \sin 2x_3 + p_2(t) \cos 2x_3 \,] f_2(x_2); \tag{5.3.3}$$

$$\varphi(x_1, x_2, x_3) = P_0(t)\psi_0(x_1, x_2) + [\, R_1(t) \sin x_3 + P_1(t) \cos x_3 \,] \psi_1(x_1, x_2)$$
$$+ [\, R_2(t) \sin 2x_3 + P_2(t) \cos 2x_3 \,] \psi_2(x_1, x_2), \tag{5.3.4}$$

where $\psi_m(x_1, x_2)$ and $f_m(x_2)$ are the leading natural sloshing modes corresponding to the eigenvalues $\varkappa_{mn}$ $(m = 0, 1, 2)$ ordered according to the Narimanov–Moiseev rules presented in Sec. 4.1. These modes satisfy the normalization condition $\psi(x_{10}, x_{20}) = 1$. By virtue of this normalization condition, the solution of the main spectral problem takes the form

$$\psi_{mn}(x_1, x_2) = \sum_{k=0}^{q} a_{nk}^{(m)} \left( \frac{x_1}{x_{10}} \right)^{\nu_{mk}} V_{\nu_{mk}}^{(m)}(x_2), \tag{5.3.5}$$

where

$$a_{nk}^{(m)} = \frac{\bar{a}_{nk}^{(m)}}{N_{mn}}; \quad N_{mn} = \psi_{mn}(x_{10}, x_{20}) = \sum_{k=0}^{q} \bar{a}_{nk}^{(m)} V_{\nu_{mk}}^{(m)}(x_{20}) \tag{5.3.6}$$

and $\bar{a}_{nk}^{(m)}$ are the Ritz–Trefftz coefficients (5.1.29).

We deduce a finite-dimensional system of nonlinear ordinary differential equations for the description of sloshing in conical tanks participating in given translational motions by using the infinite system of differential equations

$$\frac{d}{dt} A_n - \sum_k R_k A_{nk} = 0, \quad n = 1, 2, \ldots, \tag{5.3.7}$$

$$\sum_n \frac{dR_n}{dt} \frac{\partial A_n}{\partial \beta_i} + \frac{1}{2} \sum_n \sum_k \frac{\partial A_{nk}}{\partial \beta_i} R_n R_k + (\overset{*}{\mathbf{v}}_0 - \mathbf{g}) \cdot \frac{\partial \boldsymbol{l}}{\partial \beta_i} = 0, \quad i = 1, 2, \ldots \tag{5.3.8}$$

and representations (5.3.3) and (5.3.4). For the sake of brevity, we set

$$P_0(t) = Z_1(t);\ R_1(t) = Z_2(t);\ P_1(t) = Z_3(t);\ R_2(t) = Z_4(t);\ P_2(t) = Z_5(t),$$

$$\varphi_1 = \psi_0;\ \varphi_2 = \psi_1 \sin x_3;\ \varphi_3 = \psi_1 \cos x_3;\ \varphi_4 = \psi_2 \sin 2x_3;\ \varphi_5 = \psi_2 \cos 2x_3,$$

$$\beta_1(t) = p_0(t);\ \beta_2(t) = r_1(t);\ \beta_3(t) = p_1(t);\ \beta_4(t) = r_2(t);\ \beta_5(t) = p_2(t),$$

$$F_1 = f_0(x_2) = \psi_0(x_{10}, x_2);\ F_2 = f_1(x_2) \sin x_3 = \psi_1(x_{10}, x_2) \sin x_3,$$

$$F_3 = f_1(x_2) \cos x_3 = \psi_1(x_{10}, x_2) \cos x_3,$$

$$F_4 = f_2(x_2) \sin 2x_3 = \psi_2(x_{10}, x_2) \sin 2x_3,$$

$$F_5 = f_2(x_2)\cos 2x_3 = \psi_2(x_{10}, x_2)\cos 2x_3,$$

$$f_m(x_2) = a_{10}^{(m)} + \sum_{k=1}^{q} b_k^{(m)}(x_2); \; b_k^{(m)} = a_{1k}^{(m)} V_{\nu_{mk}}^{(m)}(x_2), \quad m = 0, 1, 2. \qquad (5.3.9)$$

We now reduce a finite-dimensional analog of the system of equations (5.3.7), (5.3.8) to the form

$$\frac{d}{dt}A_n - \sum_{k=1}^{5} A_{nk} Z_k = 0, \quad n = 1, \ldots, 5, \qquad (5.3.10)$$

$$\sum_{k=1}^{5} \dot{Z}_n \frac{\partial A_n}{\partial \beta_i} + \frac{1}{2} \sum_{n=1}^{5} \sum_{k=1}^{5} \frac{\partial A_{nk}}{\partial \beta_i} Z_n Z_k + \sum_{j=1}^{3} w_j \frac{\partial l_j}{\partial \beta_j} = 0, \quad i = 1, \ldots, 5, \quad (5.3.11)$$

where $w_j$ are the projections of the apparent acceleration $\overset{*}{\mathbf{v}}_0 - \mathbf{g}$ onto the axes of the moving coordinate system $Oxyz$.

By using the condition of conservation of volume

$$\int_0^{2\pi} \int_0^{x_{20}} (\alpha_1 f + \alpha_2 f^2 + \alpha_3 f^3) x_2 \, dx_2 \, dx_3 = 0, \qquad (5.3.12)$$

where

$$\alpha_1 = x_{10}^2, \; \alpha_2 = x_{10}, \; \alpha_3 = \tfrac{1}{3}; \; f(x_2, x_3, t) = \beta_0(t) + \sum_{i=1}^{5} \beta_i(t) F_i(x_2, x_3),$$

we obtain $\beta_0(t)$ in decomposition (5.3.3) as a function of the generalized coordinates $p_0(t)$, $r_1(t)$, $\ldots$, $p_2(t)$ characterizing the time evolution of the perturbed free surface.

To within the terms of the third order of smallness, we get the following relation from (5.3.12):

$$\beta_0(t) = k_1(r_1^2 + p_1^2), \qquad (5.3.13)$$

where

$$k_1 = -\frac{1}{x_{10} x_{20}^2} \int_0^{x_{20}} x_2 f_1^2(x_2) \, dx_2.$$

By using the relations

$$l_1 = \rho \int_0^{x_{20}} \int_0^{2\pi} \int_0^{x_{10}+f} x_1^3 x_2 \, dx_1 dx_2 dx_3,$$

$$l_2 = \rho \int_0^{x_{20}} \int_0^{2\pi} \int_0^{x_{10}+f} x_1^3 x_2^2 \cos x_3 \, dx_1 dx_2 dx_3, \qquad (5.3.14)$$

$$l_3 = \rho \int_0^{x_{20}} \int_0^{2\pi} \int_0^{x_{10}+f} x_1^3 x_2^2 \sin x_3 \, dx_1 dx_2 dx_3,$$

we obtain the components of the vector $\boldsymbol{l}$ specifying the position of the center-of-mass of the liquid in its perturbed state.

Assume that the terms of the third order of smallness are absent in the equations of motion. Thus, it is necessary to find the vector $\boldsymbol{l}$ to within the terms of the order of smallness higher than the third order. We find

$$l_1 = \rho[l_1^{(0)} + l_1^{(1)}(r_1^2 + p_1^2) + l_1^{(2)}p_0^2 + l_1^{(3)}(r_2^2 + p_2^2) + l_1^{(4)}(r_1^2 + p_1^2)^2 + l_1^{(5)}\left(\tfrac{1}{2}p_1^2 p_2 - \tfrac{1}{2}r_1^2 p_2 + r_1 p_1 r_2\right) + l_1^{(6)}p_0(r_1^2 + p_1^2)], \tag{5.3.15}$$

where

$$l_1^{(0)} = \pi\tfrac{1}{4}x_{10}^4 x_{20}^2,\ l_1^{(1)} = \tfrac{1}{2}x_{10}^2 e_{11},\ l_1^{(2)} = \tfrac{1}{2}x_{10}^2 e_{00},\ l_1^{(3)} = \tfrac{1}{2}x_{10}^2 e_{22},$$

$$l_1^{(4)} = \tfrac{3}{16}e_4 - \tfrac{3}{2}\frac{x_{10}^2}{s_0}e_{11}^2,\ l_1^{(5)} = 2x_{10}e_{211},\ l_1^{(6)} = 2x_{10}e_{011}. \tag{5.3.16}$$

The components of this vectors along the $Oy$- and $Oz$-axes take the form

$$l_2 = l_2^{(1)}p_1 + l_2^{(2)}p_1 p_0 + l_2^{(3)}p_1(r_1^2 + p_1^2) + l_2^{(4)}(r_1 r_2 + p_1 p_2) + l_2^{(5)}p_1 p_0^2 + l_2^{(6)}p_1(r_2^2 + p_2^2) + l_2^{(7)}p_1 p_0(r_1^2 + p_1^2) + l_2^{(8)}p_1\left(\tfrac{1}{2}p_2 p_1^2 - \tfrac{1}{2}p_2 r_1^2 + r_1 p_1 r_2\right) + l_2^{(9)}(r_1 r_2 + p_1 p_2)(r_1^2 + p_1^2) + l_2^{(10)}p_0(r_1 r_2 + p_1 p_2), \tag{5.3.17}$$

$$l_3 = l_3^{(1)}r_1 + l_3^{(2)}p_0 r_1 + l_3^{(3)}r_1(r_1^2 + p_1^2) + l_3^{(4)}(p_1 r_2 - r_1 p_2) + l_3^{(5)}r_1 p_0^2 + l_3^{(6)}r_1(r_2^2 + p_2^2) + l_3^{(7)}r_1 p_0(r_1^2 + p_1^2) + l_3^{(8)}r_1\left(\tfrac{1}{2}p_2 p_1^2 - \tfrac{1}{2}p_2 r_1^2 + r_1 p_1 r_2\right) + l_3^{(9)}(r_1^2 + p_1^2)(p_1 r_2 - r_1 p_2) + l_3^{(10)}p_0(p_1 r_2 - r_1 p_2), \tag{5.3.18}$$

where

$$l_2^{(1)} = l_3^{(1)} = \rho\, x_{10}^3 s_1,\ l_2^{(2)} = l_3^{(2)} = 3\rho\, x_{10}^2 s_{01},$$

$$l_2^{(3)} = l_3^{(3)} = 3\rho\left(x_{10}^2 k_1 s_1 + \tfrac{1}{4}x_{10}s_{13}\right),\ l_2^{(4)} = l_3^{(4)} = \tfrac{3}{2}\rho\, x_{10}^2 s_{12},$$

$$l_2^{(5)} = l_3^{(5)} = 3\rho\left(x_{10}^2 k_0 s_1 + x_{10}s_{102}\right),\ l_2^{(6)} = l_3^{(6)} = 3\rho\left(x_{10}^2 k_2 s_1 + \tfrac{1}{2}x_{10}s_{122}\right),$$

$$l_2^{(7)} = l_3^{(7)} = 3\rho\left(x_{10}^2 k_3 s_1 + 2k_1 x_{10}s_{01} + \tfrac{1}{4}s_{013}\right), \tag{5.3.19}$$

$$l_2^{(8)} = l_3^{(8)} = 3\rho\, x_{10}^2 k_4 s_1,\ l_2^{(9)} = l_3^{(9)} = 3\rho\, x_{10}s_{12}k_1,$$

$$l_2^{(10)} = l_3^{(10)} = 3\rho\, x_{10}s_{012}.$$

The quantities $e_4, e_{00}, e_{11}, \dots, e_{211}$ and $s_1, s_{01}, \dots, s_{012}$ in (5.3.16) and (5.3.19) are defined by the quadratures

$$s_1 = \pi\int_0^{x_{20}} x_2^2 f_1\,dx_2,\ s_{01} = \pi\int_0^{x_{20}} x_2^2 f_0 f_1\,dx_2,\ s_{13} = \pi\int_0^{x_{20}} x_2^2 f_1^3\,dx_2,$$

$$s_{12} = \pi \int_0^{x_{20}} x_2^2 f_1 f_2 \, dx_2, \; s_{102} = \pi \int_0^{x_{20}} x_2^2 f_0^2 f_1 \, dx_2, \; s_{122} = \pi \int_0^{x_{20}} x_2^2 f_1 f_2^2 \, dx_2,$$

$$s_{013} = \pi \int_0^{x_{20}} x_2^2 f_0 f_1^3 \, dx_2, \; s_{012} = \pi \int_0^{x_{20}} x_2^2 f_0 f_1 f_2 \, dx_2, \tag{5.3.20}$$

$$e_4 = \pi \int_0^{x_{20}} x_2 f_1^4 dx_2, \; e_{00} = 2\pi \int_0^{x_{20}} x_2 f_0^2 dx_2, \; e_{11} = \pi \int_0^{x_{20}} x_2 f_1^2 dx_2,$$

$$e_{22} = \pi \int_0^{x_{20}} x_2 f_2^2 dx_2, \; e_{011} = \pi \int_0^{x_{20}} x_2 f_0 f_1^2 dx_2, \; e_{211} = \pi \int_0^{x_{20}} x_2 f_2 f_1^2 dx_2.$$

The next step in the construction of modal equations on the basis of (5.3.10), (5.3.11) is reduced to finding the quadratures

$$A_n = \rho \int_{Q(t)} \varphi_n \, dQ \quad \text{and} \quad A_{nk} = \rho \int_{Q(t)} (\nabla\varphi_n, \nabla\varphi_k) \, dQ$$

as functions of the generalized coordinates $p_0(t)$, $r_1(t)$, ... , $p_2(t)$ specifying the position of the free surface at any time $t$.

The power dependence of the solutions $\psi_m(x_1, x_2)$ (5.3.5) on the coordinate $x_1$ and the equation of the free surface in the form (5.3.3) (solvable with respect to the coordinate $x_1$) enable us to perform this procedure in the same way as for cylindrical tanks. The quantities $A_n$ and $A_{nk}$ are found with an accuracy required for the construction of the equations of motion in terms of generalized coordinates of the system. The elements of the vector $A_n$ and the matrix $A_{nk}$ are characterized by the following dependences on the generalized coordinates $p_0$, $r_1$, ..., $p_2$:

$$\begin{aligned} A_1 &= a_0 + a_4(r_1^2 + p_1^2) + a_{17} p_0, \\ A_2 &= a_5 r_1 + a_6 r_1 (r_1^2 + p_1^2) + a_{18}(p_1 r_2 - r_1 p_2) + a_{14} r_1 p_0, \\ A_3 &= a_5 p_1 + a_6 p_1 (p_1^2 + r_1^2) + a_{18}(r_1 r_2 + p_1 p_2) + a_{14} p_1 p_0, \\ A_4 &= a_{13} r_2 - 2a_7 r_1 p_1, \quad A_5 = a_{13} p_2 + a_7 (r_1^2 - p_1^2) \end{aligned} \tag{5.3.21}$$

and

$$\begin{aligned} &A_{11} = 2a_1; \quad A_{21} = A_{12} = a_{15} r_1; \quad A_{31} = A_{13} = a_{15} p_1, \\ &A_{22} = 2a_{10} + 2a_{11} r_1^2 + 2a_{12} p_1^2 + 2a_9 p_0 - 2a_{16} p_2, \\ &A_{32} = A_{23} = a_8 r_1 p_1 + 2a_{16} r_2; \quad A_{55} = A_{44} = 2a_2, \\ &A_{33} = 2a_{10} + 2a_{11} p_1^2 + 2a_{12} r_1^2 + 2a_{16} p_2 + 2a_9 p_0, \\ &A_{42} = A_{24} = a_3 p_1; \quad A_{52} = A_{25} = -a_3 r_1; \quad A_{43} = A_{34} = a_3 r_1, \\ &A_{53} = A_{35} = a_3 p_1; \quad A_{41} = A_{14} = A_{51} = A_{15} = A_{54} = A_{45} = 0. \end{aligned} \tag{5.3.22}$$

The coefficients $a_1, \dots, a_{18}$ in relations (5.3.21) and (5.3.22) are determined by the following quadratures:

$$a_1 = \pi\rho \int_0^{x_{20}} F_0^{(0,0)}(x_2)\, x_2\, dx_2,$$

$$a_2 = \frac{\pi}{2}\rho \int_0^{x_{20}} \left( F_0^{(2,2)}(x_2) + \frac{4}{x_2^2} B_0^{(2,2)}(x_2) \right) x_2\, dx_2,$$

$$a_3 = \frac{\pi}{2}\rho \int_0^{x_{20}} \left( F_1^{(1,2)}(x_2) + \frac{2}{x_2^2} B_1^{(1,2)}(x_2) \right) f_1(x_2)\, x_2\, dx_2,$$

$$a_4 = 2\pi\rho \int_0^{x_{20}} B_0^{(2)}(x_2)\, f_1^2(x_2)\, dx_2 + 2\pi\rho k_1 \int_0^{x_{20}} B_0^{(1)}(x_2)\, x_2 dx_2,$$

$$a_5 = \pi\rho \int_0^{x_{20}} B_1^{(1)}(x_2)\, f_1(x_2)\, x_2 dx_2,$$

$$a_6 = \pi\rho \int_0^{x_{20}} \left( \frac{3}{4} B_1^{(3)}(x_2)\, f_1^2(x_2) + 2k_1 B_1^{(2)}(x_2) \right) f_1(x_2)\, x_2 dx_2,$$

$$a_7 = -\frac{\pi}{2}\rho \int_0^{x_{20}} B_2^{(2)}(x_2)\, f_1^2(x_2)\, x_2 dx_2, \tag{5.3.23}$$

$$a_8 = \frac{\pi}{2}\rho \int_0^{x_{20}} \left( F_2^{(1,1)}(x_2) - \frac{1}{x_2^2} B_2^{(1,1)}(x_2) \right) f_1^2(x_2)\, x_2 dx_2,$$

$$a_9 = \frac{\pi}{2}\rho \int_0^{x_{20}} \left( F_1^{(1,1)}(x_2) + \frac{1}{x_2^2} B_1^{(1,1)}(x_2) \right) f_0(x_2)\, x_2 dx_2,$$

$$a_{10} = \frac{\pi}{2}\rho \int_0^{x_{20}} \left( F_0^{(1,1)}(x_2) + \frac{1}{x_2^2} B_0^{(1,1)}(x_2) \right) x_2 dx_2,$$

$$a_{11} = \frac{\pi}{2}\rho \int_0^{x_{20}} \left( k_1 \left[ F_1^{(1,1)}(x_2) + \frac{1}{x_2^2} B_1^{(1,1)}(x_2) \right] \right.$$
$$\left. + \frac{3}{4} \left[ F_2^{(1,1)}(x_2) + \frac{1}{3x_2^2} B_2^{(1,1)}(x_2) \right] f_1^2(x_2) \right) x_2 dx_2,$$

$$a_{12} = \frac{\pi}{2}\rho \int_0^{x_{20}} \left( k_1 \left[ F_1^{(1,1)}(x_2) + \frac{1}{x_2^2} B_1^{(1,1)}(x_2) \right] \right.$$

$$+\frac{3}{4}\left[\frac{1}{3}F_2^{(1,1)}(x_2)+\frac{1}{x_2^2}B_2^{(1,1)}(x_2)\right]f_1^2(x_2)\Big)\,x_2dx_2,$$

$$a_{13}=\pi\rho\int_0^{x_{20}}B_2^{(1)}(x_2)\,f_2(x_2)\,x_2dx_2,$$

$$a_{14}=2\pi\rho\int_0^{x_{20}}B_1^{(2)}(x_2)\,f_0(x_2)\,f_1(x_2)\,x_2dx_2,$$

$$a_{15}=\pi\rho\int_0^{x_{20}}F_1^{(0,1)}(x_2)\,f_1(x_2)\,x_2dx_2,$$

$$a_{16}=\frac{\pi}{4}\rho\int_0^{x_{20}}\left[F_1^{(1,1)}(x_2)-\frac{1}{x_2^2}B_1^{(1,1)}(x_2)\right]f_2(x_2)\,x_2dx_2,$$

$$a_{17}=2\pi\rho\int_0^{x_{20}}B_0^{(1)}(x_2)\,f_0(x_2)\,x_2dx_2,\quad a_{18}=\pi\rho\int_0^{x_{20}}B_1^{(2)}(x_2)\,f_2(x_2)\,x_2dx_2.$$

The functions $B_0^{(1)}$, $B_0^{(2)}$, ..., $B_2^{(1,1)}$ and $F_0^{(0,0)}$, $F_0^{(1,1)}$, ..., $F_2^{(1,1)}$ in integrands (5.3.23) are expressed via $b_k^{(m)}(x_2)$ and

$$c_k^{(m)}(x_2)=a_{1k}^{(m)}\frac{dV_{\nu_{mk}}^{(m)}}{dx_2},\qquad m=0,1,2;\quad k=0,1,\dots,q.$$

They have the form

$$B_m^{(1)}(x_2)=x_{10}^2\sum_{k=0}^{q}b_k^{(m)}(x_2);\quad B_m^{(2)}(x_2)=\frac{x_{10}}{2}\sum_{k=0}^{q}(\nu_{mk}+2)b_k^{(m)}(x_2),$$

$$B_m^{(3)}(x_2)=\frac{1}{6}\sum_{k=0}^{q}(\nu_{mk}+2)(\nu_{mk}+1)b_k^{(m)}(x_2),\quad m=0,1,2,$$

$$B_0^{(m,n)}(x_2)=x_{10}\sum_{i,j=1}^{q}\frac{b_i^{(m)}(x_2)b_j^{(n)}(x_2)}{\nu_{mi}+\nu_{nj}+1};\quad B_1^{(m,n)}=\sum_{i,j=1}^{q}b_i^{(m)}(x_2)b_j^{(n)}(x_2),$$

$$B_2^{(m,n)}=\frac{1}{2x_{10}}\sum_{i,j=1}^{q}(\nu_{mi}+\nu_{nj})b_i^{(m)}b_j^{(n)},\quad m,n=1,2,\tag{5.3.24}$$

$$\Pi_{ij}^{(m,n)}(x_2)=\nu_{mi}\nu_{nj}b_i^{(m)}b_j^{(n)}-x_2(\nu_{mi}b_i^{(m)}c_j^{(n)}$$
$$+\nu_{ni}b_i^{(n)}c_j^{(m)})+(1+x_2^2)c_i^{(m)}c_j^{(n)},$$

$$F_0^{(m,n)}(x_2) = x_{10} \sum_{i,j=1}^{q} \frac{\Pi_{ij}^{(m,n)}(x_2)}{\nu_{mi} + \nu_{nj} + 1}, \quad F_1^{(m,n)}(x_2) = \sum_{i,j=1}^{q} \Pi_{ij}^{(m,n)}(x_2),$$

$$F_2^{(m,n)}(x_2) = \frac{1}{x_{10}} \sum_{i,j=1}^{q} (\nu_{mi} + \nu_{nj}) \Pi_{ij}^{(m,n)}(x_2), \quad m, n = 0, 1, 2.$$

This completes the construction of the finite-dimensional nonlinear mathematical model (5.3.10), (5.3.11).

In some cases, for the practical application of this system to the problems of dynamics of body–liquid systems, it is convenient to represent it in the form of a nonlinear system of ordinary second-order differential equations (equations of the oscillatory type). To this end, we exclude the functions $Z_k(t)$ from the first group of Eqs. (5.3.10) and express them in terms of the generalized coordinates $\beta_i(t)$ and their derivatives. To within the terms of the third order of smallness, in terms of the generalized coordinates $r_i(t)$ and $p_i(t)$, we obtain

$$\begin{aligned}
R_1(t) &= Q_1\ddot{r}_1 + C_2 r_1^2 \ddot{r}_1 + D_3 p_1^2 \ddot{r}_1 + C_1 r_1 p_1 \ddot{p}_1 + D_2(r_2\ddot{p}_1 - p_2\ddot{r}_1) \\
&\quad + C_3(p_1\ddot{r}_2 - r_1\ddot{p}_2) + B_0 p_0 \ddot{r}_1 + B_3 r_1 \ddot{p}_0, \\
P_1(t) &= Q_1\ddot{p}_1 + C_2 p_1^2 \ddot{p}_1 + D_3 r_1^2 \ddot{p}_1 + C_1 p_1 r_1 \ddot{r}_1 + D_2(r_2\ddot{r}_1 + p_2\ddot{p}_1) \\
&\quad + C_3(r_1\ddot{r}_2 + p_1\ddot{p}_2) + B_0 p_0 \ddot{p}_1 + B_3 p_1 \ddot{p}_0, \\
P_0(t) &= C_0(r_1\ddot{r}_1 + p_1\ddot{p}_1) + D_0\ddot{p}_0, \qquad (5.3.25) \\
R_2(t) &= Q_2\ddot{r}_2 - D_1(r_1\ddot{p}_1 + p_1\ddot{r}_1), \\
P_2(t) &= Q_2\ddot{p}_2 + D_1(r_1\ddot{r}_1 - p_1\ddot{p}_1),
\end{aligned}$$

where

$$C_0 = \frac{a_4}{a_1} - \frac{a_5 a_{15}}{4a_1 a_{10}}; \quad D_0 = \frac{a_{17}}{2a_1}; \quad Q_2 = \frac{a_{13}}{2a_2},$$

$$C_1 = \frac{1}{a_{10}}\left(a_6 - \frac{a_4 a_{15}}{2a_1} - \frac{a_5 a_8}{4a_{10}} + \frac{a_5 a_{15}^2}{8a_1 a_{10}}\right); \quad B_0 = \frac{1}{2a_{10}}\left(a_{14} - \frac{a_5 a_9}{a_{10}}\right),$$

$$C_3 = \frac{1}{2a_{10}}\left(a_{18} - \frac{a_3 a_{13}}{2a_2}\right); \quad D_1 = \frac{a_7}{a_2} + \frac{a_3 a_5}{4a_2 a_{10}}, \qquad (5.3.26)$$

$$Q_1 = \frac{a_5}{2a_{10}}; \quad B_3 = \frac{1}{2a_{10}}\left(a_{14} - \frac{a_{15} a_{17}}{2a_1}\right); \quad D_2 = \frac{1}{2a_{10}}\left(a_{18} - \frac{a_5 a_{16}}{a_{10}}\right),$$

$$D_3 = \frac{a_6}{2a_{10}} + Q_1\left(\frac{a_3^2}{4a_2 a_{10}} - \frac{a_{12}}{a_{10}} + \frac{a_3 a_7}{a_2 a_5}\right); \quad C_2 = D_3 + C_1.$$

Substituting the parameters $P_0(t), R_1(t), \ldots, P_2(t)$ given by relations (5.3.25) in the remaining group of equations (5.3.11), we obtain the following nonlinear system of ordinary differential equations for the description of motion of liquid in conical tanks participating in arbitrary translational motions:

$$\begin{aligned}
&\mu_1(\ddot{r}_1+\sigma_1^2 r_1)+d_1(r_1^2\ddot{r}_1+r_1\dot{r}_1^2+r_1p_1\ddot{p}_1+r_1\dot{p}_1^2)+d_2(p_1^2\ddot{r}_1+2p_1\dot{r}_1\dot{p}_1\\
&\quad -r_1p_1\ddot{p}_1-2r_1\dot{p}_1^2)-d_3(p_2\ddot{r}_1-r_2\ddot{p}_1+\dot{r}_1\dot{p}_2-\dot{p}_1\dot{r}_2)+d_4(r_1\ddot{p}_2-p_1\ddot{r}_2)\\
&\quad +d_5(p_0\ddot{r}_1+\dot{r}_1\dot{p}_0)+d_6r_1\ddot{p}_0+\underline{w_1d_1^kr_1(r_1^2+p_1^2)+2w_1d_2^k(p_1r_2-r_1p_2)}\\
&\quad +\underline{2w_1d_3^kr_1p_0+w_2(2d_5^kr_1p_1+d_6^kr_2)+w_3[\lambda+d_4^kp_0+d_5^k(p_1^2+3r_1^2)}\\
&\qquad -\underline{d_6^kp_2+d_7^kp_0]}=0;
\end{aligned}\tag{5.3.27}$$

$$\begin{aligned}
&\mu_1(\ddot{p}_1+\sigma_1^2 p_1)+d_1(p_1^2\ddot{p}_1+r_1p_1\ddot{r}_1+p_1\dot{r}_1^2+\dot{p}_1^2p_1)+d_2(r_1^2\ddot{p}_1-r_1p_1\ddot{r}_1\\
&\quad +2r_1\dot{r}_1\dot{p}_1-2p_1\dot{r}_1^2)+d_3(p_2\ddot{p}_1+r_2\ddot{r}_1+\dot{r}_1\dot{r}_2+\dot{p}_1\dot{p}_2)-d_4(p_1\ddot{p}_2+r_1\ddot{r}_2)\\
&\quad +d_5(p_0\ddot{p}_1+\dot{p}_1\dot{p}_0)+d_6p_1\ddot{p}_0+\underline{w_1[d_1^kp_1(r_1^2+p_1^2)+2d_2^k(r_1r_2+p_1p_2)}\\
&\quad +\underline{2d_3^kp_1p_0]+w_2[\lambda+d_4^kp_0+d_5^k(r_1^2+3p_1^2)+d_6^kp_2+d_7^kp_0]}\\
&\qquad +\underline{w_3[2d_5^kr_1p_1+d_6^kr_2]}=0;
\end{aligned}\tag{5.3.28}$$

$$\begin{aligned}
&\mu_0(\ddot{p}_0+\sigma_0^2p_0)+d_6(r_1\ddot{r}_1+p_1\ddot{p}_1)+d_8(\dot{r}_1^2+\dot{p}_1^2)+\underline{w_1d_3^k(r_1^2+p_1^2)}\\
&\quad \underline{+w_2[d_4^kp_1+2d_7^kp_1p_0+d_9^kp_1(r_1^2+p_1^2)+d_{12}^k(r_1r_2+p_1p_2)]}\\
&\quad \underline{+w_3[d_4^kr_1+2d_7^kr_1p_0+d_9^kr_1(r_1^2+p_1^2)+d_{12}^k(p_1r_2-r_1p_2)]}=0;
\end{aligned}\tag{5.3.29}$$

$$\begin{aligned}
&\mu_2(\ddot{r}_2+\sigma_2^2r_2)-d_4(p_1\ddot{r}_1+r_1\ddot{p}_1)-2d_7\dot{r}_1\dot{p}_1+\underline{2w_1d_2^kr_1p_1}\\
&\quad \underline{+w_2[d_6^kr_1+2d_8^kp_1r_2+2d_{10}^kr_1p_1^2+d_{11}^kr_1(r_1^2+p_1^2)+d_{12}^kr_1p_0]}\\
&\quad \underline{+w_3[d_6^kp_1+2d_8^kr_1r_2+2d_{10}^kr_1^2p_1+d_{11}^kp_1(r_1^2+p_1^2)+d_{12}^kp_1p_0]}=0;
\end{aligned}\tag{5.3.30}$$

$$\begin{aligned}
&\mu_2(\ddot{p}_2+\sigma_2^2p_2)+d_4(r_1\ddot{r}_1-p_1\ddot{p}_1)+d_7(\dot{r}_1^2-\dot{p}_1^2)-\underline{w_1d_2^k(r_1^2-p_1^2)}\\
&\quad \underline{+w_2[d_6^kp_1+2d_8^kp_1p_2+d_{10}^kp_1(p_1^2-r_1^2)+d_{11}^kp_1(r_1^2+p_1^2)+d_{12}^kp_0p_1]}\\
&\quad \underline{+w_3[2d_8^kr_1p_2-d_6^kr_1+d_{10}^kr_1(p_1^2-r_1^2)-d_{11}^kr_1(r_1^2+p_1^2)-d_{12}^kr_1p_0]}=0,
\end{aligned}\tag{5.3.31}$$

where the natural sloshing frequencies corresponding to the values $m=0,1,2$ are given by the formulas

$$\sigma_0^2=\mathrm{g}\frac{2l_1^{(2)}}{\mu_0},\quad \sigma_1^2=\mathrm{g}\frac{2l_1^{(1)}}{\mu_1},\quad \sigma_2^2=\mathrm{g}\frac{2l_1^{(3)}}{\mu_2},\tag{5.3.32}$$

$$\mu_0=a_{17}D_0,\quad \mu_1=a_5Q_1,\quad \mu_2=a_{13}Q_2,\tag{5.3.33}$$

and the other coefficients of the system (5.3.27)–(5.3.31) are equal to

$$
\begin{gathered}
d_1 = 2a_4C_0 + 2a_7D_1 + a_5C_2 + 3a_6Q_1, \\
d_2 = a_5D_3 + a_6Q_1 + 2a_7D_1; \quad d_3 = a_5D_2 + a_{18}Q_1, \\
d_4 = 2a_7Q_2 - a_5C_3; \quad d_5 = a_5B_0 + a_{14}Q_1, \quad d_6 = 2a_4D_0 + a_5B_3, \\
d_7 = d_4 + \frac{1}{2}d_3, \quad d_8 = d_6 - \frac{1}{2}d_5, \quad d_1^k = 4\,l_1^{(4)}, \quad d_2^k = \frac{1}{2}\,l_1^{(5)}, \\
d_3^k = l_1^{(6)}, \quad d_4^k = l_2^{(2)}, \quad d_5^k = l_2^{(3)}, \quad d_6^k = l_2^{(4)}, \qquad (5.3.34) \\
d_7^k = l_2^{(5)}, \quad d_8^k = l_2^{(6)}, \quad d_9^k = l_2^{(7)}, \quad d_{10}^k = \frac{1}{2}\,l_2^{(8)}, \\
d_{11}^k = l_2^{(9)}, \quad d_{12}^k = l_2^{(10)}, \quad \lambda = l_2^{(1)} = l_3^{(1)}.
\end{gathered}
$$

Equations (5.3.27)–(5.3.31) differ from the similar equations for the upright circular cylinder by the terms marked by the subscript $k$. These terms correspond to the geometric nonlinearity and appear in the equations of motion due to the curvilinearity of walls of the tank.

For the obtained system of nonlinear ordinary differential equations, we can pose the Cauchy problem of finding the configuration of the free surface and the distribution of velocities on this surface for $t = t_0$ and $v_0 = 0$ or the problem of determination of the free or forced steady-state (periodic) sloshing. To study free oscillations, it is necessary to set $w_1 = \mathrm{g}$, $w_2 = 0$, and $w_3 = 0$ in the equations of motion.

## 5.4 Nonlinear modal equations for truncated conical tanks

In [68], the results of the previous section were generalized for truncated conical tanks participating in arbitrary three-dimensional low-magnitude motions with six degrees of freedom, as shown in Fig. 5.7. These degrees of freedom are associated with the translational velocity of the tank $\boldsymbol{v}_0(t) = (\dot{\eta}_1, \dot{\eta}_2, \dot{\eta}_3)$ and its angular motions, which can be described by the instantaneous angular velocity $\boldsymbol{\omega}(t) = (\dot{\eta}_4, \dot{\eta}_5, \dot{\eta}_6)$.

The tank is partially filled with an ideal incompressible liquid with irrotational flows. The liquid motions, as well as $\boldsymbol{v}_0(t)$ and $\boldsymbol{\omega}(t)$, are considered in the tank-fixed coordinate system $Oxyz$ whose origin $O$ is placed at the artificial apex of the cone and the $Ox$-axis is directed along the axis of symmetry (Fig. 5.7). The $O'x'$-axis has the direction opposite to the direction of the vector of gravitational acceleration $\mathbf{g}$.

A seven-dimensional nonlinear modal system is constructed on the basis of the Narimanov–Moiseev multimodal asymptotics and the natural sloshing

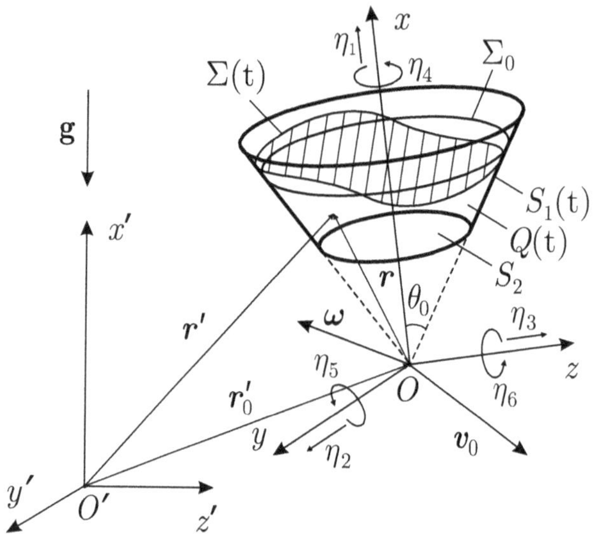


**Figure 5.7.** Schematic diagram of a tapered conical vessel and the accepted notation.

modes are computed according to the results from Sec. 5.3, where the curvilinear coordinate system $Ox_1x_2x_3$ is used in which the free surface is described by the formula

$$\begin{aligned}\zeta &= x_1 - f(x_2, x_3, t) = x_1 - f(x_2, x_3, \{p_{mi}\}, \{r_{mi}\}) \\ &= x_1 - x_{10} - \beta_0(t) - \sum_{m=0}^{\infty}\sum_{i=1}^{\infty}\Big(p_{mi}(t)\cos(mx_3) + r_{mi}(t)\sin(mx_3)\Big) f_{mi}(x_2),\end{aligned} \tag{5.4.1}$$

where $x_{10}$ is the distance between the origin and the mean free surface,

$$f_{mi}(x_2) = \frac{\sigma_{mi}}{\mathrm{g}}\psi_{mi}(x_{10}, x_2) \tag{5.4.2}$$

describes the radial natural surface profiles, and $\sigma_{mi}$ are the natural sloshing frequencies. The modal representation of the velocity potential takes the form

$$\begin{aligned}\Phi(x_1, x_2, x_3, t) &= \boldsymbol{v}_0 \cdot \boldsymbol{r} + \boldsymbol{\omega} \cdot \boldsymbol{\Omega} \\ &+ \sum_{m=0}^{\infty}\sum_{i=1}^{\infty}\Big(P_{mi}(t)\cos(mx_3) + R_{mi}(t)\sin(mx_3)\Big)\psi_{mi}(x_1, x_2).\end{aligned} \tag{5.4.3}$$

The Naimanov–Moiseev asymptotics suggests that the magnitude of dimensionless forcing has the order $\epsilon \ll 1$. For axisymmetric tanks, this means that two primary excited lowest modes that differ only by the azimuthal angle equal to $\pi/2$ and correspond to the dimensionless generalized coordinates $p_{11}$ and $r_{11}$ are predominant. They are of the order $O(\epsilon^{1/3})$. A simple trigonometric analysis in the angular coordinate leads to the following asymptotic relations for the generalized coordinates and velocities:

$$\begin{aligned}
&\frac{P_{11}}{\sigma_{11}} \sim \frac{R_{11}}{\sigma_{11}} \sim p_{11} \sim r_{11} = O(\epsilon^{1/3}),\\
&\frac{P_{2n}}{\sigma_{2n}} \sim \frac{R_{2n}}{\sigma_{2n}} \sim \frac{P_{0n}}{\sigma_{0n}} \sim p_{2n} \sim r_{2n} \sim p_{0n} = O(\epsilon^{2/3}),\\
&\frac{P_{3n}}{\sigma_{3n}} \sim \frac{R_{3n}}{\sigma_{3n}} \sim \frac{P_{1(n+1)}}{\sigma_{1(n+1)}} \sim \frac{R_{1(n+1)}}{\sigma_{1(n+1)}} \sim p_{3n}\\
&\qquad\qquad \sim r_{3n} \sim p_{1(n+1)} \sim r_{1(n+1)} = O(\epsilon), \quad n \geq 1.
\end{aligned} \tag{5.4.4}$$

The remaining dimensionless generalized coordinates and velocities are of the order $o(\epsilon)$ and can be neglected within the framework of the Moiseev asymptotics.

In constructing the asymptotic modal systems based on asymptotics (5.4.4), we neglect the terms of the $o(\epsilon)$-order. Hence, we arrive at an infinite-dimensional system of nonlinear ordinary differential equations in the generalized coordinates and velocities (5.4.4). Examples of infinite-dimensional systems of this kind can be found in [127, 57], whereas the nonlinear modal equations from the previous section contain two predominant, $r_{11}$ and $p_{11}$, and three second-order generalized coordinates and velocities corresponding to $p_{01}$, $p_{21}$, and $r_{21}$. These five-dimensional nonlinear modal equations enable us to realize the accurate approximation of steady-state sloshing caused by the resonant excitations of the lowest natural modes. This means that the weakly nonlinear modal equations of the Narimanov–Moiseev-type do not require to include a large set of generalized coordinates of the second and third orders. A physical reason for this is that the major part of kinematic energy is normally accumulated by the natural sloshing modes with lower natural sloshing frequencies. In our modal analysis, we include the corresponding above-mentioned five lowest modes and, in addition, two generalized coordinates $p_{31}$ and $r_{31}$ of the third order. The wave patterns of the adopted natural sloshing modes are shown in Fig. 5.8.

The technical details of construction of the seven-dimensional Moiseev-type modal system are outlined in [68]. For the sake of brevity, the generalized coordinates and velocities are denoted as follows:

$$p_{01} = p_0,\ r_{11} = r_1,\ p_{11} = p_1,\ \ r_{21} = r_2,\ p_{21} = p_2,\ r_{31} = r_3,\ p_{31} = p_3,$$

$$P_{01} = P_0,\ R_{11} = R_1,\ P_{11} = P_1,\ \ R_{21} = R_2,\ P_{21} = P_2,\ R_{31} = R_3,\ P_{31} = P_3.$$

As a result, we get the following system of ordinary differential equations for the dimensionless generalized coordinates:

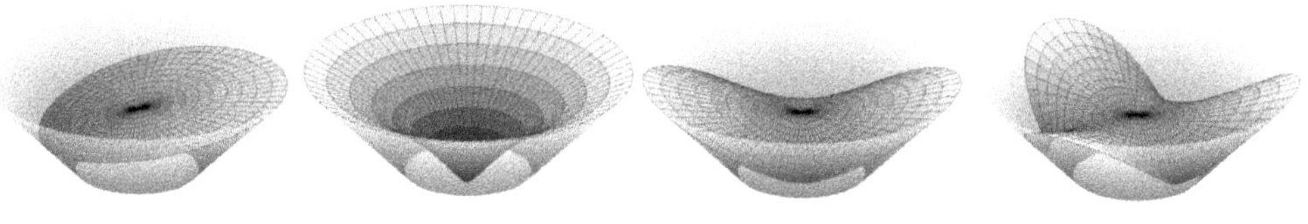

Associated with $p_{11}$ and $r_{11}$ | Associated with $p_{01}$ | Associated with $p_{21}$ and $r_{21}$ | Associated with $p_{31}$ and $r_{31}$

**Figure 5.8.** Wave patterns associated with the generalized coordinates and included in our nonlinear modal analysis. With the exception of $p_{01}$, these patterns appear to differ twice by $\pi/2$-azimuthal rotations. The angle $\theta_0 = 30°$ and the dimensionless ratio of the lower (bottom) and upper (mean free surface) radii $\mathtt{r}_1 = 0.5$.

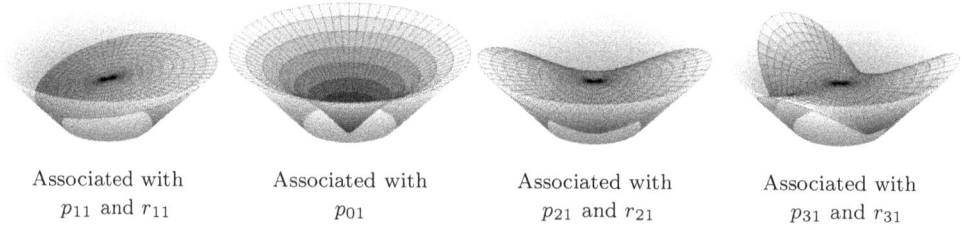

$$\ddot{p}_0 + \sigma_0^2 p_0 + \mathtt{d}_8(\dot{p}_1^2 + \dot{r}_1^2) + \mathtt{d}_{10}(\ddot{p}_1 p_1 + \ddot{r}_1 r_1) + \sigma_0^2 \mathtt{g}_0(p_1^2 + r_1^2) = 0, \tag{5.4.5}$$

$$\begin{aligned}&\ddot{r}_1 + \sigma_1^2 r_1 + \mathtt{d}_1 r_1(\ddot{p}_1 p_1 + \ddot{r}_1 r_1 + \dot{p}_1^2 + \dot{r}_1^2) + \mathtt{d}_2\big(p_1(\ddot{r}_1 p_1 - \ddot{p}_1 r_1)\\ &\quad + 2\dot{p}_1(\dot{r}_1 p_1 - \dot{p}_1 r_1)\big) + \mathtt{d}_3\,(\ddot{p}_1 r_2 - \ddot{r}_1 p_2 + \dot{p}_1 \dot{r}_2 - \dot{p}_2 \dot{r}_1) + \mathtt{d}_4(\ddot{r}_2 p_1 - \ddot{p}_2 r_1)\\ &\quad + \mathtt{d}_5(\ddot{r}_1 p_0 + \dot{r}_1 \dot{p}_0) + \mathtt{d}_6 \ddot{p}_0 r_1 + \sigma_1^2(\mathtt{g}_1 p_0 r_1 + \mathtt{g}_2(p_1 r_2 - p_2 r_1)\\ &\quad + \mathtt{g}_3(p_1^2 + r_1^2) r_1) + \Lambda(\dot{v}_{03} + g\eta_5) = 0,\end{aligned} \tag{5.4.6}$$

$$\begin{aligned}&\ddot{p}_1 + \sigma_1^2 p_1 + \mathtt{d}_1 p_1\,(\ddot{p}_1 p_1 + \ddot{r}_1 r_1 + \dot{p}_1^2 + \dot{r}_1^2) + \mathtt{d}_2\big(r_1(\ddot{p}_1 r_1 - \ddot{r}_1 p_1)\\ &\quad + 2\dot{r}_1(\dot{p}_1 r_1 - \dot{r}_1 p_1)\big) + \mathtt{d}_3(\ddot{p}_1 p_2 + \ddot{r}_1 r_2 + \dot{p}_1 \dot{p}_2 + \dot{r}_1 \dot{r}_2) + \mathtt{d}_4(\ddot{p}_2 p_1 + \ddot{r}_2 r_1)\\ &\quad + \mathtt{d}_5(\ddot{p}_1 p_0 + \dot{p}_1 \dot{p}_0) + \mathtt{d}_6 \ddot{p}_0 p_1 + \sigma_1^2(\mathtt{g}_1 p_0 p_1 + \mathtt{g}_2(p_1 p_2 + r_1 r_2)\\ &\quad + \mathtt{g}_3(p_1^2 + r_1^2) p_1) + \Lambda(\dot{v}_{02} - g\eta_6) = 0,\end{aligned} \tag{5.4.7}$$

$$\ddot{r}_2 + \sigma_2^2 r_2 + 2\mathtt{d}_7 \dot{p}_1 \dot{r}_1 + \mathtt{d}_9(\ddot{p}_1 r_1 + \ddot{r}_1 p_1) + 2\sigma_2^2 \mathtt{g}_4 p_1 r_1 = 0, \tag{5.4.8}$$

$$\ddot{p}_2 + \sigma_2^2 p_2 + \mathtt{d}_7(\dot{p}_1^2 - \dot{r}_1^2) + \mathtt{d}_9(\ddot{p}_1 p_1 - \ddot{r}_1 r_1) + \sigma_2^2 \mathtt{g}_4(p_1^2 - r_1^2) = 0, \tag{5.4.9}$$

$$\begin{aligned}&\ddot{r}_3 + \sigma_3^2 r_3 + \mathtt{d}_{11}\,(\ddot{r}_1(p_1^2 - r_1^2) + 2\ddot{p}_1 p_1 r_1) + \mathtt{d}_{12}\,(r_1(\dot{p}_1^2 - \dot{r}_1^2) + 2\dot{p}_1 \dot{r}_1 p_1)\\ &\quad + \mathtt{d}_{13}\,(\ddot{p}_1 r_2 + \ddot{r}_1 p_2) + \mathtt{d}_{14}\,(\ddot{p}_2 r_1 + \ddot{r}_2 p_1) + \mathtt{d}_{15}\,(\dot{p}_1 \dot{r}_2 + \dot{p}_2 \dot{r}_1)\\ &\quad + \sigma_3^2\,(\mathtt{g}_5\,(p_1 r_2 + p_2 r_1) + \mathtt{g}_6 r_1\,(3p_1^2 - r_1^2)) = 0,\end{aligned} \tag{5.4.10}$$

$$\begin{aligned}&\ddot{p}_3 + \sigma_3^2 p_3 + \mathtt{d}_{11}\,(\ddot{p}_1(p_1^2 - r_1^2) - 2\ddot{r}_1 p_1 r_1) + \mathtt{d}_{12}\,(p_1(\dot{p}_1^2 - \dot{r}_1^2) - 2\dot{p}_1 \dot{r}_1 r_1)\\ &\quad + \mathtt{d}_{13}\,(\ddot{p}_1 p_2 - \ddot{r}_1 r_2) + \mathtt{d}_{14}\,(\ddot{p}_2 p_1 - \ddot{r}_2 r_1) + \mathtt{d}_{15}\,(\dot{p}_1 \dot{p}_2 - \dot{r}_1 \dot{r}_2)\\ &\quad + \sigma_3^2\,(\mathtt{g}_5\,(p_1 p_2 - r_1 r_2) + \mathtt{g}_6 p_1\,(p_1^2 - 3r_1^2)) = 0,\end{aligned} \tag{5.4.11}$$

where the dimensionless hydrodynamic coefficients are functions of the parameters of mean liquid domain; the corresponding formulas and numerical values of these parameters can be found in [68].

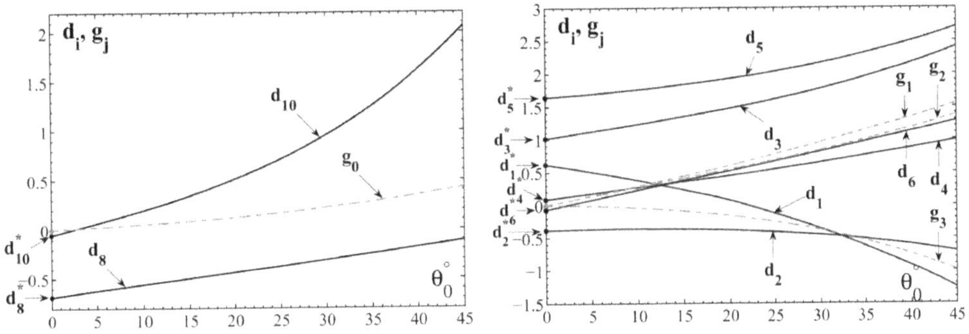


**Figure 5.9.** Coefficients $d_i$, $i = 1, \ldots, 6$, $d_8$, $d_{10}$, and $g_i$, $i = 0, 1, 2, 3$, as functions of $\theta_0$ for the dimensionless depth $h = 1$.

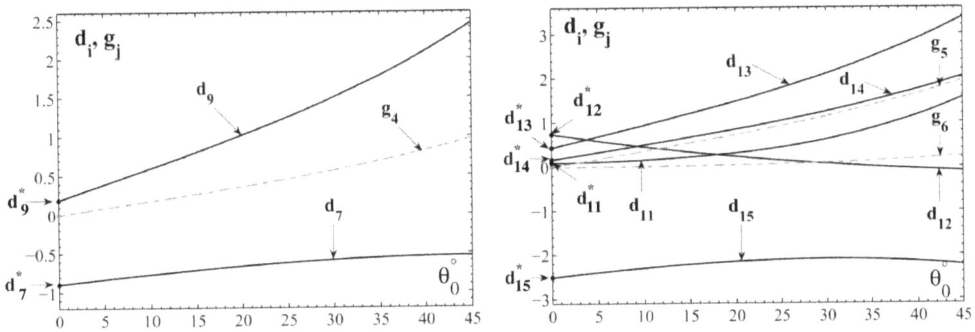


**Figure 5.10.** Coefficients $d_7$, $d_9$, $d_i$, $i = 11, \ldots, 15$, and $g_i$, $i = 4, 5, 6$, as functions of $\theta_0$ for the dimensionless depth $h = 1$.

As $r_1 \to 0$, the tank becomes nontruncated and, as expected, the dimensionless hydrodynamic coefficients approach the numerical values typical of the nontruncated conical tank. In the other limiting case, as $\theta_0 \to 0$, the tank approaches the upright circular cylindrical shape and the modal equations (5.4.5)–(5.4.11) must transform into the corresponding seven modal equations taken from the infinite-dimensional modal system presented in [127].

In Figs. 5.9–5.10, we illustrate the dependences of the dimensionless hydrodynamic coefficients $d_i$ and $g_j$ on $0 < \theta_0 < 45^o$ for a fixed dimensionless liquid depth $h = 1$. The limiting values on the vertical axis ($\theta_0 = 0$) coincide with the coefficients of the corresponding nonlinear terms computed for the upright circular cylindrical tank by using the exact natural sloshing modes [127]. The limit values are marked by $d_i^*$.

The modal equations (5.4.5)–(5.4.11) contain the hydrodynamic coefficients $g_j$ which differ from zero only for tanks with nonvertical walls. The plots confirm the fact that the limit numerical values $g_j$ vanish as the semiapex angle tends to zero.

## 5.5 Spherical vessel

As an example illustrating the algorithm used to deduce the equations of motion in the case of cavities for which the exact solutions of the spectral problem (2.2.4) do not exist, we consider the problem of free oscillations in a spherical vessel. We investigate this problem in a curvilinear coordinate system $x^1x^2x^3$ connected with the Cartesian coordinates by the formulas

$$x = x_1, \quad y = x^2\sqrt{a^2 - (x^1)^2}\cos x^3, \quad z = x^2\sqrt{a^2 - (x^1)^2}\sin x^3, \tag{5.5.1}$$

where $a$ is the radius of a sphere.

The coefficients of the reciprocal metric tensor and the nonzero Christoffel symbols are defined as follows:

$$\begin{gathered} q^{11} = 1, \quad q^{12} = q^{21} = \frac{x^1x^2}{a^2 - (x^1)^2}, \quad q^{13} = q^{31} = 0, \\ q^{22} = \frac{1}{a^2 - (x^1)^2} + \frac{(x^1x^2)^2}{[a^2 - (x^1)^2]^2}, \quad q^{23} = q^{32} = 0, \\ q^{33} = \frac{1}{(x^2)^2[a^2 - (x^1)^2]}, \quad \mu = \sqrt{q} = x^2[a^2 - (x^1)^2], \\ \Gamma_{11}^2 = -\frac{a^2x^2}{[a^2 - (x^1)^2]^2}, \quad \Gamma_{12}^2 = \Gamma_{21}^2 = -\frac{x^1}{a^2 - (x^1)^2}. \end{gathered} \tag{5.5.2}$$

We consider nonlinear sloshing in the vicinity of the natural sloshing frequency with antisymmetric modes

$$\varphi_1(x^1, x^2, x^3) = \psi_{11}(x^1, x^2)\sin x^3, \quad f_1(x^2, x^3) = \varkappa_{11}\psi_{11}(x^1, x_0^2)\sin x^3,$$
$$\varkappa_{11} = 1.560157, \tag{5.5.3}$$

and represent the equation of free surface in the form

$$x^1 = x_0^1 + \beta_0(t) + \beta(t)f_1(x^2, x^3). \tag{5.5.4}$$

In the equations of motion, we preserve the terms of the order $\beta^3(t)$ and use the following relation for the velocity potential [131]:

$$\varphi(x^1, x^2, x^3, t) = P_0(t)\psi_{10}(x^1, x^2) + R_1(t)\psi_{11}(x^1, x^2)\sin x^3 +$$
$$+ P_2(t)\psi_{12}(x^1, x^2)\cos 2x^3. \tag{5.5.5}$$

In the analyzed case, the time-dependent function $\beta_0(t)$ has the form

$$\beta_0(t) = e_1\beta^2(t), \tag{5.5.6}$$

where

$$e_1 = -\frac{x_0^1 \varkappa_{11}}{x_0^2} \int_0^{x_0^2} \psi_{11}^2(x^1, x^2) x^2 \, dx^2.$$

The eigenfunctions $\psi_{nm}(x^1, x^2)^{\cos mx^3}_{\sin mx^3}$ of the boundary-value problem (2.2.4) are obtained by the variational Ritz method in [112]:

$$\psi_{nm}(x^1, x^2) = \sum_{k=m}^{q} a_k w_k^{(m)}(x^1, x^2), \tag{5.5.7}$$

where $w_k^{(m)}(x^1, x^2)$ is the system of solutions satisfying the equation in the analyzed domain and the following recurrence relations:

$$(k+m+1)w_{k+1}^{(m)} = (2k+1)x^1 w_k^{(m)} - (k-m)[(x^1)^2 + (x^2)^2(a^2 - (x^1)^2)]w_{k-1}^{(m)},$$

$$\frac{\partial w_k^{(m)}}{\partial x^1} = \frac{1}{a^2 - (x^1)^2}\left[(k-m)a^2 w_{k-1}^{(m)} - kx^1 w_k^{(m)}\right], \tag{5.5.8}$$

$$\frac{\partial w_k^{(m)}}{\partial x^2} = \frac{1}{x^2}\left[kw_k^{(m)} - (k-m)x^1 w_{k-1}^{(m)}\right].$$

The first three functions of the system $\{w_k^{(m)}\}$, e.g., for , $m = 1$, take the form

$$w_1^{(1)} = x^2\sqrt{a^2 - (x^1)^2}, \quad w_2^{(1)} = x^1 x^2 \sqrt{a^2 - (x^1)^2},$$

$$w_3^{(1)} = x^2\sqrt{a^2 - (x^1)^2}\left[(x^1)^2 - \frac{1}{4}(x^2)^2(a^2 - (x^1)^2)\right]. \tag{5.5.9}$$

In the considered case, the system of nonlinear equations of motion (2.3.6), (2.3.7) can be represented as follows:

$$a_1\dot{P}_0\beta + a_2\dot{R}_1 + a_3\dot{R}_1\beta^2 + a_4\dot{P}_2\beta + a_5P_0^2\beta + a_6R_1^2\beta + a_7P_2^2\beta$$
$$+a_8P_0R_1 + a_9P_0P_2\beta + a_{10}R_1P_2 + \mathrm{g}a_{11}\beta + \mathrm{g}a_{12}\beta^3 = 0; \tag{5.5.10}$$

$$b_1P_0 + b_2P_0\beta^2 + b_3R_1\beta + b_4P_2\beta^2 - b_5\dot{\beta}\beta = 0,$$

$$c_1P_0\beta + c_2R_1 + c_3R_1\beta^2 + c_4P_2\beta - c_5\dot{\beta} - c_6\dot{\beta}\beta^2 = 0, \tag{5.5.11}$$

$$d_1P_2 + d_2P_2\beta^2 + d_3P_0\beta^2 + d_4R_1\beta - d_5\dot{\beta}\beta = 0.$$

We get the following relations for the coefficients of this system:

$$a_1 = b_5, \quad a_2 = c_5, \quad a_3 = c_6, \quad a_4 = d_5, \quad a_5 = b_2, \quad a_6 = c_3,$$

$$a_7 = d_2, \quad a_8 = c_1 = b_3, \quad a_9 = 2b_4 = 2d_3, \quad a_{10} = c_4 = d_4. \tag{5.5.12}$$

By analogy with the previous case, the system of equations (5.5.10), (5.5.11) can be represented as

$$\ddot{\beta} + \sigma^2\beta + \alpha(\ddot{\beta}\beta^2 + \beta\dot{\beta}^2) - s\beta^3 = 0; \tag{5.5.13}$$

$$P_0 = C_0\beta\dot{\beta}, \quad R_1 = B_4\dot{\beta} - C_2\dot{\beta}\beta^2, \quad P_2 = D_1\beta\dot{\beta}, \tag{5.5.14}$$

where

$$\sigma^2 = g\frac{a_{11}}{d_2 B_4} = g\bar{\sigma}^2, \quad C_0 = \frac{b_5 - b_3 B_4}{b_1}, \quad A_2 = \frac{d_5 - d_4 B_4}{d_1},$$

$$C_2 = \frac{a_8 a_1}{c_2 b_1} + \frac{a_{10} a_4}{c_2 d_1} + \frac{a_2 a_6}{c_2^2} - \frac{a_8^2 a_2}{c_2^2 b_1} - \frac{a_{10}^2 a_2}{c_2^2 d_1} - \frac{c_6}{c_2}, \quad B_4 = \frac{c_5}{c_2}, \tag{5.5.15}$$

$$\alpha = \frac{1}{a_2 B_4}(a_1 C_0 - a_2 C_2 + a_3 B_4 + a_4 D_1), \quad s = g\frac{a_{12}}{a_2 B_4} = g\bar{s}.$$

The numerical realization of the investigated method was performed for a spherical container of unit radius half-filled with liquid. The number $q$ of the coordinate functions for $\psi_{10}$, $\psi_{11}$, and $\psi_{12}$ is equal to 14, 16, and 14, respectively. We obtain the following values for the coefficients of nonlinear equations of motion: $a_1 = -5.793$, $a_9 = 125.9$, $C_0 = 1.143$, $a_2 = 3.142$, $a_{10} = -13.07$, $C_2 = -4.368$, $a_3 = 10.66$, $a_{11} = 4.901$, $D_1 = 2.114$, $a_4 = -6.338$, $a_{12} = -6.811$, $B_4 = 1.000$, $a_5 = 211.6$, $b_1 = 6.281$, $\sigma^2 = 1.560$, $a_6 = 40.23$, $c_2 = 3.140$, $\bar{s} = 1.390$, $a_7 = 106.6$, $d_1 = 3.139$, $\alpha = -2.167$, $a_8 = -13.10$.

On the basis of this method, one can also get a more complete system of the nonlinear equations of motion and expressions for its coefficients. The difficulties encountered in the solution of this problem are of the same order of complexity as in the solution of the corresponding problem for conical vessels.

In 2013, a general infinite-dimensional Narimanov–Moiseev-type modal system was obtained in [57]. This system is completely analogous to the system (4.1.24)–(4.1.26).

# Chapter 6

# Derivation of the nonlinear equations of space motions of the body–liquid system by the method of perturbation theory

In the present chapter, we present ideas and analytical schemes of the perturbation theory that can be applied to studying the mechanical body–liquid system. The scheme was proposed by G. S. Narimanov in [171, 172]. Later on, this analytical scheme was widely used to solve a series of problems of the dynamics of bodies with cavities containing a liquid, in particular, in the case of cavities of complex shape [112, 173]. In contrast to the variational method presented above, the perturbation methods are applied directly to the original free-boundary 'sloshing' problem. This reduces the problem to a sequence of linear boundary-value problems in the mean liquid domain. The material presented here is mainly based on works [41, 171, 173] and the results in [111, 112, 129] devoted to the Narimanov method for cavities of complicated configuration and correction of algebraic errors in the original Narimanov's derivations.

## 6.1 Reduction to a sequence of linear boundary-value problems

We consider a class of cavities of the cylindrical shape in a neighborhood of the free surface $\Sigma_0$ and construct an analytical scheme for solving the second problem of the dynamics of a rigid body with cavities containing a liquid. This problem is reduced to the nonlinear free-boundary problem (2.5.4) in which, in addition to the velocity potential $\Phi(x, y, z, t)$ and the free surface $f(y, z, t)$, the quantities $\boldsymbol{v}_0(t)$ and $\boldsymbol{\omega}(t)$ are unknown, too. We suggest a method that can be used to solve the problem of finding the velocity potential and the perturbed free surface under the assumption that $\boldsymbol{v}_0(t)$ and $\boldsymbol{\omega}(t)$ are known. As above, we represent the free surface elevations in the form of a decomposition

$$f(y, z, t) = \sum_i \beta_i(t) f_i(y, z) \tag{6.1.1}$$

in a certain system of functions $f_i(y, z)$, which, along with the constant quantity, form a complete and orthogonal system on $\Sigma_0$; here, $\beta_i(t)$ are unknown functions of time, which, in what follows, are associated with the generalized coordinates of the hydrodynamic system.

By using the nonlinear boundary conditions of the original problem (2.5.4)

$$\left.\frac{\partial \Phi}{\partial \nu}\right|_S = \boldsymbol{v}_0 \cdot \boldsymbol{\nu} + \boldsymbol{\omega} \cdot (\boldsymbol{r} \times \boldsymbol{\nu}), \tag{6.1.2}$$

$$\left.\frac{\partial \Phi}{\partial \nu}\right|_\Sigma = \boldsymbol{v}_0 \cdot \boldsymbol{\nu} + \boldsymbol{\omega} \cdot (\boldsymbol{r} \times \boldsymbol{\nu}) + \frac{1}{N}\frac{\partial f}{\partial t},$$

we present the velocity potential $\Phi(x, y, z, t)$ in the form of the sum

$$\Phi = \boldsymbol{v}_0 \cdot \boldsymbol{r} + \boldsymbol{\omega} \cdot \boldsymbol{\Omega}(x, y, z, t) + \sum_i \dot{\beta}_i(t)\varphi_i(x, y, z, t). \tag{6.1.3}$$

In view of the equalities

$$\left.\frac{\partial \boldsymbol{r}}{\partial \nu}\right|_{S+\Sigma} = \boldsymbol{\nu}, \quad \left.\frac{\partial f}{\partial t}\right|_\Sigma = \sum_i \dot{\beta}_i f_i, \tag{6.1.4}$$

relations (2.5.4) imply the following boundary-value problems for the harmonic functions $\varphi_i$ and $\boldsymbol{\Omega}_k$:

$$\Delta\varphi_i = 0, \quad \left.\frac{\partial \varphi_i}{\partial \nu}\right|_S = 0, \quad \left.\frac{\partial \varphi_i}{\partial \nu}\right|_\Sigma = \frac{f_i}{\sqrt{1 + (\nabla f)^2}}, \quad i = 1,\, 2,\, \dots\,; \tag{6.1.5}$$

$$\Delta\Omega_k = 0, \quad \left.\frac{\partial \Omega_k}{\partial \nu}\right|_{S+\Sigma} = (\boldsymbol{r} \times \boldsymbol{\nu})_k, \quad k = 1,\, 2,\, 3. \tag{6.1.6}$$

As above, we assume that the order of smallness of the squared deviation of the nonperturbed surface $f^2$ is $\varepsilon$. We neglect the quantities of the order of $\varepsilon^2$, i.e., keep the quantities $f$, $f^2$, and $f^3$. In addition, we assume that the order of smallness of the quantity $\omega^3$ is $\varepsilon^2$. Therefore, along with quantities of the type $\omega^2$, we have to keep the products $f\omega$, $f^2\omega$, and $f\omega^2$. We also impose no restrictions on the value of the translational velocity. Under these assumptions, in general, the right-hand sides of the boundary conditions of the nonlinear boundary-value problems (6.1.5) and (6.1.6) can be presented by certain polynomials in $\beta_i(t)$. This enables us to assume that the required solutions of these boundary-value problems can also be presented in the form of polynomials in $\beta_i(t)$. Up to the terms of the second order of smallness inclusively, we can write the components of the velocity potential in the form

$$\varphi_i = \varphi_{i0} + \sum_j \beta_j\varphi_{ij} + \sum_j\sum_k \beta_j\beta_k\varphi_{ijk} + \dots\,; \tag{6.1.7}$$

$$\boldsymbol{\Omega} = \boldsymbol{\Omega}_0 + \sum_j \beta_j\boldsymbol{\Omega}_i + \sum_i\sum_j \beta_i\beta_j\boldsymbol{\Omega}_{ij} + \dots\,, \tag{6.1.8}$$

where the functions $\varphi_{i0}$, $\varphi_{ij}$, $\varphi_{ijk}$, $\mathbf{\Omega}_0$, $\mathbf{\Omega}_i$, and $\mathbf{\Omega}_{ij}$ depend only on the spatial coordinates.

To proceed the development of the theory by this method, we use the Taylor series

$$F|_\Sigma = F|_{\Sigma_0} + \left.\frac{\partial F}{\partial x}\right|_{\Sigma_0} f + \tfrac{1}{2}\frac{\partial^2 F}{\partial x^2} f^2 + \tfrac{1}{6}\frac{\partial^3 F}{\partial x^3} f^3 + \dots . \tag{6.1.9}$$

This enables us to find the values of an arbitrary sufficiently smooth function on the perturbed free surface $\Sigma$ in terms of values of the function and its derivatives on the nonperturbed free surface $\Sigma_0$.

Applying relation (6.1.9) to the boundary conditions of the boundary-value problems (6.1.5) and (6.1.6) in view of (6.1.7) and (6.1.8) and equating the expressions at the same powers of the parameters $\beta_i$ to zero, we obtain the following linear boundary-value problems for the above-introduced functions $\varphi_{i0}$, $\varphi_{ij}$, ..., $\mathbf{\Omega}_{ij}$:

$$\Delta\varphi_{i0} = 0, \quad \left.\frac{\partial \varphi_{i0}}{\partial \nu}\right|_{S_0} = 0, \quad \left.\frac{\partial \varphi_{i0}}{\partial \nu}\right|_{\Sigma_0} = f_i; \tag{6.1.10}$$

$$\Delta\varphi_{ij} = 0, \quad \left.\frac{\partial \varphi_{ij}}{\partial \nu}\right|_{S_0} = 0, \quad \left.\frac{\partial \varphi_{ij}}{\partial \nu}\right|_{\Sigma_0} = \nabla' \cdot (f_i \nabla' \varphi_{i0}); \tag{6.1.11}$$

$$\Delta\varphi_{ijk} = 0, \quad \left.\frac{\partial \varphi_{ijk}}{\partial \nu}\right|_{S_0} = 0,$$

$$\left.\frac{\partial \varphi_{ijk}}{\partial \nu}\right|_{\Sigma_0} = \tfrac{1}{2}\nabla' \cdot (f_j \nabla' \varphi_{ik} + f_k \nabla' \varphi_{ij}) + \tfrac{1}{2}\nabla' \cdot \left( f_j f_k \nabla' \frac{\partial \varphi_{i0}}{\partial x} \right); \tag{6.1.12}$$

$$\Delta\mathbf{\Omega}_0 = 0, \quad \left.\frac{\partial \mathbf{\Omega}_0}{\partial \nu}\right|_{S_0+\Sigma_0} = \boldsymbol{r} \times \boldsymbol{\nu}; \tag{6.1.13}$$

$$\Delta\mathbf{\Omega}_i = 0, \quad \left.\frac{\partial \mathbf{\Omega}_i}{\partial \nu}\right|_{S_0} = 0,$$

$$\left.\frac{\partial \mathbf{\Omega}_i}{\partial \nu}\right|_{\Sigma_0} = \nabla f_i \times \boldsymbol{r} + (\nabla f_i \cdot \nabla')\mathbf{\Omega}_0 + f_i \nabla'^2 \mathbf{\Omega}_0; \tag{6.1.14}$$

$$\Delta\mathbf{\Omega}_{ij} = 0, \quad \left.\frac{\partial \mathbf{\Omega}_{ij}}{\partial \nu}\right|_{S_0} = 0,$$

$$\left.\frac{\partial \mathbf{\Omega}_{ij}}{\partial \nu}\right|_{\Sigma_0} = \tfrac{1}{2}\Big[ -\boldsymbol{i}_x \times \nabla(f_i f_j) + (\nabla f_j \cdot \nabla')\mathbf{\Omega}_i + (\nabla f_i \cdot \nabla')\mathbf{\Omega}_j + f_j \nabla'^2 \mathbf{\Omega}_i$$

$$+ f_i \nabla'^2 \mathbf{\Omega}_j + (\nabla(f_i f_j) \cdot \nabla')\frac{\partial \mathbf{\Omega}_0}{\partial x} + f_i f_j \nabla'^2 \frac{\partial \mathbf{\Omega}_0}{\partial x} \Big], \tag{6.1.15}$$

where $S_0$ is the wetted surface of the cavity in the nonperturbed state of a liquid, $\boldsymbol{i}_x$ is the unit vector of the $Ox$-axis, and $\nabla'$ is the incomplete Hamilton operator with respect to the coordinates $y$ and $z$: $\nabla' = \boldsymbol{i}_y \frac{\partial}{\partial y} + \boldsymbol{i}_z \frac{\partial}{\partial z}$. The components of the potentials $\varphi_i$ and $\boldsymbol{\Omega}$ have properties of symmetry, i.e., $\varphi_{ijk} = \varphi_{ikj}$ and $\boldsymbol{\Omega}_{ij} = \boldsymbol{\Omega}_{ji}$, which is used in what follows.

On the one hand, the boundary-value problems (6.1.10)–(6.1.15) belong to a class of well-studied linear boundary-value problems of mathematical physics. On the other hand, we now have the domain of definition of required solutions, which is a fixed domain independent of time. As above, in the frame of the considered method, for the functions $f_i(y, z)$ in (6.1.1) we choose solutions of the boundary-value problem (2.2.4) that possess the required properties. To solve these problems, we can use various methods of the linear theory [the method of separation of variables, the method of decomposition in eigenfunctions of problem (2.2.4), the variational method, etc.].

## 6.2 Vector form of the nonlinear equations of space motions of the body–liquid system

Equations of motion of the considered mechanical system can be derived by two methods. The first method is based on the use of theorems on variation in the momentum and the angular momentum and also the theorem on expansion of functions in a generalized Fourier series. For the second method, we use the equations of motion of the system in the Euler–Lagrange form in order to derive the equations of forces and moments and the Lagrange equations of the second kind to derive the equations in generalized coordinates $\beta_i(t)$. We present a general form of nonlinear equations obtained in such a way. In the body-fixed coordinate system $Oxyz$, the theorems on variation of the momentum and the angular momentum have the form [173]

$$\overset{*}{\boldsymbol{K}} + \boldsymbol{\omega} \times \boldsymbol{K} = \boldsymbol{P}^0 + \boldsymbol{P}^C; \; \overset{*}{\boldsymbol{G}} + \boldsymbol{\omega} \times \boldsymbol{G} + \boldsymbol{v}_0 \times \boldsymbol{K} = \boldsymbol{M}^0 + \boldsymbol{M}^C, \quad (6.2.1)$$

where $\boldsymbol{P}^0$ and $\boldsymbol{M}^0$ are, respectively, the principal vector and the principal momentum of external forces applied at the point $O$; $\boldsymbol{P}^C$ and $\boldsymbol{M}^C$ are, respectively, the principal vector and the principal momentum of static mass forces that affect the rigid body and the liquid. By definition, the vectors of momentum $\boldsymbol{K}$ of the system, the angular momentum $\boldsymbol{G}$, and the static moment $\boldsymbol{L}$ are written in the form

$$\boldsymbol{K} = \int_{\tau_0} (\boldsymbol{v}_0 + \boldsymbol{\omega} \times \boldsymbol{r}) \rho_0 dQ + \rho \int_Q \Big[ \boldsymbol{v}_0 + \nabla(\boldsymbol{\omega} \cdot \boldsymbol{\Omega}) + \sum_i \dot{\beta}_i(t) \nabla \varphi_i \Big] dQ,$$

$$\boldsymbol{G} = \int_{\tau_0} \boldsymbol{r} \times (\boldsymbol{v}_0 + \boldsymbol{\omega} \times \boldsymbol{r}) \rho_0 dQ + \rho \int_Q \boldsymbol{r} \times \Big[ \boldsymbol{v}_0 + \nabla(\boldsymbol{\omega} \cdot \boldsymbol{\Omega}) + \sum_i \dot{\beta}_i(t) \nabla \varphi_i \Big] dQ,$$

$$\boldsymbol{L} = \int\limits_{\tau_0} \boldsymbol{r}\rho_0 dQ + \rho \int\limits_{Q} \boldsymbol{r} dQ, \tag{6.2.2}$$

where $\tau_0$ and $Q$ are the domains of rigid body and liquid, respectively; $\rho_0$ is the rigid body density; $\rho$ is the liquid density.

By using the integral theorems and relations (6.1.7), (6.1.8), we can transform relations (6.2.2). By using the results presented in Sec. 2.8, we obtain

$$\begin{aligned}\boldsymbol{L} = M\boldsymbol{r}_C = m_0\boldsymbol{r}_C^0 + m_1\boldsymbol{r}_{1C} &= m_0\boldsymbol{r}_C^0 + \boldsymbol{i}_x\left(l_{10} + \tfrac{1}{2}\rho\int\limits_{\Sigma_0} f^2 dx\right)\\ &+\boldsymbol{i}_y\left(l_{20} + \rho\int\limits_{\Sigma_0} fy dS\right) + \boldsymbol{i}_z\left(l_{30} + \rho\int\limits_{\Sigma_0} fz dS\right)\\ &= m_0\boldsymbol{r}_C^0 + \boldsymbol{i}_x\left(l_{10} + \tfrac{1}{2}\sum_i \lambda_{i1}\beta_1^2\right) + \boldsymbol{i}_y\left(l_{20} + \sum_i \lambda_{i2}\beta_i\right)\\ &+\boldsymbol{i}_z\left(l_{30} + \sum_i \lambda_{i3}\beta_i\right);\end{aligned} \tag{6.2.3}$$

$$\boldsymbol{K} = M\boldsymbol{v}_0 + \boldsymbol{\omega} \times \boldsymbol{L} + \overset{*}{\boldsymbol{L}} = M(\boldsymbol{v}_0 + \boldsymbol{\omega} \times \boldsymbol{r}_C) + m_1\overset{*}{\vec{\boldsymbol{r}}}_{1C}; \tag{6.2.4}$$

$$\boldsymbol{G} = M\boldsymbol{r}_C \times \boldsymbol{v}_0 + \boldsymbol{J} \cdot \boldsymbol{\omega} + \sum_i \boldsymbol{R}_i\beta_i, \tag{6.2.5}$$

where $\boldsymbol{r}_C$, $\boldsymbol{r}_C^0$, and $\boldsymbol{r}_{1C}$ are the radius vectors of the entire system, rigid body, and liquid, respectively; $\boldsymbol{J}$ is the tensor of inertia of the mechanical system that consists of the tensor of inertia of rigid body $\boldsymbol{J}^0$ and the tensor of inertia of the liquid $\boldsymbol{J}^1 = \rho\int_{\Sigma_0} \boldsymbol{\Omega} \cdot \frac{\partial \boldsymbol{\Omega}}{\partial \nu} dS$;

$$l_{10} = \rho\int\limits_{Q_0} x dQ, \quad l_{20} = \rho\int\limits_{Q_0} y dQ, \quad l_{30} = \rho\int\limits_{Q_0} z dQ,$$

$$N_i^2 = \int\limits_{\Sigma_0} f_i^2 dS, \tag{6.2.6}$$

$$\lambda_{i1} = \rho N_i^2, \quad \lambda_{i2} = \rho\int\limits_{\Sigma_0} y f_i dS, \quad \lambda_{i3} = \rho\int\limits_{\Sigma_0} z f_i dS.$$

We define the vector $\boldsymbol{R}_i$ in (6.2.5) by the formula [173]

$$\boldsymbol{R}_i = \rho\int\limits_{Q} \boldsymbol{r} \times \nabla\varphi_i dQ = \int\limits_{S+\Sigma} (\boldsymbol{r} \times \boldsymbol{\nu})\varphi_i dS = \int\limits_{S+\Sigma} \boldsymbol{\Omega}\frac{\partial \varphi_i}{\partial \nu} dS = \int\limits_{\Sigma} \boldsymbol{\Omega}\frac{f_i}{N} dS$$

$$= \int\limits_{\Sigma_0} \boldsymbol{\Omega}\Big|_{\Sigma_0} f_i dS = \boldsymbol{\lambda}_{0i} + \sum_j \boldsymbol{\lambda}_{ij}\beta_j + \sum_j\sum_k \boldsymbol{\lambda}_{ijk}\beta_j\beta_k, \tag{6.2.7}$$

where

$$\boldsymbol{\lambda}_{0i} = \rho \int\limits_{\Sigma_0} \boldsymbol{\Omega}_0 f_i dS;$$

$$\boldsymbol{\lambda}_{ij} = \rho \int\limits_{\Sigma_0} \left[\boldsymbol{\Omega}_0 \frac{\partial \varphi_{ij}}{\partial \nu} + f_j \frac{\partial \varphi_{i0}}{\partial \nu}(\boldsymbol{r} \times \boldsymbol{\nu}) - \varphi_{i0}(\boldsymbol{r} \times \nabla f_i)\right] dS$$

$$= \rho \int\limits_{\Sigma_0} \left(\boldsymbol{\Omega}_j + f_j \frac{\partial \boldsymbol{\Omega}_0}{\partial x}\right) f_i dS; \tag{6.2.8}$$

$$\boldsymbol{\lambda}_{ijk} = \rho \int\limits_{\Sigma_0} \Bigg\{\boldsymbol{\Omega}_0 \frac{\partial \varphi_{ijk}}{\partial \nu} + f_k \left(\frac{\partial \varphi_{ij}}{\partial x} + \tfrac{1}{2} f_j \frac{\partial^2 \varphi_{i0}}{\partial x^0}\right)(\boldsymbol{r} \times \boldsymbol{\nu})$$

$$- \left(\varphi_{ij} + f_j \frac{\partial \varphi_{i0}}{\partial x}\right)(\boldsymbol{r} \times \nabla f_k) - \tfrac{1}{2}\varphi_{i0}[\boldsymbol{i}_x \times \nabla(f_j f_k)]\Bigg\} dS$$

$$= \rho \int\limits_{\Sigma_0} \left(\boldsymbol{\Omega}_{jk} + f_k \frac{\partial \boldsymbol{\Omega}_j}{\partial x} + \tfrac{1}{2} f_j f_k \frac{\partial^2 \boldsymbol{\Omega}_0}{\partial x^2}\right) f_i dS.$$

We derive the expression for the inertia tensor of a liquid $\boldsymbol{J}^1$ by using the following chain of transformations:

$$\rho \int\limits_{Q} \boldsymbol{r} \times \nabla(\boldsymbol{\omega} \cdot \boldsymbol{\Omega}) dQ = \rho \int\limits_{S+\Sigma} (\boldsymbol{r} \times \boldsymbol{\nu})(\boldsymbol{\omega} \cdot \boldsymbol{\Omega}) dS$$

$$= \rho \int\limits_{S+\Sigma_0} (\boldsymbol{\omega} \cdot \boldsymbol{\Omega}) \frac{\partial \boldsymbol{\Omega}}{\partial \nu} dS = \boldsymbol{J}^1 \cdot \boldsymbol{\omega}. \tag{6.2.9}$$

To find the components of the tensor of inertia, it makes sense to divide the liquid domain into two subdomains: the nonperturbed domain $Q_0$ and the domain $Q_1$ whose element can be presented in the form of a cylinder with the base $dS$ and height $f$. If, by using the Taylor formula in the domain $Q_1$, we present the integrand in the form of the power series

$$F(x, y, z) = F|_{\Sigma_0} + (x - a)\frac{\partial F}{\partial x}\Big|_{\Sigma_0} + \tfrac{1}{2}(x - a)^2 \frac{\partial^2 F}{\partial x^2} + \dots$$

up to $f^3$ inclusively, then we obtain

$$\int\limits_{Q_1} F(x, y, z) dQ = \int\limits_{\Sigma_0} \int\limits_{a}^{a+f} \left[F|_{\Sigma_0} + (x - a)\frac{\partial F}{\partial x}\Big|_{\Sigma_0} + \tfrac{1}{2}(x - a)^2 \frac{\partial^2 F}{\partial x^2}\Big|_{\Sigma_0}\right] dx dS$$

$$= \int\limits_{\Sigma_0} \left[ fF|_{\Sigma_0} + \tfrac{1}{2} f^2 \frac{\partial F}{\partial x}\bigg|_{\Sigma_0} + \tfrac{1}{6} f^3 \frac{\partial^2 F}{\partial x^2}\bigg|_{\Sigma_0} \right] dS. \tag{6.2.10}$$

The component of the tensor of inertia associated with the integration over the nonperturbed liquid domain is defined by the relation

$$\rho \int\limits_{Q_0} \boldsymbol{r} \times \nabla(\boldsymbol{\omega} \cdot \boldsymbol{\Omega}) dQ = \rho \int\limits_{S_0+\Sigma_0} (\boldsymbol{r} \times \boldsymbol{\nu})(\boldsymbol{\omega} \cdot \boldsymbol{\Omega}) dS$$

$$= \rho \int\limits_{S_0+\Sigma_0} \frac{\partial \boldsymbol{\Omega}_0}{\partial \nu} (\boldsymbol{\omega} \cdot \boldsymbol{\Omega}) dS = \rho \int\limits_{S_0+\Sigma_0} \boldsymbol{\Omega}_0 \left( \boldsymbol{\omega} \cdot \frac{\partial \boldsymbol{\Omega}}{\partial \nu} \right) dS$$

$$= \rho \int\limits_{S_0+\Sigma_0} \left\{ \boldsymbol{\Omega}_0 \left[ \boldsymbol{\omega} \cdot \left( \frac{\partial \boldsymbol{\Omega}_0}{\partial \nu} + \sum_i \beta_i \frac{\partial \boldsymbol{\Omega}_i}{\partial \nu} + \sum_i \sum_j \beta_i \beta_j \frac{\partial \boldsymbol{\Omega}_{ij}}{\partial \nu} \right) \right] \right\} dS$$

$$= \rho \int\limits_{S_0+\Sigma_0} \left( \boldsymbol{\omega} \cdot \frac{\partial \boldsymbol{\Omega}_0}{\partial \nu} \right) \boldsymbol{\Omega}_0 dS + \sum_i \beta_i \rho \int\limits_{\Sigma_0} \left( \boldsymbol{\omega} \cdot \frac{\partial \boldsymbol{\Omega}_i}{\partial \nu} \right) \boldsymbol{\Omega}_0 dS$$

$$+ \sum_i \sum_j \beta_i \beta_j \rho \int\limits_{\Sigma_0} \left( \boldsymbol{\omega} \cdot \frac{\partial \boldsymbol{\Omega}_{ij}}{\partial \nu} \right) \boldsymbol{\Omega}_0 dS$$

$$= \rho \left\{ \int\limits_{S_0+\Sigma_0} \left( \boldsymbol{\Omega}_0 \,;\, \frac{\partial \boldsymbol{\Omega}_0}{\partial \nu} \right) dS + \sum_i \beta_i \int\limits_{\Sigma_0} \left( \boldsymbol{\Omega}_0 \,;\, \frac{\partial \boldsymbol{\Omega}_i}{\partial \nu} \right) dS \right.$$

$$\left. + \sum_i \sum_j \beta_i \beta_j \int\limits_{\Sigma_0} \left( \boldsymbol{\Omega}_0 \,;\, \frac{\partial \boldsymbol{\Omega}_{ij}}{\partial \nu} \right) dS \right\} \boldsymbol{\omega} = \boldsymbol{J}_0^1 \boldsymbol{\omega}, \tag{6.2.11}$$

where $(\boldsymbol{a}; \boldsymbol{b})$ denotes the dyadic product of vectors, i.e.,

$$(\boldsymbol{a}; \boldsymbol{b}) = \begin{pmatrix} a_x b_x & a_x b_y & a_x b_z \\ a_y b_x & a_y b_y & a_y b_z \\ a_z b_x & a_z b_y & a_z b_z \end{pmatrix}.$$

The components of the inertia tensor of a liquid $\boldsymbol{J}_1^1$, which correspond to a perturbation of the domain $Q_1$, can be derived directly from the relation

$$\boldsymbol{J}_1^1 \boldsymbol{\omega} = \rho \int\limits_{Q_1} \boldsymbol{r} \times \nabla(\boldsymbol{\omega} \cdot \boldsymbol{\Omega}) dQ \tag{6.2.12}$$

by using relation (6.2.10). To this end, we can use expressions for the components of the inertia tensor $J_{ij}^1$ in the form of integration over a domain that were defined by relations (2.6.21).

The total vectors of the static forces and moments induced by the forces, which affect the body and the deformable liquid domain, are determined by the relations

$$\boldsymbol{P}^C = (m_0 + m_1)\mathbf{g}, \quad \boldsymbol{M}^C = \boldsymbol{L} \times \mathbf{g}. \tag{6.2.13}$$

Substituting (6.2.4), (6.2.5), and (6.2.13) into (6.2.1), we obtain the following two vector equations that describe the motion of the mechanical system under consideration:

$$\boldsymbol{M}(\overset{*}{\boldsymbol{v}}_0 + \boldsymbol{\omega} \times \boldsymbol{v}_0 - \mathbf{g}) + \dot{\boldsymbol{\omega}} \times \boldsymbol{L} + 2\boldsymbol{\omega} \times \overset{*}{\boldsymbol{L}} + \boldsymbol{\omega} \times (\boldsymbol{\omega} \times \boldsymbol{L}) + \overset{**}{\boldsymbol{L}} = \boldsymbol{P}^0,$$

$$\begin{aligned} &\boldsymbol{J} \cdot \dot{\boldsymbol{\omega}} + \overset{*}{\boldsymbol{J}}{}^1 \cdot \boldsymbol{\omega} + \boldsymbol{\omega} \times (\boldsymbol{J} \cdot \boldsymbol{\omega}) + \boldsymbol{L} \times (\overset{*}{\boldsymbol{v}}_0 + \boldsymbol{\omega} \times \boldsymbol{v}_0 - \mathbf{g}) \\ &+ \sum_i \boldsymbol{R}_i \ddot{\beta}_i + \sum_i (\overset{*}{\boldsymbol{R}}_i + \boldsymbol{\omega} \times \boldsymbol{R}_i)\dot{\beta}_i = \boldsymbol{M}_0. \end{aligned} \tag{6.2.14}$$

The system of equations (6.2.14) is not complete because it contains an infinite number of the generalized coordinates $\beta_i$ in addition to $\boldsymbol{v}_0$ and $\boldsymbol{\omega}$. The last group of equations, which associates the generalized coordinates $\beta_i$ with quasivelocities $\boldsymbol{v}_0$ and $\boldsymbol{\omega}$, is derived from the dynamical condition on the free surface $\Sigma$ [the fourth condition in (2.5.4))]. Substituting the velocity potential $\boldsymbol{\Phi}$ in the form of (6.1.3) into this condition with regard to relations (6.1.7), (6.1.8), and (6.1.1), and applying the Taylor formula (6.1.9) to the result, we obtain an approximate expression for the dynamical condition $L(\boldsymbol{\Phi}, f) = 0$ on the perturbed surface $\Sigma$ that depends only on solutions of the boundary-value problems (6.1.10)–(6.1.15) for the domain $Q_0$ with fixed boundaries.

Multiplying $L(\boldsymbol{\Phi})$ by $\rho f_i$ and integrating over the nonperturbed surface $\Sigma_0$ up to the terms of the third order of smallness inclusively with regard to the orthogonality of the system of functions $f_i$, we obtain the following system of nonlinear ordinary differential equations [173]:

$$\begin{aligned} &\mu_i \ddot{\beta}_i + w_1 N_i^2 \beta_i + \boldsymbol{\lambda}_i \cdot \overset{*}{\boldsymbol{w}} + \sum_j \sum_k \mu_{jik} \ddot{\beta}_i \beta_k + \boldsymbol{\lambda}_{0i} \cdot \dot{\boldsymbol{\omega}} + \sum_j \sum_k E_{jik} \dot{\beta}_j \dot{\beta}_k \\ &+ \sum_j \sum_k \sum_l \mu_{jikl} \beta_k \beta_l \ddot{\beta}_j + \sum_j \sum_k \sum_l E_{jikl} \beta_l \dot{\beta}_j \dot{\beta}_k + \sum_j \boldsymbol{\lambda}_{ij} \cdot \dot{\boldsymbol{\omega}} \beta_j \\ &+ \sum_j \boldsymbol{B}_{ij} \cdot \boldsymbol{\omega} \dot{\beta}_j + \boldsymbol{\omega} \cdot \boldsymbol{F}_i \cdot \boldsymbol{\omega} + \sum_j \sum_k \boldsymbol{\lambda}_{ijk} \cdot \dot{\boldsymbol{\omega}} \beta_j \beta_k + \sum_j \sum_k \boldsymbol{B}_{ijk} \cdot \boldsymbol{\omega} \beta_k \dot{\beta}_j \\ &+ \sum_j \boldsymbol{\omega} \cdot \boldsymbol{F}_{ij} \cdot \boldsymbol{\omega} \beta_j = 0, \quad i = 1, 2, \ldots . \end{aligned} \tag{6.2.15}$$

In system (6.2.15), we use the following notation for the components of vector and tensor quantities:

$$E_{jik} = \rho \int_{\Sigma_0} \left(\varphi_{jk} + \tfrac{1}{2}\nabla\varphi_{j0} \cdot \nabla\varphi_{k0}\right) f_i dS,$$

$$E_{jikl} = \rho \int_{\Sigma_0} \left[ 2\varphi_{jkl} + f_l \frac{\partial \varphi_{jk}}{\partial x} + \nabla\varphi_{j0} \cdot \nabla\varphi_{kl} + \nabla\varphi_{k0} \cdot \nabla\varphi_{jl} \right.$$

$$\left. + \tfrac{1}{2} f_l \frac{\partial}{\partial x}(\nabla\varphi_{j0} \cdot \nabla\varphi_{k0}) \right] f_i dS,$$

$$B_{ij}^{(n)} = \rho \int_{\Sigma_0} [\Omega_j^{(n)} + \nabla\Omega_0^{(n)} \cdot \nabla\varphi_{j0} - (\boldsymbol{r} \times \nabla\varphi_{j0})^{(n)}] f_i dS,$$

$$F_i^{(n,m)} = \rho \int_{\Sigma_0} \left[ \tfrac{1}{2} \nabla\Omega_0^{(n)} \cdot \nabla\Omega_0^{(m)} - (\boldsymbol{r} \times \nabla\Omega_0^{(m)})^{(n)} \right] f_i dS, \qquad (6.2.16)$$

$$B_{ijk}^{(n)} = \rho \int_{\Sigma_0} \left\{ 2\Omega_{jk}^{(n)} + f_k \frac{\partial \Omega_j^{(n)}}{\partial x} + \nabla\Omega_k^{(n)} \cdot \nabla\varphi_{j0} + \nabla\Omega_0^{(n)} \cdot \nabla\varphi_{jk} - (\boldsymbol{r} \times \nabla\varphi_{jk})^{(n)} \right.$$

$$\left. - f_k \frac{\partial}{\partial x} [\nabla\Omega_0^{(n)} \cdot \nabla\varphi_{j0} - (\boldsymbol{r} \times \nabla\varphi_{j0})^{(n)}] \right\} f_i f_j dS,$$

$$F_{ij}^{(mn)} = \rho \int_{\Sigma_0} \left\{ \nabla\Omega_0^{(n)} \cdot \nabla\Omega_j^{(m)} - (\boldsymbol{r} \times \nabla\Omega_j^{(m)})^{(n)} \right.$$

$$\left. + f_j \frac{\partial}{\partial x} \left[ \tfrac{1}{2} \nabla\Omega_0^{(n)} \cdot \nabla\Omega_0^{(m)} - (\boldsymbol{r} \times \nabla\Omega_0^{(m)})^{(n)} \right] \right\} f_i dS.$$

Under the assumption that the velocity potential and the shape of the free surface are known, equations of motion (6.2.14) and (6.2.15) can also be derived by the Lagrange method [173]. To this end, we have to write down the expression for the kinetic energy of the system

$$T = \frac{1}{2} \int_{\tau_0} (\boldsymbol{v}_0 + \boldsymbol{\omega} \times \boldsymbol{r})^2 \rho_0 d\tau + \frac{1}{2} \rho \int_Q \left[ \boldsymbol{v}_0 + \nabla(\boldsymbol{\omega} \cdot \boldsymbol{\Omega}) + \sum_i \dot{\beta}_i \nabla\varphi_i \right]^2 dQ. \qquad (6.2.17)$$

To derive this expression, we use the integral transformation

$$\mu_{ij} = \rho \int_Q \nabla\varphi_i \cdot \nabla\varphi_j dQ. \qquad (6.2.18)$$

By using the integral theorems, we reduce relation (6.2.18) to the form

$$\mu_{ij} = \rho \int_Q \nabla\varphi_i \cdot \nabla\varphi_j dQ = \mu_{ij0} + \sum_k \mu_{ijk}\beta_k + \sum_k \sum_i \mu_{ijkl}\beta_k\beta_l, \qquad (6.2.19)$$

where

$$\mu_{ij0} = \rho \int\limits_{\Sigma_0} \varphi_{i0} f_j dS,$$

$$\mu_{ijk} = \rho \int\limits_{\Sigma_0} \left( \varphi_{ik} + f_k \frac{\partial \varphi_{i0}}{\partial x} \right) f_j dS, \tag{6.2.20}$$

$$\mu_{ijkl} = \rho \int\limits_{\Sigma_0} \left[ \varphi_{j0} \frac{\partial \varphi_{ikl}}{\partial x} + \left( \frac{\partial \varphi_{ik}}{\partial x} + \tfrac{1}{2} f_k \frac{\partial^2 \varphi_{i0}}{\partial x^2} \right) f_l f_j \right] dS.$$

As soon as the kinetic energy is derived, the equations for forces and moments can be rewritten in the Euler–Lagrange form

$$\frac{d}{dt}\left(\frac{\partial T}{\partial v_x}\right) + \frac{\partial T}{\partial v_z}\omega_y - \frac{\partial T}{\partial v_y}\omega_z = P_x^C + P_x^0,$$

$$\frac{d}{dt}\left(\frac{\partial T}{\partial \omega_x}\right) + \frac{\partial T}{\partial \omega_z}\omega_y - \frac{\partial T}{\partial \omega_y}\omega_z + \frac{\partial T}{\partial v_z}v_y - \frac{\partial T}{\partial v_y}v_z = M_x^C + M_x^0. \tag{6.2.21}$$

By performing the cyclic permutation of the indices $x$, $y$, and $z$, we obtain the other four equations of the system. In these equations, $M_x^C$ and $M_x^0$ are the generalized forces defined by the work of external and mass forces on virtual displacements of the mechanical system in the directions of corresponding quasicoordinates.

Due to relation (6.1.3) for the velocity potential, the virtual displacement of liquid particles can be presented in the form

$$\delta \boldsymbol{r}' = \delta \boldsymbol{r}_0' + \nabla(\delta \boldsymbol{\theta} \times \boldsymbol{\Omega}) + \sum_i \delta \beta_i \nabla \varphi_i. \tag{6.2.22}$$

In view of relation (2.9.1) for the virtual displacement of points of a rigid body, we obtain the following relation for the elementary work of mass forces on the virtual displacement:

$$\delta' A = \int\limits_{\tau_0} (\delta \boldsymbol{r}_0' + \delta \boldsymbol{\theta} \times \boldsymbol{r}) \cdot \mathbf{g} \rho_0 d\tau$$

$$+ \rho \int\limits_{G} \left[ \delta \boldsymbol{r}_0' + \nabla(\delta \boldsymbol{\theta} \times \boldsymbol{\Omega}) + \sum_i \delta \beta_i \nabla \varphi_i \right] \cdot \mathbf{g} dQ$$

$$= (m_0 + m_1)\mathbf{g} \cdot \delta \boldsymbol{r}_0' + (\boldsymbol{L} \times \mathbf{g}) \cdot \delta \boldsymbol{\theta} + \sum_i \frac{\partial \boldsymbol{L}}{\partial \beta_i} \cdot \mathbf{g} \delta \beta_i. \tag{6.2.23}$$

In this formula, the generalized forces of variations of the quasicoordinates $\delta \boldsymbol{r}_0'$ and $\delta \boldsymbol{\theta}$ are vectors of the static force and the static moment defined by relations (6.2.13). Writing down the partial derivatives of the kinetic energy with respect to projections of the translational $\boldsymbol{v}_0$ and angular $\boldsymbol{\omega}$ velocities, we rewrite the Euler–Lagrange equations in the form

$$\frac{d}{dt}(\text{grad}_{v_0} T) + \boldsymbol{\omega} \times \text{grad}_{v_0} T = \boldsymbol{P}^C + \boldsymbol{P}^0,$$

$$\frac{d}{dt}(\text{grad}_{\omega} T) + \boldsymbol{\omega} \times \text{grad}_{\omega} T + \boldsymbol{v}_0 \times \text{grad}_{v_0} T = \boldsymbol{M}^C + \boldsymbol{M}_0. \tag{6.2.24}$$

Noting that $\text{grad}_{v_0} T = \boldsymbol{K}$ and $\text{grad}_{\omega} T = \boldsymbol{G}$, we conclude that systems (6.2.1), (6.2.14) and (6.2.24) coincide identically. Equations (6.2.24) should be supplemented with a system of equations for the generalized coordinates $\beta_i$, which is derived by using the Lagrange equations of the second kind

$$\frac{d}{dt}\left(\frac{\partial T}{\partial \dot\beta_i}\right) - \frac{\partial T}{\partial \beta_i} = B_i, \quad B_i = \frac{\partial \boldsymbol{L}}{\partial \beta_i} \cdot \mathbf{g} \tag{6.2.25}$$

within the framework of the considered method. In the expanded form, this system is presented as follows:

$$\begin{aligned}
&\mu_i \ddot\beta_i + w_1 N_i^2 \beta_i + \boldsymbol{\lambda}_i \cdot \overset{*}{\boldsymbol{w}} + \boldsymbol{\lambda}_{0i} \cdot \dot{\boldsymbol{\omega}} + \sum_j \sum_k \mu_{jik} \beta_k \ddot\beta_j \\
&+ \tfrac{1}{2} \sum_j \sum_k (2\mu_{jik} - \mu_{jki}) \dot\beta_j \dot\beta_k + \sum_j \sum_k \sum_l \mu_{ijkl} \beta_k \beta_l \ddot\beta_j \\
&+ \tfrac{1}{2} \sum_j \sum_k \sum_l (2\mu_{jikl} + 2\mu_{jilk} - \mu_{jkil} - \mu_{jkli}) \beta_l \dot\beta_j \dot\beta_k + \sum_j \boldsymbol{\lambda}_{ij} \cdot \dot{\boldsymbol{\omega}} \beta_j \\
&+ \sum_j (\boldsymbol{\lambda}_{ij} - \boldsymbol{\lambda}_{ji}) \cdot \boldsymbol{\omega} \dot\beta_j - \tfrac{1}{2} \boldsymbol{\omega} \cdot J_i \cdot \boldsymbol{\omega} + \sum_j \sum_k \boldsymbol{\lambda}_{ijk} \cdot \dot{\boldsymbol{\omega}} \beta_j \beta_k \\
&+ \sum_j \sum_k (\boldsymbol{\lambda}_{ijk} + \boldsymbol{\lambda}_{ikj} - \boldsymbol{\lambda}_{jik} - \boldsymbol{\lambda}_{jki}) \cdot \boldsymbol{\omega} \beta_k \dot\beta_j \\
&- \tfrac{1}{2} \sum_j \boldsymbol{\omega} \cdot (J_{ij} + J_{ji}) \cdot \boldsymbol{\omega} \beta_j = 0, \quad i = 1, 2, \ldots .
\end{aligned} \tag{6.2.26}$$

To derive (6.2.26), we present the tensor of inertia $\boldsymbol{J}$ of the mechanical system in the form

$$\boldsymbol{J} = \boldsymbol{J}^0 + \boldsymbol{J}^{\text{Zh}} + \sum_i J_i \beta_i + \sum_i \sum_j J_{ij} \beta_i \beta_j, \tag{6.2.27}$$

where $\boldsymbol{J}^0$ is the tensor of inertia with respect to the point $O$ for the rigid body without any liquid; $\boldsymbol{J}^0 + \boldsymbol{J}^{\text{Zh}}$ is the tensor of inertia for an equivalent body introduced by N. E. Zhukovsky in the case of a body with liquid under the condition of nondeformability of the boundaries. The system of equations

(6.2.26) coincides structurally with (6.2.15). For complete coincidence of these systems, we have to assume the additional relations [173]

$$E_{jik} = \mu_{jik} - \tfrac{1}{2}\mu_{jki}, \quad E_{jikl} = \mu_{jikl} + \mu_{jilk} - \tfrac{1}{2}\mu_{jkil} - \tfrac{1}{2}\mu_{jkli},$$
$$\boldsymbol{B}_{ij} = \boldsymbol{\lambda}_{ij} - \boldsymbol{\lambda}_{ji}, \quad F_i = -\tfrac{1}{2}J_i, \quad \boldsymbol{B}_{ijk} = \boldsymbol{\lambda}_{ijk} + \boldsymbol{\lambda}_{ikj} - \boldsymbol{\lambda}_{jik} - \boldsymbol{\lambda}_{jki}, \tag{6.2.28}$$
$$F_{ij} = -\tfrac{1}{2}(J_{ij} + J_{ji}).$$

## 6.3 Main boundary-value problems for axially symmetric cylindrical cavities near the free surface

Assume that a cavity of cylindrical shape in a neighborhood of the free surface $\Sigma_0$ is a surface of revolution. In the cylindrical coordinate system $x$, $\xi$, $\eta$, we divide a set of functions $\varphi_{no}$ into four subsets with the following properties:

(a) even with respect to $y$ and $z$;

(b) odd with respect to $y$ and even with respect to $z$;

(c) even with respect to $y$ and $z$;

(d) odd with respect to $y$ and $z$.

Conventionally, these systems of functions can be written in the form

$$\begin{aligned}
\varphi_{i0}^{(r)} &= \varphi_{mn}^{(\alpha)} = \psi_{mn}(x,\xi)\sin m\eta, \quad m = 1,\ 3,\ 5,\ \ldots,\\
\varphi_{i0}^{(p)} &= \varphi_{mn}^{(\beta)} = \psi_{mn}(x,\xi)\cos m\eta, \quad m = 1,\ 3,\ 5,\ \ldots,\\
\varphi_{i0}^{(s)} &= \varphi_{mn}^{(s)} = \psi_{mn}(x,\xi)\cos m\eta, \quad m = 0,\ 2,\ 4,\ \ldots,\\
\varphi_{i0}^{(q)} &= \varphi_{mn}^{(q)} = \psi_{mn}(x,\xi)\sin m\eta, \quad m = 0,\ 2,\ 4,\ \ldots.
\end{aligned} \tag{6.3.1}$$

Further, by setting

$$f_{mn}(\xi,\eta) = \varkappa_{mn}\varphi_{mn}(x,\xi,\eta)\Big|_{\Sigma_0} = \varkappa_{mn}\psi_{mn}(x,\xi)|_{\Sigma_0}\,{\sin m\eta \atop \cos m\eta}, \tag{6.3.2}$$

we can present the free surface in the form

$$\begin{aligned}
f(r,\eta,t) = s(t)f_{01}(\xi) &+ [\alpha(t)\sin\eta + \beta(t)\cos\eta]f_{11}(\xi)\\
&+[q(t)\sin 2\eta + p(t)\cos 2\eta]f_{21}(\xi),
\end{aligned} \tag{6.3.3}$$

i.e., in accordance with (4.3.4), we set $s = p_0$, $\alpha = r_1$, $\beta = p_1$, $q = r_2$, and $p = p_2$. With regard to harmonics $\varphi_{i0}$ in (6.1.7) (for the sake of brevity, we denote them by $\varphi_{\alpha 0} = \psi_{11}(x,\xi)\sin\eta$, $\varphi_{\beta 0} = \psi_{11}(x,\xi)\cos\eta$, $\varphi_{s0} = \psi_{01}(x,\xi)$, $\varphi_{p0} = \psi_{21}(x,\xi)\cos 2\eta$, and $\varphi_{q0} = \psi_{21}\sin 2\eta$), for the boundary-value problem (6.1.11), we present the boundary conditions on the nonperturbed free surface $\Sigma_0$ as follows:

$$\frac{\partial\varphi_{\alpha\alpha}}{\partial\nu} = G_1\sin^2\eta + G_2\cos^2\eta,\quad \frac{\partial\varphi_{\alpha\beta}}{\partial\nu} = \frac{\partial\varphi_{\beta\alpha}}{\partial\nu} = (G_1 - G_2)\sin\eta\cos\eta,$$

$$\frac{\partial\varphi_{\beta\beta}}{\partial\nu} = G_1\cos^2\eta + G_2\sin^2\eta,\quad \frac{\partial\varphi_{\alpha s}}{\partial\nu} = G_{01}\sin\eta,\quad \frac{\partial\varphi_{s\alpha}}{\partial\nu} = G_{10}\sin\eta,$$

$$\begin{aligned}
\frac{\partial\varphi_{\alpha p}}{\partial\nu} &= G_{21}\cos 2\eta\sin\eta - 2G'_{21}\sin 2\eta\cos\eta,\\
\frac{\partial\varphi_{p\alpha}}{\partial\nu} &= G_{12}\cos 2\eta\sin\eta - 2G'_{12}\sin 2\eta\cos\eta,\\
\frac{\partial\varphi_{\alpha q}}{\partial\nu} &= G_{21}\sin 2\eta\sin\eta + 2G'_{21}\cos 2\eta\cos\eta,\\
\frac{\partial\varphi_{q\alpha}}{\partial\nu} &= G_{12}\sin 2\eta\sin\eta + 2G'_{12}\cos 2\eta\cos\eta,
\end{aligned}$$

$$\frac{\partial\varphi_{\beta s}}{\partial\nu} = G_{01}\cos\eta,\quad \frac{\partial\varphi_{s\beta}}{\partial\nu} = G_{10}\cos\eta, \tag{6.3.4}$$

$$\begin{aligned}
\frac{\partial\varphi_{\beta p}}{\partial\nu} &= G_{21}\cos 2\eta\cos\eta + 2G'_{21}\sin 2\eta\sin\eta,\\
\frac{\partial\varphi_{p\beta}}{\partial\nu} &= G_{12}\cos 2\eta\cos\eta + 2G'_{21}\sin 2\eta\sin\eta,\\
\frac{\partial\varphi_{\beta q}}{\partial\nu} &= G_{21}\sin 2\eta\cos\eta - 2G'_{21}\cos 2\eta\sin\eta,\\
\frac{\partial\varphi_{q\beta}}{\partial\nu} &= G_{12}\sin 2\eta\cos\eta - 2G'_{21}\cos 2\eta\sin\eta,
\end{aligned}$$

where

$$G_1 = \frac{\partial}{\partial\xi}\left(f_{11}\frac{\partial\psi_{11}}{\partial\xi}\right) + \frac{f_{11}}{\xi}\frac{\partial\psi_{11}}{\partial\xi} - \frac{f_{11}}{\xi^2}\psi_{11} = \frac{\partial f_{11}}{\partial\xi}\frac{\partial\psi_{11}}{\partial\xi} - f_{11}\frac{\partial^2\psi_{11}}{\partial x^2},$$

$$G_2 = \frac{1}{\xi^2}f_{11}\psi_{11},\quad G_{01} = \frac{\partial f_{01}}{\partial\xi}\frac{\partial\psi_{11}}{\partial\xi} - f_{01}\frac{\partial^2\psi_{11}}{\partial x^2},$$

$$G_{10} = \frac{\partial f_{11}}{\partial\xi}\frac{\partial\psi_{01}}{\partial\xi} - f_{11}\frac{\partial^2\psi_{01}}{\partial x^2}, \tag{6.3.5}$$

$$G_{21} = \frac{\partial f_{21}}{\partial\xi}\frac{\partial\psi_{11}}{\partial\xi} - f_{21}\frac{\partial^2\psi_{11}}{\partial x^2},\quad G'_{21} = \frac{1}{\xi^2}f_{21}\psi_{11},$$

$$G_{12} = \frac{\partial f_{11}}{\partial\xi}\frac{\partial\psi_{21}}{\partial\xi} - f_{11}\frac{\partial^2\psi_{21}}{\partial x^2},\quad G'_{12} = \frac{1}{\xi^2}f_{11}\psi_{21}.$$

It is convenient to present a solution of the boundary-value problems for the functions $\varphi_{\alpha\alpha}$, $\varphi_{\alpha\beta}$, and $\varphi_{\beta\beta}$ in the form

$$\varphi_{\alpha\alpha} = \Psi_0 - \Psi_2 \cos 2\eta, \quad \varphi_{\alpha\beta} = \varphi_{\beta\alpha} = \Psi_2 \sin 2\eta, \tag{6.3.6}$$
$$\varphi_{\beta\beta} = \Psi_0 + \Psi_2 \cos 2\eta.$$

This enables us to reduce the corresponding three-dimensional boundary-value problems to the following boundary-value problems in the plane of the meridional section of the cavity:

$$L_0(\Psi) \equiv \xi^2 \frac{\partial^2 \Psi_0}{\partial x^2} + \xi \frac{\partial \Psi_0}{\partial \xi} + \xi^2 \frac{\partial^2 \Psi_0}{\partial \xi^2} = 0, \quad (x, \xi) \in G, \tag{6.3.7}$$
$$\left.\frac{\partial \Psi_0}{\partial \nu}\right|_{L_1} = 0, \quad \left.\frac{\partial \Psi_0}{\partial \nu}\right|_{L_0} = \tfrac{1}{2}(G_1 + G_2);$$
$$L_2(\Psi) \equiv \xi^2 \frac{\partial^2 \Psi_2}{\partial x^2} + \xi \frac{\partial \Psi_2}{\partial \xi} + \xi^2 \frac{\partial^2 \Psi_2}{\partial \xi^2} - 4\Psi_2 = 0, \quad (x, \xi) \in G, \tag{6.3.8}$$
$$\left.\frac{\partial \Psi_2}{\partial \nu}\right|_{L_1} = 0, \quad \left.\frac{\partial \Psi_2}{\partial \nu}\right|_{L_0} = \tfrac{1}{2}(G_1 - G_2).$$

Methods for solving these problems for cavities of various geometric shapes are given in [112, 173].

For the remaining boundary-value problems of the type (6.1.12) associated with the potentials $\varphi_i$ in the form (6.1.7), the boundary conditions on $\Sigma_0$ can be reduced to relations

$$\begin{aligned}
\frac{\partial \varphi_{\alpha\alpha\alpha}}{\partial \nu} &= G_3 \sin^3 \eta + G_5 \sin \eta - G_6 \sin \eta \cos 2\eta + (G_4 + G_7) \cos \eta \sin 2\eta, \\
\frac{\partial \varphi_{\alpha\alpha\beta}}{\partial \nu} &= G_3 \sin^2 \eta \cos \eta + G_4 \cos \eta \cos 2\eta + \tfrac{1}{2} G_5 \cos \eta - \tfrac{1}{2}(G_6 - G_7) \cos 3\eta, \\
\frac{\partial \varphi_{\alpha\beta\beta}}{\partial \nu} &= G_3 \sin \eta \cos^2 \eta - (G_4 - G_6) \cos \eta \sin 2\eta - G_7 \sin \eta \cos 2\eta, \\
\frac{\partial \varphi_{\beta\alpha\alpha}}{\partial \nu} &= G_3 \sin^2 \eta \cos \eta - (G_4 - G_6) \sin \eta \sin 2\eta + G_7 \cos \eta \cos 2\eta, \\
\frac{\partial \varphi_{\beta\alpha\beta}}{\partial \nu} &= G_3 \sin\eta \cos^2\eta - G_4 \sin\eta \cos 2\eta + \tfrac{1}{2} G_5 \sin\eta + \tfrac{1}{2}(G_6 - G_7) \sin 3\eta, \\
\frac{\partial \varphi_{\beta\beta\beta}}{\partial \nu} &= G_3 \cos^3 \eta + G_5 \cos \eta + G_6 \cos \eta \cos 2\eta + (G_4 + G_7) \sin \eta \cos 2\eta,
\end{aligned} \tag{6.3.9}$$

where

$$\begin{aligned}
G_3 &= \frac{1}{2}\left[\frac{\partial}{\partial \xi}\left(f_{11}^2 \frac{\partial^2 \psi_{11}}{\partial x \partial \xi}\right) + \frac{f_{11}^2}{\xi} \frac{\partial^2 \psi_{11}}{\partial x \partial \xi} - \frac{f_{11}^2}{\xi^2} \frac{\partial \psi_{11}}{\partial x}\right], \\
G_4 &= \frac{f_{11}^2}{2\xi^2} \frac{\partial \psi_{11}}{\partial x}, \quad G_5 = \frac{\partial}{\partial \xi}\left(f_{11} \frac{\partial \Psi_0}{\partial \xi}\right) + \frac{f_{11}}{\xi} \frac{\partial \Psi_0}{\partial \xi}, \\
G_6 &= \frac{\partial}{\partial \xi}\left(f_{11} \frac{\partial \Psi_2}{\partial \xi}\right) + \frac{f_{11}}{\xi} \frac{\partial \Psi_2}{\partial \xi} - 4 \frac{f_{11}}{\xi^2} \Psi_2, \quad G_7 = \frac{2 f_{11}}{\xi^2} \Psi_2.
\end{aligned} \tag{6.3.10}$$

We now present boundary-value problems for components of the Stokes–Zhukovsky potential. In accordance with the assumptions accepted in this section, the potential takes the form

$$\boldsymbol{\Omega} = \boldsymbol{\Omega}_0 + \alpha\boldsymbol{\Omega}_\alpha + \beta\boldsymbol{\Omega}_\beta + s\boldsymbol{\Omega}_s + p\boldsymbol{\Omega}_p + q\boldsymbol{\Omega}_q + \alpha^2\boldsymbol{\Omega}_{\alpha\alpha} + 2\alpha\beta\boldsymbol{\Omega}_{\alpha\beta} + \beta^2\boldsymbol{\Omega}_{\beta\beta}. \quad (6.3.11)$$

Since the components of the normal vector on $S$ are defined by the relations

$$\nu_x = \nu_x, \quad \nu_y = \nu_\xi \cos\eta, \quad \nu_z = \nu_\xi \sin\eta, \quad (6.3.12)$$

we can present a solution of the boundary-value problem (6.1.13) for $\boldsymbol{\Omega}_0$ in the form

$$\Omega_0^x = 0, \quad \Omega_0^y = -F(x,\xi)\sin\eta, \quad \Omega_0^z = F(x,\xi)\cos\eta. \quad (6.3.13)$$

To define the function $F(x,\xi)$, we obtain the following two-dimensional boundary-value problem:

$$\xi^2 F_{xx} + \xi F_\xi + \xi^2 F_{\xi\xi} - F = 0, \quad (x,\xi) \in G, \quad \left.\frac{\partial F}{\partial \nu}\right|_{L_1+L_0} = x\nu_\xi - \xi\nu_x. \quad (6.3.14)$$

By analyzing the boundary conditions of the boundary-value problems (6.1.14) and (6.1.15), we obtain the following relations for the $x$-components of the vector functions:

$$\Omega_\alpha^x = -\varphi_{\beta 0}, \quad \Omega_\beta^x = \varphi_{\alpha 0}, \quad \Omega_{\alpha\alpha}^x = -\varphi_{\beta\alpha}, \quad (6.3.15)$$
$$\Omega_{\alpha\beta}^x = \tfrac{1}{2}(\varphi_{\alpha\alpha} - \varphi_{\beta\beta}), \quad \Omega_{\beta\beta}^x = \varphi_{\alpha\beta}.$$

The boundary conditions for components $\Omega_\alpha^y$, $\Omega_\beta^y$, $\Omega_\alpha^z$, and $\Omega_\beta^z$ of the vector functions in (6.1.14) on the nonperturbed free surface $\Sigma_0$ are presented as follows:

$$\frac{\partial \Omega_\alpha^y}{\partial \nu} = H_1 \sin^2\eta + H_2 \cos^2\eta,$$
$$\frac{\partial \Omega_\beta^y}{\partial \nu} = -\frac{\partial \Omega_\alpha^z}{\partial \nu} = (H_1 - H_2)\sin\eta\cos\eta, \quad (6.3.16)$$
$$\frac{\partial \Omega_\beta^z}{\partial \nu} = -H_1 \cos^2\eta - H_2 \sin^2\eta,$$

where

$$H_1 = \frac{\partial f_{11}}{\partial \xi}\left(x - \frac{\partial F}{\partial \xi}\right) + f_{11}\frac{\partial^2 F}{\partial x^2}; \quad H_2 = \frac{f_{11}}{\xi^2}(x\xi - F). \quad (6.3.17)$$

Hence, these components can be represented in the form

$$\Omega_\alpha^y = F_0 + F_2\cos 2\eta, \quad \Omega_\alpha^z = -\Omega_\beta^z = F_2 \sin 2\eta, \quad (6.3.18)$$
$$\Omega_\beta^y = -F_0 + F_2\cos 2\eta.$$

Further, in order to determine the functions $F_0$ and $F_2$, we obtain the following boundary-value problems, which are equivalent to (6.3.7) and (6.3.8):

$$L_0(F_0) = 0, \quad (x,\xi) \in G; \quad \left.\frac{\partial F_0}{\partial \nu}\right|_{L_1} = 0, \quad \left.\frac{\partial F_0}{\partial \nu}\right|_{L_0} = \tfrac{1}{2}(H_1 + H_2); \tag{6.3.19}$$

$$L_2(F_2) = 0, \quad (x,\xi) \in G; \quad \left.\frac{\partial F_2}{\partial \nu}\right|_{L_1} = 0, \quad \left.\frac{\partial F_2}{\partial \nu}\right|_{L_0} = \tfrac{1}{2}(H_2 - H_1). \tag{6.3.20}$$

Now, the boundary conditions on $\Sigma_0$ for the remaining components of vector functions in (6.1.14) and (6.1.15) take the following form, which is similar to (6.3.4) and (6.3.9):

$$\begin{gathered}
\frac{\partial \Omega_s^x}{\partial \nu} = 0, \quad \frac{\partial \Omega_s^y}{\partial \nu} = H_{01}\sin\eta, \quad \frac{\partial \Omega_s^z}{\partial \nu} = -H_{01}\cos\eta, \quad \frac{\partial \Omega_p^x}{\partial \nu} = 2f_{21}\sin 2\eta, \\
\frac{\partial \Omega_q^x}{\partial \nu} = -2f_{21}\cos 2\eta, \quad \frac{\partial \Omega_p^y}{\partial \nu} = H_{21}\cos 2\eta\sin\eta - 2H'_{21}\sin 2\eta\cos\eta, \\
\frac{\partial \Omega_p^z}{\partial \nu} = -H_{21}\cos 2\eta\cos\eta - 2H'_{21}\sin 2\eta\sin\eta, \\
\frac{\partial \Omega_q^y}{\partial \nu} = H_{21}\sin 2\eta\sin\eta + 2H'_{21}\cos 2\eta\cos\eta, \\
\frac{\partial \Omega_q^z}{\partial \nu} = -H_{21}\sin 2\eta\cos\eta + 2H'_{21}\cos 2\eta\sin\eta, \\
\frac{\partial \Omega_{\alpha\alpha}^y}{\partial \nu} = H_3\sin^3\eta + (H_4 - H_7)\cos\eta\sin 2\eta + H_5\sin\eta + H_6\sin\eta\cos 2\eta, \\
\frac{\partial \Omega_{\alpha\beta}^y}{\partial \nu} = H_3\sin^2\eta\cos\eta + H_4\cos\eta\cos 2\eta + \tfrac{1}{2}H_5\cos\eta + \tfrac{1}{2}(H_6 - H_7)\cos 3\eta, \\
\frac{\partial \Omega_{\beta\beta}^y}{\partial \nu} = H_3\sin\eta\cos^2\eta - (H_4 + H_6)\cos\eta\sin 2\eta + H_7\sin\eta\cos 2\eta, \\
\frac{\partial \Omega_{\alpha\alpha}^z}{\partial \nu} = -H_3\sin^2\eta\cos\eta + (H_4 + H_6)\sin\eta\sin 2\eta + H_7\cos\eta\cos 2\eta, \\
\frac{\partial \Omega_{\alpha\beta}^z}{\partial \nu} = -H_3\sin\eta\cos^2\eta + H_4\sin\eta\cos 2\eta - \tfrac{1}{2}H_5\sin\eta + \tfrac{1}{2}(H_6 - H_7)\sin 3\eta, \\
\frac{\partial \Omega_{\beta\beta}^z}{\partial \nu} = -H_3\cos^3\eta - (H_4 - H_7)\sin\eta\sin 2\eta - H_5\cos\eta + H_6\cos\eta\cos 2\eta,
\end{gathered} \tag{6.3.21}$$

where

$$H_{01} = x\frac{\partial f_{01}}{\partial \xi} - \frac{\partial f_{01}}{\partial \xi}\frac{\partial F}{\partial \xi} + f_{01}\frac{\partial^2 F}{\partial x^2},$$

$$H_{21} = x\frac{\partial f_{21}}{\partial \xi} - \frac{\partial f_{21}}{\partial \xi}\frac{\partial F}{\partial \xi} + f_{21}\frac{\partial^2 F}{\partial x^2}, \quad H'_{21} = \frac{f_{21}}{\xi^2}(x\xi - F),$$

$$H_3 = f_{11}\frac{\partial f_{11}}{\partial \xi}\left(1 - \frac{\partial^2 F}{\partial x \partial \xi}\right) + \frac{f_{11}^2}{2}\frac{\partial^3 F}{\partial x^3}, \tag{6.3.22}$$

$$H_4 = \frac{f_{11}^2}{2\xi}\left(1 - \frac{1}{\xi}\frac{\partial F}{\partial x}\right), \quad H_5 = \frac{\partial f_{11}}{\partial \xi}\frac{\partial F_0}{\partial \xi} - f_{11}\frac{\partial^2 F_0}{\partial x^2},$$

$$H_6 = \frac{\partial f_{11}}{\partial \xi}\frac{\partial F_2}{\partial \xi} - f_{11}\frac{\partial^2 F_2}{\partial x^2}, \quad H_7 = \frac{2 f_{11}}{\xi^2}F_2.$$

Let us recall once again that the boundary-value problems, which are described in this section, are related to the case where axially symmetric containers have the cylindrical shape only in a neighborhood of the free surface, while outside this neighborhood their shape can be determined by an arbitrary surface of revolution. It is assumed that deformations of the free surface are finite since two parameters, which correspond to two degenerating antisymmetric sloshing modes, are not small. It will be shown in Chap. 8 that the generalized coordinates $s$, $p$, and $q$ make a dominant contribution to qualitative and quantitative variations of nonlinear phenomena.

## 6.4 Scalar form of the nonlinear equations and their hydrodynamic coefficients in special cases

In the case of cavities of revolution considered in the previous section, we define the coefficients of expressions for dynamical characteristics of liquid assuming that solutions of the main boundary-value problems are known. Due to symmetry properties of these solutions, a lot of scalar coefficients and components of the vector and tensor values vanish.

First, we define coefficients $\lambda_{i1}$, $\lambda_{i2}$, and $\lambda_{i3}$ in the expression for the vector of static moment $\boldsymbol{L}$ (6.2.3). By using relations (6.2.6), we get

$$\lambda_{\alpha 1} \equiv \lambda_\alpha^x = \lambda_{\beta 1} \equiv \lambda_\beta^x = \rho\pi \int\limits_{L_0} \xi f_{11}^2 d\xi = \rho N_{11}^2 = \lambda_1^1,$$

$$\lambda_{\alpha 3} \equiv \lambda_\alpha^z = \lambda_{\beta 2} = \lambda_\beta^y = \rho\pi \int\limits_{L_0} \xi^2 f_{11} d\xi = \lambda. \tag{6.4.1}$$

The other coefficients of the generalized coordinates $s$, $p$, and $q$ are equal to zero.

The nonzero components of the vectors $\boldsymbol{\lambda}_{0i}$, $\boldsymbol{\lambda}_{ij}$, and $\boldsymbol{\lambda}_{ijk}$ in the definition of $\boldsymbol{R}_i$ (6.2.7) are presented in the form

$$\lambda_{0\beta}^z = -\lambda_{0\alpha}^y = \pi\rho \int\limits_{L_0} \xi f_{11} F d\xi = \lambda_0,$$

$$\lambda_{\alpha\beta}^x = -\lambda_{\beta\alpha}^x = \pi\rho \int\limits_{L_0} \xi f_{11}\psi_{11} d\xi = \mu,$$

$$\lambda^y_{\alpha s} = -\lambda^z_{\beta s} = \pi\rho\int\limits_{L_0}\left(\psi_{11}H_{01} - f_{11}f_{01}\frac{\partial F}{\partial x}\right)\xi d\xi = \lambda_{10}, \tag{6.4.2}$$

$$\lambda^y_{s\alpha} = -\lambda^z_{s\beta} = \pi\rho\int\limits_{L_0}\left[\psi_{01}(H_1 + H_2) - f_{11}f_{01}\frac{\partial F}{\partial x}\right]\xi d\xi = \lambda'_{10},$$

$$-\lambda^y_{\alpha p} = -\lambda^z_{\beta p} = \lambda^y_{\beta q} = -\lambda^z_{\alpha q} = \frac{\pi}{2}\rho\int\limits_{L_0}\left[\psi_{11}(H_{21}+2H'_{21}) - f_{11}f_{21}\frac{\partial F}{\partial x}\right]\xi d\xi = \lambda_{12},$$

$$-\lambda_{p\alpha} = -\lambda^z_{p\beta} = \lambda^p_{q\beta} = -\lambda^z_{q\alpha} = \frac{\pi}{2}\rho\int\limits_{L_0}\left[\psi_{21}(H_1 - H_2) - f_{11}f_{21}\frac{\partial F}{\partial x}\right]\xi d\xi = \lambda'_{12},$$

$$\lambda^y_{\alpha\alpha\alpha} = -\lambda^z_{\beta\beta\beta} = \lambda_1, \ \ \lambda^z_{\beta\alpha\alpha} = -\lambda^y_{\alpha\beta\beta} = \lambda_2,$$

$$\lambda^y_{\beta\alpha\beta} = -\lambda^z_{\alpha\beta\alpha} = \lambda_3, \ \ \lambda^y_{\beta\beta\alpha} = -\lambda^z_{\alpha\alpha\beta} = \lambda_4,$$

where

$$\lambda_1 = \frac{\pi}{4}\rho\int\limits_{L_0}\Big[(3H_1 + H_2)f_{11}^2$$
$$+(3H_3 + 2H_4 + 4H_5 - 2H_6 - 2H_7)\psi_{11} - \tfrac{3}{2}\frac{\partial^2 F}{\partial x^2}f_{11}^3\Big]\xi d\xi,$$

$$\lambda_2 = \frac{\pi}{4}\rho\int\limits_{L_0}\Big[(-H_1 + H_2)f_{11}^2$$
$$+(-H_3 + 2H_4 + 2H_6 + 2H_7)\psi_{11} + \frac{1}{2}\frac{\partial^2 F}{\partial x^2}f_{11}^3\Big]\xi d\xi, \tag{6.4.3}$$

$$\lambda_3 = \frac{\pi}{4}\rho\int\limits_{L_0}\left[(H_1 + 3H_2)f_{11}^2 + (H_3 + 2H_4 + 2H_5)\psi_{11} - \frac{1}{2}\frac{\partial^2 F}{\partial x^2}f_{11}^3\right]\xi d\xi,$$

$$\lambda_4 = \frac{\pi}{4}\rho\int\limits_{L_0}\left[(H_1 - H_2)f_{11}^2 + (H_3 + 2H_4 + 2H_5)\psi_{11} - \frac{1}{2}\frac{\partial^2 F}{\partial x^2}f_{11}^3\right]\xi d\xi.$$

Among the entire collection of coefficients $\mu_{ijk}$ and $\mu_{ijkl}$ of the nonlinear ordinary differential equations of motion (6.2.26), the following coefficients are not equal to zero:

$$\mu_\alpha = \mu_\beta = \lambda^x_{\alpha\beta} = -\lambda^x_{\beta\alpha} = \mu,$$

$$\mu_s = 2\pi\rho\int\limits_{L_0} f_{01}\psi_{01}\xi d\xi = \mu_{10}, \ \ N_s^2 = 2\pi\int\limits_{L_0}\xi f_{01}^2 d\xi = N_{10}^2,$$

$$\mu_p = \mu_q = \pi\rho\int\limits_{L_0} f_{21}\psi_{21}\xi d\xi = \mu_{12}, \ \ N_q^2 = N_p^2 = \pi\rho\int\limits_{L_0} f_{21}^2\xi d\xi = N_{12}^2,$$

$$\mu_{\alpha\alpha s} = \mu_{\beta\beta s} = \mu_{101}, \quad \mu_{\alpha s\alpha} = \mu_{\beta s\beta} = \mu'_{110},$$

$$\mu_{s\alpha\alpha} = \mu_{s\beta\beta} = \mu_{110}, \quad \mu_{\beta\alpha q} = \mu_{\alpha\beta q} = \mu_{\beta\beta p} = -\mu_{\alpha\alpha p} = \mu_{121},$$

$$\mu_{\beta p\beta} = \mu_{\beta q\alpha} = \mu_{\alpha q\beta} = -\mu_{\alpha p\alpha} = \mu'_{112},$$

$$\mu_{q\beta\alpha} = \mu_{q\alpha\beta} = \mu_{p\beta\beta} = -\mu_{p\alpha\alpha} = \mu_{112},$$

$$\mu_{\alpha\alpha\alpha\alpha} = \mu_{\beta\beta\beta\beta} = \mu_1, \quad \mu_{\alpha\alpha\beta\beta} = \mu_{\beta\beta\alpha\alpha} = \mu_2,$$

$$\mu_{\alpha\beta\alpha\beta} = \mu_{\beta\alpha\beta\alpha} = \mu_3, \quad \mu_{\alpha\beta\beta\alpha} = \mu_{\beta\alpha\alpha\beta} = \mu_4,$$

$$\begin{aligned}
\mu_1 &= \frac{\pi}{4}\rho \int\limits_{L_0} \Big(3G_1 f_{11}^2 + 2G_2 f_{11}^2 + 3G_3\psi_{11} \\
&\qquad + 4G_5\psi_{11} + 2G_6\psi_{11} + 2G_7\psi_{11} + \frac{3}{2}\frac{\partial^2\psi_{11}}{\partial x^2} f_{11}^3\Big)\xi d\xi, \\
\mu_2 &= \frac{\pi}{4}\rho \int\limits_{L_0} \Big(G_1 f_{11}^2 - 2G_2 f_{11}^2 + G_3\psi_{11} \\
&\qquad + 2G_6\psi_{11} + 2G_7\psi_{11} + \frac{1}{2}\frac{\partial^2\psi_{11}}{\partial x^2} f_{11}^3\Big)\xi d\xi, \\
\mu_3 &= \frac{\pi}{4}\rho \int\limits_{L_0} \Big(G_1 f_{11}^2 + 4G_2 f_{11}^2 + G_3\psi_{11} + 2G_5\psi_{11} + \frac{1}{2}\frac{\partial^2\psi_{11}}{\partial x^2} f_{11}^3\Big)\xi d\xi, \\
\mu_4 &= \frac{\pi}{4}\rho \int\limits_{L_0} \Big(G_1 f_{11}^2 + G_3\psi_{11} + 2G_5\psi_{11} + \frac{1}{2}\frac{\partial^2\psi_{11}}{\partial x^2} f_{11}^3\Big)\xi d\xi, \\
\mu_{101} &= \pi\rho \int\limits_{L_0} (\psi_{11}G_{01} + f_{11}^2 f_{01})\xi d\xi, \\
\mu'_{110} &= \pi\rho \int\limits_{L_0} [\psi_{10}(G_1 + G_2) + f_{11}^2 f_{01}]\xi d\xi, \\
\mu_{110} &= \pi\rho \int\limits_{L_0} (\psi_{11}G_{01} + f_{11}^2 f_{01})\xi d\xi, \\
\mu_{121} &= \frac{\pi}{2}\rho \int\limits_{L_0} [\psi_{11}(G_{21} + 2G'_{21}) + f_{11}^2 f_{21}]\xi d\xi, \\
\mu'_{112} &= \frac{\pi}{2}\rho \int\limits_{L_0} [\psi_{21}(G_1 - G_2) + f_{11}^2 f_{21}]\xi d\xi, \\
\mu_{112} &= \frac{\pi}{2}\rho \int\limits_{L_0} [\psi_{11}(G_{12} + 2G'_{12}) + f_{11}^2 f_{21}]\xi d\xi.
\end{aligned} \tag{6.4.4}$$

We now define components of the inertia tensor of a liquid on the base of its presentation in the form (6.2.27). For components of the Zhukovsky inertia tensor $\boldsymbol{J}^{\mathrm{Zh}}$, we obtain the known relation

$$J_{yy}^{\mathrm{Zh}} = J_{zz}^{\mathrm{Zh}} = \pi\rho \int\limits_{L_0+L_1} (x\nu_\xi - \xi\nu_x)F\xi d\xi = J, \tag{6.4.5}$$

which is widely used in the linear theory. The other components of the inertia tensor in relation (6.2.27) have the form

$$\begin{aligned}
&J_\alpha^{xz} = J_\alpha^{zx} = J_\beta^{xy} = J_\beta^{yx} = -\lambda_0,\\
&J_s^{zz} = J_s^{yy} = J_{10} = \pi\rho \int\limits_{L_0} \left\{ \left[ \xi^2 + \left(\frac{\partial F}{\partial \xi}\right)^2 + \frac{1}{\xi^2}F^2 \right] f_{01} - 2FH_{01} \right\} \xi d\xi,\\
&J_p^{yy} = -J_p^{zz} = J_q^{yz} = J_q^{zy} = J_{12}\\
&\quad = \pi\rho \int\limits_{L_0} \left\{ (2H_{21}' + H_{21})F + \tfrac{1}{2}\left[ \frac{1}{\xi^2}F^2 - \left(\frac{\partial F}{\partial \xi}\right)^2 - \xi^2 \right] f_{21} \right\} \xi d\xi,\\
&J_{\alpha\alpha}^{yy} = J_{\beta\beta}^{zz} = J_1, \quad J_{\alpha\alpha}^{zz} = J_{\beta\beta}^{yy} = J_2, \quad J_{\alpha\beta}^{yz} = J_{\beta\alpha}^{zy} = J_{\beta\alpha}^{yz} = J_{\alpha\beta}^{zy} = J_3,
\end{aligned} \tag{6.4.6}$$

where

$$\begin{aligned}
J_1 = \pi\rho \int\limits_{L_0} \Bigg[ & F\left(-\frac{3}{2}H_3 - H_4 + H_7 - 2H_5 + H_6\right) + 2F_0\frac{\partial F_0}{\partial x} + F_2\frac{\partial F_2}{\partial x}\\
& + 2f_{11}\left(\xi\frac{\partial F_0}{\partial x} - \frac{\xi}{2}\frac{\partial F_2}{\partial x} - \frac{\partial F}{\partial \xi}\frac{\partial F_0}{\partial \xi} + \frac{1}{2}\frac{\partial F}{\partial \xi}\frac{\partial F_2}{\partial \xi} + \frac{1}{\xi^2}FF_2\right)\\
& + \tfrac{3}{4}f_{11}^2\left(\frac{\partial F}{\partial \xi}\frac{\partial^2 F}{\partial x\partial \xi} - \xi\frac{\partial^2 F}{\partial x^2} - \frac{1}{3\xi}F\right)\Bigg]\xi d\xi,\\
J_2 = \pi\rho \int\limits_{L_0} \Bigg[ & F\left(-\tfrac{1}{2}H_3 + H_4 + H_6 + H_7\right) + F_2\frac{\partial F_2}{\partial x}\\
& + f_{11}\left(\frac{\partial F}{\partial \xi}\frac{\partial F_2}{\partial \xi} - \xi\frac{\partial F_2}{\partial x} + \frac{2}{\xi^2}FF_2\right)\\
& + \tfrac{1}{4}f_{11}^2\left(\frac{\partial F}{\partial \xi}\frac{\partial^2 F}{\partial x\partial \xi} - \xi\frac{\partial^2 F}{\partial x^2} - \frac{3}{\xi}F\right)\Bigg]\xi d\xi,\\
J_3 = \pi\rho \int\limits_{L_0} \Bigg[ & F\left(H_3 + 2H_4 + 2H_5\right) - 2F_0\frac{\partial F_0}{\partial x}\\
& + 2f_{11}\left(\frac{\partial F}{\partial \xi}\frac{\partial F_0}{\partial \xi} - \xi\frac{\partial F_0}{\partial \xi}\right) + \tfrac{1}{2}f_{11}^2\left(\xi\frac{\partial^2 F}{\partial x^2} - \frac{\partial F}{\partial \xi}\frac{\partial^2 F}{\partial x\partial \xi} - \frac{1}{\xi}F\right)\Bigg]\xi d\xi.
\end{aligned} \tag{6.4.7}$$

By using relations (6.4.3), (6.4.4), and (6.4.7), we obtain the following obvious relations between the above defined coefficients:

$$\lambda_1 + \lambda_2 = \lambda_3 + \lambda_4, \quad \mu_1 - \mu_2 = \mu_3 + \mu_4, \quad J_2 - J_1 = J_3. \tag{6.4.8}$$

Now, by using results obtained above, we present relations for dynamical characteristics of the body–liquid system. We impose restrictions on the choice of the rigidly-fixed coordinate system $Oxyz$ accepted in Sec. 4.6. Namely, we assume that the point $O$ coincides with the center of mass of the body in the case of nonperturbed free surface and the coordinate axes coincide with the principal axes of inertia of the rigid body. Then the static moment of the solidified system and the centrifugal moments of inertia become equal to zero. According to relations (6.4.1), the projections of the static moment (6.2.3) and the momentum vector (6.2.4) onto the principal axes of the rigidly-fixed coordinate system are presented as follows:

$$\begin{aligned}
L_x &= \tfrac{1}{2}\lambda_1^1(\alpha^2+\beta^2), \quad L_y = \lambda\beta, \quad L_z = \lambda\alpha,\\
K_x &= Mv_x + \lambda(\alpha\omega_y - \beta\omega_z) + \tfrac{1}{2}\lambda_1^1(\alpha^2+\beta^2)^{\cdot},\\
K_y &= Mv_y + \tfrac{1}{2}\lambda_1^1(\alpha^2+\beta^2)\omega_z - \lambda\alpha\omega_x + \lambda\dot{\beta},\\
K_z &= Mv_z - \tfrac{1}{2}\lambda_1^1(\alpha^2+\beta^2)\omega_y + \lambda\beta\omega_x + \lambda\dot{\alpha}.
\end{aligned} \tag{6.4.9}$$

By virtue of relations (6.4.2) and (6.4.6), the projections of vector (6.2.7) and the inertia tensor of the system take the form

$$\begin{aligned}
&R_\alpha^x = \mu\beta, \quad R_\alpha^y = -\lambda_0 + \lambda_1\alpha^2 - \lambda_2\beta^2 + \lambda_{10}s - \lambda_{12}p,\\
&R_\alpha^z = -(\lambda_1+\lambda_2)\alpha\beta - \lambda_{12}q, \quad R_\beta^y = (\lambda_1+\lambda_2)\alpha\beta + \lambda_{12}q,\\
&R_\beta^x = -\mu\alpha, \quad R_\beta^z = \lambda_0 + \lambda_2\alpha^2 - \lambda_1\beta^2 - \lambda_{10}s - \lambda_{12}p,\\
&R_s^x = R_p^x = R_q^x = 0, \quad R_s^y = \lambda_{10}'\alpha; \quad R_s^z = -\lambda_{10}'\beta,\\
&R_p^y = R_q^z = -\lambda_{12}'\alpha, \quad R_q^y = -R_p^z = -\lambda_{12}'\beta,\\
&J^0 = \begin{pmatrix} J_x^0 & 0 & 0\\ 0 & J_y^0 & 0\\ 0 & 0 & J_r^0\end{pmatrix}, \quad J^{\mathrm{Zh}} = \begin{pmatrix} 0 & 0 & 0\\ 0 & J & 0\\ 0 & 0 & J\end{pmatrix},\\
&J_\alpha = \begin{pmatrix} 0 & 0 & -\lambda_0\\ 0 & 0 & 0\\ -\lambda_0 & 0 & 0\end{pmatrix}, \quad J_\beta = \begin{pmatrix} 0 & -\lambda_0 & 0\\ -\lambda_0 & 0 & 0\\ 0 & 0 & 0\end{pmatrix}.
\end{aligned} \tag{6.4.10}$$

$$J_q = \begin{pmatrix} 0 & 0 & 0\\ 0 & 0 & J_{12}\\ 0 & J_{12} & 0\end{pmatrix}, \quad J_s = \begin{pmatrix} 0 & 0 & 0\\ 0 & J_{10} & 0\\ 0 & 0 & J_{10}\end{pmatrix},$$

$$J_p = \begin{pmatrix} 0 & 0 & 0 \\ 0 & J_{12} & 0 \\ 0 & 0 & -J_{12} \end{pmatrix},$$

$$J_{\alpha\alpha} = \begin{pmatrix} \mu & 0 & 0 \\ 0 & J_1 & 0 \\ 0 & 0 & J_2 \end{pmatrix}, \quad J_{\alpha\beta} = \begin{pmatrix} 0 & 0 & 0 \\ 0 & 0 & J_3 \\ 0 & J_3 & 0 \end{pmatrix},$$

$$J_{\beta\alpha} = \begin{pmatrix} 0 & 0 & 0 \\ 0 & 0 & J_3 \\ 0 & J_3 & 0 \end{pmatrix}, \quad J_{\beta\beta} = \begin{pmatrix} \mu & 0 & 0 \\ 0 & J_2 & 0 \\ 0 & 0 & J_1 \end{pmatrix}.$$

By using relations (6.4.4), we reduce relation (6.2.10) to the form

$$\begin{gathered}
\mu_{\alpha\alpha} = \mu + \mu_1\alpha^2 + \mu_2\beta^2 + \mu_{101}s - \mu_{121}p, \\
\mu_{\alpha s} = \mu_{110}'\alpha, \quad \mu_{\alpha\beta} = \mu_{\beta\alpha} = (\mu_1 - \mu_2)\alpha\beta + \mu_{121}q, \\
\mu_{\alpha p} = -\mu_{\beta q} = -\mu_{112}'\alpha, \quad \mu_{\beta\beta} = \mu + \mu_2\alpha^2 + \mu_1\beta^2 + \mu_{101}s + \mu_{121}p, \\
\mu_{\beta s} = \mu_{110}'\beta, \quad \mu_{\alpha q} = \mu_{\beta p} = \mu_{112}'\beta, \quad \mu_{s\alpha} = \mu_{110}\alpha, \\
\mu_{s\beta} = \mu_{110}\beta, \quad \mu_{ss} = \mu_{10}, \quad \mu_{p\alpha} = -\mu_{q\beta} = -\mu_{112}\alpha, \\
\mu_{p\beta} = \mu_{q\alpha} = \mu_{112}\beta, \quad \mu_{qq} = \mu_{pp} = \mu_{12}.
\end{gathered} \tag{6.4.11}$$

By using the obtained results, we can write down the expressions for the projections of the vector of angular momentum (6.2.5) onto the axes of the moving coordinate system and also the relation for the kinetic energy of the system.

Substituting (6.4.9) and (6.4.10) into (6.2.5), we obtain

$$\begin{aligned}
G_x &= J_x^0\omega_x + \lambda(\beta v_z - \alpha v_y) + \mu[(\alpha^2 + \beta^2)\omega_x + \beta\dot\alpha - \alpha\dot\beta] - \lambda_0(\beta\omega_y + \alpha\omega_z), \\
G_y &= (J_y^0 + J)\omega_y + \lambda\alpha v_x - \tfrac{1}{2}\lambda_1^1(\alpha^2 + \beta^2)v_z - \lambda_0(\beta\omega_x + \dot\alpha) \\
&\quad + (J_{10}s + J_{12}p + J_1\alpha^2 + J_2\beta^2)\omega_y - (J_1 - J_2)\alpha\beta\omega_z \\
&\quad + (\lambda_1\alpha^2 - \lambda_2\beta^2 + \lambda_{10}s - \lambda_{12}p)\dot\alpha + [(\lambda_1 + \lambda_2)\alpha\beta + \lambda_{12}q]\dot\beta \\
&\quad + \lambda_{10}'\alpha\dot s - \lambda_{12}'(\alpha\dot p + \beta\dot q), \\
G_z &= (J_z^0 + J)\omega_z - \lambda\beta v_z + \tfrac{1}{2}\lambda_1^1(\alpha^2 + \beta^2)v_y - \lambda_0(\alpha\omega_x - \dot\beta) \\
&\quad + (J_2\alpha^2 + J_1\beta^2 + J_{10}s - J_{12}p)\omega_2 - (J_1 - J_2)\alpha\beta\omega_y \\
&\quad - [(\lambda_1 + \lambda_2)\alpha\beta + \lambda_{12}q]\dot\alpha + (\lambda_2\alpha^2 - \lambda_1\beta^2 - \lambda_{10}s - \lambda_{12}p)\dot\beta \\
&\quad - \lambda_{10}'\beta\dot s + \lambda_{12}'(\beta\dot p - \alpha\dot q).
\end{aligned} \tag{6.4.12}$$

In accordance with relations (6.4.11), the relation for the kinetic energy (6.2.17) takes the form

$$
\begin{aligned}
T = {} & \tfrac{1}{2}M(v_x^2 + v_y^2 + v_z^2) + \tfrac{1}{2}J_x^0\omega_x^2 + \tfrac{1}{2}(J_y^0 + J)\omega_y^2 + \tfrac{1}{2}(J_z^0 + J)\omega_z^2 \\
& +\lambda[\alpha(v_x\omega_y - v_y\omega_x) + \beta(v_z\omega_x - v_x\omega_z) + v_y\dot\beta + v_z\dot\alpha] \\
& +\tfrac{1}{2}\lambda_1^1[(\alpha^2+\beta^2)(v_y\omega_z - v_z\omega_y) + v_x(\alpha^2+\beta^2)^{\cdot}] \\
& -\lambda_0[(\alpha\omega_z + \beta\omega_y)\omega_x + \omega_y\dot\alpha - \omega_z\dot\beta] \\
& +\mu\left[\tfrac{1}{2}(\alpha^2+\beta^2)\omega_x^2 + (\beta\dot\alpha - \alpha\dot\beta)\omega_x + \tfrac{1}{2}(\dot\alpha^2+\dot\beta^2)\right] \\
& +(\lambda_1\alpha^2 + \lambda_{10}s - \lambda_{12}p - \lambda_2\beta^2)\omega_y\dot\alpha + (\lambda_2\alpha^2 - \lambda_{10}s - \lambda_{12}p - \lambda_1\beta^2)\omega_z\dot\beta \\
& +[(\lambda_1+\lambda_2)\alpha\beta + \lambda_{12}q](\omega_y\dot\beta - \omega_z\dot\alpha) + \tfrac{1}{2}(J_1\alpha^2 + J_{10}s + J_{12}p + J_2\beta^2)\omega_y^2 \\
& -(J_1 - J_2)\alpha\beta\omega_y\omega_z + \tfrac{1}{2}(J_2\alpha^2 + J_{10}s - J_{12}p + J_1\beta^2)\omega_z^2 \\
& +[\lambda_{10}'\alpha\dot s - \lambda_{12}'(\alpha\dot p + \beta\dot q]\omega_y + [\lambda_{12}'(\beta\dot p - \alpha\dot q) - \lambda_{10}'\beta\dot s]\omega_z \\
& +\tfrac{1}{2}(\mu + \mu_1\alpha^2 + \mu_2\beta^2 + \mu_{101}s - \mu_{121}p)\dot\alpha^2 + [(\mu_1 - \mu_2)\alpha\beta + \mu_{121}q]\dot\alpha\dot\beta \\
& +\tfrac{1}{2}(\mu_{110}' + \mu_{110})(\dot s\dot\alpha\alpha + \beta\dot s\dot\beta) + \tfrac{1}{2}(\mu_{112} + \mu_{112}')(\beta\dot\alpha\dot q - \alpha\dot\alpha\dot p + \beta\dot\beta\dot p + \alpha\dot\beta\dot q) \\
& +\tfrac{1}{2}(\mu + \mu_2\alpha^2 + \mu_1\beta^2 + \mu_{101}s + \mu_{121}p)\dot\beta^2 \\
& +\tfrac{1}{2}(\mu_{10}\dot s^2 + \mu_{12}\dot q^2 + \mu_{12}\dot p^2). \qquad (6.4.13)
\end{aligned}
$$

By using the mentioned relations for the static and angular momentum and the kinetic energy, we can obtain nonlinear differential equations of spatial motion of the rigid body–liquid mechanical system in terms of projections on the axes of the rigidly-fixed coordinate system $Oxyz$. In the scalar form, the equation of forces of system (6.2.14) takes the form

$$
\begin{aligned}
& Mw_x + \lambda_1^1(\ddot\alpha\alpha + \ddot\beta\beta + \dot\alpha^2 + \dot\beta^2) \\
& \qquad +\lambda[2(\omega_y\dot\alpha - \omega_z\dot\beta) + \alpha(\dot\omega_y + \omega_x\omega_z) - \beta(\dot\omega_z - \omega_x\omega_y)] = P_x^0, \\
& Mw_y + \lambda[\ddot\beta - 2\omega_x\dot\alpha - \alpha(\dot\omega_x - \omega_y\omega_z) - \beta(\omega_x^2 + \omega_z^2)] \\
& \qquad + \tfrac{1}{2}\lambda_1^1[2\omega_z(\alpha^2+\beta^2)^{\cdot} + (\alpha^2+\beta^2)\dot\omega_z] = P_y^0, \qquad (6.4.14) \\
& Mw_z + \lambda[\ddot\alpha + 2\omega_x\dot\beta + \beta(\dot\omega_x + \omega_y\omega_z) - \alpha(\omega_x^2 + \omega_z^2)] \\
& \qquad - \tfrac{1}{2}\lambda_1^1[2\omega_y(\alpha^2+\beta^2)^{\cdot} + (\alpha^2+\beta^2)\dot\omega_y] = P_z^0.
\end{aligned}
$$

In view of the introduced notation, the equations of moments relative to the principal central axes have the form

$$
\begin{aligned}
& J_x^0\dot\omega_x + (J_z^0 - J_y^0)\omega_y\omega_z + \mu[(\alpha^2+\beta^2)\dot\omega_x + 2(\alpha\dot\alpha + \beta\dot\beta)\omega_x - \alpha\ddot\beta + \ddot\alpha\beta] \\
& \qquad -\lambda_0(\dot\omega_y\beta + \dot\omega_z\alpha + \omega_x\omega_y\alpha - \omega_x\omega_z\beta) + \lambda(\beta w_z - \alpha w_y) = M_x^0, \\
& (J_y^0 + J)\dot\omega_y + \lambda w_x\alpha - \lambda_0\ddot\alpha - \tfrac{1}{2}\lambda_1^1(\alpha^2+\beta^2)w_z + (\lambda_1\alpha^2 - \lambda_2\beta^2)\ddot\alpha + 2\lambda_1\alpha\dot\alpha^2 + a_2\alpha(\beta\dot\beta)^{\cdot} \\
& \qquad +a_3\beta\dot\alpha\dot\beta + a_1\omega_x\omega_z - \lambda_0[2\omega_x\dot\beta + \beta(\dot\omega_x + \omega_y\omega_z) + \alpha(\omega_z^2 - \omega_x^2)]
\end{aligned}
$$

$$+[(J_1\alpha^2+J_2\beta^2)\omega_y]^{\cdot}-a_4\alpha\beta\dot{\omega}_z-(a_5\alpha\dot{\beta}+a_6\beta\dot{\alpha})\omega_z+(\lambda_{10}s-\lambda_{12}p)\ddot{\alpha}$$
$$+\lambda_{12}q\ddot{\beta}+(a_7\dot{s}+a_8\dot{p})\dot{\alpha}+a_8\dot{q}\dot{\beta}+(\lambda'_{10}\ddot{s}-\lambda'_{12}\ddot{p})\alpha+\lambda'_{12}\ddot{q}\beta$$
$$+[(J_{10}s+J_{12}p)\omega_y+J_{12}q\omega_z]^{\cdot}=M_y^0, \qquad (6.4.15)$$
$$(J_z^0+J)\dot{\omega}_x-\lambda w_1\beta+\lambda_0\ddot{\beta}+\tfrac{1}{2}\lambda_1^1(\alpha^2+\beta^2)w_y-a_2\beta(\alpha\dot{\alpha})^{\cdot}-2\lambda_1\beta\dot{\beta}^2$$
$$+(\lambda_2\alpha^2-\lambda_1\beta^2)\ddot{\beta}-a_3\alpha\dot{\alpha}\dot{\beta}-a_9\omega_x\omega_y$$
$$-\lambda_0[2\omega_x\dot{\alpha}+\alpha(\dot{\omega}_x-\omega_y\omega_z)+\beta(\omega_x^2-\omega_y^2)]+[(J_2\alpha^2+J_1\beta^2)\omega_z]^{\cdot}$$
$$-a_4\alpha\beta\dot{\omega}_y-(a_5\beta\dot{\alpha}+a_6\alpha\dot{\beta})\omega_y-(\lambda_{10}s+\lambda_{12}p)\ddot{\beta}-\lambda_{12}q\ddot{\alpha}-(a_7\dot{s}+a_8\dot{p})\dot{\beta}$$
$$-(\lambda_{12}+\lambda'_{12})\dot{q}\dot{\alpha}-(\lambda'_{10}\ddot{s}+\lambda'_{12}\ddot{p})\beta-\lambda'_{12}\ddot{q}\alpha$$
$$+[(J_{10}s-J_{12}p)\omega_z+J_{12}q\omega_y]^{\cdot}=M_z^0.$$

The last group of equations, which associate the generalized coordinates $\alpha$, $\beta$, $s$, $p$, and $q$ with the quasivelocities $v_i$ and $\omega_i$, takes the form

$$\mu(\ddot{\alpha}+\varkappa_{11}w_x\alpha)+\lambda w_z-\lambda_0(\dot{\omega}_y-\omega_x\omega_z)+\tfrac{1}{2}\mu_1\alpha(\alpha^2+\beta^2)^{\cdot\cdot}$$
$$+\mu_2(\beta^2\ddot{\alpha}-\alpha\beta\ddot{\beta}+2\beta\dot{\beta}\dot{\alpha}-2\alpha\dot{\beta}^2)+2\mu\omega_x\dot{\beta}^2+(\mu\dot{\omega}_x+a_4\omega_y\omega_z)\beta$$
$$+(\lambda_1\alpha^2-\lambda_2\beta^2)\dot{\omega}_y-b_1\alpha\beta\dot{\omega}_z-b_2(\beta\omega_y+\alpha\omega_z)\dot{\beta}-(\mu\omega_x^2+J_1\omega_y^2+J_2\omega_z^2)\alpha$$
$$+(\lambda_{10}s-\lambda_{12}p)\dot{\omega}_y-\lambda_{12}q\dot{\omega}_z+(b_3\dot{s}-b_4\dot{p})\omega_y-b_4\dot{q}\omega_z+\tfrac{1}{2}b_5\alpha\ddot{s}+\tfrac{1}{2}b_6(\beta\ddot{q}-\alpha\ddot{p})$$
$$+[(\mu_{101}s-\mu_{121}p)\dot{\alpha}+\mu_{121}q\dot{\beta}]^{\cdot}=0,$$

$$\mu(\ddot{\beta}+\varkappa_{11}w_x\beta)+\lambda w_y+\lambda_0(\dot{\omega}_z+\omega_x\omega_y)+\tfrac{1}{2}\mu_1\beta(\alpha^2+\beta^2)^{\cdot\cdot}$$
$$+\mu_2(\alpha^2\ddot{\beta}-\alpha\beta\ddot{\alpha}+2\alpha\dot{\alpha}\dot{\beta}-2\beta\dot{\alpha}^2)-2\mu\omega_x\dot{\alpha}-(\mu\dot{\omega}_x-a_4\omega_y\omega_z)\alpha$$
$$+(\lambda_2\alpha^2-\lambda_1\beta^2)\dot{\omega}_z+b_1\alpha\beta\dot{\omega}_y+b_2(\alpha\omega_z+\beta\omega_y)\dot{\alpha}-(\mu\omega_x^2+J_2\omega_y^2+J_1\omega_z^2)\beta$$
$$-(\lambda_{10}s+\lambda_{12}p)\dot{\omega}_z+\lambda_{12}q\dot{\omega}_y-(b_3\dot{s}+b_4\dot{p})\omega_z+b_4\dot{q}\omega_y+\tfrac{1}{2}b_5\beta\ddot{s} \qquad (6.4.16)$$
$$+\tfrac{1}{2}b_6(\beta\ddot{p}+\alpha\ddot{q})+[(\mu_{101}s+\mu_{121}p)\dot{\beta}+\mu_{121}q\dot{\alpha}]^{\cdot}=0,$$

$$\mu_{10}(\ddot{s}+\varkappa_{01}w_xs)+\tfrac{1}{2}b_5(\alpha\ddot{\alpha}+\beta\ddot{\beta})+\tfrac{1}{2}b_7(\dot{\alpha}^2+\dot{\beta}^2)-\tfrac{1}{2}J_{10}(\omega_y^2+\omega_z^2)$$
$$+\lambda'_{10}(\alpha\dot{\omega}_y-\beta\dot{\omega}_z)+b_3(\dot{\alpha}\omega_y-\dot{\beta}\omega_z)=0,$$

$$\mu_{12}(\ddot{p}+\varkappa_{21}w_xp)-\tfrac{1}{2}b_6(\alpha\ddot{\alpha}-\beta\ddot{\beta})-\tfrac{1}{2}b_8(\dot{\alpha}^2-\dot{\beta}^2)-\tfrac{1}{2}J_{12}(\omega_z^2-\omega_y^2)$$
$$-\lambda'_{12}(\alpha\dot{\omega}_y+\beta\dot{\omega}_z)-b_4(\dot{\alpha}\omega_y+\dot{\beta}\omega_z)=0,$$

$$\mu_{12}(\ddot{q}+\varkappa_{21}w_xq)+\tfrac{1}{2}b_6(\beta\ddot{\alpha}+\alpha\ddot{\beta})+b_8\dot{\alpha}\dot{\beta}-J_{12}\omega_y\omega_z$$
$$+\lambda'_{12}(\beta\dot{\omega}_y-\alpha\dot{\omega}_z)-b_4(\omega_y\dot{\beta}-\omega_z\dot{\alpha})=0.$$

In addition to the notation in (6.4.15) and (6.4.16), we introduce the notation

$$a_1 = J_x^0 - J_z^0 - J, \quad a_2 = \lambda_1 + \lambda_2, \quad a_3 = \lambda_1 - \lambda_2,$$
$$a_4 = J_1 - J_2, \quad a_5 = a_4 + \mu, \quad a_6 = a_4 - \mu,$$
$$a_7 = \lambda_{10} + \lambda_{10}', \quad a_8 = \lambda_{12} + \lambda_{12}', \quad a_9 = J_x^0 - J_y^0 - J, \tag{6.4.17}$$
$$b_1 = \lambda_1 + \lambda_2, \quad b_2 = \lambda_1 + 3\lambda_2, \quad b_3 = \lambda_{10} - \lambda_{10}',$$
$$b_4 = \lambda_{12} - \lambda_{12}', \quad b_5 = \mu_{110} + \mu_{110}', \quad b_6 = \mu_{112} + \mu_{112}',$$
$$b_7 = b_5 - \mu_{101}, \quad b_8 = b_6 - \mu_{121}.$$

The system of nonlinear differential equations (6.4.14), (6.4.15), and (6.4.16), which is derived by the methods presented above, coincides almost completely with similar equations presented in Chap. 4. The system presented here does not contain the terms, which are underlined in (4.7.8)–(4.7.16). For complete coincidence of these equations, it is necessary to know the components of the inertia tensor $\boldsymbol{J}^1$ along with the components of the vectors $\boldsymbol{R}_i$ and the coefficients $\mu_{ij}$ in (6.4.11) up to the terms of the third order inclusively. In addition to $J_{\alpha\alpha}$, $J_{\alpha\beta}$, and $J_{\beta\beta}$, the quantities $J_{s\alpha}$, $J_{\alpha s}$, $J_{\alpha p}$, etc. have to be known, too.

## 6.5 The case of an upright cylindrical cavity with circular or annular cross section

We apply the above-presented method to the case of a cylindrical cavity where the nonperturbed surface $\Sigma_0$ has a circular or ring shape. For the coordinate system $Oxyz$, we assume that its origin coincides with the center of mass of the rigid body–liquid system in the nonperturbed state and the longitudinal $Ox$-axis coincides with the axis of the cylinder. The coordinate of the center of the cavity bottom $S_0$ is denoted by $x_0$ and the coordinate of the center of the nonperturbed surface $\Sigma_0$ is denoted by $x_0 + h$, where $h$ is the depth of the liquid in the container. Denote the radius of the lateral surface $S_1$ by $R_1$ and the radius of the interior surface by $R_0$.

Solutions of the type (6.3.1) can be obtained from the eigenvalue boundary problem

$$\frac{\partial^2\psi}{\partial x^2} + \frac{\partial^2\psi}{\partial \xi^2} + \frac{1}{\xi}\frac{\partial\psi}{\partial\xi} - \frac{m^2}{\xi^2}\psi = 0, \quad (x,\xi) \in G,$$
$$\left.\frac{\partial\psi}{\partial x}\right|_{x=x_0+h} = \varkappa\psi, \quad \left.\frac{\partial\psi}{\partial x}\right|_{x=x_0} = 0, \quad \left.\frac{\partial\psi}{\partial \xi}\right|_{\xi=R_1} = 0, \quad \left.\frac{\partial\psi}{\partial \xi}\right|_{\xi=R_0} = 0. \tag{6.5.1}$$

By applying the method of separation of variables, we obtain

$$\psi_m(x,\xi) = X_m(x)Y_m(\xi), \tag{6.5.2}$$

and moreover, for each factor, we obtain a single-dimensional boundary-value problem leading to the following result:

$$X_{mn}(x) = \frac{\cosh k_{mn}(x - x_0)}{k_{mn} \sinh k_{mn} h}, \quad \varkappa_{mn} = k_{mn} \tanh k_{mn} h, \tag{6.5.3}$$

$$Y_m(k_{mn}\xi) = \frac{N'_m(\zeta_{mn}) J_m(k_{mn}\xi) - N_m(k_{mn}\xi) J'_m(\zeta_{mn})}{J_m(\zeta_{mn}) N'_m(\zeta_{mn}) - N_m(\zeta_{mn}) J'_m(\zeta_{mn})}, \tag{6.5.4}$$

where $J_m(\xi)$ and $N_m(\xi)$ are, respectively, the Bessel and Neumann functions of the order $m$; $\xi_{mn} = k_{mn} R_1$ is the $n$th root of the equation

$$J'_m(\delta\zeta) N'_m(\zeta) - N'_m(\delta\zeta) J'_m(\zeta) = 0, \quad \delta = R_0/R_1. \tag{6.5.5}$$

Values of the first roots of this equation are given in Tables 11–13 (see Sec. 4.1).

In addition to the boundary conditions, solutions (6.5.3) and (6.5.4) satisfy the normalization conditions

$$X'_{mn}(x_0 + h) = 1, \quad Y_m(k_{mn} R_1) = 1. \tag{6.5.6}$$

By virtue of these conditions, the generalized coordinates $\alpha(t)$, $\beta(t)$, ..., $q(t)$ present the deviation of the free surface from the shape of the corresponding sloshing mode. The square of the natural sloshing frequency for the $(m, n)$th mode is determined by the equality $\sigma_{mn}^2 = \varkappa_{mn} w_x$, where $w_x$ is the apparent acceleration of the mechanical system along the $Ox$-axis.

Boundary-value problem (6.3.14) for the function $F(x, \xi)$, which is related to the Stokes–Zhukovsky potential in the linear approximation, is relatively simple and can be solved by the method of decomposition in eigenfunctions of boundary-value problems of type (6.5.1). As it is known, such a solution takes the form

$$F(x, \xi) = x\xi - \sum_n 2 d_{1n} Z_{1n}(x) Y_1(k_{1n}\xi),$$

$$Z_{1n}(x) = \frac{\sinh k_{1n}\left(x - x_0 - \frac{h}{2}\right)}{k_{1n} \cosh\left(k_{1n}\frac{h}{2}\right)}, \quad d_{1n} = \frac{E_{1n}}{N_{1n}^2}, \tag{6.5.7}$$

$$E_{1n} = \int \xi^2 Y_1(k_{1n}\xi) d\xi, \quad N_{1n}^2 = \int \xi Y_1^2(k_{1n}\xi) d\xi.$$

Here and in what follows, quadratures of various combinations of Bessel functions are taken from $R_0$ to $R_1$. Solutions (6.5.2) and (6.5.7) provide a possibility to compute a group of coefficients of a linear system of equations derived from system (6.4.14), (6.4.15), and (6.4.16). Among these are

$$\lambda = \pi\rho E_{11}, \quad \mu = \frac{\pi\rho}{\varkappa_{11}} N_{11}^2, \quad \lambda_0 = \pi\rho\left[(x_0 + h)E_{11} - \frac{2d_{11}}{\varepsilon_{11}} N_{11}^2\right], \tag{6.5.8}$$

$$J = \pi\rho\left[\left(x_0^2 h + x_0 h^2 + \tfrac{1}{3}h^3\right)\left(R_1^2 - R_0^2\right) - \tfrac{3}{4}h(R_1^4 - R_0^4)\right] + 8\pi\rho\sum_n \frac{s_n}{k_{1n}} d_{1n}E_{1n},$$

$$\frac{1}{\varepsilon_{1n}} = \frac{1}{k_{1n}}\tanh\left(k_{1n}\frac{h}{2}\right).$$

These solutions also define several coefficients of nonlinear equations of motion. In order to obtain the entire collection of hydrodynamic coefficients, it is necessary to solve the boundary-value problems (6.3.7), (6.3.8), (6.3.19), and (6.3.20).

To this end, we define functions $G_1$, $G_2$, $H_1$, and $H_2$ on the right-hand sides of these boundary-value problems. By using (6.3.5) and (6.3.17), we obtain

$$G_1 = \frac{1}{\varkappa_{11}}\left({Y'_{11}}^2 - k_{11}^2 Y_{11}^2\right), \quad G_2 = \frac{1}{\varkappa_{11}}\frac{Y_{11}^2}{\xi^2},$$

$$H_1 = \sum_n \frac{2d_{1n}}{\varepsilon_{1n}}\left(Y'_{1n}Y'_{11} - k_{1n}^2 Y_{1n}Y_{11}\right), \quad H_2 = \sum_n \frac{2d_{1n}}{\varepsilon_{1n}}\frac{1}{\xi^2}Y_{1n}Y_{11}. \tag{6.5.9}$$

By using the method of decomposition in eigenfunctions of the eigenvalue boundary problem with parameters in the boundary condition (6.5.1), we present the required functions $\Psi_0$, $\Psi_2$, $F_0$, and $F_2$ in the form of decompositions

$$\Psi_m(x,\xi) = \sum_n c_{mn}X_{mn}(x)Y_{mn}(\xi), \quad m = 0,\, 2,$$

$$F_m(x,\xi) = \sum_n d_{mn}X_{mn}(x)Y_{mn}(\xi), \quad m = 0,\, 2;\; Y_{mn}(\xi) = Y_m(k_{mn}\xi). \tag{6.5.10}$$

The coefficients $c_{mn}$ and $d_{mn}$ are defined by the relations

$$c_{0n} = \frac{1}{2\varkappa_{11}N_{0n}^2}\int\left[{Y'_{11}}^2 + \left(\frac{1}{\xi^2} - k_{11}^2\right)Y_{11}^2\right]Y_{0n}\xi d\xi,$$

$$c_{2n} = \frac{1}{2\varkappa_{11}N_{2n}^2}\int\left[{Y'_{11}}^2 - \left(\frac{1}{\xi^2} + k_{11}^2\right)Y_{11}^2\right]Y_{2n}\xi d\xi,$$

$$d_{0n} = \frac{1}{N_{0n}^2}\sum_i \frac{d_{1i}}{\varepsilon_{1i}}\int Y_{0n}\left[Y'_{11}Y'_{1i} + \left(\frac{1}{\xi^2} - k_{1i}^2\right)Y_{11}Y_{1i}\right]\xi d\xi, \tag{6.5.11}$$

$$d_{2n} = -\frac{1}{N_{2n}^2}\sum_i \frac{d_{1i}}{\varepsilon_{1i}}\int\left[Y'_{11}Y'_{1i} - \left(\frac{1}{\xi^2} + k_{1i}^2\right)Y_{11}Y_{1i}\right]Y_{2n}\xi d\xi;$$

$$N_{mn}^2 = \int Y_{mn}^2 \xi d\xi.$$

By using solutions (6.5.2), (6.5.7), and (6.5.10), we define the values of $G_i$, $H_i$, $G_{ik}$, and $H_{ik}$ on the free surface $\Sigma_0$ from (6.3.5) and (6.3.10). These values are necessary for computing the coefficients of nonlinear equations. They have the form

$$G_3 = Y_{11}\left({Y'_{11}}^2 - \tfrac{1}{2}k_{11}^2 Y_{11}^2\right), \quad G_4 = \frac{1}{2}\frac{Y_{11}^3}{\xi^2},$$

$$G_5 = \sum_n \frac{c_{0n}}{\varkappa_{0n}}(Y'_{0n}Y'_{11} - k_{0n}^2 Y_{0n}Y_{11}), \quad G_6 = \sum_n \frac{c_{2n}}{\varkappa_{2n}}(Y'_{2n}Y'_{11} - k_{2n}^2 Y_{2n}Y_{11}),$$

$$G_7 = 2\sum_n \frac{c_{2n}}{\varkappa_{2n}} Y_{2n}Y_{11},$$

$$H_3 = \sum_n d_{1n}(2Y'_{1n}Y'_{11} - k_{1n}^2 Y_{1n}Y_{11})Y_{11}, \quad H_4 = \sum_n d_{1n}\frac{1}{\xi^2}Y_{1n}Y_{11}^2,$$

$$H_5 = \sum_n \frac{d_{0n}}{\varkappa_{0n}}(Y'_{0n}Y'_{11} - k_{0n}^2 Y_{0n}Y_{11}), \quad H_6 = \sum_n \frac{d_{2n}}{\varkappa_{2n}}(Y'_{2n}Y'_{11} - k_{2n}^2 Y_{2n}Y_{11}),$$

$$H_7 = \sum_n \frac{2d_{2n}}{\varkappa_{2n}}\frac{Y_{2n}Y_{11}}{\xi^2}, \tag{6.5.12}$$

$$G_{01} = \frac{1}{\varkappa_{11}}(Y'_{11}Y'_{01} - k_{11}^2 Y_{11}Y_{01}), \quad G_{10} = \frac{1}{\varkappa_{01}}(Y'_{01}Y'_{11} - k_{01}^2 Y_{01}Y_{11}),$$

$$G_{21} = \frac{1}{\varkappa_{11}}(Y'_{11}Y'_{21} - k_{11}^2 Y_{11}Y_{21}), \quad G_{12} = \frac{1}{\varkappa_{21}}(Y'_{21}Y'_{11} - k_{21}^2 Y_{21}Y_{11}),$$

$$G'_{21} = \frac{1}{\varkappa_{11}\xi^2}Y_{1n}Y_{21}, \quad G'_{12} = \frac{1}{\varkappa_{21}\xi^2}Y_{2n}Y_{11},$$

$$H'_{21} = \sum_n \frac{2d_{1n}}{\varepsilon_{1n}\xi^2}Y_{1n}Y_{21}, \quad H_{01} = \sum_n \frac{2d_{1n}}{\varepsilon_{1n}}(Y'_{1n}Y'_{01} - k_{1n}^2 Y_{1n}Y_{01}),$$

$$H_{21} = \sum_n \frac{2d_{1n}}{\varepsilon_{1n}}(Y'_{1n}Y'_{21} - k_{1n}^2 Y_{1n}Y_{21}).$$

In conclusion, we suggest expressions for main parameters that characterize nonlinear interaction of the rigid body with liquid and are included in the system of nonlinear equations of motion (6.4.14)–(6.4.16) as coefficients of this system:

$$\mu_1 = \frac{\pi}{4\varkappa_{11}}\rho\int\left\{\left[6{Y'_{11}}^2 + \left(\frac{2}{\xi^2} - 3k_{11}^2\right)Y_{11}^2\right]Y_{11} + \sum_n \frac{4c_{0n}}{\varkappa_{0n}}(Y'_{0n}Y'_{11} - k_{0n}^2 Y_{0n}Y_{11})\right.$$
$$\left. + \sum_n \frac{2c_{2n}}{\varkappa_{2n}}\left[Y'_{2n}Y'_{11} + \left(\frac{2}{\xi^2} - k_{2n}^2\right)Y_{2n}Y_{11}\right]\right\}\xi d\xi,$$

$$\mu_2 = \frac{\pi}{4\varkappa_{11}}\rho\int\left\{\left[2{Y'}_{11}^2 - \left(\frac{2}{\xi^2} + k_{11}^2\right)Y_{11}^2\right]Y_{11}\right.$$
$$\left. + \sum_n \frac{2c_{2n}}{\varkappa_{2n}}\left[Y'_{2n}Y'_{11} + \left(\frac{2}{\xi^2} - k_{2n}^2\right)Y_{2n}Y_{11}\right]\right\}Y_{11}\xi d\xi,$$

$$\lambda_1 = \frac{\pi}{4\varkappa_{11}}\rho\int\left\{\sum_n \frac{d_{1n}(\varepsilon_{1n} + \varkappa_{11})}{\varepsilon_{1n}}\left[6Y'_{1n}Y_{11} + \left(\frac{2}{\xi^2} - 3k_{1n}^2\right)Y_{1n}Y_{11}\right]Y_{11}\right.$$

$$+\sum_n \frac{4d_{0n}}{\varkappa_{0n}}(Y'_{0n}Y'_{11} + k_{0n}^2 Y_{0n}Y_{11})$$

$$-\sum_n \frac{2d_{2n}}{\varkappa_{2n}}\left[Y'_{2n}Y'_{11} + \left(\frac{2}{\xi^2} - k_{2n}^2\right) Y_{2n}Y_{11}\right]\Bigg\}Y_{11}\xi d\xi,$$

$$\lambda_2 = \frac{\pi}{4\varkappa_{11}}\rho\int\Bigg\{\sum_n \frac{d_{1n}(\varepsilon_{1n} + \varkappa_{11})}{\varepsilon_{1n}}\left[-2Y'_{1n}Y'_{11} + \left(\frac{2}{\xi^2} + k_{1n}^2\right) Y_{1n}Y_{11}\right]Y_{11}$$

$$+\sum_n \frac{2d_{2n}}{\varkappa_{2n}}\left[Y'_{2n}Y'_{11} + \left(\frac{2}{\xi^2} - k_{2n}^2\right) Y_{2n}Y_{11}\right]\Bigg\}Y_{11}\xi d\xi,$$

$$\mu_{10} = \frac{2\pi}{\varkappa_{01}}\rho\int Y_{01}^2\xi d\xi, \quad \mu_{12} = \frac{\pi}{\varkappa_{21}}\rho\int Y_{21}^2\xi d\xi,$$

$$\lambda_{10} = \pi\rho\int\left[\xi Y_{01} + \frac{1}{\varkappa_{11}}\sum_n \frac{2d_{1n}}{\varepsilon_{1n}}(Y'_{1n}Y'_{01} - k_{1n}^2 Y_{1n}Y_{01})\right]Y_{11}\xi d\xi,$$

$$\lambda'_{10} = \pi\rho\int\Bigg\{\xi Y_{11} + \frac{1}{\varkappa_{01}}\sum_n \frac{2d_{1n}}{\varepsilon_{1n}}\left[Y'_{1n}Y'_{11} + \left(\frac{1}{\xi^2} - k_{1n}^2\right) Y_{1n}Y_{11})\right]\Bigg\}Y_{11}\xi d\xi,$$

$$\lambda_{12} = \frac{\pi}{2}\rho\int\Bigg\{\xi Y_{21} + \frac{1}{\varkappa_{11}}\sum_n \frac{2d_{1n}}{\varepsilon_{1n}}\left[Y'_{1n}Y'_{21} + \left(\frac{2}{\xi^2} - k_{1n}^2\right) Y_{1n}Y_{21})\right]\Bigg\}Y_{11}\xi d\xi,$$

$$\lambda'_{12} = \frac{\pi}{2}\rho\int\Bigg\{Y_{11}\xi + \frac{1}{\varkappa_{21}}\sum_n \frac{2d_{1n}}{\varepsilon_{1n}}\left[Y'_{1n}Y'_{11} - \left(\frac{1}{\xi^2} + k_{1n}^2\right) Y_{1n}Y_{11})\right]\Bigg\}Y_{21}\xi d\xi,$$

$$\mu_{101} = \frac{\pi}{\varkappa_{11}^2}\rho\int[Y'_{11}Y'_{01} + (\varkappa_{11}^2 - k_{11}^2)Y_{11}Y_{01}]Y_{11}\xi d\xi, \tag{6.5.13}$$

$$\mu_{110} = \mu'_{110} = \frac{\pi}{\varkappa_{11}\varkappa_{01}}\rho\int(\varkappa_{11}\varkappa_{01}Y_{01}Y_{11} - Y'_{01}Y'_{11})Y_{11}\xi d\xi,$$

$$\mu_{121} = \frac{\pi}{2\varkappa_{11}^2}\rho\int\left[Y'_{11}Y'_{21} + \left(\frac{2}{\xi^2} + \varkappa_{11}^2 - k_{11}^2\right) Y_{11}Y_{21}\right]Y_{11}\xi d\xi,$$

$$\mu_{112} = \mu'_{112} = \frac{\pi\rho}{2\varkappa_{11}\varkappa_{21}}\int\left[\left(\varkappa_{11}\varkappa_{21} - \frac{2}{\xi^2}\right) Y_{11}^2 Y_{21} - Y_{11}Y'_{11}Y'_{21}\right]\xi d\xi,$$

$$J_{10} = \pi\rho\int\left[\xi^2 - \sum_n\sum_k \frac{4d_{1n}d_{1k}}{\varepsilon_{1n}\varepsilon_{1k}}\left(\frac{1}{\xi^2}Y_{1n}Y_{1k} + Y'_{1n}Y'_{1k}\right)\right]Y_{01}\xi d\xi,$$

$$J_{12} = \pi\rho\int\left[\sum_n\sum_k \frac{2d_{1n}d_{1k}}{\varepsilon_{1n}\varepsilon_{1k}}\left(Y'_{1n}Y'_{1k} - \frac{1}{\xi^2}Y_{1n}Y_{1k}\right) - \frac{\xi^2}{2}\right]Y_{21}\xi d\xi,$$

$$J_1 = \pi\rho\int\Bigg\{\sum_k \frac{2d_{1k}}{\varepsilon_{1k}}\Big[-\tfrac{1}{2}Y_{11}^2Y_{1k} - \tfrac{3}{4}Y_{11}^2Y'_{1k}\xi d\xi - \tfrac{3}{4}\sum_n d_{1n}\xi Y_{11}^2 Y'_{1n}Y'_{1k}$$

$$+\sum_n \frac{d_{2n}}{\varkappa_{2n}} \left( \frac{1}{\xi} Y_{2n} Y_{1k} + \frac{\xi}{2} Y'_{2n} Y'_{1k} \right) Y_{11} - \sum_n \frac{d_{0n}}{\varkappa_{0n}} Y_{11} Y'_{0n} Y'_{1k} \xi \Big] \Big\} d\xi$$

$$+(x_0 + h) \frac{\pi \rho}{4} \int \left( 1 + 3 \sum_n d_{1n} Y'_{1n} \right) Y_{11}^2 \xi d\xi,$$

$$J_2 = \pi \rho \int \Big\{ \sum_k \frac{2 d_{1k}}{\varepsilon_{1k}} \Big[ \left( -\tfrac{3}{2} Y_{1k} - \frac{\xi}{4} Y'_{1k} - \tfrac{1}{4} \sum_n d_{1n} Y'_{1n} Y'_{1k} \right) Y_{11}^2$$

$$+\sum_n \frac{d_{2n}}{\frac{1}{\xi} \varkappa_{2n}} \left( Y_{2n} Y_{1k} + \frac{\xi}{2} Y'_{2n} Y'_{1k} \right) Y_{11} \Big] \Big\} d\xi$$

$$+\frac{(x_0 + h)}{4} \pi \rho \int \left( 3 + \sum_n d_{1n} Y'_{1n} \right) Y_{11}^2 \xi d\xi.$$

In relations (6.5.13), we eliminate the errors found in published works devoted to the study of the nonlinear sloshing in cylindrical containers.

By comparing the equations of motion (6.4.14)–(6.4.16) with analogous equations from Chap. 4, we establish their complete structural coincidence except for the underlined terms that are absent in (6.4.14)–(6.4.16). However, these terms can be recovered by applying the method presented in this chapter provided that we have the relations for the inertia tensor given in Chap. 4 and, additionally, the quantities $\mu_{ij}$ and the projections of the vector $\boldsymbol{R}_i$ up to the terms of the third order of smallness. We also reveal the formal coincidence of coefficients of these systems, except for the coefficients $c_1$, $c_2$, $G_{22}^3$, and $G_{22}^4$. Nevertheless, numerical values of all corresponding coefficients computed by both analytical schemes are in a satisfactory agreement (up to the third significant digit).

## 6.6 The case of cavities of complex geometric shapes

The presented method for solving the nonlinear boundary-value problems of liquid sloshing in fixed or mobile cavities is based on the assumption of the cylindrical shape of a cavity in a neighborhood of the free surface. Since it is not often occurs in practice, we have to develop an analytical scheme for finite-dimensional models in more complicated cases.

Consider a class of cavities of revolution that can be parametrized by using relations (1.5.7) and (1.5.8) introduced in Chap. 1:

$$x = x^1, \quad y = \xi(x^1, x^2) \cos x^3, \quad z = \xi(x^1, x^2) \sin x^3, \tag{6.6.1}$$

$$x^1 = x, \quad x^2 = x^2(x, y, z), \quad x^3 = \text{arctg}\,(z/y)\,. \tag{6.6.2}$$

The nonperturbed liquid domain $Q_0$, which has the shape of a body of revolution, is transformed into a cylindrical domain $Q_0^*$ in the space of variables $x^1x^2x^3$. As it was mentioned above, the nonperturbed free surface $\Sigma_0$, which is the horizontal plane $x = h$, is transformed by relations (6.6.1) into a plane circular domain $\Sigma_0^*$ whose element is defined by

$$d\sigma_2^{(1)} = \left(\xi \frac{\partial \xi}{\partial x^2}\right)_{x^1=h} dx^2 dx^3. \tag{6.6.3}$$

The perturbed surface $\Sigma$ in the parameter space $x^1x^2x^3$ is transformed into a certain surface $\Sigma^*$ whose projection at any time coincides with the nonperturbed surface $\Sigma_0^*$. In this case, the equation of the perturbed free surface has the form

$$x^1 = x_0^1 + f(x^2, x^3, t) = x_0^1 + f_0(t) + \sum_i \beta_i(t) f_i(x^2, x^3), \tag{6.6.4}$$

where $x_0^1$ is the coordinate of the center of the nonperturbed surface $\Sigma_0^*$ on the $Ox^1$-axis; $f_0(t)$ is the mean value of the function $f(x^2, x^3, t)$ at time $t$; $\beta_i(t)$ are the Fourier coefficients that are beforehand unknown and play the role of generalized coordinates of the liquid motion; $f_i(x^2, x^3)$ is a system of functions defined on $\Sigma_0^*$, which is complete and orthogonal along with a constant:

$$\int_{\Sigma_0} f_i dS = \int_{\Sigma_0^*} f_i d\sigma_0^{(1)} = 0,$$

$$\int_{\Sigma_0} f_i f_j dS = \int_{\Sigma_0^*} f_i f_j d\sigma_0^{(1)} = \begin{cases} 0 \text{ for } i \neq j, \\ N_i^2 \text{ for } i = j. \end{cases} \tag{6.6.5}$$

Having defined constructively the free surface of a liquid by (6.6.4), we perform the subsequent construction of an analytical scheme of solving the nonlinear problems with the use of the perturbation theory methods based on their tensor form. In this procedure, the generalized Taylor series

$$F(x_0^1 + f, x^2, x^3, t) = F|_{x^1=x_0^1} + \nabla_1 F|_{x^1=x_0^1} f + \tfrac{1}{2}\nabla_{11} F|_{x^1=x_0^1} f^2 + \dots \tag{6.6.6}$$

plays the main role because it provides a possibility to reduce the original problem to a sequence of linear problems in a domain with fixed boundaries. Here, as it was mentioned above, the covariant derivatives $\nabla_i$ and $\nabla_{ij}$ are associated with the corresponding partial derivatives by the relations

$$\nabla_i F = \frac{\partial F}{\partial x^i}, \quad \nabla_{ij} = \frac{\partial^2 F}{\partial x^i x^j} - \Gamma_{ij}^{\nu} \frac{\partial F}{\partial x^\nu}. \tag{6.6.7}$$

We present the problem of free liquid oscillations in an invariant form by assuming that the free surface $\Sigma^*$ is described by the equation

$$\zeta(x^1, x^2, x^3, t) \equiv x^1 - x_0^1 - f(x^2, x^3, t) = 0. \tag{6.6.8}$$

The problem is formulated as follows:

$$\nabla^2\varphi = 0, \;\; x^1, x^2, x^3 \in Q^*, \;\; \left.\frac{\partial\varphi}{\partial\nu}\right|_{S^*} = 0, \;\; \left.\frac{\partial\varphi}{\partial\nu}\right|_{\Sigma^*} = -\left.\frac{\partial\zeta}{\partial t}\frac{1}{\sqrt{(\nabla\zeta)^2}}\right|_{\Sigma^*}, \tag{6.6.9}$$

where the action of the differential operator $\nabla$ and the Laplace operator $\nabla^2$ on any scalar function $\psi$ is understood as follows:

$$\nabla\psi = \boldsymbol{e}_i q^{ij}\frac{\partial\psi}{\partial x^j}, \quad \nabla^2\psi = q^{ij}\left(\frac{\partial^2\psi}{\partial x^i\partial x^j} - \Gamma^k_{ij}\frac{\partial\psi}{\partial x^k}\right),$$

$$\frac{\partial\psi}{\partial\nu} = \nabla\psi\cdot\boldsymbol{\nu} = q^{ij}\frac{\partial\psi}{\partial x^j}\nu_i. \tag{6.6.10}$$

The condition for solvability of boundary-value problem (6.6.9) is

$$\int\limits_{\Sigma^*}\frac{\partial\zeta}{\partial t}\frac{dS}{\sqrt{(\nabla\zeta)^2}} = -\int\limits_{\Sigma^*_0}\left(\xi\frac{\partial\xi}{\partial x^2}\right)_0\frac{\partial f}{\partial t}dx^2dx^3 = 0, \tag{6.6.11}$$

where the subscript "0" means that the expression in brackets is taken for $x^1 = x^1_0 + f(x^2, x^3, t)$.

In what follows, we assume that parameters $\beta_k(t)$ and $\beta_l(t)$ play an essential role among all parameters that characterize small but finite deformations of free surface (6.6.4). Deriving equations of the liquid motion, we restrict our consideration by the terms of the third order of smallness in $\beta_k(t)$ and $\beta_l(t)$ and assume that the terms $\beta_k^2(t)$ and $\beta_l^2(t)$ have the order of smallness of other parameters that characterize the position of the free surface. (The necessity to consider two parameters $\beta_k$ and $\beta_l$, which are not small, occurs in some practically important problems, e.g., when we have to solve the problem of spatial motion of a liquid in a container that performs a harmonic oscillation.) In what follows, it is convenient to present the kinematic boundary condition on the free surface in the form

$$\left.\frac{\partial\varphi}{\partial\nu}\right|_{\Sigma^*} = \left.\frac{f_t}{N}\right|_{\Sigma^*}, \quad N = \sqrt{(\nabla x^1 - \nabla f)^2}. \tag{6.6.12}$$

Further, we use condition (6.6.11) in order to compute the mean value $f_0(t)$ of the function $f(x^2, x^3, t)$ at time $t$ and, taking into account the orthogonality of the system of functions $f_i(x^2, x^3)$ and the order of smallness of the parameters that characterize deformation of the free surface, we obtain

$$f_0(t) = \beta_k^2 e_{kk} + \beta_l^2 e_{ll} + \beta_k^3\varepsilon_{kkk} + \beta_k^2\beta_l\varepsilon_{kkl} + \beta_k\beta_l^2\varepsilon_{kll} + \beta_l^3\varepsilon_{lll}, \tag{6.6.13}$$

where the constants $e_{kk}$, $e_{ll}$, …, $\varepsilon_{lll}$ are defined by the relations

$$e_{kk} = -\frac{\delta_1}{S_0}\int\limits_{\Sigma^*_0} f_k^2 d\sigma_0^{(1)}, \quad e_{ll} = -\frac{\delta_1}{S_0}\int\limits_{\Sigma^*_0} f_l^2 d\sigma_0^{(1)},$$

$$\varepsilon_{kkk} = -\frac{\delta_2}{S_0}\int\limits_{\Sigma_0^*} f_k^2 d\sigma_0^{(1)}, \quad \varepsilon_{kkl} = -\frac{3\delta_2}{S_0}\int\limits_{\Sigma_0^*} f_k^2 f_l d\sigma_0^{(1)}, \tag{6.6.14}$$

$$\varepsilon_{kll} = -\frac{3\delta_2}{S_0}\int\limits_{\Sigma_0^*} f_k f_l^2 d\sigma_0^{(1)}, \quad \varepsilon_{lll} = -\frac{\delta_2}{S_0}\int\limits_{\Sigma_0^*} f_l^3 d\sigma_0^{(1)}.$$

The constants $\delta_1$ and $\delta_2$ are defined by the geometric parameters of the container. In view of (6.6.13), by using (6.6.4) we have

$$f_t = \sum_{i=1}^{\infty} \dot{\beta}_i [f_i + \delta_{ik} F_{ik}(\beta_k, \beta_l) + \delta_{il} F_{il}(\beta_k, \beta_l)], \tag{6.6.15}$$

where

$$F_{ik} = 2e_{ik}\beta_k + 3\varepsilon_{ikk}\beta_k^2 + 2\varepsilon_{ikl}\beta_k\beta_l + \varepsilon_{ill}\beta_l^2,$$
$$F_{il} = 2e_{il}\beta_l + \varepsilon_{ikk}\beta_k^2 + 2\varepsilon_{ikl}\beta_k\beta_l + 3\varepsilon_{ill}\beta_l^2, \tag{6.6.16}$$
$$\delta_{ik} = \begin{cases} 1 & \text{for } i = k, \\ 0 & \text{for } i \neq k, \end{cases} \qquad \delta_{il} = \begin{cases} 1 & \text{for } i = l, \\ 0 & \text{for } i \neq l. \end{cases}$$

By considering relations (6.6.12) and (6.6.15), we present the velocity potential $\varphi$ in the form

$$\varphi = \sum_{i=1}^{\infty} \dot{\beta}_i \varphi_i, \tag{6.6.17}$$

where the system of functions $\varphi_i$, which depends on time and spatial coordinates, is defined by the solution of the boundary-value problem

$$\nabla^2 \varphi_i = 0; \quad x^1, x^2, x^3 \in Q^*, \quad \left.\frac{\partial \varphi_i}{\partial \nu}\right|_{S^*} = 0; \tag{6.6.18}$$

$$\left.\frac{\partial \varphi_i}{\partial \nu}\right|_{\Sigma^*} = \frac{1}{N}[f_i + \delta_{ik} F_{ik} + \delta_{il} F_{il}].$$

Since the right-hand side of the boundary condition for $\Sigma^*$ in the boundary-value problem (6.6.18) depends on the parameters $\beta_i(t)$ and, generally speaking, can be presented by a certain polynomial in $\beta_i(t)$, we assume that the function $\varphi_i$ admits an expansion in the form of a series in powers of $\beta_i(t)$. Keeping the terms of the third order of smallness in $\beta_k$ and $\beta_l$ in the relation for the velocity potential (6.6.17), we seek the function $\varphi_i(x^1, x^2, x^3, t)$ in the form

$$\varphi_i = \varphi_{i0} + \sum_{j=1}^{\infty} \beta_j \varphi_{ij} + \beta_k^2 \varphi_{ik}^{(1)} + \beta_k \beta_l \varphi_{ikl}^{(1)} + \beta_l^2 \varphi_{il}^{(1)}. \tag{6.6.19}$$

Therefore, the velocity potential $\varphi$ is completely defined by the harmonic functions $\varphi_{i0}, \varphi_{ij}, \ldots, \varphi_{il}^{(1)}$ that depend only on spatial coordinates. To construct these functions, we use the generalized Taylor series (6.6.6) whose Christoffel symbols are defined by the relations

$$\Gamma_{11}^1 = 0, \quad \Gamma_{11}^2 = \frac{\partial^2 \xi}{\partial x^{1\,2}} \Big/ \frac{\partial \xi}{\partial x^2}, \quad \Gamma_{11}^3 = 0. \tag{6.6.20}$$

By applying relation (6.6.6) to the boundary conditions of the boundary-value problem (6.6.18) in view of expressions (6.6.4) and (6.6.19) and equating the coefficients at the same powers of the parameters $\beta$ to zero, we obtain the following sequence of the boundary-value problems with respect to the functions $\varphi_{i0}$, $\varphi_{ij}$, $\overset{(1)}{\varphi}_{ik}$, $\overset{(1)}{\varphi}_{il}$, and $\overset{(1)}{\varphi}_{ikl}$:

$$\nabla^2 \varphi_{i0} = 0, \quad x^1, x^2, x^3 \in Q^*, \quad \left.\frac{\partial \varphi_{i0}}{\partial \nu}\right|_{S^*} = 0, \quad \left.\frac{\partial \varphi_{i0}}{\partial \nu}\right|_{\Sigma_0^*} = f_i; \tag{6.6.21}$$

$$\nabla^2 \varphi_{ij} = 0, \quad x^1, x^2, x^3 \in Q^*, \quad \left.\frac{\partial \varphi_{ij}}{\partial \nu}\right|_{S^*} = 0,$$

$$\left.\frac{\partial \varphi_{ij}}{\partial \nu}\right|_{\Sigma_0^*} = (\nabla \varphi_{i0}, \nabla f_j) - f_j \nabla_1 (\nabla \varphi_{i0}, \nabla x^1) + 2\delta_{ij} e_{ij}; \tag{6.6.22}$$

$$\nabla^2 \overset{(1)}{\varphi}_{1k} = 0, \quad x^1, x^2, x^3 \in Q^*, \quad \left.\frac{\partial \overset{(1)}{\varphi}_{ik}}{\partial \nu}\right|_{S^*} = 0,$$

$$\left.\frac{\partial \overset{(1)}{\varphi}_{ik}}{\partial \nu}\right|_{\Sigma_0^*} = (\nabla \varphi_{ik}, \nabla f_k) + f_k \nabla_1 (\nabla \varphi_{i0}, \nabla f_k) - \tfrac{1}{2} f_k^2 \nabla_{11} (\nabla \varphi_{i0}, \nabla x^1)$$
$$- f_k \nabla_1 (\nabla \varphi_{ik}, \nabla x^1) - e_{kk} \nabla_1 (\nabla \varphi_{i0}, \nabla x^1) + 3\delta_{ik} \varepsilon_{ikk} + \delta_{il} \varepsilon_{ikk}; \tag{6.6.23}$$

$$\nabla^2 \overset{(1)}{\varphi}_{1kl} = 0, \quad x^1, x^2, x^3 \in Q^*, \quad \left.\frac{\partial \overset{(1)}{\varphi}_{ikl}}{\partial \nu}\right|_{S^*} = 0,$$

$$\left.\frac{\partial \overset{(1)}{\varphi}_{ikl}}{\partial \nu}\right|_{\Sigma_0^*} = (\nabla \varphi_{ik}, \nabla f_l) + (\nabla \varphi_{il}, \nabla f_k) + f_l \nabla_1 (\nabla \varphi_{i0}, \nabla f_k) + f_k \nabla_1 (\nabla \varphi_{i0}, \nabla f_l)$$
$$- f_l \nabla_1 (\nabla \varphi_{ik}, \nabla x^1) - f_k \nabla_1 (\nabla \varphi_{il}, \nabla x^1) - f_k f_l \nabla_{11} (\nabla \varphi_{i0}, \nabla x^1)$$
$$- 2 e_{lk} \nabla_1 (\nabla \varphi_{i0}, \nabla x^1) + 2(\delta_{ik} \varepsilon_{ikl} + \delta_{il} \varepsilon_{kli}); \tag{6.6.24}$$

$$\nabla^2 \overset{(1)}{\varphi}_{il} = 0, \quad x^1, x^2, x^3 \in Q^*, \quad \left.\frac{\partial \overset{(1)}{\varphi}_{il}}{\partial \nu}\right|_{S^*} = 0,$$

$$\left.\frac{\partial \overset{(1)}{\varphi}_{il}}{\partial \nu}\right|_{\Sigma_0^*} = (\nabla \varphi_{il}, \nabla f_l) + f_l \nabla_1 (\nabla \varphi_{i0}, \nabla f_l) - \tfrac{1}{2} f_l^2 \nabla_{11} (\nabla \varphi_{i0}, \nabla x^1)$$
$$- f_l \nabla_1 (\nabla \varphi_{il}, \nabla x^1) - e_{ll} \nabla_1 (\nabla \varphi_{i0}, \nabla x^1) + 3\delta_{il} \varepsilon_{ill} + \delta_{ik} \varepsilon_{ill}. \tag{6.6.25}$$

Thus, we obtain a sequence of linear boundary-value problems (6.6.21)–(6.6.25) that have an advantage over the boundary-value problem (6.6.9) because, in this case, the domain of definition of the required functions is a domain with fixed boundaries. Up to this point, we assumed that the system of functions $f_i$ is known. Now, we can define it by solving the following eigenvalue boundary problem:

$$\nabla^2 \varphi_{i0} = 0, \quad x^1,\, x^2,\, x^3 \in Q^*, \quad \left.\frac{\partial \varphi_{i0}}{\partial \nu}\right|_{\Sigma_0^*} = \varkappa_i \varphi_{i0}, \quad \left.\frac{\partial \varphi_{i0}}{\partial \nu}\right|_{S^*} = 0, \tag{6.6.26}$$

i.e., $f_i = \varkappa_i \varphi_{i0}\Big|_{\Sigma_0^*}$.

To determine the velocity potential in the form (6.6.17), (6.6.19) and the free surface (6.6.4) completely, it is necessary to find the time-dependent parameters $\beta_i(t)$ that characterize the liquid motion. To this end, we use the dynamical condition on the free surface. To find the value of the operator

$$L(\Phi) = \frac{\partial \Phi}{\partial t} + \tfrac{1}{2}(\nabla \Phi)^2 + \mathrm{g}x = 0 \tag{6.6.27}$$

on the perturbed free surface, we use the Taylor formula (6.6.6) again and apply it to the function $L(\Phi)$. The expression for the function $L(\Phi)|_{\Sigma^*}$, which is obtained as a result, depends only on the variables $x^2$ and $x^3$. We expand this expression in a generalized Fourier series with respect to the system of functions $f_i$ on $\Sigma_0^*$. Since the system of functions is orthogonal on $\Sigma_0^*$ to an arbitrary constant, the coefficients of this expansion are equal to zero, i.e.,

$$\int\limits_{\Sigma_0^*} L(\Phi) f_i \, d\sigma_0^{(1)} = 0, \quad i = 1,\, 2,\, \ldots . \tag{6.6.28}$$

This condition leads to the following system of nonlinear ordinary differential equations for the parameters $\beta_i(t)$:

$$\begin{aligned}
&\mu_i(\ddot\beta_i + \sigma_i^2 \beta_i) + \sum_{j=1}^{\infty}{}'' \big[(\dot\beta_j \beta_k)^{\cdot} a_{jk}^{(i)} + (\dot\beta_j \beta_l)^{\cdot} a_{jl}^{(i)} + (\dot\beta_k \beta_j)^{\cdot} a_{kj}^{(i)} + (\dot\beta_l \beta_j)^{\cdot} a_{lj}^{(i)}\big] \\
&\quad + (\dot\beta_k \beta_l)^{\cdot} a_{kl}^{(i)} + (\dot\beta_k \beta_k)^{\cdot} a_{kk}^{(i)} + (\dot\beta_l \beta_l)^{\cdot} a_{ll}^{(i)} + (\dot\beta_l \beta_k)^{\cdot} a_{lk}^{(i)} + (\dot\beta_k \beta_k^2)^{\cdot} b_{kk}^{(i)} \\
&\quad + (\dot\beta_k \beta_k \beta_l)^{\cdot} b_{kkl}^{(i)} + (\dot\beta_k \beta_l^2)^{\cdot} b_{kl}^{(i)} + (\dot\beta_l \beta_k^2)^{\cdot} b_{lk}^{(i)} + (\dot\beta_l \beta_k \beta_l)^{\cdot} b_{lkl}^{(i)} + (\dot\beta_l \beta_l^2)^{\cdot} b_{ll}^{(i)} \\
&\quad + c_{kk}^{(i)} \dot\beta_k^2 + c_{kl}^{(i)} \dot\beta_k \dot\beta_l + c_{ll}^{(i)} \dot\beta_l^2 + \dot\beta_k \sum_{j=1}^{\infty}{}'' c_{kj}^{(i)} \dot\beta_j + \dot\beta_l \sum_{j=1}^{\infty}{}'' c_{lj}^{(i)} \dot\beta_j + d_{kk}^{(i)} \dot\beta_k^2 \beta_k \\
&\quad + d_{kl}^{(i)} \dot\beta_k^2 \beta_l + d_{lk}^{(i)} \dot\beta_l^2 \beta_k + d_{ll}^{(i)} \dot\beta_l^2 \beta_l + d_{klk}^{(i)} \dot\beta_k \dot\beta_l \beta_k + d_{kll}^{(i)} \dot\beta_k \dot\beta_l \beta_l = 0.
\end{aligned} \tag{6.6.29}$$

Primes at the summation sign mean that the summation should be performed with omitted indices $k$ and $l$. The coefficients of the system of differential equations (6.6.29) are defined by the relations

$$\mu_i = \varkappa_i \int\limits_{\Sigma_0^*} \varphi_{i0}^2 d\sigma_0^{(1)}, \quad \sigma_i^2 = g\varkappa_i, \quad a_{jk}^{(i)} = \int\limits_{\Sigma_0^*} (\varphi_{jk} + f_k \nabla_1 \varphi_{j0}) f_i d\sigma_0^{(1)},$$

$$a_{jl}^{(i)} = \int\limits_{\Sigma_0^*} (\varphi_{ji} + f_l \nabla_1 \varphi_{j0}) f_i d\sigma_0^{(1)}, \quad a_{kj}^{(i)} = \int\limits_{\Sigma_0^*} (\varphi_{kj} + f_j \nabla_1 \varphi_{k0}) f_i d\sigma_0^{(1)},$$

$$a_{lj}^{(i)} = \int\limits_{\Sigma_0^*} (\varphi_{lj} + f_j \nabla_1 \varphi_{l0}) f_i d\sigma_0^{(1)},$$

$$b_{kk}^{(i)} = \int\limits_{\Sigma_0^*} \left(\varphi_{kk}^{(1)} + f_k \nabla_1 \varphi_{kk} + \tfrac{1}{2} f_k^2 \nabla_{11} \varphi_{k0} + e_{kk} \nabla_1 \varphi_{k0}\right) f_i d\sigma_0^{(1)},$$

$$b_{kkl}^{(i)} = \int\limits_{\Sigma_0^*} \left(\varphi_{kkl}^{(1)} + f_l \nabla_1 \varphi_{kk} + f_k \nabla_1 \varphi_{kl} + f_k f_l \nabla_{11} \varphi_{k0}\right) f_i d\sigma_0^{(1)},$$

$$b_{kl}^{(i)} = \int\limits_{\Sigma_0^*} \left(\varphi_{kl}^{(1)} + f_l \nabla_1 \varphi_{kl} + \tfrac{1}{2} f_l^2 \nabla_{11} \varphi_{k0} + e_{ll} \nabla_1 \varphi_{k0}\right) f_i d\sigma_0^{(1)},$$

$$b_{lk}^{(i)} = \int\limits_{\Sigma_0^*} \left(\varphi_{lk}^{(1)} + f_k \nabla_1 \varphi_{lk} + \tfrac{1}{2} f_k^2 \nabla_{11} \varphi_{l0} + e_{kk} \nabla_1 \varphi_{l0}\right) f_i d\sigma_0^{(1)},$$

$$b_{lkl}^{(i)} = \int\limits_{\Sigma_0^*} \left(\varphi_{lkl}^{(1)} + f_l \nabla_1 \varphi_{lk} + f_k \nabla_1 \varphi_{ll} + f_k f_l \nabla_{11} \varphi_{l0}\right) f_i d\sigma_0^{(1)},$$

$$b_{ll}^{(i)} = \int\limits_{\Sigma_0^*} \left(\varphi_{ll}^{(1)} + f_l \nabla_1 \varphi_{ll} + \tfrac{1}{2} f_l^2 \nabla_{11} \varphi_{l0} + e_{ll} \nabla_1 \varphi_{l0}\right) f_i d\sigma_0^{(1)}, \qquad (6.6.30)$$

$$c_{kk}^{(i)} = \frac{1}{2} \int\limits_{\Sigma_0^*} [(\nabla \varphi_{k0})^2 - 2 f_k \nabla_1 \varphi_{k0}] f_i d\sigma_0^{(1)},$$

$$c_{kl}^{(i)} = \int\limits_{\Sigma_0^*} [(\nabla \varphi_{k0}, \nabla \varphi_{l0}) - f_l \nabla_1 \varphi_{k0} - f_k \nabla_1 \varphi_{l0}] f_i d\sigma_0^{(1)},$$

$$c_{ll}^{(i)} = \frac{1}{2} \int\limits_{\Sigma_0^*} [(\nabla \varphi_{l0})^2 - 2 f_l \nabla_1 \varphi_{l0}] f_i d\sigma_0^{(1)},$$

$$c_{kj}^{(i)} = \int\limits_{\Sigma_0^*} [(\nabla \varphi_{k0}, \nabla \varphi_{j0}) - f_k \nabla_1 \varphi_{j0} - f_j \nabla_1 \varphi_{k0}] f_i d\sigma_0^{(1)},$$

$$c_{lj}^{(i)} = \int\limits_{\Sigma_0^*} [(\nabla\varphi_{l0}, \nabla\varphi_{j0}) - f_l\nabla_1\varphi_{j0} - f_j\nabla_1\varphi_{l0}]f_i d\sigma_0^{(1)},$$

$$d_{kk}^{(i)} = \int\limits_{\Sigma_0^*} [(\nabla\varphi_{k0}, \nabla\varphi_{kk}) + (\nabla\varphi_{k0}, \nabla_1(\nabla\varphi_{k0}))f_k$$
$$- f_k\nabla_1\varphi_{kk} - f_k^2\nabla_{11}\varphi_{k0} - 2e_{kk}\nabla_1\varphi_{k0}]f_i d\sigma_0^{(1)},$$

$$d_{kl}^{(i)} = \int\limits_{\Sigma_0^*} [(\nabla\varphi_{k0}, \nabla\varphi_{kl}) + (\nabla\varphi_{k0}, \nabla_1(\nabla\varphi_{k0}))f_l - f_k\nabla_1\varphi_{kl} - f_k f_l\nabla_{11}\varphi_{k0}]f_i d\sigma_0^{(1)},$$

$$d_{lk}^{(i)} = \int\limits_{\Sigma_0^*} [(\nabla\varphi_{l0}, \nabla\varphi_{lk}) + (\nabla\varphi_{l0}, \nabla_1(\nabla\varphi_{l0}))f_k - f_l\nabla_1\varphi_{lk} - f_l f_k\nabla_{11}\varphi_{l0}]f_i d\sigma_0^{(1)},$$

$$d_{ll}^{(i)} = \int\limits_{\Sigma_0^*} [(\nabla\varphi_{l0}, \nabla\varphi_{ll}) + (\nabla\varphi_{l0}, \nabla_1(\nabla\varphi_{l0}))f_l$$
$$- f_l\nabla_1\varphi_{ll} - f_l^2\nabla_{11}\varphi_{l0} - 2e_{ll}\nabla_1\varphi_{l0}]f_i d\sigma_0^{(1)},$$

$$d_{klk}^{(i)} = \int\limits_{\Sigma_0^*} [(\nabla\varphi_{k0}, \nabla\varphi_{lk}) + (\nabla\varphi_{l0}, \nabla\varphi_{kk}) + (\nabla\varphi_{k0}, \nabla_1(\nabla\varphi_{l0}))f_k$$
$$+ (\nabla\varphi_{l0}, \nabla_1(\nabla\varphi_{k0}))f_k - f_l\nabla_1\varphi_{kk} - f_k\nabla_1\varphi_{lk}$$
$$- f_k f_l\nabla_{11}\varphi_{k0} - f_k^2\nabla_{11}\varphi_{l0} - 2e_{kk}\nabla_1\varphi_{l0}]f_i d\sigma_0^{(1)},$$

$$d_{kll}^{(i)} = \int\limits_{\Sigma_0^*} [(\nabla\varphi_{k0}, \nabla\varphi_{ll}) + (\nabla\varphi_{l0}, \nabla\varphi_{kl}) + (\nabla\varphi_{k0}, \nabla_1(\nabla\varphi_{l0}))f_l$$
$$+ (\nabla\varphi_{l0}, \nabla_1(\nabla\varphi_{k0}))f_l - f_k\nabla_1\varphi_{ll} - f_l\nabla_1\varphi_{kl}$$
$$- f_k f_l\nabla_{11}\varphi_{l0} - f_l^2\nabla_{11}\varphi_{k0} - 2e_{ll}\nabla_1\varphi_{k0}]f_i d\sigma_0^{(1)}.$$

By using the Green formula, these coefficients can be presented in another form. For example, integrals of the type

$$\int\limits_{\Sigma_0^*} \varphi_{kl}^{(1)} f_i d\sigma_0^{(1)}$$

can be presented in the form

$$\int\limits_{\Sigma_0^*} \varphi_{kl}^{(1)} f_i d\sigma_0^{(1)} = \int\limits_{\Sigma_0^*} \varphi_{kl}^{(1)} \frac{\partial\varphi_{i0}}{\partial\nu} d\sigma_0^{(1)} = \int\limits_{\Sigma_0^*} \varphi_{i0} \frac{\partial\varphi_{kl}^{(1)}}{\partial\nu} d\sigma_0^{(1)}.$$

With this presentation, in order to obtain coefficients (6.6.30), it is not necessary to construct actual expressions for the functions $\varphi_{kl}^{(1)}$, $\varphi_{lk}^{(1)}$, $\varphi_{ll}^{(1)}$, $\varphi_{lkl}^{(1)}$, etc., it is sufficient to use only the values of normal derivatives of these functions on $\Sigma_0^*$. Therefore, to obtain the complete set of coefficients of equations of motion (6.6.29), it suffices to solve the boundary-value problems (6.6.22) and (6.6.26). At present, fairly efficient methods are developed for solving these boundary-value problems. Since the main results for cavities of revolution are obtained in terms of the cylindrical coordinate system, it makes sense at this stage of our study to rewrite accordingly the boundary conditions of the boundary-value problems (6.6.22) and (6.6.26) and the expressions for the coefficients of equations of motion. The main relations used for this transition are the following:

$$
\begin{aligned}
&\frac{\partial\varphi}{\partial x^1}=\frac{\partial\varphi}{\partial x}+n_1\frac{\partial\varphi}{\partial\xi},\quad \frac{\partial\varphi}{\partial x^2}=n\frac{\partial\varphi}{\partial\xi},\quad \frac{\partial\varphi}{\partial x^3}=\frac{\partial\varphi}{\partial\eta},\\
&(\nabla\varphi,\nabla f)=q^{ij}\frac{\partial\varphi}{\partial x^i}\frac{\partial f}{\partial x^j}=-n_1\frac{\partial\varphi}{\partial x}\frac{\partial\varphi}{\partial\xi}+\frac{\partial\varphi}{\partial\xi}\frac{\partial f}{\partial\xi}+\frac{1}{\xi^2}\frac{\partial f}{\partial\eta}\frac{\partial\varphi}{\partial\eta},\\
&\frac{\partial}{\partial x_1}(\nabla\varphi,\nabla x^1)=\frac{\partial^2\varphi}{\partial x^2}+n_1\frac{\partial^2\varphi}{\partial x\partial\xi},\\
&\nabla_{11}\varphi=\frac{\partial^2\varphi}{\partial x^2}+2n_1\frac{\partial^2\varphi}{\partial x\partial\xi}+n_1^2\frac{\partial^2\varphi}{\partial\xi^2},\\
&(\nabla\varphi,\nabla\psi)=q^{ij}\frac{\partial\varphi}{\partial x^i}\frac{\partial\psi}{\partial x^j}=\frac{\partial\varphi}{\partial x}\frac{\partial\psi}{\partial x}+\frac{\partial\varphi}{\partial\xi}\frac{\partial\psi}{\partial\xi}+\frac{1}{\xi^2}\frac{\partial f}{\partial\eta}\frac{\partial\psi}{\partial\eta},\\
&(\nabla\varphi,\nabla_1(\nabla\varphi))=\frac{\partial^2\varphi}{\partial x^2}\frac{\partial\varphi}{\partial x}+n_1\frac{\partial^2\varphi}{\partial x\partial\xi}\frac{\partial\varphi}{\partial x}+\frac{\partial\varphi}{\partial\xi}\frac{\partial^2\varphi}{\partial x\partial\xi}\\
&\qquad+n_1\frac{\partial^2\varphi}{\partial\xi^2}\frac{\partial\varphi}{\partial\xi}+\frac{1}{\xi^2}\frac{\partial\varphi}{\partial\eta}\left(\frac{\partial^2\varphi}{\partial x\partial\eta}+n_1\frac{\partial^2\varphi}{\partial\xi\partial\eta}-\frac{n_1}{\xi}\frac{\partial\varphi}{\partial\eta}\right),\\
&\nabla_1(\nabla\varphi,\nabla f)=\frac{\partial}{\partial x^1}\left(q^{ij}\frac{\partial\varphi}{\partial x^i}\frac{\partial f}{\partial x^j}\right)\\
&\quad=\left[-n_1\frac{\partial^2\varphi}{\partial x^2}+(1-n_1^2)\frac{\partial^2\varphi}{\partial x\partial\xi}+n_1\frac{\partial^2\varphi}{\partial\xi^2}-n_2\frac{\partial\varphi}{\partial\xi}-n_3\frac{\partial\varphi}{\partial x}\right]\frac{\partial f}{\partial\xi}\\
&\qquad-\frac{2n_1}{\xi^3}\frac{\partial\varphi}{\partial\eta}\frac{\partial f}{\partial\eta}+\frac{1}{\xi^2}\left(\frac{\partial^2\varphi}{\partial x\partial\eta}+n_1\frac{\partial^2\varphi}{\partial\xi\partial\eta}\right)\frac{\partial f}{\partial\eta},\\
&\nabla_{11}(\nabla\varphi,\nabla x^1)=\frac{\partial^3\varphi}{\partial x^3}+2n_1\frac{\partial^3\varphi}{\partial x^2\partial\xi}+n_1^2\frac{\partial^3\varphi}{\partial x\partial\xi^2}\\
&\qquad+2n_3\frac{\partial^2\varphi}{\partial x\partial\xi}+2n_1n_3\frac{\partial^2\varphi}{\partial\xi^2}+2n_2n_3\frac{\partial\varphi}{\partial\xi},
\end{aligned}
\tag{6.6.31}
$$

where

$$n = \frac{\partial \xi}{\partial x^2}, \quad n_1 = \frac{\partial \xi}{\partial x^1}, \quad n_2 = \frac{\partial^2 \xi}{\partial x^1 \partial x^2} \frac{\partial x^2}{\partial \xi}, \quad n_3 = \frac{\partial^2 \xi}{\partial x^{1 2}} - n_1 n_2. \tag{6.6.32}$$

Relations (6.6.31) will be used later in order to derive expressions for the coefficients of equations of motion in some specific cases.

We give a more detailed analysis of the obtained system of nonlinear equations of motion (6.6.29) by using spectral properties of the eigenvalue boundary problem (6.6.26) and specifying the sense of the parameters $\beta_k$ and $\beta_l$, which are not small. As it is known, for containers that have the shape of bodies of revolution, a solution of the boundary-value problem (6.6.26) can be written in the cylindrical coordinate system in the form $\varphi_{i0} = \psi_{mn}(x, \xi) f_m(\eta)$. As above, we divide the collection of all eigenfunctions of the boundary-value problem (6.6.26) into four subsystems that completely exhaust this collection:

$$\begin{aligned}
\varphi_{i0}^{(\alpha)} &= \psi_{mn}(x, \xi) \sin m\eta, \quad m = 1, 3, 5, \dots, \\
\varphi_{i0}^{(\beta)} &= \psi_{mn}(x, \xi) \cos m\eta, \quad m = 1, 3, 5, \dots, \\
\varphi_{i0}^{(s)} &= \psi_{mn}(x, \xi) \cos m\eta, \quad m = 0, 2, 4, \dots, \\
\varphi_{i0}^{(q)} &= \psi_{mn}(x, \xi) \sin m\eta, \quad m = 0, 2, 4, \dots .
\end{aligned} \tag{6.6.33}$$

According to (6.6.33), the equation of the perturbed free surface (6.6.4) can be presented as follows:

$$x^1 - x_0^1 = f_0(t) + \sum_{i=1}^{\infty} \alpha_i(t) f_{\alpha i} + \sum_{i=1}^{\infty} \beta_i(t) f_{\beta i} + \sum_{i=1}^{\infty} s_i(t) f_{si} + \sum_{i=1}^{\infty} q_i(t) f_{qi}, \tag{6.6.34}$$

where the parameters $\alpha_i(t)$, $\beta_i(t)$, $s_i(t)$, and $q_i(t)$ characterize the deformation of the free surface in time for the natural sloshing modes that possess certain symmetry and antisymmetry properties in spatial coordinates. An analysis of the coefficients of equations (6.6.29) shows that they become equal to zero after substituting solutions of the boundary-value problems (6.6.22)–(6.6.26) for all indices $m$ different from $m = 0, 1, 2$.

For further study, we determine the parameters $\beta_k$ and $\beta_l$ by assuming that their sense is the same as the sense of the parameters $\alpha_1(t)$ and $\beta_1(t)$ in (6.6.34). The obtained nonlinear equations of motion provide a possibility to investigate the motion of a liquid in containers of considered types in a neighborhood of the main resonance of the parameters $\alpha_1(t)$ and $\beta_1(t)$. We now define the general form of solutions $\varphi_{kk} = \varphi_{11}^{(\alpha)}$, $\varphi_{ll} = \varphi_{11}^{(p)}$, and $\varphi_{kl} = \varphi_{11}^{(\alpha\beta)}$ of the boundary-value problem (6.6.22), which are encountered in what follows. By substituting $\varphi_{11}^{(\alpha)} = \psi_{11}(x, r) \sin \eta$ and $f_{\alpha_1} = \varkappa_{11} \psi_{11}(x, \xi)|_{x=x_0} \sin \eta$ into the boundary condition on the free surface of problem (6.6.22), in view of relations (6.6.31) we obtain

$$\frac{\partial \varphi_{11}^{(\alpha)}}{\partial \nu} = \tfrac{1}{2}\varkappa_{11}F_1(\xi) - \tfrac{1}{2}\varkappa_{11}F(\xi)\cos 2\eta + 2e_{11}, \tag{6.6.35}$$

where

$$F_1(\xi) = \left[-n_1\frac{\partial\psi_{11}}{\partial x}\frac{\partial\psi_{11}}{\partial\xi} + \left(\frac{\partial\psi_{11}}{\partial\xi}\right)^2 + \frac{1}{\xi^2}\psi_{11}^2 - \psi_{11}\left(\frac{\partial^2\psi_{11}}{\partial x^2} + n\frac{\partial^2\psi_{11}}{\partial x\partial\xi}\right)\right]_{x=x_0},$$

$$F(\xi) = F_1(\xi) - \frac{2}{\xi^2}\psi_{11}^2\bigg|_{x=x_0}, \quad e_{11} = -\frac{\delta_1\varkappa_{11}^2}{r_0^2}\int\limits_0^{r_0}\xi\psi_{11}^2\,d\xi \tag{6.6.36}$$

($r_0$ is the radius of the nonperturbed free surface). By using (6.6.35), we write the solution of the boundary-value problem for the function $\varphi_{11}^{(\alpha)}$ in the form

$$\varphi_{11}^{(\alpha)} = \Psi_0(x,\xi) - \Psi_2(x,\xi)\cos 2\eta, \tag{6.6.37}$$

where the functions $\Psi_0$ and $\Psi_2$ are defined as solutions of the following boundary-value problems:

$$L_0(\Psi_0) = 0, \quad x,\,\xi \in G, \quad m = 0,$$

$$\frac{\partial\Psi_0}{\partial\nu}\bigg|_{L_0} = \tfrac{1}{2}\varkappa_{11}F_1(\xi) + 2e_{11}, \quad \frac{\partial\Psi_0}{\partial\nu}\bigg|_{L} = 0; \tag{6.6.38}$$

$$L_2(\Psi_2) = 0, \ x,\xi \in G, \ m = 2, \ \frac{\partial\Psi_2}{\partial\nu}\bigg|_{L_0} = \tfrac{1}{2}\varkappa_{11}F(\xi), \ \frac{\partial\Psi_2}{\partial\nu}\bigg|_{L} = 0. \tag{6.6.39}$$

In addition, the functions $\Psi_0$ and $\Psi_2$ must satisfy the condition of boundedness at $\xi = 0$.

By analogy, we get

$$\varphi_{11}^{(\beta)} = \Psi_0(x,\xi) + \Psi_2(x,\xi)\cos 2\eta, \quad \varphi_{11}^{(\alpha\beta)} = \Psi_2(x,\xi)\sin 2\eta. \tag{6.6.40}$$

Further, we successively substitute functions $f_{\alpha i}$, $f_{\beta i}$, $f_{si}$, and $f_{qi}$ into relations (6.6.30) taking their symmetry and antisymmetry properties into account. This corresponds to the decomposition of functions (6.6.27) in a generalized Fourier series in systems of functions $f_{\alpha i}$, $f_{\beta i}$, $f_{si}$, and $f_{qi}$ that leads to nonlinear systems of ordinary differential equations for the parameters $\alpha_i$, $\beta_i$, $s_i$, and $q_i$. As a result, instead of the system of differential equations (6.6.29), we obtain

$$\mu_i^{(\alpha)}(\ddot\alpha_i + \sigma_{i\alpha}^2\alpha_i) + \sum_{j=1}^{\infty}[a_{js\alpha}^{(i\alpha)}(\dot s_j\alpha_1)^{\cdot} + a_{\alpha js}^{(i\alpha)}(\dot\alpha_1 s_j)^{\cdot} + a_{jq\beta}^{(i\alpha)}(\dot q_j\beta_1)^{\cdot} + a_{\beta jq}^{(i\alpha)}(\dot\beta_1 q_j)^{\cdot}]$$

$$+b_{\alpha\alpha}^{(i\alpha)}(\dot\alpha_1\alpha_1^2)^{\cdot} + b_{\alpha\beta}^{(i\alpha)}(\dot\alpha_1\beta_1^2)^{\cdot} + b_{\beta\alpha\beta}^{(i\alpha)}(\dot\beta_1\alpha_1\beta_1)^{\cdot} + \dot\alpha_1\sum_{j=1}^{\infty}c_{\alpha js}^{(i\alpha)}\dot s_j + \dot\beta_1\sum_{j=1}^{\infty}c_{pjq}^{(i\alpha)}\dot q_j$$

$$+d_{\alpha\alpha}^{(i\alpha)}\dot\alpha_1^2\alpha_1 + d_{\beta\alpha}^{(i\alpha)}\dot\beta_1^2\alpha_1 + d_{\alpha\beta\beta}^{(i\alpha)}\dot\alpha_1\dot\beta_1\beta_1 = 0, \quad i = 1,\,2,\,\ldots\,; \tag{6.6.41}$$

$$\mu_i^{(\beta)}(\ddot{\beta}_i+\sigma_{i\beta}^2\beta_i)+\sum_{j=1}^{\infty}[a_{iq\alpha}^{(i\beta)}(\dot{q}_j\alpha_1)^{\cdot}+a_{\alpha jq}^{(i\beta)}(\dot{\alpha}_1 q_j)^{\cdot}+a_{js\beta}^{(i\beta)}(\dot{s}_j\beta_1)^{\cdot}+a_{\beta js}^{(i\beta)}(\dot{\beta}_1 s_j)^{\cdot}]$$

$$+b_{\beta\alpha}^{(i\beta)}(\dot{\beta}_1\alpha_1^2)^{\cdot}+b_{\beta\beta}^{(i\beta)}(\dot{\beta}_1\beta_1^2)^{\cdot}+b_{\alpha\alpha\beta}^{(i\beta)}(\dot{\alpha}_1\alpha_1\beta_1)^{\cdot}+\dot{\alpha}_1\sum_{j=1}^{\infty}c_{\alpha jq}^{(i\beta)}\dot{q}_j+\dot{\beta}_1\sum_{j=1}^{\infty}c_{\beta js}^{(i\beta)}\dot{s}_j$$

$$+d_{\alpha\beta}^{(i\beta)}\dot{\alpha}_1^2\beta_1+d_{\beta\beta}^{(i\beta)}\dot{\beta}_1^2\beta_1+d_{\alpha\beta\alpha}^{(i\beta)}\dot{\alpha}_1\dot{\beta}_1\alpha_1=0,\quad i=1,\,2,\,\ldots; \tag{6.6.42}$$

$$\mu_i^{(s)}(\ddot{s}_i+\sigma_{is}^2 s_i)+\sum_{j=2}^{\infty}[a_{j\alpha\alpha}^{(is)}(\dot{\alpha}_j\alpha_1)^{\cdot}+a_{j\beta\beta}^{(is)}(\dot{\beta}_j\beta_1)^{\cdot}+a_{\alpha j\alpha}^{(is)}(\dot{\alpha}_1\alpha_j)^{\cdot}+a_{\beta j\beta}^{(is)}(\dot{\beta}_1\beta_j)^{\cdot}]$$

$$+a_{\alpha\alpha}^{(is)}(\dot{\alpha}_1\alpha_1)^{\cdot}+a_{\beta\beta}^{(is)}(\dot{\beta}_1\beta_1)^{\cdot}+c_{\alpha\alpha}^{(is)}\dot{\alpha}_1^2+c_{\beta\beta}^{(is)}\dot{\beta}_1^2+\dot{\alpha}_1\sum_{j=2}^{\infty}c_{\alpha j\alpha}^{(is)}\dot{\alpha}_j$$

$$+\dot{\beta}_1\sum_{j=2}^{\infty}c_{\beta j\beta}^{(is)}\dot{\beta}_j=0,\quad i=1,\,2,\,\ldots; \tag{6.6.43}$$

$$\mu_i^{(q)}(\ddot{q}_i+\sigma_{iq}^2 q_i)+\sum_{j=2}^{\infty}[a_{j\beta\alpha}^{(iq)}(\dot{\beta}_j\alpha_1)^{\cdot}+a_{\alpha j\beta}^{(iq)}(\dot{\alpha}_1\beta_j)^{\cdot}+a_{j\alpha\beta}^{(iq)}(\dot{\alpha}_j\beta_1)^{\cdot}+a_{\beta j\alpha}^{(iq)}(\dot{\beta}_1\alpha_j)^{\cdot}]$$

$$+a_{\alpha\beta}^{(iq)}(\dot{\alpha}_1\beta_1)^{\cdot}+a_{\beta\alpha}^{(iq)}(\dot{\beta}_1\alpha_1)^{\cdot}+c_{\alpha\beta}^{(iq)}(\dot{\alpha}_1\dot{\beta}_1)+\dot{\alpha}_1\sum_{j=2}^{\infty}c_{\alpha j\beta}^{(iq)}\dot{\beta}_j$$

$$+\dot{\beta}_1\sum_{j=2}^{\infty}c_{\beta j\alpha}^{(iq)}\dot{\alpha}_j=0,\quad i=1,\,2,\,\ldots. \tag{6.6.44}$$

The coefficients of (6.6.41)–(6.6.44), as compared with expressions (6.6.30), have additional indices $\alpha$, $\beta$, $s$, and $q$ indicating that the functions in (6.6.30) belong to the corresponding subsystem (6.6.33). Moreover, the indices $k$ and $l$ in relations (6.6.30) are replaced by the sign of unity in accordance with the results presented above. Therefore, the nonlinear equations (6.6.41)–(6.6.44) can be used, in general, for the description of nonlinear sloshing in the range of natural frequencies of the generalized coordinates $\alpha_1$ and $\beta_1$. For other special cases of sloshing, the corresponding equations can be derived from (6.6.29) in a similar way. In this case, the parameters $\beta_k$ and $\beta_l$ should be treated as parameters $\alpha$, $\beta$, $s$, or $q$ that are changing essentially during motion under consideration. Studying the systems of nonlinear differential equations (6.6.41)–(6.6.44) is an extremely complicated problem. However, there is a lot of practically important problems, for which the number of considered parameters $\alpha_i$, $\beta_i$, $s_i$, and $q_i$ can be restricted by analyzing solutions of the linear approximation of the problem. For example, if the motion of a container is oscillatory and the frequency of oscillations is close to the frequency of natural oscillations of the parameters $\alpha_1$ and $\beta_1$, then it is quite natural to neglect variations of all parameters $\alpha_i$ and $\beta_i$ for $i > 1$. The same reasoning is applicable to the parameters $s_i$ and $q_i$ ($s_{mn}$ and $q_{mn}$) whose frequencies of natural oscillations are not smaller than the frequencies of natural oscillations of the parameters $\alpha_2$ ($\alpha_{12}$) and $\beta_2$ ($\beta_{12}$). A similar conclusion can be obtained on the base of analysis of the spectrum of eigenvalues of the boundary-value problem (6.6.26).

The simplified system of equations of motion of a liquid takes the form

$$L_1(\alpha,\beta,s,q) = \mu_{11}^{(1)}(\ddot{\alpha}+\sigma_{1\alpha}^2\alpha) + b_{11}^{(1)}(\dot{\alpha}\alpha^2)^{\cdot} + b_{12}^{(1)}(\dot{\alpha}\beta^2)^{\cdot} + b_{212}^{(1)}(\dot{\beta}\alpha\beta)^{\cdot}$$
$$+d_{11}^{(1)}\dot{\alpha}^2\alpha + d_{21}^{(1)}\dot{\beta}^2\alpha + d_{122}^{(1)}\alpha\dot{\beta}\beta + \sum_{i=0,2}\left[a_{i31}^{(1)}(\dot{s}_{1i}\alpha)^{\cdot} + a_{i13}^{(1)}(\dot{\alpha}s_{1i})^{\cdot}\right]$$
$$+a_{42}^{(1)}(\dot{q}_{12}\beta)^{\cdot} + a_{24}^{(1)}(\dot{\beta}q_{12})^{\cdot} + \dot{\alpha}\sum_{i=0,2} c_{13i}^{(1)}\dot{s}_{1i} + c_{24}^{(1)}\dot{\beta}q_{12} = 0; \qquad (6.6.45)$$

$$L_2(\alpha,\beta,s,q) = \mu_{11}^{(2)}(\ddot{\beta}+\sigma_{1\beta}^2\beta) + b_{21}^{(2)}(\dot{\beta}\alpha^2)^{\cdot} + b_{22}^{(2)}(\dot{\beta}\beta^2)^{\cdot} + b_{112}^{(2)}(\dot{\alpha}\alpha\beta)^{\cdot}$$
$$+d_{12}^{(2)}\dot{\alpha}^2\beta + d_{22}^{(2)}\dot{\beta}^2\beta + d_{121}^{(2)}\dot{\alpha}\beta\alpha + a_{41}^{(2)}(\dot{q}_{12}\alpha)^{\cdot} + a_{14}^{(2)}(\dot{\alpha}q_{12})^{\cdot}$$
$$+\sum_{i=0,2}\left[a_{i32}^{(2)}(\dot{s}_{1i}\beta)^{\cdot} + a_{23i}^{(2)}(\dot{\beta}s_{1i})^{\cdot}\right] + c_{14}^{(2)}\dot{\alpha}\dot{q}_{12} + \dot{\beta}\sum_{i=0,2} c_{23i}^{(2)}\dot{s}_{1i} = 0; \qquad (6.6.46)$$

$$L_3(\alpha,\beta,s,q) = \mu_{1i}^{(3)}(\ddot{s}_{1i}+\sigma_{is}^2 s_{1i}) + a_{1i}^{(3)}(\dot{\alpha}\alpha)^{\cdot} + a_{2i}^{(3)}(\dot{\beta}\beta)^{\cdot}$$
$$+c_{1i}^{(3)}\dot{\alpha}^2 + c_{2i}^{(3)}\dot{\beta}^2 = 0, \qquad i = 0,\ 2; \qquad (6.6.47)$$

$$L_4(\alpha,\beta,s,q) = \mu_{11}^{(4)}(\ddot{q}_{12}+\sigma_{1q}^2 q_{12}) + a_{12}^{(4)}(\dot{\alpha}\beta)^{\cdot} + a_{21}^{(4)}(\dot{\beta}\alpha)^{\cdot} + c_{12}^{(4)}\dot{\alpha}\dot{\beta} = 0, \quad (6.6.48)$$

where $\alpha = \alpha_{11}$ and $\beta = \beta_{11}$.

The coefficients of (6.6.45)–(6.6.48) are defined by the relations

$$\mu_{11}^{(1)} = \pi\varkappa_{11}\int_{L_0}\xi\psi_{11}^2 d\xi,$$

$$b_{11}^{(1)} = \pi\varkappa_{11}\int_{L_0}\Big[K(\xi) + \tfrac{1}{2}L(\xi) + \tfrac{3}{4}\varkappa_{11}T_1(\xi) + \frac{\varkappa_{11}}{4\xi^2}\psi_{11}^2 T_2(\xi) - \frac{\varkappa_{11}}{2\xi^3}n_1\psi_{11}^4$$
$$-\tfrac{3}{8}\varkappa_{11}T_3(\xi) - \frac{e_{11}}{\varkappa_{11}}T_5(\xi) + \tfrac{3}{8}\varkappa_{11}^2 T_4(\xi) + e_{11}T_2(\xi)\Big]\xi d\xi,$$

$$b_{12}^{(1)} = \pi\varkappa_{11}\int_{L_0}\Big[\tfrac{1}{2}L(\xi) + \tfrac{1}{4}\varkappa_{11}T_1(\xi) - \frac{\varkappa_{11}}{4\xi^2}\psi_{11}^2 T_2(\xi) + \frac{\varkappa_{11}}{2\xi^3}n_1\psi_{11}^4$$
$$-\frac{\varkappa_{11}}{8}T_3(\xi) - \frac{e_{22}}{\varkappa_{11}}T_5(\xi) + \frac{\varkappa_{11}}{8}T_4(\xi) + e_{22}T_2(\xi)\Big]\xi d\xi,$$

$$b_{212}^{(1)} = \pi\varkappa_{11}\int\limits_{L_0}\Big[K(\xi) + \frac{\varkappa_{11}}{2}T_1(\xi) + \frac{\varkappa_{11}}{2\xi^2}\psi_{11}^2T_2(\xi) - \frac{\varkappa_{11}}{\xi^3}n_1\psi_{11}^4 - \frac{\varkappa_{11}}{4}T_3(\xi) + \frac{\varkappa_{11}^2}{4}T_4(\xi)\Big]\xi d\xi,$$

$$d_{11}^{(1)} = \pi\varkappa_{11}\int\limits_{L_0}\Big[P_0(\xi) + \tfrac{1}{2}Q_2(\xi) + \tfrac{3}{4}\varkappa_{11}T_6(\xi) + \tfrac{1}{4}\varkappa_{11}T_7(\xi) - \tfrac{3}{4}\varkappa_{11}^2T_4(\xi) - \frac{2e_{11}}{\varkappa_{11}}T_2(\xi)\Big]\xi d\xi,$$

$$d_{21}^{(1)} = \pi\varkappa_{11}\int\limits_{L_0}\Big[\tfrac{1}{2}Q_2(\xi) + \tfrac{1}{4}\varkappa_{11}T_6(\xi) + \tfrac{3}{4}\varkappa_{11}T_7(\xi) - \tfrac{1}{4}\varkappa_{11}^2T_4(\xi)\Big]\xi d\xi,$$

$$d_{122}^{(1)} = \pi\varkappa_{11}\int\limits_{L_0}\Big[P_0(\xi) + \frac{\varkappa_{11}}{2}T_6(\xi) - \frac{\varkappa_{11}}{2}T_7(\xi) - \frac{\varkappa_{11}^2}{2}T_4(\xi) - \frac{2e_{11}}{\varkappa_{11}}T_2(\xi)\Big]\xi d\xi,$$

$$a_{031}^{(1)} = \pi\varkappa_{11}\int\limits_{L_0}K_1(\xi)\xi d\xi, \quad a_{231}^{(1)} = -\frac{\pi}{2}\varkappa_{11}\int\limits_{L_0}L_1(\xi)\xi d\xi,$$

$$a_{013}^{(1)} = \pi\varkappa_{01}\int\limits_{L_0}M(\xi)\xi d\xi, \quad a_{213}^{(1)} = -\frac{\pi}{2}\varkappa_{21}\int\limits_{L_0}N(\xi)\xi d\xi,$$

$$a_{42}^{(1)} = -a_{231}^{(1)}, \quad a_{24}^{(1)} = -a_{213}^{(1)},$$

$$c_{130}^{(1)} = \pi\varkappa_{11}\int\limits_{L_0}P_1(\xi)\xi d\xi, \quad c_{132}^{(1)} = -c_{24}^{(1)} = -\frac{\pi}{2}\varkappa_{11}\int\limits_{L_0}Q_1(\xi)\xi d\xi,$$

$$\mu_{11}^{(2)} = \mu_{11}^{(1)}, \quad \sigma_{1\alpha}^2 = \sigma_{1\beta}^2, \quad b_{21}^{(2)} = b_{12}^{(1)}, \quad b_{22}^{(2)} = b_{11}^{(1)}, \quad b_{112}^{(2)} = b_{212}^{(1)},$$

$$d_{12}^{(2)} = d_{21}^{(1)}, \quad d_{22}^{(2)} = d_{11}^{(1)}, \quad d_{121}^{(2)} = d_{122}^{(1)}, \quad a_{41}^{(2)} = a_{42}^{(1)}, \quad a_{14}^{(2)} = a_{24}^{(1)},$$

$$a_{032}^{(2)} = a_{031}^{(1)}, \quad a_{232}^{(2)} = -a_{231}^{(1)}, \quad a_{230}^{(2)} = a_{013}^{(1)}, \quad a_{232}^{(2)} = a_{213}^{(1)},$$

$$c_{14}^{(2)} = c_{24}^{(1)}, \quad c_{230}^{(2)} = c_{130}^{(1)}, \quad c_{232}^{(2)} = -c_{132}^{(1)},$$

$$\mu_{i1}^{(3)} = \delta\varkappa_{i1}\int\limits_{L_0}\xi\psi_{i1}^2 d\xi, \quad \delta = \begin{cases} 2\pi \text{ for } i = 0, \\ \pi \text{ for } i = 2, \end{cases} \tag{6.6.49}$$

$$a_{10}^{(3)} = \pi\varkappa_{11}\int\limits_{L_0}[F_1(\xi) + \varkappa_{01}T_2(\xi)]\psi_{01}\xi d\xi,$$

$$a_{12}^{(3)} = -\frac{\pi}{2}\varkappa_{11}\int\limits_{L_0}[F(\xi) + \varkappa_{21}T_2(\xi)]\psi_{21}\xi d\xi,$$

$$a_{20}^{(3)} = a_{10}^{(3)}, \quad a_{22}^{(3)} = -a_{12}^{(3)}, \quad c_{10}^{(3)} = \frac{\pi}{2}\varkappa_{01} \int\limits_{L_0} R(\xi)\xi d\xi,$$

$$c_{12}^{(3)} = -\frac{\pi}{4}\varkappa_{21} \int\limits_{L_0} S(\xi)\xi d\xi, \quad c_{20}^{(3)} = c_{10}^{(3)}, \quad c_{22}^{(3)} = -c_{12}^{(3)},$$

$$\mu_{12}^{(4)} = \pi\varkappa_{21} \int\limits_{L_0} \xi\psi_{12}^2 d\xi, \quad a_{12}^{(4)} = -a_{12}^{(3)}, \quad a_{21}^{(4)} = a_{12}^{(4)}, \quad c_{12}^{(4)} = -2c_{12}^{(3)},$$

where

$$T_1(\xi) = \psi_{11}^2 \frac{\partial\psi_{11}}{\partial\xi}\left[-n_1\frac{\partial^2\psi_{11}}{\partial x^2} + (1-n_1^2)\frac{\partial^2\psi_{11}}{\partial x\partial\xi} + n_1\frac{\partial^2\psi_{11}}{\partial\xi^2} - n_2\frac{\partial\psi_{11}}{\partial\xi} - n_3\frac{\partial\psi_{11}}{\partial x}\right],$$

$$T_2(\xi) = \psi_{11}T_8, \quad T_3(\xi) = \psi_{11}^3\left[\frac{\partial^3\psi_{11}}{\partial x^3} + 2n_1\frac{\partial^3\psi_{11}}{\partial x^2\partial\xi} + n_1^2\frac{\partial^3\psi_{11}}{\partial x\partial\xi^2}\right],$$

$$T_4(\xi) = \psi_{11}^3\left[\frac{\partial^2\psi_{11}}{\partial x^2} + 2n_1\frac{\partial^2\psi_{11}}{\partial x\partial\xi} + n_1^2\frac{\partial^2\psi_{11}}{\partial\xi^2}\right],$$

$$T_5(\xi) = \psi_{11}\left[\frac{\partial^2\psi_{11}}{\partial x^2} + n_1\frac{\partial^2\psi_{11}}{\partial x\partial\xi}\right],$$

$$T_6(\xi) = \psi_{11}^2\left(\frac{\partial^2\psi_{11}}{\partial x^2}\frac{\partial\psi_{11}}{\partial x} + n_1\frac{\partial^2\psi_{11}}{\partial x\partial\xi}\frac{\partial\psi_{11}}{\partial x} + \frac{\partial\psi_{11}}{\partial x}\frac{\partial^2\psi_{11}}{\partial x\partial\xi} + n_1\frac{\partial^2\psi_{11}}{\partial\xi^2}\frac{\partial\psi_{11}}{\partial\xi}\right),$$

$$T_7(\xi) = \frac{1}{\xi^2}\psi_{11}^3\left(T_8 - \frac{n_1}{\xi}\psi_{11}\right), \quad T_8(\xi) = \frac{\partial\psi_{11}}{\partial x} + n_1\frac{\partial\psi_{11}}{\partial\xi},$$

$$T_9(\xi) = \frac{\partial\Psi_0}{\partial x} + n_1\frac{\partial\Psi_0}{\partial\xi}, \quad T_{10}(\xi) = \frac{\partial\Psi_2}{\partial x} + n_1\frac{\partial\Psi_2}{\partial\xi},$$

$$T_{11}(\xi) = \frac{\partial\psi_{01}}{\partial x} + n_1\frac{\partial\psi_{01}}{\partial\xi}, \quad T_{12}(\xi) = \frac{\partial\psi_{21}}{\partial x} + n_1\frac{\partial\psi_{21}}{\partial\xi},$$

$$K(\xi) = \psi_{11}\left[-n_1\frac{\partial\Psi_0}{\partial x}\frac{\partial\psi_{11}}{\partial\xi} + \frac{\partial\Psi_0}{\partial\xi}\frac{\partial\psi_{11}}{\partial\xi}\right.$$
$$\left. -\psi_{11}\left(\frac{\partial^2\Psi_0}{\partial x^2} + n_1\frac{\partial^2\Psi_0}{\partial x\partial\xi}\right) + \varkappa_{11}\psi_{11}T_9(\xi)\right],$$

$$L(\xi) = \psi_{11}\left[-n_1\frac{\partial\Psi_2}{\partial x}\frac{\partial\psi_{11}}{\partial\xi} + \frac{\partial\Psi_2}{\partial\xi}\frac{\partial\psi_{11}}{\partial\xi} + \frac{2}{\xi^2}\Psi_2\psi_{11}\right.$$
$$\left. -\psi_{11}\left(\frac{\partial^2\Psi_2}{\partial x^2} + n_1\frac{\partial^2\Psi_2}{\partial x\partial\xi}\right) + \varkappa_{11}\psi_{11}T_{10}(\xi)\right],$$

$$F_1(\xi) = -n_1 \frac{\partial \psi_{11}}{\partial x}\frac{\partial \psi_{11}}{\partial \xi} + \left(\frac{\partial \psi_{11}}{\partial \xi}\right)^2 + \frac{1}{\xi^2}\psi_{11}^2 - T_5(\xi), \tag{6.6.50}$$

$$F(\xi) = F_1(\xi) - \frac{2}{\xi^2}\psi_{11}^2,$$

$$P_1(\xi) = \psi_{11}\left[\frac{\partial \psi_{11}}{\partial x}\frac{\partial \psi_{01}}{\partial x} + \frac{\partial \psi_{11}}{\partial \xi}\frac{\partial \psi_{01}}{\partial \xi} - \varkappa_{11}\psi_{11}T_{11}(\xi) - \varkappa_{01}\psi_{01}T_8(\xi)\right],$$

$$Q_1(\xi) = \psi_{11}\left[\frac{\partial \psi_{11}}{\partial x}\frac{\partial \psi_{21}}{\partial x} + \frac{\partial \psi_{11}}{\partial \xi}\frac{\partial \psi_{21}}{\partial \xi} + \frac{2}{\xi^2}\psi_{11}\psi_{21} - \varkappa_{11}\psi_{11}T_{12}(\xi) - \varkappa_{21}\psi_{21}T_8(\xi)\right],$$

$$R(\xi) = \psi_{01}\left[\left(\frac{\partial \psi_{11}}{\partial x}\right)^2 + \left(\frac{\partial \psi_{11}}{\partial \xi}\right)^2 + \frac{1}{\xi^2}\psi_{11}^2 - 2\varkappa_{11}\psi_{11}T_8(\xi)\right],$$

$$S(\xi) = \psi_{21}\left[\left(\frac{\partial \psi_{11}}{\partial x}\right)^2 + \left(\frac{\partial \psi_{11}}{\partial \xi}\right)^2 - \frac{1}{\xi^2}\psi_{11}^2 - 2\varkappa_{11}\psi_{11}T_8(\xi)\right],$$

$$P_0(\xi) = \psi_{11}\left[\frac{\partial \psi_{11}}{\partial x}\frac{\partial \Psi_0}{\partial x} + \frac{\partial \psi_{11}}{\partial \xi}\frac{\partial \Psi_0}{\partial \xi} - \varkappa_{11}\psi_{11}T_9(\xi)\right],$$

$$Q_2(\xi) = \psi_{11}\left[\frac{\partial \psi_{11}}{\partial x}\frac{\partial \Psi_2}{\partial x} + \frac{\partial \psi_{11}}{\partial \xi}\frac{\partial \Psi_2}{\partial \xi} + \frac{2}{\xi^2}\psi_{11}\Psi_2 - \varkappa_{11}\psi_{11}T_{10}(\xi)\right],$$

$$K_1(\xi) = \psi_{11}\left[-n_1\frac{\partial \psi_{01}}{\partial x}\frac{\partial \psi_{11}}{\partial \xi} + \frac{\partial \psi_{10}}{\partial \xi}\frac{\partial \psi_{11}}{\partial \xi} \right.$$
$$\left. -\psi_{11}\left(\frac{\partial^2 \psi_{01}}{\partial x^2} + n_1\frac{\partial^2 \psi_{01}}{\partial x \partial \xi}\right) + \varkappa_{11}\psi_{11}T_{11}(\xi)\right],$$

$$L_1(\xi) = \psi_{11}\left[-n_1\frac{\partial \psi_{21}}{\partial x}\frac{\partial \psi_{11}}{\partial \xi} + \frac{\partial \psi_{21}}{\partial \xi}\frac{\partial \psi_{11}}{\partial \xi} + \frac{2}{\xi^2}\psi_{21}\psi_{11} \right.$$
$$\left. -\psi_{11}\left(\frac{\partial^2 \psi_{21}}{\partial x^2} + n_1\frac{\partial^2 \psi_{21}}{\partial x \partial \xi}\right) + \varkappa_{11}\psi_{11}T_{12}(\xi)\right],$$

$$M(\xi) = \psi_{11}\left[-n_1\frac{\partial \psi_{11}}{\partial x}\frac{\partial \psi_{01}}{\partial \xi} + \frac{\partial \psi_{11}}{\partial \xi}\frac{\partial \psi_{01}}{\partial \xi} \right.$$
$$\left. -\psi_{01}\left(\frac{\partial^2 \psi_{11}}{\partial x^2} + n_1\frac{\partial^2 \psi_{11}}{\partial x \partial \xi}\right) + \varkappa_{11}\psi_{01}T_8(\xi)\right],$$

$$N(\xi) = \psi_{11}\left[-n_1\frac{\partial \psi_{11}}{\partial x}\frac{\partial \psi_{21}}{\partial \xi} + \frac{\partial \psi_{11}}{\partial \xi}\frac{\partial \psi_{21}}{\partial \xi} + \frac{2}{\xi^2}\psi_{11}\psi_{21} \right.$$
$$\left. -\psi_{21}\left(\frac{\partial^2 \psi_{11}}{\partial x^2} + n_1\frac{\partial^2 \psi_{11}}{\partial x \partial \xi}\right) + \varkappa_{11}\psi_{21}T_8(\xi)\right].$$

Relations (6.6.49) and (6.6.50) imply that the actual values of coefficients of the equations of motion can be obtained only after obtaining the functions $\psi_{m1}$,

($m = 0, 1, 2$), $\Psi_0$, and $\Psi_2$. For some special cases, this problem is considered in [112, 173].

It is worth noting that, in the most cases of practically important problems, we are forced to seek approximate solutions of the main boundary-value problems (6.6.22)–(6.6.26). If these solutions are among the functions that do not satisfy the boundary condition on the wetted surface $S^*(t)$, then we have to impose additional conditions on these solutions at the contour $l$ given by the intersection of the nonperturbed surface $\Sigma_0^*$ and the surface $S^*$. Usually, they are obtained from the Taylor expansion of the normal derivative of required functions at boundary points of the surface $S^*$ of the perturbed domain in a neighborhood of the contour $l$. Up to the terms of the third order, these additional conditions have the form [173]

$$\nabla_1\left(\frac{\partial\varphi_{i0}}{\partial\nu}\right) = 0 \text{ for } x = x_0^1, \quad \nabla_{11}\left(\frac{\partial\varphi_{i0}}{\partial\nu}\right) = 0 \text{ for } x^1 = x_0^1,$$

$$\nabla_1\left(\frac{\partial\varphi_{il}}{\partial\nu}\right) = 0 \text{ for } x = x_0^1, \quad \nabla_{11}\left(\frac{\partial\varphi_{il}}{\partial\nu}\right) = 0 \text{ for } x^1 = x_0^1. \tag{6.6.51}$$

An application of the variational method with the use of conditions (6.6.51) is described in [173]. For cylindrical containers, the above-presented results follow from the results obtained by the considered method as a special case.

# Chapter 7

# Equivalent mechanical systems in the dynamics of a rigid body with liquid

In this chapter, on the basis of modern results of studying the dynamics of liquid-filled systems by the multimodal method, we suggest a nonlinear mathematical model in the form of equations of motion of a mechanical system, which is equivalent to the body–liquid system in a certain sense from the viewpoint of the analytical mechanics. As is known, equivalent mechanical systems of this type in the linear theory were proposed by several authors quite long ago. These systems are either a rigid body with mathematical or spherical pendulums attached to the rigid body in a special way or a rigid body with spring-mass oscillators. Principles of construction of such equivalent mechanical systems are described in [86, 149, 59]. The parameters of equivalent mechanical systems (lengths of pendulums, points of suspension, masses and spring constants, etc.) are chosen depending on hydrodynamic characteristics of the hydrodynamic subsystem defined by solutions of the corresponding free-boundary problems. Therefore, differential equations of motion of a rigid body with liquid are completely equivalent to differential equations of motion of a rigid body with mathematical pendulums or spring masses. However, from an engineering point of view, a mathematical analog of the original mechanical system is more informative if we are interested in estimating the role of each element of the system when we study the specific motion of the system as a whole. As a rule, it is hidden rather deeply in equations of motion of the hydrodynamic type because of a certain arbitrariness in solutions, e.g., of homogeneous boundary-value problems. Since a mechanical analog of a rigid body–liquid system is the subject of the analytical mechanics, which is an old and the most developed part of the mechanics in the whole, related physical processes can be predicted by using the well developed methods of analytical mechanics. As a consequence, mechanical analogs are widely adopted in the practice of engineering (first of all, in the aerospace science). The achievements in developing and using the mechanical analogs in nonlinear dynamics of a rigid body with liquid are much less successful because of considerable difficulties related to studying nonlinear problems of liquid dynamics of a bounded volume.

First attempts of using nonlinear analogs in the dynamics of bodies with liquid were performed in the 1960–1970s [108, 160, 173, 213, 11]. In [108], a mechanical analog of the linear theory is used to derive the equations of perturbed motion of a body with a cavity partially filled with liquid under the assumption of finite deviations of the free surface. The delinearization, i.e., partial account of nonlinearities in equations of motion of a body with liquid, were performed by using a mechanical system equivalent to the initial one in a linear approximation. These nonlinear equations of motion were verified in the study of the first problem of dynamics of a body–liquid system, namely, the problem of forced sloshing in a neighborhood of the main resonance for a cylindrical tank oscillating in a prescribed way by the harmonic time law.

A simple analytic mechanical model in the form of a point mass attached to a parabolic surface by a nonlinear spring is proposed in [11]. For a spring with cubic characteristics and two generalized coordinates related to two main sloshing modes, this model makes it possible to describe both planar and swirling motions in a vibrating cylindrical container. In this case, the elasticity constant is determined for each depth of the liquid from the experimental results.

By using the notion of a surface of centers in the case where the perturbed surface is always planar, a mechanical analog is proposed in [160]; the related nonlinear ordinary differential equation of motion has just one parameter $\beta(t)$, which corresponds to the angle of rotation of the mirror of the nonperturbed free surface. For small but finite values of $\beta(t)$, this model provides a qualitative description of planar sloshing in a neighborhood of the main resonance with respect to the nodal diameter. A similar model under the same assumptions is proposed in [213] for the spatial liquid motion in a sphere. A mechanical analog in the form of a rigid body with a suspended spherical pendulum whose equations of motion in their linear approximation are equivalent to the equations of motion of a body with liquid is described in [173]. In this case, the generalized coordinates of the pendulum are chosen to be the direction cosines of the pendulum bar, and such a choice provides certain advantages in the simulation of spatial motions of a body with liquid.

In this chapter, by restricting ourselves to the theory of the third order of smallness presented in Chap. 4 of this book, we derive an invariant form of nonlinear equations of plane-parallel motion of a rigid body with a cylindrical cavity partially filled with liquid by using a seven-mode approximation of the free surface. To this end, we suggest a mechanical analog in the form of a rigid body with suspended spherical pendulums whose lengths and masses are defined by coefficients of the initial nonlinear system of equations of motion.

## 7.1 Nonlinear equations of translational motion of a rigid body with a cylindrical cavity partially filled with liquid

Let a mechanical system consist of a rigid body with ideal incompressible liquid and perform a translational motion in the gravity field under the action of the external forces $\boldsymbol{F}_1, \boldsymbol{F}_2, \ldots, \boldsymbol{F}_n$ applied to the body. We characterize the motion of the mechanical system in the inertial coordinate system $O'x'y'z'$ by the vector of translational velocity $\boldsymbol{v}_0$ of an arbitrarily chosen point $O$ of the rigid body. This point is chosen to be the origin of the coordinate system $Oxyz$ rigidly attached to the rigid body whose $Ox$-axis is directed, as above, oppositely to the gravity force vector $\mathbf{g}$. In what follows, we present equations of motion of a rigid body with liquid in the moving coordinate system $Oxyz$ by using the principles of the mechanics of relative motions. The hydrodynamic part of the problem assumes finding the velocity and pressure fields in a moving liquid domain and the hydrodynamic loads. Here, we consider the case of irrotational flow of ideal liquid. This means that the vector of absolute velocity of particles of a liquid is defined by the scalar function $\Phi(x, y, z, t)$, $\boldsymbol{v}_a = \nabla\Phi$. As it is known, the velocity potential $\Phi(x, y, z, t)$ and the instant shape of the perturbed free surface can be defined as solutions of the following free-boundary problem:

$$\begin{aligned} &\Delta\Phi = 0, \quad \boldsymbol{r} \in Q, \\ &\frac{\partial\Phi}{\partial\nu} = \boldsymbol{v}_0 \cdot \boldsymbol{\nu} - \frac{\zeta_t}{\sqrt{|\nabla\zeta|^2}}, \quad \boldsymbol{r} \in \Sigma, \\ &\frac{\partial\Phi}{\partial\nu} = \boldsymbol{v}_0 \cdot \boldsymbol{\nu}, \quad \boldsymbol{r} \in S, \\ &\frac{\partial'\Phi}{\partial t} + \frac{1}{2}|\nabla\Phi|^2 - \nabla\Phi \cdot \boldsymbol{v}_0 + U = 0, \quad \boldsymbol{r} \in \Sigma, \end{aligned} \tag{7.1.1}$$

where $U$ is the gravity potential; $\boldsymbol{\nu}$ is the unit vector of the outward normal to the surface of the domain $Q$; $\boldsymbol{r}$ is the radius vector of points of the domain $Q$ in the tank-fixed coordinate system; $S$ and $\Sigma$ are the rigid walls of the cavity and the perturbed free surface, respectively; the prime sign at the derivative with respect to time $t$ means that the differentiation is performed in the moving coordinate system (in what follows, the prime sign is omitted); $\zeta(x, y, z, t) = 0$ is the equation of the perturbed free surface. The pressure $p$ is determined by using the Bernoulli equation rewritten in the body-fixed coordinate system.

For the considered mechanical system, the differential equations of motion can be derived in a several ways. Traditionally used method of derivation of the equations suggests to consider the momentum of a body–liquid mechanical system and, subsequently, to apply the dynamic theorem on variation of the momentum of the system. Other methods are based on the use of variational principles of the mechanics.

In the present book, we use a nonlinear system of differential equations of motion derived earlier by using the Bateman–Luke variational principle. For a prescribed motion of a rigid body, i.e., in the case where the translational velocity of motion $\boldsymbol{v}_0$ is assumed to be known, the problem of solving the free-boundary problem (7.1.1) is equivalent, in a certain sense, to studying the variational problem with an action $W$ for the Lagrangian

$$L = \int\limits_{Q(t)} p\, dQ = -\rho \int\limits_{Q(t)} \left[ \frac{\partial \Phi}{\partial t} + \frac{1}{2}|\nabla\Phi|^2 - \nabla\Phi \cdot \boldsymbol{v}_0 + U \right] dQ, \tag{7.1.2}$$

where $\rho$ is the density of liquid. In the potential field of gravity forces, the force function $U$ can be presented in the form

$$U = -\mathbf{g} \cdot \boldsymbol{r}' \tag{7.1.3}$$

and, additionally, $\boldsymbol{r}' = \boldsymbol{r}_0' + \boldsymbol{r}$, where $\boldsymbol{r}'$ is the radius vector of the points of the body–liquid system relative to the origin $O'$ of the absolute coordinate system; $\boldsymbol{r}_0'$ is the radius vector of the point $O$ relative to the fixed point $O'$; $\boldsymbol{r}$ is the radius vector of an arbitrary point of the system relative to the point $O$. For real motions that are wave motions of a liquid in a moving cavity having free surface, the action takes stationary values, i.e.,

$$\delta W = \delta \int\limits_{t_1}^{t_2} L\, dt = 0. \tag{7.1.4}$$

Since the free-boundary problem (7.1.1) is equivalent to the variational problem for an action, the method of approximate solving the problem of liquid sloshing in a moving container can be reformulated in terms of the direct method of solving the variational problem [119]. In this case, the most essential ideas are obtaining a constructive presentation of the perturbed free surface and transformations in the domain of the velocity potential, which is unknown in advance.

Note at first that, under the condition of translational motions of the body–liquid system, the velocity potential $\Phi(x, y, z, t)$ can be divided into two components

$$\Phi(x, y, z, t) = \boldsymbol{v}_0 \cdot \boldsymbol{V} + \varphi, \tag{7.1.5}$$

for which, due to the kinematic boundary conditions of the boundary problem (7.1.1), we have $\boldsymbol{V} = \boldsymbol{r}$ up to an arbitrary function of time, and the harmonic function $\varphi$ satisfies the boundary conditions

$$\left.\frac{\partial \varphi}{\partial \nu}\right|_S = 0, \quad \left.\frac{\partial \varphi}{\partial \nu}\right|_\Sigma = -\frac{\zeta_t}{\sqrt{|\nabla\zeta|^2}} \tag{7.1.6}$$

and, hence, describes sloshing in a motionless cavity. Further, if the cavity has cylindrical shape in a neighborhood of the free surface, then the equation $\zeta(x, y, z, t) = 0$ can be presented in the form

$$x = x_0 + h + f(y, z, t), \tag{7.1.7}$$

where $x_0$ is the coordinate of the center of the cavity bottom and $h$ is the depth of liquid.

If the motion law for a rigid body is unknown and the body with liquid moves under the influence of a system of external forces that includes also nonpotential ones, then equations of motion of the system can be derived by using the generalized rational principle

$$\delta S + \delta W + \int_{t_1}^{t_2} \delta' A dt = \int_{t_1}^{t_2} (\delta L + \delta L_1 + \delta' A) dt = 0, \tag{7.1.8}$$

where $L_1 = T - \Pi$ is the kinetic potential of a rigid body; $S = \int_{t_1}^{t_2} L_1\, dt$ is the action; and $\delta' A$ is an elementary work of nonpotential forces.

If the cavity has the shape of an upright circular cylinder (or it is formed by two cylindrical surfaces), the free surface and the velocity potential in the cylindrical coordinate system $Ox\xi\eta$ are defined by the relations

$$x = f(\xi, \eta, t) = \sum_{m,n} [r_{mn}(t) \sin m\eta + p_{mn}(t) \cos m\eta] f_{mn}(\xi), \tag{7.1.9}$$

$$\varphi(x, \xi, \eta, t) = \sum_{m,n} [R_{mn}(t) \sin m\eta + P_{mn}(t) \cos m\eta]\, \psi_{mn}(x, \xi), \tag{7.1.10}$$

where

$$Y_m(k_{mn}\xi) = \frac{J_m(k_{mn}\xi) N_m'(\zeta_{mn}) - N_m(k_{mn}\xi) J_m'(\zeta_{mn})}{J_m(\zeta_{mn}) N_m'(\zeta_{mn}) - N_m(\zeta_{mn}) J_m'(\zeta_{mn})}, \tag{7.1.11}$$

$$f_{mn} = Y_m(k_{mn}\xi), \quad \psi_{mn}(x, \xi) = \frac{\cosh k_{mn} x}{\cosh k_{mn} h} Y_m(k_{mn}\xi);$$

$J_m(k_{mn}\xi)$ and $N_m(k_{mn}\xi)$ are the Bessel and Neumann functions of order $m$, respectively, $\zeta_{mn} = k_{mn} R_1$ ($m = 0, 1, 2, \ldots$ and $n = 1, 2, \ldots$) are roots of the equation

$$J_m'(\delta\zeta) N_m'(\zeta) - N_m'(\delta\zeta) J_m'(\zeta) = 0, \quad \delta = \frac{R_0}{R_1}, \tag{7.1.12}$$

where $R_0$ is the radius of the inner cylinder; $R_1$ is the radius of the external cylinder; $h$ is the depth of liquid; $x = x$, $y = \xi \cos \eta$, and $z = \xi \sin \eta$ are relations between the Cartesian and cylindrical coordinates. The eigenvalues of the boundary-value problem (1.5.34) are defined by the relation

$$\varkappa_{mn} = k_{mn} \tanh(k_{mn} h), \quad k_{mn} = \frac{\zeta_{mn}}{R_1}. \tag{7.1.13}$$

Some numerical values of the parameter $\zeta_{mn}$ are given in Table 7.1 (for $\delta = 0$).

Note separately that, under the chosen normalization of solutions of the eigenvalue boundary problem (1.5.34) [$\varphi(h, y, z) = 1$], the generalized coordinates $r_{mn}(t)$ and $p_{mn}(t)$ in (7.1.9) correspond to deviations of the free surface at points of the contour ($\xi = R_1$) from the equilibrium state ($\Sigma_0$) for the related

**Table 7.1**

| $n/\zeta_{mn}$ | 1 | 2 | 3 | 4 | 5 |
|---|---|---|---|---|---|
| $\zeta_{0n}$ | 3.831706 | 7.015587 | 10.173468 | 13.323692 | 16.470630 |
| $\zeta_{1n}$ | 1.841184 | 5.331443 | 8.536316 | 11.706005 | 14.863589 |
| $\zeta_{2n}$ | 3.054237 | 6.706133 | 9.969468 | 13.170371 | 16.347522 |
| $\zeta_{3n}$ | 4.201192 | 8.015243 | 11.345996 | 14.585846 | 17.788675 |

modes of oscillation. Sometimes, for problem (1.5.34), other normalization rules for solutions can be used, however, all these normalization rules require the condition

$$\int_{\Sigma_0} \varphi_n(x, y, z)dS = 0, \tag{7.1.14}$$

which is a corollary of the volume conservation condition.

The linear theory of motion of a body with liquid is developed under the assumption of smallness of the kinematic parameters of motion of a liquid. This means that the equations of motion keep only the linear terms with parameters $r_{mn}(t)$, $p_{mn}(t)$, $R_{mn}(t)$, and $P_{mn}(t)$ that appear in decompositions (7.1.9) and (7.1.10). The nonlinear dynamics of a rigid body with cavities partially filled with liquid is based on the principle of "delinearization." According to this principle, the nonlinearity is partially recovered in agreement with the priority of natural sloshing modes with respect to specific physical processes under study. If the problem under consideration includes, among others, e.g., uncontrolled oscillatory processes that may be observed in resonance zones, then the system of equations of the perturbed motion for the corresponding mechanical system has to provide a possibility of "delinearization" of the system for natural sloshing modes, at least in domains of main resonances that are dangerous for the system operation. A method of such a "delinearization" was proposed by G. S. Narimanov and it was applied above. With the use of the proposed variational method, we obtain the most complete nonlinear system of differential equations that describes the spatial motion of a mechanical body–liquid system. Here, we present only equations of the plane-parallel motion of the system that follows from system (2.9.31)–(2.9.34). In the vector notation, they have the form

$$M(\overset{*}{\boldsymbol{v}}_0 - \mathbf{g}) + m_1 \overset{**}{\boldsymbol{r}}_{C_1} = \boldsymbol{P}, \tag{7.1.15}$$

$$\frac{d}{dt} A_n - \sum_k A_{nk} R_k = 0, \quad n = 1, 2, \ldots, \tag{7.1.16}$$

$$\sum_n \frac{\partial A_n}{\partial \beta_i} \dot{R}_n + \frac{1}{2} \sum_n \sum_k \frac{\partial A_{nk}}{\partial \beta_i} R_n R_k + (\overset{*}{\boldsymbol{v}}_0 - \mathbf{g}) \frac{\partial \boldsymbol{l}}{\partial \beta_i} = 0, \; i = 1, 2, \ldots, \tag{7.1.17}$$

where $\beta_i(t)$ is the set of generalized coordinates $r_{mn}(t)$ and $p_{mn}(t)$ of the problem; $M = m_0 + m_1$ is the mass of the mechanical system; $\boldsymbol{P}$ is the principal vector of all active forces relative to the pole $O$; $\boldsymbol{r}_{C_1}$ is the radius vector of the liquid mass center;

$$A_n = \rho \int\limits_{Q(t)} \varphi_n \, dQ, \quad A_{nk} = A_{kn} = \rho \int\limits_{Q(t)} (\nabla\varphi_n, \nabla\varphi_k) dQ,$$

$$m_1 = \rho \int\limits_{Q(t)} dQ, \quad l_1 = m_1 x_{C_1} = \rho \int\limits_{Q(t)} x \, dQ; \tag{7.1.18}$$

$$l_2 = m_1 y_{C_1} = \rho \int\limits_{Q(t)} y \, dQ; \quad l_3 = m_1 z_{C_1} = \rho \int\limits_{Q(t)} z \, dQ.$$

All quantities in (7.1.18) depend on time by means of the generalized coordinates $\beta_i(t)$ that define a part of the domain $Q(t)$, namely, the free surface $\Sigma$. The scalar form of nonlinear ordinary differential equations of motion of a rigid body of mass $m_0$ with a cylindrical cavity filled with liquid of depth $h$, which are derived from system (7.1.15)–(7.1.17) under certain additional restricting conditions (the assumptions of the theory of the third order of smallness), has the form

$$(m_0 + m_1)(w_{01} + \mathrm{g}) + \lambda_{11x}(\ddot{r}_{11} r_{11} + \ddot{p}_{11} p_{11} + \dot{r}_{11}^2 + \dot{p}_{11}^2) = P_1, \tag{7.1.19}$$

$$(m_0 + m_1) w_{02} + \sum_{n=1}^{\infty} \lambda_{1n} \ddot{p}_{1n} = P_2, \tag{7.1.20}$$

$$(m_0 + m_1) w_{03} + \sum_{n=1}^{\infty} \lambda_{1n} \ddot{r}_{1n} = P_3, \tag{7.1.21}$$

$$\begin{aligned} \mu_{11}(\ddot{r}_{11} + \sigma_{11}^2 r_{11}) &+ w_{01} \varkappa_{11} \mu_{11} r_{11} + \lambda_{11} w_{03} + d_1(r_{11}^2 \ddot{r}_{11} + r_{11} \dot{r}_{11}^2 \\ &+ r_{11} p_{11} \ddot{p}_{11} + r_{11} \dot{p}_{11}^2) + d_2(p_{11}^2 \ddot{r}_{11} + 2 p_{11} \dot{r}_{11} \dot{p}_{11} - r_{11} p_{11} \ddot{p}_{11} \\ &- 2 r_{11} \dot{p}_{11}^2) - d_3(p_{21} \ddot{r}_{11} - r_{21} \ddot{p}_{11} - \dot{r}_{11} \dot{p}_{21} - \dot{p}_{11} \dot{r}_{21}) \\ &+ d_4(r_{11} \ddot{p}_{21} - p_{11} \ddot{r}_{21}) + d_5(p_{01} \ddot{r}_{11} + \dot{r}_{11} \dot{p}_{01}) \\ &+ d_6 r_{11} \ddot{p}_{01} = 0, \end{aligned} \tag{7.1.22}$$

$$\mu_{1n}(\ddot{r}_{1n} + \sigma_{1n}^2 r_{1n}) + w_{01} \varkappa_{1n} \mu_{1n} r_{1n} + \lambda_{1n} w_{03} = 0, \quad n = 2, 3, \ldots, \tag{7.1.23}$$

$$\begin{aligned} \mu_{11}(\ddot{p}_{11} + \sigma_{11}^2 p_{11}) &+ w_{01} \varkappa_{11} \mu_{11} p_{11} + \lambda_{11} w_{02} + d_1(p_{11}^2 \ddot{p}_{11} + r_{11} p_{11} \ddot{r}_{11} \\ &+ p_{11} \dot{r}_{11}^2 + p_{11} \dot{p}_{11}^2) + d_2(r_{11}^2 \ddot{p}_{11} + 2 r_{11} \dot{r}_{11} \dot{p}_{11} - r_{11} p_{11} \ddot{r}_{11} \\ &- 2 p_{11} \dot{r}_{11}^2) + d_3(p_{21} \ddot{p}_{11} + r_{21} \ddot{r}_{11} + \dot{r}_{11} \dot{r}_{21} + \dot{p}_{11} \dot{p}_{21}) \\ &- d_4(p_{11} \ddot{p}_{21} + r_{11} \ddot{r}_{21}) + d_5(p_{01} \ddot{p}_{11} + \dot{p}_{11} \dot{p}_{01}) \\ &+ d_6 p_{11} \ddot{p}_{01} = 0, \end{aligned} \tag{7.1.24}$$

$$\mu_{1n}(\ddot{p}_{1n} + \sigma_{1n}^2 p_{1n}) + w_{01} \varkappa_{1n} \mu_{1n} p_{1n} + \lambda_{1n} w_{02} = 0, \quad n = 2, 3, \ldots, \tag{7.1.25}$$

$$\mu_{01}(\ddot{p}_{01}+\sigma_{01}^2 p_{01})+w_{01}\varkappa_{01}\mu_{01}p_{01}+d_6(r_{11}\ddot{r}_{11}+p_{11}\ddot{p}_{11})$$
$$+d_8(\dot{r}_{11}^2+\dot{p}_{11}^2)=0, \qquad (7.1.26)$$
$$\mu_{21}(\ddot{r}_{21}+\sigma_{21}^2 r_{21})+w_{01}\varkappa_{21}\mu_{21}r_{21}-d_4(p_{11}\ddot{r}_{11}+r_{11}\ddot{p}_{11})$$
$$-2d_7\dot{r}_{11}\dot{p}_{11}=0, \qquad (7.1.27)$$
$$\mu_{21}(\ddot{p}_{21}+\sigma_{21}^2 p_{21})+w_{01}\varkappa_{21}\mu_{21}p_{21}+d_4(r_{11}\ddot{r}_{11}-p_{11}\ddot{p}_{11})$$
$$+d_7(\dot{r}_{11}^2-\dot{p}_{11}^2)=0, \qquad (7.1.28)$$
$$\mu_{31}(\ddot{r}_{31}+\sigma_{31}^2 r_{31})+w_{01}\varkappa_{31}\mu_{31}r_{31}+d_9(-r_{11}\dot{p}_{11}^2+r_{11}\dot{r}_{11}^2-2p_{11}\dot{p}_{11}\dot{r}_{11})$$
$$+d_{10}(-2p_{11}r_{11}\ddot{p}_{11}-p_{11}^2\ddot{r}_{11}+r_{11}^2\ddot{r}_{11})+d_{11}(\dot{p}_{21}\dot{r}_{11}+\dot{p}_{11}\dot{r}_{21})$$
$$+d_{12}(r_{21}\ddot{p}_{11}+p_{21}\ddot{r}_{11})+d_{13}(r_{11}\ddot{p}_{21}+p_{11}\ddot{r}_{21})=0, \qquad (7.1.29)$$
$$\mu_{31}(\ddot{p}_{31}+\sigma_{31}^2 p_{31})+w_{01}\varkappa_{31}\mu_{31}p_{31}+d_9(-p_{11}\dot{p}_{11}^2+2r_{11}\dot{p}_{11}\dot{r}_{11}+p_{11}\dot{r}_{11}^2)$$
$$+d_{10}(-p_{11}^2\ddot{p}_{11}+r_{11}^2\ddot{p}_{11}+2p_{11}r_{11}\ddot{r}_{11})+d_{11}(\dot{p}_{11}\dot{p}_{21}-\dot{r}_{11}\dot{r}_{21})$$
$$+d_{12}(p_{21}\ddot{p}_{11}-r_{21}\ddot{r}_{11})+d_{13}(p_{11}\ddot{p}_{21}-r_{11}\ddot{r}_{21})=0. \qquad (7.1.30)$$

In this system of differential equations, in accordance with the notation used above, symbols $w_{0i}$ denote projections of the acceleration of the pole $O$ onto axes of the moving coordinate system, $\mu_{mn}$, $\lambda_{mn}$, $\lambda_{mnx}$, $d_1$, $d_2$, … , $d_{13}$ are the added masses, and $\sigma_{mn}^2 = \mathrm{g}\varkappa_{mn} = \mathrm{g}k_{mn} \tanh k_{mn}h$ is the square of the natural sloshing frequencies.

Properties of the eigenvalues $\varkappa_{mn}$ (the frequency parameters) and the natural sloshing modes $\varphi_{mn}$ of the special boundary problem (1.5.34) are studied in many scientific works and monographs [59, 124]. From the spectral theory viewpoint, representations (7.1.9) and (7.1.10) of the free surface and the velocity potential are quite reasonable. However, the problem of reduction of an infinite system of nonlinear ordinary differential equations (7.1.15)–(7.1.17) or (7.1.19)–(7.1.30) to a system with a finite number of degrees of freedom remains unsolved. By using certain transformations, the nonlinear system of ordinary differential equations (7.1.15)–(7.1.17) can be reduced to the system of Lagrange-type equations of the second kind (7.1.22)–(7.1.30) under assumptions proposed by G.S. Narimanov. These assumptions are derived from an experimental study of dynamical properties of the system under the influence of harmonic external forces in a neighborhood of resonance values of eigenfrequencies $\sigma_{mn}$. From the point of view of interaction of a body with liquid, the neighborhood of the natural sloshing frequency $\sigma_{11}$ is the most interesting and dangerous. In this connection, the generalized coordinates $r_{11}(t)$ and $p_{11}(t)$ are preferred in the nonlinear theory over the other parameters since the parameters, which are the closest to them with respect to eigenvalues, have the second or third order of smallness in parameters $r_{11}(t)$ and $p_{11}(t)$: these parameters are $p_{01}(t)$ ($\zeta_{01} = 3.831706$), $r_{21}(t)$ ($\zeta_{21} = 3.054237$), and $p_{21}(t)$ ($\zeta_{21} = 3.054237$), which are quantities of the second order of smallness ($p_{01} \sim r_{11}^2$, $r_{21} \sim r_{11}^2$, and $p_{21} \sim r_{11}^2$), and $r_{31}(t)$ ($\zeta_{31} = 4.20119$) and $p_{31}(t)$ ($\zeta_{31} = 4.20119$), which are quantities of the third order of smallness ($r_{31} \sim r_{11}^3$ and $p_{31} \sim r_{11}^3$). This hypothesis is confirmed in the process of obtaining the equations of motion of the system (up to the terms of the third order of smallness) that can be verified by analyzing the nonlinear part of (7.1.26)–(7.1.30).

The nonlinear system of equations (7.1.19)–(7.1.30), which is derived under these assumptions, is a basis of the so-called theory of the third order of smallness for the motion of a rigid body with a cavity partially filled with liquid. To construct the theory of the fifth order of smallness (and obtain equations of motion containing the terms up to $r_{11}^5$ inclusively), we have to consider additional modes that correspond to the parameters $r_{41}$, $p_{41}$, $r_{51}$, and $p_{51}$, and, possibly, some additional parameters, which are close to them. This theory would be useful not only for establishing the bounds of applicability of the theory of the third order of smallness but also for solving the problem of mathematical simulation of infinite-dimensional systems where we replace them by finite-dimensional systems. The verification of the theory of the third order of smallness in the form of model (7.1.19)–(7.1.28) in the case of solving some specific problems of dynamics (including an investigation of stability of corresponding physical processes) shows its high efficiency and good agreement of its results with experimental data [119, 141, 173, 53, 69].

The above-presented modal method makes it possible to find the coordinates of the center of the liquid mass for an arbitrarily perturbed free surface. In the body-fixed coordinate system $Oxyz$ with its origin at the center of the cylinder bottom, by using relations (7.1.18) and also taking into account (7.1.7) and (7.1.9), we obtain

$$m_1 x_{C_1} = \tfrac{1}{2} m_1 h + \tfrac{1}{2} \sum_{m,n}^{\infty} \lambda_{mnx} (r_{mn}^2 + p_{mn}^2),$$

$$m_1 y_{C_1} = \sum_{n=1}^{\infty} \lambda_{1n} p_{1n}, \quad m_1 z_{C_1} = \sum_{n=1}^{\infty} \lambda_{1n} r_{1n}, \tag{7.1.31}$$

where

$$\lambda_{0nx} = 2\pi\rho N_{0n}^2, \quad \lambda_{mnx} = \pi\rho N_{mn}^2, \quad m = 1, 2, \ldots,$$

$$\lambda_{1n} = \pi\rho E_{1n}, \quad E_{1n} = \int_{R_0}^{R_1} \xi^2 Y_m(k_{1n}\xi) d\xi, \tag{7.1.32}$$

$$N_{mn}^2 = \int_{R_0}^{R_1} \xi Y_m^2(k_{mn}\xi) d\xi.$$

Relations (7.1.32) are very important for obtaining the dynamical characteristics of a rigid body–liquid system.

**Table 7.2**

| $n$ | $\lambda_{1n}$ | $\mu_{1n}$ | $\mu_{0n}$ | $\mu_{2n}$ | $\mu_{3n}$ |
|---|---|---|---|---|---|
| 1 | 0.9267 | 0.6015 | 0.8199 | 0.2937 | 0.1832 |
| 2 | 0.1105 | 0.2843 | 0.4478 | 0.2134 | 0.1685 |
| 3 | 0.04311 | 0.1815 | 0.3088 | 0.1512 | 0.1288 |
| 4 | 0.02293 | 0.13321 | 0.2358 | 0.1165 | 0.1031 |
| 5 | 0.01422 | 0.1052 | 0.1907 | 0.09465 | 0.08579 |

**Table 7.3**

| $\bar{h}$ | $\mu_{01}$ | $\mu_{11}$ | $\mu_{21}$ | $\mu_{31}$ | $d_1$ | $d_2$ |
|---|---|---|---|---|---|---|
| 0.2 | 1.272 | 1.707 | 0.5393 | 0.2671 | 9.287 | 4.992 |
| 0.4 | 0.9001 | 0.9593 | 0.3497 | 0.1964 | 1.541 | 0.2769 |
| 0.6 | 0.8366 | 0.7498 | 0.3092 | 0.1856 | 0.6918 | −0.1336 |
| 0.8 | 0.8235 | 0.6682 | 0.2982 | 0.1855 | 0.4713 | −0.2171 |
| 1.0 | 0.8207 | 0.6325 | 0.2951 | 0.1833 | 0.3920 | −0.2416 |
| 1.2 | 0.8201 | 0.6161 | 0.2942 | 0.1832 | 0.3590 | −0.2503 |
| 1.6 | 0.8199 | 0.6048 | 0.2938 | 0.1832 | 0.3375 | −0.2554 |
| 2.0 | 0,8199 | 0.6022 | 0.2938 | 0.1832 | 0.3328 | −0.2564 |
| 3.0 | 0.8199 | 0.6015 | 0.2938 | 0.1832 | 0.3315 | −0.2567 |

Finally, we present the linearized system of equations of the translational motion of a rigid body with a cylindrical cavity partially filled with liquid, which is used in what follows. It has the form

$$(m_0 + m_1)(w_{01} + \mathrm{g}) = P_1, \tag{7.1.33}$$

$$(m_0 + m_1)w_{02} + \sum_{n=1}^{\infty} \lambda_{1n}\ddot{p}_{1n} = P_2, \tag{7.1.34}$$

$$(m_0 + m_1)w_{03} + \sum_{n=1}^{\infty} \lambda_{1n}\ddot{r}_{1n} = P_3, \tag{7.1.35}$$

$$\mu_{1n}(\ddot{r}_{1n} + \sigma_{1n}^2 r_{1n}) + w_{01}\varkappa_{1n}\mu_{1n}r_{1n} + \lambda_{1n}w_{03} = 0, \quad n = 1, 2, \ldots, \tag{7.1.36}$$

$$\mu_{1n}(\ddot{p}_{1n} + \sigma_{1n}^2 p_{1n}) + w_{01}\varkappa_{1n}\mu_{1n}p_{1n} + \lambda_{1n}w_{02} = 0, \quad n = 1, 2, \ldots, \tag{7.1.37}$$

where

$$\mu_{1n} = \frac{\pi\rho N_{1n}^2}{\varkappa_{1n}}; \quad \varkappa_{1n} = k_{1n}\tanh(k_{1n}h). \tag{7.1.38}$$

For a cavity that has the shape of upright circular cylinder of radius $r_0$, we have

$$\mu_{1n} = \frac{\pi\rho r_0^3(\zeta_{1n}^2 - 1)}{2\zeta_{1n}^3 \tanh \zeta_{1n}\bar{h}}; \quad \lambda_{1n} = \frac{\pi\rho r_0^3}{\zeta_{1n}^2}; \quad \bar{h} = \frac{h}{r_0}. \tag{7.1.39}$$

**Table 7.4**

| $\bar{h}$ | $d_3$ | $d_4$ | $d_5$ | $d_6$ | $d_7$ | $d_8$ |
|---|---|---|---|---|---|---|
| 0.2 | 1.821 | 1.502 | 4.714 | −1.625 | 2.4125 | −3.982 |
| 0.4 | 0.8847 | 0.2598 | 1.800 | −0.3732 | 0.7022 | −1.2732 |
| 0.6 | 0.7165 | 0.0399 | 1.276 | −0.1469 | 0.3982 | −0.7849 |
| 0.8 | 0.6621 | −0.0292 | 1.107 | −0.0731 | 0.3019 | −0.6266 |
| 1.0 | 0.6403 | −0.0560 | 1.039 | −0.0433 | 0.2642 | −0.5628 |
| 1.2 | 0.6307 | −0.0675 | 1.009 | −0.0300 | 0.2479 | −0.5345 |
| 1.6 | 0.6242 | −0.0750 | 0.9890 | −0.0210 | 0.2371 | −0.5155 |
| 2.0 | 0.6227 | −0.0767 | 0.9845 | −0.0190 | 0.2347 | −0.5113 |
| 3.0 | 0,6223 | −0.0771 | 0.9832 | −0.0184 | 0.2341 | −0.5100 |

**Table 7.5**

| $\bar{h}$ | $d_9$ | $d_{10}$ | $d_{11}$ | $d_{12}$ | $d_{13}$ |
|---|---|---|---|---|---|
| 0.2 | −7.5998 | −2.3789 | −2.6736 | −0.8288 | −0.5678 |
| 0.4 | −1.0141 | −0.2831 | −0.8941 | −0.1144 | −0.06716 |
| 0.6 | −0.3470 | −0.07804 | −0.3877 | 0.01712 | 0.00776 |
| 0.8 | −0.1881 | −0.03124 | −0.4850 | 0.06104 | 0.02512 |
| 1.0 | −0.1353 | −0.0164 | −0.4602 | 0.07914 | 0.02978 |
| 1.2 | −0.1147 | −0.01082 | −0.4456 | 0.08728 | 0.03094 |
| 1.6 | −0.1020 | −0.00752 | −0.4361 | 0.09286 | 0.03140 |
| 2.0 | −0.0993 | −0.00686 | −0.4341 | 0.09412 | 0.03144 |
| 3.0 | −0.0986 | −0.00668 | −0.4337 | 0.09441 | 0.03144 |

Actual values of the added masses $\lambda_{1n}$, $\mu_{mn}$, $d_1$, $d_2$, ... , $d_{13}$ for a cylindrical tank with radius $r_0 = 1$ of the free surface for $\rho = 1$ are given in Tables 7.2–7.5. Numerical values of the added masses for the case of deep liquid are given in Tables 7.2–7.3. Numerical values of the hydrodynamic coefficients versus the liquid depth $\bar{h}$ are presented in Tables 7.4–7.5.

The numeral values of the added masses for coaxial cylindrical tanks are obtained for a wide range of geometric parameters (depending on $\delta = R_0/R_1$ and $\bar{h} = h/R_1$). The passage from the dimensionless values given in the tables to the dimensional values is performed by the relations

$$\begin{gathered} h = \bar{h}R_1, \quad \mu_{mn} = \bar{\mu}_{mn}R_1^3, \quad \varkappa_{mn} = \frac{\bar{\varkappa}_{mn}}{R_1}, \quad d_1 = \bar{d}_1 R_1, \\ d_2 = \bar{d}_2 R_1, \quad d_3 = \bar{d}_3 R_1^2, \quad d_4 = \bar{d}_4 R_1^2, \quad d_5 = \bar{d}_5 R_1^2, \\ d_6 = \bar{d}_6 R_1^2, \quad d_7 = \bar{d}_7 R_1^2, \quad d_8 = \bar{d}_8 R_1^2, \quad d_9 = \bar{d}_9 R_1, \\ d_{10} = \bar{d}_{10} R_1, \quad d_{11} = \bar{d}_{11} R_1^2, \quad d_{12} = \bar{d}_{12} R_1^2, \quad d_{13} = \bar{d}_{13} R_1^2. \end{gathered} \tag{7.1.40}$$

The coefficients $\bar{d}_7$ and $\bar{d}_8$ are found from the relations

$$\bar{d}_7 = \bar{d}_4 + \tfrac{1}{2}\bar{d}_3, \quad \bar{d}_8 = \bar{d}_6 - \tfrac{1}{2}\bar{d}_5. \tag{7.1.41}$$

## 7.2 Equivalent mechanical systems for a rigid body with liquid

It follows from the results presented above that the modal approach makes it possible to describe the motion of the mechanical system rigid body–liquid, which is not homogeneous from the viewpoint of the aggregate state, in terms of an infinite system of linear or nonlinear differential equations. Presentation of the free surface, which is unknown in advance, and of the velocity potential in the form of decompositions (7.1.9) and (7.1.10) is a very important point of this approach. The corresponding relations have a reliable mathematical foundation on the base of the spectral theory of the related linear eigenvalue problem with a parameter in boundary conditions (1.5.34). It is worth noting especially that the basis of decompositions (7.1.9) and (7.1.10) is formed by the natural sloshing modes and, in the linear theory, this leads to an infinite collection of independent linear oscillators that have the type of a mathematical pendulum or spring mass. Frequencies of free oscillations of these modes form the increasing sequence $\sigma_1^2 < \sigma_2^2 < \sigma_3^2 < \ldots < \sigma_n^2$, so the lowest natural sloshing frequency is dominant.

The coordinates of the center of mass of a liquid, which is in its arbitrary perturbed state, are completely defined by the generalized Fourier coefficients (the generalized coordinates) that can be presented in the form (7.1.31). Note that the projections of the radius vector of the center of liquid mass onto $Oy$- and $Oz$-axes depend on the generalized coordinates linearly, and a similar projection onto the $Ox$-axis has the quadratic dependence. This determines properties of important concepts of the analytical mechanics: the static moment $\boldsymbol{L} = \rho \int_Q \boldsymbol{r}\, dQ$ and the potential energy $\Pi$ of a liquid in the gravitational field. This shows that the mechanical system under consideration and other objects of the analytical mechanics have common properties.

We also consider the nature of interaction forces of a rigid body with a liquid in the problem in its linear and nonlinear statements. Comparing the equations of motion (7.1.19)–(7.1.21) and (7.1.33)–(7.1.35), we can see that the projections of hydrodynamic forces onto $Oy$- and $Oz$-axes are linear in acceleration with respect to the generalized coordinates only for $m = 1$, therefore, the other modes with $m \neq 1$ do not interact with body directly. However, equations (7.1.22)–(7.1.30) of the nonlinear theory imply that these modes (with $m \neq 1$) interact with other modes in the case of considerable perturbation of the generalized coordinates $r_{11}(t)$ and $p_{11}(t)$ and strongly affect them both quantitatively and qualitatively. The modes with generalized coordinates

$p_{01}(t)$, $r_{21}(t)$, $p_{21}(t)$, $r_{13}(t)$, and $p_{31}(t)$ interact with the rigid body via the generalized coordinates $r_{11}(t)$ and $p_{11}(t)$. In the linear theory, the projection of interaction forces onto the $Ox$-axis is absent, whereas such a projection in the nonlinear theory is revealed in the form of nonlinear terms of the second order of smallness in accelerations of the generalized coordinates $r_{11}(t)$ and $p_{11}(t)$.

Significant experience of using the mathematical models of type (7.1.19)–(7.1.30) and (7.1.33)–(7.1.37) in the practice of engineering shows that it is necessary to have an interpretation of physical results obtained by these models in terms of their equivalent mechanical analogs. The mechanical analog in the form of a rigid body with attached mathematical pendulums, which was proposed by B. I. Rabinovich [149] and developed later in [108], proved to be an efficient model in the linear theory.

We start with a mathematical model of plane-parallel motion in the plane $Oxz$ for the case of a rigid body with mathematical pendulums attached to the $Ox$-axis by using cylindrical hinges. Let $m^0$ denote the mass of the rigid body, let $m_n$ denote masses of pendulums, and let $l_n$ and $\alpha_n$ be, accordingly, the lengths and deflection angles with respect to the vertical direction for these pendulums. The kinetic energy of the considered system has the form

$$T = \tfrac{1}{2}[(m^0 + \sum_n m_n)(v_{01}^2 + v_{03}^2) + 2v_{01}\sum_n m_n l_n \dot{\alpha}_n \sin\alpha_n +$$

$$+2v_{03}\sum_n m_n l_n \dot{\alpha}_n \cos\alpha_n + \sum_n m_n l_n^2 \dot{\alpha}_n^2. \tag{7.2.1}$$

Due to (7.1.3), the potential energy of this system can be reduced to the form

$$\Pi = -M\mathbf{g}\cdot \boldsymbol{r}'_C; \quad \boldsymbol{r}'_C = \boldsymbol{r}'_0 + \boldsymbol{r}_C, \tag{7.2.2}$$

where $\boldsymbol{r}'_C$ and $\boldsymbol{r}_C$ are the radius vectors of the center of mass of the system relative to the points $O'$ and $O$, respectively. By choosing the generalized coordinates of the system equal to the components of the displacement vector $\boldsymbol{u}_0$ of the point $O$ and the deflection angles of the mathematical pendulums, we obtain the following nonlinear system of ordinary differential equations of motion of the body–pendulum system, which is derived from the Euler–Lagrange equations of the second kind [108]:

$$(m^0 + \sum_n m_n)\ddot{u}_{01} + \sum_n m_n l_n(\sin\alpha_n\ddot{\alpha}_n + \dot{\alpha}_n^2\cos\alpha_n) = P_1, \tag{7.2.3}$$

$$(m^0 + \sum_n m_n)\ddot{u}_{03} + \sum_n m_n l_n(\cos\alpha_n\ddot{\alpha}_n - \dot{\alpha}_n\sin\alpha_n) = P_3, \tag{7.2.4}$$

$$m_n l_n^2\ddot{\alpha}_n + \mathrm{g} m_n l_n\sin\alpha_n + \ddot{u}_{01}m_n l_n\sin\alpha_n + \ddot{u}_{03}m_n l_n\cos\alpha_n = 0, \quad n = 1, 2, \ldots. \tag{7.2.5}$$

The system of equations (7.2.3)–(7.2.5) can be generalized by considering additionally the three-dimensional displacement vector $\boldsymbol{u}_0(u_{01},\ u_{02},\ u_{03})$ and replacing the cylindrical hinge at the suspension point of pendulums by a spherical hinge. This increases the number of degrees of freedom of the system due to both the projection $u_{02}$ of the displacement of the point $O$ and the deflection angles $\beta_n$ of spherical pendulums in the plane $Oxy$. A further generalization of the mathematical model of the mechanical body–pendulum system is based on considering angular displacements of a rigid body [108, 149], although we do not dwell upon this problem here.

Returning to the analysis of mathematical model (7.2.3)–(7.2.5), we note a specific feature, which is important for applications [108]. For a series of dynamical problems, the horizontal displacement of a pendulum $z_n = l_n \sin\alpha_n$ referred to a certain characteristic size of the system (e.g., to the tank radius $r_0$) is a quantity of the first order of smallness in spite of the fact that the deflection angle $\alpha_n$ is not small. In this case, the vertical displacement of a pendulum, up to the terms of order $\sin^4\alpha_n$, can be defined by the relation

$$l_n \cos\alpha_n = l_n\sqrt{1-\sin^2\alpha_n} = l_n\left(1-\tfrac{1}{2}\sin^2\alpha_n - \ldots\right), \tag{7.2.6}$$

which is used in what follows.

First, we compare qualitatively the mathematical hydrodynamic-type models (7.1.19)–(7.1.21) and (7.1.33)–(7.1.35) with the corresponding mechanical analog model (7.2.3)–(7.2.5). The equation of forces (7.1.21) projected onto the $Ox$-axis can be reduced to the equation of motion (7.2.4) for a mechanical analog by introducing the generalized coordinates $\alpha_{1n}$ defined by the relation

$$r_{1n} = a_{1n}\sin\alpha_{1n}, \tag{7.2.7}$$

where $a_{1n}$ is the normalization factor that has the dimension of length. Substituting (7.2.7) into (7.1.21), we obtain the equation

$$(m_0+m_1)w_{03} + \sum_n \lambda_{1n}a_{1n}(\ddot{\alpha}_{1n}\cos\alpha_{1n} - \dot{\alpha}_{1n}^2\sin\alpha_{1n}) = P_3. \tag{7.2.8}$$

This equation and equality (7.2.4) lead to the relation

$$m_{1n}l_{1n} = \lambda_{1n}a_{1n}. \tag{7.2.9}$$

A similar relation for spherical pendulums can be obtained from (7.1.20) by using the following substitution formula:

$$p_{1n} = a_{1n}\sin\beta_{1n}. \tag{7.2.10}$$

In a similar way, by using substitutions (7.2.7) and (7.2.10), we transform the equations of forces (7.1.19) of the hydrodynamic theory. For the projection of the hydrodynamic force onto the $Ox$-axis, we use (7.2.6) in order to obtain

$$\lambda_{11x}(\ddot{r}_{11}r_{11} + \ddot{p}_{11}p_{11} + \dot{r}_{11}^2 + \dot{p}_{11}^2) = \lambda_{11x}a_{11}^2\left(\tfrac{1}{2}\ddot{\alpha}_{11}\sin 2\alpha_{11} + \dot{\alpha}_{11}^2\cos 2\alpha_{11}\right.$$

$$\left. + \tfrac{1}{2}\ddot{\beta}_{11}\sin 2\beta_{11} + \dot{\beta}_{11}^2\cos 2\beta_{11}\right)$$

$$\approx \lambda_{11x}a_{11}^2(\ddot{\alpha}_{11}\sin\alpha_{11} + \dot{\alpha}_{11}^2\cos\alpha_{11} + \ddot{\beta}_{11}\sin\beta_{11} + \dot{\beta}_{11}^2\cos\beta_{11}). \tag{7.2.11}$$

Relation (7.2.11) is presented up to the terms of smallness of $\ddot{\alpha}_{11}\sin^3\alpha_{11}$, etc., i.e., up to the terms of the fourth order of smallness in $\alpha_{11}$ and $\beta_{11}$. Comparing relations (7.2.11) with analogous relations of the theory of pendulum, we obtain

$$\lambda_{11x}a_{11}^2 = m_{11}l_{11} \tag{7.2.12}$$

up to the same accuracy. Since eigenfrequencies of pendulum oscillators should coincide with eigenfrequencies of hydrodynamic oscillators, we can use the relations for added liquid masses (7.1.32), (7.1.38), and (7.1.39) in order to obtain the following relations from (7.2.9) and (7.2.12):

$$l_{11} = \frac{1}{\varkappa_{11}}; \quad a_{11} = \frac{\lambda_{11}}{\varkappa_{11}\mu_{11}}; \quad m_{11} = \frac{\lambda_{11}^2}{\mu_{11}}. \tag{7.2.13}$$

Under the restrictions accepted to derive relation (7.2.11), we can obtain a more general relation for the projection of the hydrodynamic force onto the $Ox$-axis. This relation is derived from the hydrodynamic equation (7.1.15) and relations for projections of the center of mass of liquid (7.1.31), and it has the form

$$\lambda_{mnx}(\ddot{r}_{mn}r_{mn} + \ddot{p}_{mn}p_{mn} + \dot{r}_{mn}^2 + \dot{p}_{mn}^2) \approx \lambda_{mnx}a_{mn}^2(\ddot{\alpha}_{mn}\sin\alpha_{mn}$$

$$+ \dot{\alpha}_{mn}^2\cos\alpha_{mn} + \ddot{\beta}_{mn}\sin\beta_{mn} + \dot{\beta}_{mn}^2\cos\beta_{mn}). \tag{7.2.14}$$

This relation implies that the equations

$$l_{1n} = \frac{1}{\varkappa_{1n}}; \quad a_{1n} = \frac{\lambda_{1n}}{\varkappa_{1n}\mu_{1n}}; \quad m_{1n} = \frac{\lambda_{1n}^2}{\mu_{1n}} \tag{7.2.15}$$

are true for all (nonsymmetric) modes of interaction with rigid body via the added mass $\lambda_{1n}$. For other oscillators ($m \neq 1$), which appear in the hydrodynamic theory, relation (7.2.14) yields

$$l_{mn} = \frac{1}{\varkappa_{mn}}; \quad m_{mn} = a_{mn}^2\mu_{mn}\varkappa_{mn}^2. \tag{7.2.16}$$

These oscillators contribute only to projections of forces onto the $Ox$-axis. They are not present in (7.2.11) since the assumptions of the theory of the third order of smallness are supposed to be satisfied. Within the framework of this theory, they affect the generalized coordinates $r_{11}(t)$ and $p_{11}(t)$ (and, hence, $\alpha_{11}$ and $\beta_{11}$) and, thereby, interact with the rigid body indirectly.

Thus, provided that the parameters of spherical pendulums ($l_{1n}$ and $m_{1n}$) and the normalization factor $a_{1n}$ in accordance with relations (7.2.15) are chosen, the equations of forces (7.1.19)–(7.1.21) of the nonlinear hydrodynamic theory of the third order of smallness can be presented in the form of a nonlinear mathematical model of motion of a rigid body with spherical pendulums. An important advantage of this analog is the invariance of the quantity $m_{1n} = \lambda_{1n}^2/\mu_{1n}$ with respect to the normalization of solutions of the boundary-value problem (1.5.34). As it was mentioned above, this provides a possibility to estimate the contribution of each mode to the dynamical processes under study.

Pendulums in model (7.2.3)–(7.2.5) are primary. Their masses (7.2.15) are determined by the added inertial masses $\lambda_{1n}$ and the masses of hydrodynamic oscillators $\mu_{1n}$ as a result of the interaction of modes, which have generalized coordinates $r_{1n}$ and $p_{1n}$, with a rigid body. By analogy, we introduce lengths and masses of "secondary" pendulums that are used to construct the mathematical model of nonlinear hydrodynamic theory (7.1.26)–(7.1.30). The generalized forces in the Lagrange-type equations of the second kind (7.1.26)–(7.1.30) for the generalized coordinates $r_{mn}$ and $p_{mn}$, $m \neq 1$, are present in the nonlinear theory due to the energy exchange between the modes. By analyzing (7.1.26)–(7.1.30), we conclude that these forces have a complicated nonlinear structure.

Similarly to transformations of the equations of forces (7.1.19)–(7.1.21), we perform corresponding transformations in (7.1.22)–(7.1.30) by using the substitution formulas (7.2.7) and (7.2.10) and relations (7.2.15).

We start with equations (7.1.19)–(7.1.22) that define a system of Lagrange-type equations of the second kind for the generalized coordinates $r_{1n}(t)$ and $p_{1n}(t)$ of modes, which interact with the rigid body directly. To bring the structure of these equations into accordance with the structure of equations of motion of a mechanical analog (in the form of mathematical or spherical pendulums), we multiply each equation by the factor $a_{1n}\cos\alpha_{1n}$ and $a_{1n}\cos\beta_{1n}$, respectively. For the group of equations (7.1.23), we obtain

$$a_{1n}^2\mu_{1n}(\cos^2\alpha_{1n}\ddot{\alpha}_{1n} - \dot{\alpha}_{1n}^2\sin\alpha_{1n}\cos\alpha_{1n}) + a_{1n}^2\mu_{1n}g\varkappa_{1n}w_{01}\sin\alpha_{1n}\cos\alpha_{1n}$$
$$+w_{03}\lambda_{1n}a_{1n}\cos\alpha_{1n} = 0, \quad n = 2, 3, \dots. \tag{7.2.17}$$

Due to relations (7.2.15) and

$$a_{1n}^2\cos^2\alpha_{1n} = a_{1n}^2(1 - \sin^2\alpha_{1n}),$$
$$a_{1n}^2\sin\alpha_{1n}\cos\alpha_{1n} = a_{1n}^2\sin\alpha_{1n}\left(1 - \tfrac{1}{2}\sin^2\alpha_{1n} - \tfrac{1}{8}\sin^4\alpha_{1n} - \dots\right)$$

(neglecting the terms of the order of $a_{1n}^2\sin^3\alpha_{1n}$ and higher), we obtain

$$m_{1n}l_{1n}^2\ddot{\alpha}_{1n} + gm_{1n}l_{1n}\sin\alpha_{1n} + w_{01}l_{1n}m_{1n}\sin\alpha_{1n}$$
$$+w_{03}m_{1n}l_{1n}\cos\alpha_{1n} = 0, \quad n = 2, 3, \dots. \tag{7.2.18}$$

These restrictions are admissible from the viewpoint of the theory of the third order of smallness. By similar transformations, equations (7.1.25) can be reduced to the following equations of motion of a pendulum analog in the plane $Oxy$

$$m_{1n}l_{1n}^2\ddot{\beta}_{1n} + gm_{1n}l_{1n}\sin\beta_{1n} + w_{01}l_{1n}m_{1n}\sin\beta_{1n}$$
$$+w_{02}m_{1n}l_{1n}\cos\beta_{1n} = 0 \quad (n = 2, 3, \ldots). \tag{7.2.19}$$

In the process of transformation of equations (7.1.22) and (7.1.24), we should take into account that, for the generalized coordinates $r_{mn}(t)$ and $p_{mn}(t)$ in the theory of the third order of smallness, the following subordination of generalized coordinates is assumed:

$$p_{01} \sim r_{11}^2, \quad p_{21} \sim r_{11}^2, \quad r_{21} \sim r_{11}^2, \quad r_{31} \sim r_{11}^3, \quad p_{31} \sim r_{11}^3, \quad r_{11} \sim p_{11}. \tag{7.2.20}$$

For $r_{mn} = a_{mn}\sin\alpha_{mn}$ and $p_{mn} = a_{mn}\cos\beta_{mn}$, this yields

$$a_{01}\sin\beta_{01} \sim a_{11}^2\sin^2\alpha_{11}, \quad a_{21}\sin\alpha_{21} \sim a_{11}^2\sin^2\alpha_{11},$$

$$a_{01}\cos\beta_{01} = a_{01}\left(1 - \tfrac{1}{2}\sin^2\beta_{01}\right) \sim a_{01} - \tfrac{1}{2}a_{11}^2\sin\beta_{01}\sin^2\alpha_{11}, \quad \text{etc.}$$

Thus, the nonlinear terms of the type

$$d_1(r_{11}^2\ddot{r}_{11} + r_{11}\dot{r}_{11}^2)a_{11}\cos\alpha_{11}, \quad d_4a_{11}\cos\alpha_{11}r_{11}\ddot{p}_{21}$$

in (7.1.22)–(7.1.24) and of the type

$$d_6a_{01}\cos\beta_{01}r_{11}\ddot{r}_{11}$$

in (7.1.26)–(7.1.28) are transformed as follows:

$$d_1(r_{11}^2\ddot{r}_{11} + r_{11}\dot{r}_{11}^2)a_{11}\cos\alpha_{11} = d_1a_{11}^4\cos\alpha_{11}\sin\alpha_{11}\left(\tfrac{1}{2}\sin 2\alpha_{11}\ddot{\alpha}_{11}\right.$$
$$\left.-\dot{\alpha}^2\cos 2\alpha_{11}\right) = d_1a_{11}^4\sin\alpha_{11}(\ddot{\alpha}_{11}\sin\alpha_{11} - \dot{\alpha}_{11}^2),$$
$$d_4a_{11}^2a_{21}\cos\alpha_{11}\sin\alpha_{11}(\ddot{\beta}_{21}\cos\beta_{21} - \dot{\beta}_{21}^2\sin\beta_{21}) = d_4a_{11}^2a_{21}\sin\alpha_{11} \tag{7.2.21}$$
$$\times\left(1 - \tfrac{1}{2}\sin^2\alpha_{11} - \tfrac{1}{8}\sin^4\alpha_{11} - \ldots\right)\left(\ddot{\beta}_{21}\cos\beta_{21} - \dot{\beta}_{21}^2\sin\beta_{21}\right)$$
$$= d_4a_{11}^2a_{21}\ddot{\beta}_{21}\sin\alpha_{11},$$
$$d_6a_{01}\cos\beta_{01}a_{11}^2\sin\alpha_{11}(\ddot{\alpha}_{11}\cos\alpha_{11} - \dot{\alpha}_{11}^2\sin\alpha_{11}) = d_6a_{11}^2a_{01}\ddot{\alpha}_{11}\sin\alpha_{11}.$$

In relations (7.2.21), the terms of the fourth order of smallness and higher are omitted.

For the "secondary" pendulums that correspond to oscillators of hydrodynamic type (7.1.26)–(7.1.30), we now introduce parameters similar to (7.2.15). Choosing the lengths of these pendulums, it is logical to use the principle of conservation of frequencies of natural oscillations. Since the normalization coefficients $a_{mn}$ have dimensions of length, we choose them by analogy with the previous case, taking into account that the role of "perturbation" forces of the type $\lambda_{1n}w_{03}$ in (7.1.26)–(7.1.30) is played by inertial forces caused by dominant sloshing modes with generalized coordinates $r_{11}(t)$ and $p_{11}(t)$. By using the theory of dimensions and the results from Table 7.3, we set

$$a_{01} = \frac{a_{11}d_8}{\varkappa_{01}\mu_{01}}; \quad a_{21} = \frac{a_{11}d_7}{\varkappa_{21}\mu_{21}}; \quad a_{31} = \frac{a_{11}^2 d_9^*}{\varkappa_{31}\mu_{31}}; \quad l_{mn} = \frac{1}{\varkappa_{mn}}. \tag{7.2.22}$$

In accordance with relation (7.2.16), we obtain the following relations for masses of "secondary" pendulums:

$$m_{01} = a_{11}^2 \frac{d_8^2}{\mu_{01}}; \quad m_{21} = a_{11}^2 \frac{d_7^2}{\mu_{21}}; \quad m_{31} = \frac{a_{11}^4 d_9^{*2}}{\mu_{31}}; \quad d_9^* = d_9 - \frac{d_7 d_{13}}{\mu_{21}}. \tag{7.2.23}$$

Summarizing the above-presented results, we can write down the following equations of translational motion of a rigid body with a partially filled cylindrical cavity, which present a nonlinear system of ordinary differential equations in an invariant form of the type of motion equations of an equivalent mechanical system:

$$(m_0 + m_1)(w_{01} + \mathrm{g}) + m_{11}l_{11}(\sin\alpha_{11}\ddot{\alpha}_{11} + \dot{\alpha}_{11}^2\cos\alpha_{11}) = P_1, \tag{7.2.24}$$

$$(m_0 + m_1)w_{02} + \sum_n m_{1n}l_{1n}(\cos\beta_{1n}\ddot{\beta}_{1n} + \dot{\beta}_{1n}^2\sin\beta_{1n}) = P_2, \tag{7.2.25}$$

$$(m_0 + m_1)w_{03} + \sum_n m_{1n}l_{1n}(\cos\alpha_{1n}\ddot{\alpha}_{1n} - \dot{\alpha}_{1n}^2\sin\alpha_{1n}) = P_3, \tag{7.2.26}$$

$$\begin{aligned}
&m_{11}l_{11}^2\ddot{\alpha}_{11} + \mathrm{g}l_{11}m_{11}\sin\alpha_{11} + w_{01}l_{11}m_{11}\sin\alpha_{11} + w_{03}m_{11}l_{11}\cos\alpha_{11} \\
&\quad - m_{11}l_{11}\left(\ddot{\alpha}_{11}\sin^2\alpha_{11} + \dot{\alpha}_{11}^2\sin\alpha_{11} + \tfrac{1}{2}\mathrm{g}\sin^3\alpha_{11} + \tfrac{1}{2}w_{01}\sin^3\alpha_{11}\right) \\
&\quad + b_1\sin\alpha_{11}(\ddot{\alpha}_{11}\sin\alpha_{11} + \dot{\alpha}_{11}^2\cos\alpha_{11} + \ddot{\beta}_{11}\sin\beta_{11} + \dot{\beta}_{11}^2\cos\beta_{11}) \\
&\quad + b_2(\ddot{\alpha}_{11}\sin^2\beta_{11} + 2\dot{\alpha}_{11}\dot{\beta}_{11}\sin\beta_{11} - \ddot{\beta}_{11}\sin\alpha_{11}\sin\beta_{11} - 2\dot{\beta}_{11}^2\sin\alpha_{11}) \\
&\quad - b_3(\ddot{\alpha}_{11}\sin\beta_{21} - \ddot{\beta}_{11}\sin\alpha_{21} + \dot{\alpha}_{11}\dot{\beta}_{21} - \dot{\beta}_{11}\dot{\alpha}_{21}) \\
&\quad + b_4(\ddot{\beta}_{21}\sin\alpha_{11} - \ddot{\alpha}_{21}\sin\beta_{11}) + b_5(\ddot{\alpha}_{11}\sin\beta_{01} + \dot{\alpha}_{11}\dot{\beta}_{01}) \\
&\quad + d_6 a_{11}^2 a_{01}\ddot{\beta}_{01}\sin\alpha_{11} = 0,
\end{aligned} \tag{7.2.27}$$

$$\begin{aligned}
&m_{1n}l_{1n}^2\ddot{\alpha}_{1n} + \mathrm{g}l_{1n}m_{1n}\sin\alpha_{1n} + w_{01}l_{1n}m_{1n}\sin\alpha_{1n} \\
&\quad + w_{03}m_{1n}l_{1n}\cos\alpha_{1n} = 0, \quad n = 2, 3, \dots,
\end{aligned} \tag{7.2.28}$$

$$\begin{aligned}
&m_{11}l_{11}^2\ddot{\beta}_{11} + \mathrm{g}l_{11}m_{11}\sin\beta_{11} + w_{01}l_{11}m_{11}\sin\beta_{11} + w_{02}m_{11}l_{11}\cos\beta_{11} \\
&\quad - m_{11}l_{11}\left(\ddot{\beta}_{11}\sin^2\beta_{11} + \dot{\beta}_{11}^2\sin\beta_{11} + \tfrac{1}{2}\mathrm{g}\sin^3\beta_{11} + \tfrac{1}{2}w_{01}\sin^3\beta_{11}\right) \\
&\quad + b_1\sin\beta_{11}(\ddot{\beta}_{11}\sin\beta_{11} + \ddot{\alpha}_{11}\sin\alpha_{11} + \dot{\beta}_{11}^2 + \dot{\alpha}_{11}^2) \\
&\quad + b_2(\ddot{\beta}_{11}\sin^2\alpha_{11} + 2\dot{\alpha}_{11}\dot{\beta}_{11}\sin\alpha_{11} - \ddot{\alpha}_{11}\sin\alpha_{11}\sin\beta_{11} - 2\dot{\alpha}_{11}^2\sin\beta_{11})
\end{aligned}$$

$$+b_3(\ddot{\beta}_{11}\sin\beta_{21}+\ddot{\alpha}_{11}\sin\alpha_{21}+\dot{\alpha}_{11}\dot{\alpha}_{21}+\dot{\beta}_{11}\dot{\beta}_{21})$$
$$-d_4a_{11}^2a_{21}(\ddot{\beta}_{21}\sin\beta_{11}+\ddot{\alpha}_{21}\sin\alpha_{11})+d_5a_{11}^2a_{01}(\ddot{\beta}_{11}\sin\beta_{01}+\dot{\beta}_{11}\dot{\beta}_{01})$$
$$+d_6a_{11}^2a_{01}\ddot{\beta}_{01}\sin\beta_{11}=0, \tag{7.2.29}$$

$$m_{1n}l_{1n}^2\ddot{\beta}_{1n}+gl_{1n}m_{1n}\sin\beta_{1n}+w_{01}l_{1n}m_{1n}\sin\beta_{1n}$$
$$+w_{02}m_{1n}l_{1n}\cos\beta_{1n}=0,\quad n=2,3,\ldots, \tag{7.2.30}$$

$$m_{01}l_{01}^2\ddot{\beta}_{01}+gm_{01}l_{01}\sin\beta_{01}+w_{01}m_{01}l_{01}\sin\beta_{01}+b_8(\dot{\alpha}_{11}^2+\dot{\beta}_{11}^2)$$
$$+b_6(\ddot{\alpha}_{11}\sin\alpha_{11}+\ddot{\beta}_{11}\sin\beta_{11})=0, \tag{7.2.31}$$

$$m_{21}l_{21}^2\ddot{\alpha}_{21}+gm_{21}l_{21}\sin\alpha_{21}+w_{01}m_{21}l_{21}\sin\alpha_{21}-2b_7\dot{\alpha}_{11}\dot{\beta}_{11}$$
$$-b_4(\ddot{\alpha}_{11}\sin\beta_{11}+\ddot{\beta}_{11}\sin\alpha_{11})=0, \tag{7.2.32}$$

$$m_{21}l_{21}^2\ddot{\beta}_{21}+gm_{21}l_{21}\sin\beta_{21}+w_{01}m_{21}l_{21}\sin\beta_{21}+b_7(\dot{\alpha}_{11}^2-\dot{\beta}_{11}^2)$$
$$+b_4(\ddot{\alpha}_{11}\sin\alpha_{11}-\ddot{\beta}_{11}\sin\beta_{11})=0, \tag{7.2.33}$$

$$m_{31}l_{31}^2\ddot{\alpha}_{31}+gm_{31}l_{31}\sin\alpha_{31}+w_{01}m_{31}l_{31}\sin\alpha_{31}$$
$$+b_9(\dot{\beta}_{11}^2\sin\alpha_{11}+2\dot{\alpha}_{11}\dot{\beta}_{11}\sin\beta_{11}-\alpha_{11}^2\sin\alpha_{11})$$
$$+b_{10}(2\ddot{\beta}_{11}\sin\alpha_{11}\sin\beta_{11}+\ddot{\alpha}_{11}\sin^2\beta_{11}-\ddot{\alpha}_{11}\sin^2\alpha_{11})$$
$$+b_{11}(\dot{\beta}_{21}\dot{\alpha}_{11}+\dot{\beta}_{11}\dot{\alpha}_{21})+b_{12}(\alpha_{21}\ddot{\beta}_{11}+\beta_{21}\ddot{\alpha}_{11})$$
$$+b_{13}(\alpha_{11}\ddot{\beta}_{21}+\beta_{11}\ddot{\alpha}_{21})=0, \tag{7.2.34}$$

$$m_{31}l_{31}^2\ddot{\beta}_{31}+gm_{31}l_{31}\sin\beta_{31}+w_{01}m_{31}l_{31}\sin\beta_{31}$$
$$+b_9(\dot{\beta}_{11}^2\sin\beta_{11}-2\dot{\alpha}_{11}\dot{\beta}_{11}\sin\alpha_{11}-\dot{\alpha}_{11}^2\sin\beta_{11})$$
$$+b_{10}(\ddot{\beta}_{11}\sin^2\beta_{11}-\ddot{\beta}_{11}\sin^2\alpha_{11}-2\ddot{\alpha}_{11}\sin\beta_{11}\sin\alpha_{11})$$
$$+b_{11}(\dot{\beta}_{11}\dot{\beta}_{21}-\dot{\alpha}_{11}\dot{\alpha}_{21})+b_{12}(\ddot{\beta}_{11}\beta_{21}-\alpha_{21}\ddot{\alpha}_{11})$$
$$+b_{13}(\beta_{11}\ddot{\beta}_{21}-\alpha_{11}\ddot{\alpha}_{21})=0, \tag{7.2.35}$$

where $b_1=d_1a_{11}^4$, $b_2=d_2a_{11}^4$, $b_3=d_3c_{21}$, $b_4=d_4c_{21}$, $b_5=d_5c_{01}$, $b_6=d_6c_{01}$, $b_7=d_7c_{21}$, $b_8=d_8c_{01}$, $b_9=-d_9c_{31}$, $b_{10}=-d_{10}c_{31}$, $b_{11}=d_{11}c_{23}$, $b_{12}=d_{12}c_{23}$, $b_{13}=d_{13}c_{23}$, $c_{01}=a_{11}^2a_{01}$, $c_{21}=a_{11}^2a_{21}$, $c_{31}=a_{11}^3a_{31}$, and $c_{23}=a_{11}a_{21}a_{31}$.

This system of differential equations can be called the system of differential equations of invariants of motion of a body–liquid system. All coefficients of these equations are independent of the method of solution of basic hydrodynamic eigenvalue boundary problems (1.5.34) with a parameter in boundary conditions. We note also that, in addition to relations (7.2.15) in the linear theory, the following relation for the mass of an equivalent rigid body is used:

$$m^0=m_0+m_1-\sum_n\frac{\lambda_{1n}^2}{\mu_{1n}}.$$

**Table 7.6**

| $n$ | $m_{1n}$ | $l_{1n}$ | $a_{1n}$ | $m_{1n}l_{1n}^2$ |
|---|---|---|---|---|
| 1 | 1.428 | 0.5431 | 0.8368 | 0.4212 |
| 2 | 0.0431 | 0.1876 | 0.07315 | $0.1516.10^{-2}$ |
| 3 | 0.01024 | 0.1172 | 0.02783 | $0.1405.10^{-3}$ |
| 4 | 0.003937 | 0.08543 | 0.01467 | $0.2873.10^{-4}$ |
| 5 | 0.002274 | 0.06728 | 0.01076 | $0.1029.10^{-4}$ |

To understand more comprehensively the considered mechanical system as an object of the analytical mechanics, we provide numerical values of the parameters of an equivalent mechanical system for a specific case, namely, for a cylindrical tank with the nonperturbed free surface of radius $r_0 = 1.0$ m and the filling depth $h = 3.0$ m (Table 7.6).

The numerical values of the related parameters of mathematical pendulums for symmetric modes of oscillations of the free surface are the following: $m_{01} = 0.1555$; $m_{21} = 0.1306$; $m_{31} = 0.07009$; $l_{01} = 0.2610$; $l_{21} = 0.3274$; $l_{31} = 0.2380$; $a_{01} = -0.1358$; $a_{21} = 0.2183$; $a_{31} = -0.1472$; $m_{01}l_{01}^2 = 0.01059$; $m_{21}l_{21}^2 = 0.01400$; $m_{31}l_{31}^2 = 0.00397$.

The application of this model to the study of conventional problems of forced liquid sloshing in a neighborhood of the main resonance of the perturbing force confirms a high efficiency of the model. Theoretical results agree both qualitatively and quantitatively with experimental data [119]. In our opinion, models of this type may be very useful in studying the dynamical processes in body–liquid systems under the condition of vibrating motion of the rigid body and, especially, in studying the dynamical shapes of equilibrium of the free surface of a liquid in a vibrating tank.

# Chapter 8

# Forced finite-amplitude liquid sloshing in moving vessels

In the present chapter, we study nonlinear sloshing phenomena caused by harmonic excitations of a rigid container with liquid. In practice, the most interesting problem is the case of resonant forcing in the vicinity of the lowest natural frequency. In this case, the typical nonlinear phenomena are dependence of the sloshing frequency on wave amplitude, boundedness of wave elevations in the resonant mode, mobility of nodal curves of the free surface, passage to three-dimensional wave patterns (e.g., swirling), etc. These nonlinear phenomena were discovered experimentally long ago (see, e.g., [5, 6, 21, 147, 148, 29, 196]) and studied theoretically at various levels of completeness (see, e.g., [24, 103, 112, 119, 133, 173, 210, 3, 80]). Early in these studies, a special attention was paid to sloshing in a cylindrical container, and the problem was analyzed by examining the resonance interaction of two lowest degenerating natural sloshing modes. This made it possible to describe the phenomena qualitatively.

The forthcoming studies employ more complicated (modal) mathematical models for upright cylindrical and conic containers discussed in the previous chapters. We formulate and solve problems for the steady-state forced sloshing regimes and their dynamic stability. For the problem of dynamic stability, we consider the principal wave motion and additional motions that appear only for certain values of input parameters. Due to coupling these motions in multimodal nonlinear models, we provide a detailed study of the nonlinear wave phenomena including three-dimensional ones. As compared with the results of [121, 139, 132], we are able to better clarify the issues related to the influence of liquid damping on nonlinear dynamic wave regimes. Comparing our theoretical results with experimental model data, we found that the accepted mathematical model describes all physical processes also on the quantitative level. The discrepancy between theoretical and experimental results is usually not greater than 5% for steady-state sloshing amplitudes and 2% for amplitudes of the hydrodynamic force, which is satisfactory for applications. The results of the present chapter are based on the studies presented in [87, 112, 130, 125, 142, 139, 140, 138, 173, 184, 186, 69, 73].

## 8.1 Dynamic stability of resonant liquid sloshing under horizontal harmonic excitations

The nonlinear models constructed in the previous chapter enable one to consider a series of special problems of dynamics for a body with liquid. At present, the problems of resonance interaction of a liquid with a container that performs translational or angular harmonic oscillations are studied well enough. These problems were attractive for researchers due to interesting nonlinear phenomena that are not described by linear theories of sloshing.

We briefly discuss main models related to the problem of forced sloshing in a container perpendicular to the $Ox$-axis and associated problem of dynamic stability of the free surface. This problem was thoroughly studied in experiments for containers of various geometric shapes. An analysis of experimental data shows that the original form of liquid motion (steady-state forced oscillations in the plane of the perturbation force) loses stability for certain values of parameters of the perturbation force. The subsequent dynamical process is accompanied by the appearance of new types of wave patterns caused by the so-called parametric resonance. The interaction of forced and parametric components of liquid sloshing leads to complex spatial motions of the free surface. For axisymmetric containers, this is a swirling wave.

In the present chapter, a theoretical analysis of these processes for cylindrical containers is carried out by using the nonlinear differential equations (4.7.8), (4.7.9)–(4.7.11), and (4.7.12)–(4.7.16) under the assumption that the cavity performs prescribed motions along the $Oz$-axis governed by the law

$$U(t) = H\cos\omega t. \tag{8.1.1}$$

By setting in these equations

$$w_1 = \mathrm{g}, \quad w_2 = 0, \quad w_3 = -H\omega^2\cos\omega t, \tag{8.1.2}$$

we obtain

$$\begin{aligned}
L_1(r_1, p_1, p_0, r_2, p_2) &= \ddot{r}_1 + \sigma_1^2 r_1 + d_1^*(r_1^2\ddot{r}_1 + \ddot{r}_1\dot{r}_1^2 + r_1 p_1\ddot{p}_1 + r_1\dot{p}_1^2) \\
&+ d_2^*(p_1^2\ddot{r}_1 + 2p_1\dot{r}_1\dot{p}_1 - r_1p_1\ddot{p}_1 - 2r_1\dot{p}_1^2) - d_3^*(p_2\ddot{r}_1 - r_2\ddot{p}_1 + \dot{r}_1\dot{p}_2 - \dot{p}_1\dot{r}_2) \\
&+ d_4^*(r_1\ddot{p}_2 - p_1\ddot{r}_2) + d_5^*(p_0\ddot{r}_1 + \dot{r}_1\dot{p}_0) + d_6^* r_1\ddot{p}_0 - \omega^2 P_1\cos\omega t = 0, \\
L_2(r_1, p_1, p_0, r_2, p_2) &= \ddot{p}_1 + \sigma_1^2 p_1 + d_1^*(p_1^2\ddot{p}_1 + r_1p_1\ddot{r}_1 + p_1\dot{r}_1^2 + p_1\dot{p}_1^2) \\
&+ d_2^*(r_1^2\ddot{p}_1 - r_1p_1\ddot{r}_1 + 2r_1\dot{r}_1\dot{p}_1 - 2p_1\dot{r}_1^2) + d_3^*(p_2\ddot{p}_1 + r_2\ddot{r}_1 + \dot{r}_1\dot{r}_2 + \dot{p}_1\dot{p}_2) \\
&- d_4^*(p_1\ddot{p}_2 + r_1\ddot{r}_2) + d_5^*(p_0\ddot{p}_1 + \dot{p}_1\dot{p}_0) + d_6^* p_1\ddot{p}_0 = 0, \\
L_3(r_1, p_0, p_1) &= \ddot{p}_0 + \sigma_0^2 p_0 + d_{10}^*(r_1\ddot{r}_1 + p_1\ddot{p}_1) + d_8^*(\dot{r}_1^2 + \dot{p}_1^2) = 0, \\
L_4(r_1, p_1, p_2) &= \ddot{p}_2 + \sigma_2^2 p_2 + d_9^*(r_1\ddot{r}_1 - p_1\ddot{p}_1) + d_7^*(\dot{r}_1^2 - \dot{p}_1^2) = 0, \\
L_5(r_1, p_1, r_2) &= \ddot{r}_2 + \sigma_2^2 r_2 - d_9^*(r_1\ddot{p}_1 + p_1\ddot{r}_1) - 2d_7^*\dot{r}_1\dot{p}_1 = 0,
\end{aligned} \tag{8.1.3}$$

where

$$d_1^* = \frac{d_1}{\mu_1}, \quad d_2^* = \frac{d_2}{\mu_1}, \quad d_3^* = \frac{d_3}{\mu_1}, \quad d_4^* = \frac{d_4}{\mu_1},$$
$$d_5^* = \frac{d_5}{\mu_1}, \quad d_6^* = \frac{d_6}{\mu_1}, \quad d_7^* = \frac{d_7}{\mu_2}, \quad d_8^* = \frac{d_8}{\mu_0}, \tag{8.1.4}$$
$$d_9^* = \frac{d_4}{\mu_2}, \quad d_{10}^* = \frac{d_6}{\mu_0}, \quad P_1 = \frac{H\lambda}{\mu_1}.$$

From now on, the symbol "$*$" at the coefficients $d_i$ is omitted.

When deducing the nonlinear modal equations, we assume that the perturbations of the free surface have five degrees of freedom. The generalized coordinates $r_1(t)$, $p_0(t)$, and $p_2(t)$ characterize the liquid motion in the plane of motion of the container unlike coordinates $p_1(t)$ and $r_2(t)$ which characterize the cross waves perpendicular to the plane of excitation.

Considering the forced nonlinear sloshing described by system (8.1.3), we distinguish the primary-excited wave regime that occurs for all values of $H$ and $\omega$ and an additional motion that appears only for certain values of these parameters. The investigation of the stability of these regimes is split into the following stages: finding a periodic solution that describes forced steady-state regime in the excitation plane and studying its stability; constructing domains of dynamic instability of the steady-state regime (domains of parametrically perturbed sloshing); finding the other steady-state regimes in domains of dynamic stability and instability.

In these stages of the study, we encounter numerous difficulties of technical and principal character. The most serious difficulties are caused by the multidimensionality of the original system of equations and the corresponding variational equations in the nonlinear theory. When we construct approximate periodic solutions of the system of equations (8.1.3), it makes sense to use analytical methods since, otherwise, the rigorous analysis of stability of these solutions seems to be a quite difficult task.

In the present chapter, we mainly present descriptions of the construction of approximate particular solutions to nonlinear equations in an analytic form and derive approximate solutions of variational equations on the basis of these approximate solutions. The physical interpretation of these results is expressed by the construction of the domains of stability and instability, the study of the evolution of free surface, and the analysis of hydrodynamic forces.

## 8.2 Stability of forced and parametric sloshing in the simplest case

We start with the simplest case of the problem of forced sloshing in a cylindrical container where the generalized coordinates $p_0(t)$, $r_2(t)$, and $p_2(t)$ in the equations of motion (8.1.3) are neglected. In this case, the corresponding system of nonlinear equations has the form

$$\begin{aligned} L_1(r,p) &= \ddot{r} + \sigma^2 r + d_1(r^2\ddot{r} + \ddot{r}\dot{r}^2 + rp\ddot{p} + r\dot{p}^2) \\ &\quad + d_2(p^2\ddot{r} + 2p\dot{r}\dot{p} - rp\ddot{p} - 2r\dot{p}^2) - \omega^2 P_1 \cos\omega t = 0; \\ L_2(r,p) &= \ddot{p} + \sigma^2 p + d_1(p^2\ddot{p} + rp\ddot{r} + p\dot{r}^2 + p\dot{p}^2) \\ &\quad + d_2(r^2\ddot{p} - rp\ddot{r} + 2r\dot{r}\dot{p} - 2p\dot{r}^2) = 0. \end{aligned} \tag{8.2.1}$$

This system describes forced and parametrically excitable sloshing approximately. In the case where parametric oscillations in the system do not appear ($p \equiv 0$), forced oscillations are described by the nonlinear differential equation

$$L_1(r) = \ddot{r} + \sigma^2 r + d_1(\ddot{r}r^2 + \dot{r}^2 r) = \omega^2 P_1 \cos\omega t. \tag{8.2.2}$$

We construct an approximate periodic solution of this equation by the Bubnov–Galerkin method. To this end, we represent the solution by the Fourier series

$$r(t) = \alpha_0 + \sum_{k=1}^{\infty} (\alpha_k \cos k\omega t + \bar{\alpha}_k \sin k\omega t), \tag{8.2.3}$$

where $\alpha_k$ and $\bar{\alpha}_k$ are unknown constants. Keeping only the main harmonics

$$r(t) = A\cos\omega t + \bar{A}\sin\omega t \tag{8.2.4}$$

in (8.2.3) and using the Fredholm alternative

$$\int_0^{\frac{2\pi}{\omega}} L_1(r)\cos\omega t dt = 0, \quad \int_0^{\frac{2\pi}{\omega}} L_1(r)\sin\omega t dt = 0,$$

we obtain

$$(\bar{\sigma}^2 - 1)A - m_1 A^3 = P, \quad \bar{A} \equiv 0, \tag{8.2.5}$$

where $m_1 = \frac{1}{2}d_1$ and $\bar{\sigma}^2 = \sigma^2/\omega^2$. Equation (8.2.5) is used for computing response amplitude versus $P$ and $\omega$. By setting $P = 0$ in (8.2.5), we obtain the relation

$$(\bar{\sigma}^2 - 1) - m_1 A^2 = 0 \tag{8.2.6}$$

that enables us to find the so-called skeleton line.

In order to find the steady-state sloshing regimes and boundaries of domains of their stability, we can use main terms in decompositions of the type (8.2.3). If necessary, we can increase the number of terms in (8.2.3), although, it is clear that this makes derivations more complicated. It is shown below for a particular case that additional harmonics lead to a correction of amplitude-frequency characteristics only at the third significant digit. Therefore, the error rate of the obtained approximate solution (8.2.4) is quite acceptable within the framework of the considered theory of the third order of smallness.

We dwell on the problem of stability of the periodic solution (8.2.4). To this end, we have to find parameters $H$ and $\omega$ such that the steady-state regime

$$r(t) = A\cos\omega t, \quad p \equiv 0, \tag{8.2.7}$$

which is described by the system of nonlinear equations (8.2.1), is realizable physically. Following the first Lyapunov method, along with motion (8.2.7), which is chosen to be unperturbed, we also consider motions

$$r = \bar{r}(t) + \xi(t), \quad p = \bar{p}(t) + \eta(t) \tag{8.2.8}$$

whose initial conditions differ slightly enough from the initial conditions (8.2.7) for $\bar{r}(t)$ and $\bar{p}(t)$. These motions (8.2.7) are close to the unperturbed motion. We say that these motions are perturbed and the quantities $\xi(t)$ and $\eta(t)$ are perturbations. The periodic solution is stable if small variations in initial conditions slightly deviate the system from the unperturbed motion. By decreasing initial perturbations, we can made deviations arbitrarily small for $t > t_0$.

According to the first Lyapunov method, we compose variational equations by substituting (8.2.7) into (8.2.1) given that $\bar{r}(t)$ is nonzero. As a result, we obtain equations of perturbed motions

$$(d_1\bar{r}^2 + 1)\ddot{\xi} + 2d_1\dot{\bar{r}}\,\bar{r}\dot{\xi} + (\sigma^2 + 2d_1\bar{r}\ddot{\bar{r}} + d_1\dot{\bar{r}}^2)\xi + F_1(\xi, p, \dot{\xi}, \dot{p}, \ddot{\xi}, \ddot{p}) = 0; \tag{8.2.9}$$

$$(1 + d_2)\ddot{\eta} + 2d_2\bar{r}\,\dot{\bar{r}}\dot{\eta} + (\sigma^2 + c\bar{r}\ddot{\bar{r}} + k_3\dot{\bar{r}}^2)\eta + F_2(\xi, p, \dot{\xi}, \dot{p}, \ddot{\xi}, \ddot{p}) = 0, \tag{8.2.10}$$

where $F_1$ and $F_2$ are functions that contain perturbations and their derivatives raised to powers higher than the first power. Keeping only linear terms in (8.2.9) and (8.2.10) and taking (8.2.7) into account, we obtain

$$L_3(\xi) = (\alpha + \beta\cos\theta t)\ddot{\xi} - \varepsilon_1\dot{\xi}\sin\theta t + (\gamma - \delta\cos\theta t)\xi = 0; \tag{8.2.11}$$

$$L_4(\xi) = (\bar{\alpha} + \bar{\beta}\cos\theta t)\ddot{\eta} - \varepsilon\dot{\eta}\sin\theta t + (\bar{\gamma} - \bar{\delta}\cos\theta t)\eta = 0, \tag{8.2.12}$$

which are called equations of the first approximation. Here, we use the following notation:

$$\beta = \tfrac{1}{2}d_1A^2, \quad \alpha = 1 + \beta, \quad \varepsilon = d_1\omega A^2, \quad \gamma = \left(\bar{\sigma}^2 - \tfrac{1}{2}d_1A^2\right)\omega^2,$$
$$\delta = \tfrac{3}{2}A^2\omega^2 d_1, \quad \bar{\alpha} = 1 + \bar{\beta}, \quad \bar{\beta} = \tfrac{1}{2}d_2A^2, \tag{8.2.13}$$
$$\bar{\gamma} = \left(\bar{\sigma}^2 - \tfrac{1}{2}d_2A^2\right)\omega^2, \quad \bar{\delta} = A^2\omega^2(2d_1 - 3d_2), \quad \varepsilon = d_2\omega A^2, \quad \theta = 2\omega.$$

The variational equations (8.2.9) and (8.2.10) are linear equations with periodic coefficients. The main information about solutions of these equations is provided by the Floquet theory. According to this theory, it is usual that these equations are reduced to the form

$$y'' + \Omega^2[1 - 2\mu\Phi(t)]y = 0, \tag{8.2.14}$$

where $\Phi(t)$ is a $T$-periodic function. In scientific publications, (8.2.14) is called the Hill equation. There are three types of solutions to this equation: (1) an "unstable" solution that increases infinitely as $t \to \infty$; (2) a "stable" solution that remains bounded as $t \to \infty$; and (3) a solution that has a period $T$ or $2T$ and is called neutral. This solution can be regarded as a special case of the stable solution.

In the plane of parameters of (8.2.14), unstable solutions fill up certain domains completely. Moreover, domains of instability are separated from domains of stability by periodic solutions of periods $T$ and $2T$. Two solutions of equal periods bound a domain of instability, whereas two solutions of different periods bound a domain of stability. Therefore, the problem of detecting the boundaries of domains of instability reduces to the problem of finding the conditions under which a differential equation has periodic solutions of periods $T$ and $2T$.

Most often, domains of instability are determined by using the method of trigonometric series. To apply this method to the Hill equations, periodic solutions of period $2T$ are presented in the form [25, 77]

$$y(t) = \sum_{k=1,3,5,\ldots}^{\infty} \left( a_k \sin \frac{k\theta t}{2} + b_k \cos \frac{k\theta t}{2} \right). \tag{8.2.15}$$

Substituting series (8.2.15) into (8.2.14) and grouping the coefficients at the same terms $\sin \frac{k\theta t}{2}$ and $\cos \frac{k\theta t}{2}$, we obtain a homogeneous algebraic system of equations for $a_k$ and $b_k$. The condition for solvability of this system leads to equations for boundaries of domains of instability. Similar results can be also obtained by the Bubnov–Galerkin method used in what follows.

Let us proceed with the construction of instability domains for solutions of (8.2.11) and (8.2.12). First, consider equation (8.2.11) for the function $\xi(t)$ that characterizes a perturbation of the periodic solution $\bar{r}(t) = A \cos \omega t$. The amplitude $A$ of this solution is defined by (8.2.5). We have to find out which pair of values $A$ and $\omega$ satisfying (8.2.5) leads to stable solutions and which pair of these values leads to unstable solutions. In accordance with the construction of the main domain of instability of a solution of (8.2.11), we write down it in the form

$$\xi(t) = a_1 \cos \frac{\theta t}{2} + b_1 \sin \frac{\theta t}{2}. \tag{8.2.16}$$

By using the Fredholm alternative

$$\int_0^{\frac{4\pi}{\theta}} L_3(\xi) \cos \frac{\theta t}{2} dt = 0, \quad \int_0^{\frac{4\pi}{\theta}} L_3(\xi) \sin \frac{\theta t}{2} dt = 0,$$

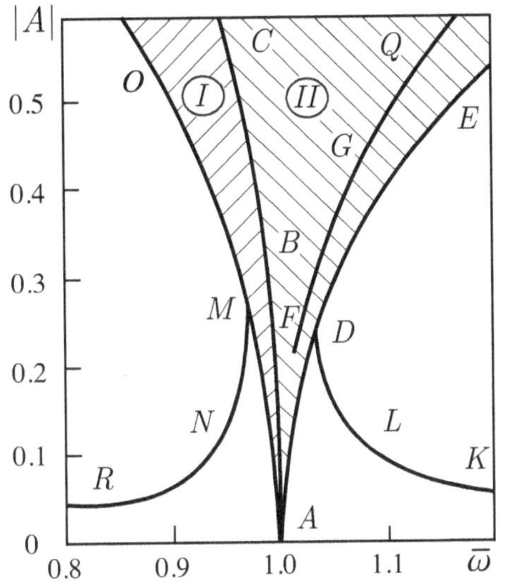


**Figure 8.1**

we obtain the following relations defining boundaries of domains of instability:

$$(\bar{\sigma}^2 - 1) - 3m_1A^2 = 0; \tag{8.2.17}$$

$$(\bar{\sigma}^2 - 1) - m_1A^2 = 0. \tag{8.2.18}$$

Equation (8.2.18) coincides with the equation of skeleton line (8.2.6). In Fig. 8.1, this line is depicted by the curve $ABC$. Equation (8.2.17) gives the curve $AMO$. Comparing (8.2.17) with the equation of response curves (8.2.5), we see that the stable branch of the curve $RNM$ is separated from the unstable branch by the point $M$ at which the amplitude curves have the vertical tangent. On the stable part of the branch of the response curves, which is located to the left of the skeleton line, the derivative is necessarily positive. At the point $M$, the derivative tends to infinity. On the unstable left branch, the derivative is negative. In the domain $I$ bounded by the curves $AMO$ and $ABC$, the solution $\bar{r}(t) = A\cos\omega t$ is unstable. From the physical point of view, for a periodic solution with the period of the perturbation force, the condition of stability means that the amplitude of forced sloshing increases when the external force $P$ increases.

Further, consider the variational equation (8.2.12) for the perturbation $\eta(t)$ of the trivial solution $p(t) \equiv 0$. By studying solutions of (8.2.12), we should obtain an answer to the question of stability of this trivial solution. According to the results presented above, the domain of instability of (8.2.12) is associated with domains of parametrically perturbed oscillations ($p \not\equiv 0$), i.e., domains of dynamic instability of motion (8.2.7). To determine the main domain of instability, we set

$$\eta(t) = a_0 \sin\frac{\theta t}{2} + b_0 \cos\frac{\theta t}{2}. \tag{8.2.19}$$

Substituting (8.2.19) into the Fredholm alternative

$$\int_0^{\frac{4\pi}{\theta}} L_4(\eta) \sin\frac{\theta t}{2} dt = 0, \quad \int_0^{\frac{4\pi}{\theta}} L_4(\eta) \cos\frac{\theta t}{2} dt = 0,$$

we obtain the following equations for finding the boundaries of domains of dynamic stability:

$$\bar{\sigma}^2 = m_2 A^2 + 1; \tag{8.2.20}$$

$$\bar{\sigma}^2 = m_1 A^2 + 1, \tag{8.2.21}$$

where $m_2 = \frac{1}{2}(4d_2 - d_1)$. In Fig. 8.1, the curve $ADE$ corresponds to (8.2.20) and the curve $ABC$ corresponds to (8.2.21). Therefore, the domains of instability of (8.2.11) and (8.2.12) go into each other continuously.

In domain $II$ bounded by the curves $ABC$ and $ADE$, the solution $p \equiv 0$ is unstable. Thus, both domains $I$ and $II$ are domains of instability of forced steady-state sloshing in the excitation plane of the perturbation force. In domain $II$, this instability is caused by instability of the trivial solution $p \equiv 0$, i.e., a parametric perturbation of the generalized coordinate $p$ is possible. The second stable branch of the response curve $KLD$ adjoins domain $II$ from the right and is separated from it by the point $D$. In the domain of dynamic instability $II$, a steady-state regime, if it exists, is described by the nonlinear system of equations (8.2.1).

We now consider the problem of construction of steady-state regimes in the main domain of dynamic instability $II$. As was mentioned above, on the boundaries of the odd domains of instability, solutions of the linear equations (8.2.11) and (8.2.12) take the form (8.2.15). It is the presentation of stable motions that is used in order to find such motions in domains of dynamic instability.

Suppose that an approximate solution of the system of nonlinear equations (8.2.1) can be represented in the form

$$r(t) = A\cos\omega t + \bar{A}\sin\omega t, \quad p(t) = \bar{B}\cos\omega t + B\sin\omega t, \tag{8.2.22}$$

in the domain of the main resonance. By using the Bubnov–Galerkin method for the solution of this system, we obtain the following algebraic relations for the constants $A$, $\bar{A}$, $B$, and $\bar{B}$:

$$(\bar{\sigma}^2 - 1)A - m_1 A^3 - m_2 A B^2 = P, \tag{8.2.23}$$

$$(\bar{\sigma}^2 - 1) - m_1 B^2 - m_2 A^2 = 0, \quad \bar{A} = 0, \quad \bar{B} = 0. \tag{8.2.24}$$

Eliminating $B$ from (8.2.24) and substituting the result into (8.2.23), we obtain the following equation for the response curve in domain $II$:

$$(\bar{\sigma}^2 - 1)A - m_4 A^3 = m_5 P, \tag{8.2.25}$$

where

$$m_4 = 2d_2, \quad m_5 = m_1/(m_1 - m_2). \tag{8.2.26}$$

Solution (8.2.22) corresponds to the experimentally observed swirling mode. The resonance curve corresponding to this solution is depicted by the curve $FGQ$ in Fig. 8.1.

It is a point of interest to find parameters $A$, $B$, and $\omega$, for which there exists a stable steady-state regime (8.2.22). To this end, we derive variational equations for the nonlinear system (8.2.1) by choosing

$$\bar{r}(t) = A\cos\omega t, \quad \bar{p}(t) = B\sin\omega t \tag{8.2.27}$$

as an unperturbed solution. Substituting the perturbed solutions

$$r(t) = \bar{r}(t) + \xi(t), \quad p(t) = \bar{p}(t) + \eta(t) \tag{8.2.28}$$

into (8.2.1) and assuming that the perturbations $\xi$ and $\eta$ are small, we obtain equations in variations

$$(\alpha + \beta\cos\theta t)\ddot{\xi} + \beta_1\ddot{\eta}\sin\theta t - \varepsilon\dot{\xi}\sin\theta t - \varepsilon_1(1 + \cos\theta t)\dot{\eta}$$
$$+(\gamma - \delta\cos\theta t)\xi - \delta_1\eta\sin\theta t = 0; \tag{8.2.29}$$

$$(\bar{\alpha} + \bar{\beta}\cos\theta t)\ddot{\eta} + \beta_1\ddot{\xi}\sin\theta t - \bar{\varepsilon}\dot{\eta}\sin\theta t - \varepsilon_1(\cos\theta t - 1)\dot{\xi}$$
$$+(\bar{\gamma} - \bar{\delta}\cos\theta t)\eta - \delta_1\xi\sin\theta t = 0, \tag{8.2.30}$$

where

$$\begin{gathered}
\alpha = 1 + \tfrac{1}{2}d_1A^2 + \tfrac{1}{2}d_2B^2, \quad \beta = \tfrac{1}{2}(d_1A^2 - d_2B^2), \quad \beta_1 = \tfrac{1}{2}cAB, \\
\varepsilon = (d_1A^2 - d_2B^2)\omega, \quad \varepsilon_1 = \tfrac{1}{2}k_4AB\omega, \quad c = d_1 - d_2, \\
\gamma = \left(\bar{\sigma}^2 - \tfrac{1}{2}d_1A^2 - \tfrac{1}{2}d_2B^2\right)\omega^2, \quad \delta_1 = \tfrac{1}{2}\omega^2(3d_2 + d_1)AB, \\
\delta = \tfrac{1}{2}[3d_1A^2 - (2d_1 - 3d_2)B^2]\omega^2, \quad \bar{\alpha} = 1 + \tfrac{1}{2}d_1B^2 + \tfrac{1}{2}d_2A^2, \\
\bar{\beta} = \tfrac{1}{2}(d_2A^2 - d_1B^2), \quad \bar{\varepsilon} = \tfrac{1}{2}(-2d_1A^2 + 2d_2A^2)\omega, \quad \bar{\gamma} = \gamma, \\
\bar{\delta} = \tfrac{1}{2}[(2d_1 - 3d_2)A^2 - 3d_1B^2]\omega^2, \quad k_4 = 2(3d_2 - d_1), \quad \bar{\sigma}^2 = \sigma^2/\omega^2.
\end{gathered} \tag{8.2.31}$$

Unlike the variational equations (8.2.11) and (8.2.12), the considered system (8.2.29), (8.2.30) is a system of equations of the fourth order. Solving this system is a fairly complicated problem. In accordance with the theory of systems of differential equations with periodic coefficients [225], the fundamental system of solutions of (8.2.29) and (8.2.30) has solutions of the form

$$\xi(t) = e^{\lambda t}\varphi_1(t), \quad \eta(t) = e^{\lambda t}\varphi_2(t), \tag{8.2.32}$$

where $\varphi_1(t)$ and $\varphi_2(t)$ are periodic functions with period $\frac{2\pi}{\omega}$; $\lambda$ are characteristic numbers of the system.

The problem of stability of solutions of variational equations and, hence, the problem of stability of solutions (8.2.27) of the original nonlinear system of equations (8.2.1) is solved by computing the characteristic numbers $\lambda$. To find an approximate solution of the problem, we present the periodic functions $\varphi_1(t)$ and $\varphi_2(t)$ in the form of the Fourier series [225]

$$\begin{aligned}\varphi_1(t) &= a_0 + \sum_{n=1}^{\infty}(a_k \cos n\omega t + b_n \sin n\omega t),\\ \varphi_2(t) &= \bar{a}_0 + \sum_{n=1}^{\infty}(\bar{a}_k \cos n\omega t + \bar{b}_n \sin n\omega t).\end{aligned} \tag{8.2.33}$$

To find the unknown constants, we use the Bubnov–Galerkin method. In the first approximation, we obtain the following system of homogeneous algebraic equations for $a_1$, $\bar{a}_1$, $b_1$, and $\bar{b}_1$:

$$\begin{aligned}(A_{11} + \bar{\lambda}^2 B_{11})b_1 + \bar{\lambda}B_{12}a_1 + \bar{\lambda}B_{13}\bar{b}_1 + (A_{14} + \bar{\lambda}^2 B_{14})\bar{a}_1 &= 0,\\ \bar{\lambda}B_{21}b_1 + (A_{22} + \bar{\lambda}^2 B_{22})a_1 + (A_{23} + \bar{\lambda}^2 B_{23})\bar{b}_1 + \bar{\lambda}B_{24}\bar{a}_1 &= 0,\\ \bar{\lambda}B_{31}b_1 + (A_{32} + \bar{\lambda}^2 B_{32})a_1 + (A_{33} + \bar{\lambda}^2 B_{33})\bar{b}_1 + \bar{\lambda}B_{34}\bar{a}_1 &= 0,\\ (A_{41} + \bar{\lambda}^2 B_{41})b_1 + \bar{\lambda}B_{42}a_1 + \bar{\lambda}B_{43}\bar{b}_1 + (A_{44} + \bar{\lambda}^2 B_{44})\bar{a}_1 &= 0.\end{aligned} \tag{8.2.34}$$

The quantities $A_{ij}$ and $B_{ij}$ are defined by the parameters $\omega$, $A$, and $B$ as follows:

$$\begin{gathered}A_{11} = \bar{\sigma}^{-1}\left[(\bar{\sigma}^2 - 1) - \tfrac{1}{2}d_1(A^2 + B^2)\right] = A_{44},\\ A_{14} = -\bar{\sigma}^{-1}(d_1 - 2d_2)AB = A_{41},\\ A_{33} = \bar{\sigma}^{-1}\left[(\bar{\sigma}^2 - 1) - \tfrac{3}{2}d_1 B^2 - m_2 A^2\right], \quad A_{32} = -2\bar{\sigma}^{-1}m_2 AB = A_{23},\\ A_{22} = \bar{\sigma}^{-1}[(\bar{\sigma}^2 - 1) - 3m_1 A^2 - m_2 B^2], \quad B_{11} = 1 + \tfrac{1}{4}d_1 A^2 + \tfrac{3}{4}d_2 B^2,\\ B_{12} = -2 - d_1 A^2 - d_2 B^2, \quad B_{13} = (d_1 - 3d_2)AB, \quad B_{21} = -B_{12},\\ B_{31} = -B_{13}, \quad B_{34} = -2 - d_2 A^2 - d_1 B^2, \quad B_{24} = -B_{31},\\ B_{42} = -B_{13}, \quad B_{43} = -B_{34}, \quad B_{22} = 1 + \tfrac{3}{4}d_1 A^2 + \tfrac{1}{4}d_2 B^2,\\ B_{33} = 1 + \tfrac{3}{4}d_1 B^2 + \tfrac{1}{4}d_2 A^2,\\ B_{14} = B_{41} = B_{23} = B_{32} = \tfrac{1}{4}(d_1 - d_2)AB, \quad B_{44} = 1 + \tfrac{1}{4}d_1 B^2 + \tfrac{3}{4}d_2 A^2.\end{gathered} \tag{8.2.35}$$

By using the condition for solvability of the system of homogeneous algebraic equations (8.2.34), we obtain the characteristic equation with respect to $\bar{\lambda}$. Further, on the basis of a solution of this equation, in the plane of the parameters $A$ and $\omega$, we find their critical values for which an unstable branch of the response curves (8.2.25) becomes stable. In Fig. 8.1, this point is denoted by $F$.

The method of slowly varying amplitudes [25, 81, 57] can also be used successfully to analyze the stability of solutions of the nonlinear system of equations (8.2.1). According to this method, a solution of this system has the form

$$r(t) = f_1(t)\sin\omega t + f_2(t)\cos\omega t, \;\; p(t) = f_3(t)\sin\omega t + f_4(t)\cos\omega t. \tag{8.2.36}$$

Under known assumptions on the amplitudes $f_i(t)$ from (8.2.1), we obtain the following system of equations analogous to equations in [87]:

$$\begin{aligned}
\frac{df_1}{d\tau} &= -\nu f_2 + m_1 f_2 \sum_{i=1}^{4} f_i^2 - k f_3(f_2 f_3 - f_1 f_4) + P,\\
\frac{df_2}{d\tau} &= \nu f_1 - m_1 f_1 \sum_{i=1}^{4} f_i^2 - k f_4(f_2 f_3 - f_1 f_4),\\
\frac{df_3}{d\tau} &= -\nu f_4 + m_1 f_4 \sum_{i=1}^{4} f_i^2 + k f_1(f_2 f_3 - f_1 f_4),\\
\frac{df_4}{d\tau} &= \nu f_3 - m_1 f_3 \sum_{i=1}^{4} f_i^2 + k f_2(f_2 f_3 - f_1 f_4),
\end{aligned} \tag{8.2.37}$$

where $\tau = \frac{1}{2}\omega t$; $\nu = \bar{\sigma}^2 - 1$, and $k = m_1 - m_2$.

Setting the left-hand sides of (8.2.37) equal to zero, we come to the steady-state regimes (8.2.4) and (8.2.22) obtained earlier. Note that the system of nonlinear equations (8.2.37) with respect to $f_i(\tau)$ is autonomous unlike the original equations (8.2.1). We represent the perturbed solution of (8.2.37) as

$$\tilde{f}_i(\tau) = f_i^{(0)} + \eta_i(\tau), \tag{8.2.38}$$

where $f_i^{(0)}$ are amplitudes of the steady-state regime and $\eta_i(\tau)$ are small perturbations. By using this representation, we derive from (8.2.37) the following system of linear differential equations with constant coefficients for $\eta_i(\tau)$ :

$$\begin{aligned}
\frac{d\eta_1}{d\tau} &= \eta_2(-\nu + 3m_1 f_2^{(0)^2} + m_2 f_3^{(0)^2}) - 2m_2 f_2^{(0)} f_3^{(0)} \eta_3,\\
\frac{d\eta_2}{d\tau} &= \eta_1[\nu - m_1(f_2^{(0)^2} + f_3^{(0)^2})] - k f_2^{(0)} f_3^{(0)} \eta_4,\\
\frac{d\eta_3}{d\tau} &= \eta_4[-\nu + m_1(f_2^{(0)^2} + f_3^{(0)^2})] + k f_2^{(0)} f_3^{(0)} \eta_1,\\
\frac{d\eta_4}{d\tau} &= \eta_3[\nu - 3m_1 f_3^{(0)^2} - m_2 f_2^{(0)^2}] - 2m_2 f_2^{(0)} f_3^{(0)} \eta_2.
\end{aligned} \tag{8.2.39}$$

The fundamental system of solutions of these equations has the form $\eta_i(t) = c_i e^{\lambda\tau_0}$, $i = 1, 2, 3, 4$. To find the constants $c_i$, we derive the following system of algebraic equations from (8.2.39):

$$(A + \lambda E)C = 0. \tag{8.2.40}$$

The condition for solvability of this system gives the following secular equation

$$\det |A + \lambda E| = 0. \tag{8.2.41}$$

Setting $\mu = \lambda^2$, we rewrite this equation in the form

$$\mu^2 + C\mu + D = 0, \tag{8.2.42}$$

where the coefficients $C$ and $D$ are expressed in terms of the amplitudes $f_i^{(0)}$, the coefficients of the nonlinear system (8.2.1), and the frequency $\omega$. The numerical results obtained by both methods coincide.

The response curves at the primary resonance for the upright circular cylindrical tank with $h = 1$, $R_1 = 1$, and $H = 0.01$ are depicted in Fig. 8.1. These curves are typical for tanks of revolution.

As the forcing frequency $\omega$ increases from a certain value $\omega_R$ to $\omega_M$, a 'planar' steady-state sloshing occurs. For the 'planar' sloshing, the generalized coordinate $p$ is not excited (the response curve $RNM$). In the frequency range $\omega_M < \omega < \omega_F$, no stable steady-state regime exists. An increase of $\omega$ from a certain value $\omega_F$, which corresponds to the point $F$ on the response curve, results in cross-waves, which are perpendicular to the excitation plane ($p \neq 0$). The interaction of the primary excited 'planar' sloshing and parametrically excitable cross waves leads to the so-called swirling (the curve $FGQ$). For a certain value of the frequency $\omega_D$, quenching of spatial oscillations happens, and the steady-state regime corresponding to steady-state sloshing in the excitation plane occurs in the system (the curve $DLK$). Whereas $\omega$ decreases, the described phenomena are mainly repeated. In the domain of dynamic instability bounded by the curves $ABC$ and $ADE$, chaotic sloshing is possible [196]. In the present monograph, these motions are not discussed. Quantitative characteristics of the studied processes are given in what follows.

## 8.3 Construction the periodic solutions of (8.2.1) by alternative approximate methods

In the previous section, we studied the forced sloshing and its stability by using the Bubnov–Galerkin method applied to the single-dimensional model. We arrive at the problem of accuracy in predicting the response curves on the basis of (8.2.5), (8.2.23), and (8.2.24) and thresholds of instability on the basis of (8.2.17), (8.2.18), and (8.2.20). In this section, we solve (8.2.1) and (8.2.2) by other approximate methods. Comparing the obtained results, we reveal the limits of applicability of the above-obtained solution [185]. This is an important problem since, in what follows, we have to compare computed amplitudes of the free surface and the pressure on tank walls with the experimental data. On the basis of this comparison, we have to determine the boundaries of applicability of the constructed mathematical model defined by (8.1.3) to the problem of quantitative estimates of nonlinear physical phenomena under consideration.

I. Consider (8.2.2):

$$\ddot{r}_1 + \sigma_1^2 r_1 + d_1(r_1^2\ddot{r}_1 + r_1\dot{r}_1^2) = \omega^2 P_1 \cos\omega t. \tag{8.3.1}$$

To solve this equation, we successively use three approximate methods: the Bubnov–Galerkin method with several harmonics in decomposition (8.2.3), the averaging method, and the Poincaré method.

A. We present a periodic solution of (8.3.1) in the form

$$r_1(t) = A_0 + A_1\cos\omega t + \bar{A}_1 \sin\omega t + \ldots + A_3\cos 3\omega t + \bar{A}_3 \sin 3\omega t, \tag{8.3.2}$$

where $A_0$, $A_1$, …, $\bar{A}_3$ are unknown constants to be found. By using the Bubnov–Galerkin method applied to (8.3.1), we arrive at the following relations coupling the amplitudes $A_0$, $A_1$, …, $\bar{A}_3$:

$$\begin{gathered} A_0 = 0, \;\; \bar{A}_1 = 0, \;\; \bar{A}_2 = 0, \;\; \bar{A}_3 = 0, \\ A_1\left[\bar{\sigma}_1^2 - 1 - a_1\left(\tfrac{1}{2}A_1^2 + \tfrac{3}{2}A_1A_3 + 5A_3^2\right)\right] = P_1, \\ (\bar{\sigma}_1^2 - 9)A_3 - a_1\left(\tfrac{1}{2}A_1^3 + 5A_1^2A_3 + \tfrac{9}{2}A_3^3\right) = 0. \end{gathered} \tag{8.3.3}$$

Thus, relation (8.3.2) takes the form

$$r_1(t) = A_1\cos\omega t + A_3\cos 3\omega t. \tag{8.3.4}$$

The results of calculations for typical input parameters show that $|A_3|$ is at most 0.1% of the maximum value of $|A_1|$. Therefore, the first harmonic in decomposition (8.3.2) gives the main contribution to the generalized coordinate $r_1(t)$ and, hence, the other terms of the decomposition can be neglected. After elimination of higher harmonics in decomposition (8.3.2), the error in estimating the generalized coordinates $p_k(t)$ is even smaller because they are of the order of $O(r_1^2)$.

B. We present (8.3.1) in the form

$$\ddot{r}_1 + \sigma_1^2 r_1 = \varepsilon f(r_1, \dot{r}_1, \ddot{r}_1, t), \tag{8.3.5}$$

where $\varepsilon = d_1, f(r_1, \dot{r}_1, \ddot{r}_1, t) = \omega^2/d_1 P_1 \cos\omega t - r_1^2\ddot{r}_1 - r_1\dot{r}_1^2$, and solve this equation by the averaging (Struble) method [168]. For a small value of $\varepsilon$, we look for an asymptotic solution

$$r_1(t) = A\cos\psi + \varepsilon x_1(t) + \varepsilon^2 x_2(t) + \dots, \tag{8.3.6}$$

where $\psi = \sigma_1(t) - \theta$; $A$ and $B$ are functions slowly varying with time.
Substituting (8.3.6) into (8.3.5), we have

$$\begin{aligned}
&\left[\frac{d^2A}{dt^2} + A\frac{d\theta}{dt}\left(2\sigma_1 - \frac{d\theta}{dt}\right)\right]\cos\psi \\
&+\left[A\frac{d^2\theta}{dt^2} + 2\frac{dA}{dt}\left(\frac{d\theta}{dt} - \sigma_1\right)\right]\sin\psi + \varepsilon\left(\frac{d^2x_1}{dt^2} + \sigma_1^2 x_1\right) + \dots \\
&= \varepsilon\frac{P_1}{a_1}\cos\omega t - \varepsilon\frac{A}{4}\left\{3A\frac{d^2A}{dt^2} + 3\left(\frac{dA}{dt}\right)^2 - 2A^2\left(\frac{d\theta}{dt} - \sigma_1\right)^2\right\}\cos\psi \\
&\quad -\varepsilon A^2\left\{\frac{1}{4}A\frac{d^2\theta}{dt^2} - \frac{dA}{dt}\left(\frac{d\theta}{dt} - \sigma_1\right)\right\}\sin\psi \\
&\quad -\varepsilon\frac{A}{4}\left\{A\frac{d^2A}{dt^2} + \left(\frac{dA}{dt}\right)^2 - 2A^2\left(\frac{d\theta}{dt} - \sigma_1\right)^2\right\}\cos 3\psi \\
&\quad -\varepsilon A^2\left\{\frac{1}{4}A\frac{d^2A}{dt^2} + \frac{dA}{dt}\left(\frac{d\theta}{dt} - \sigma_1\right)\right\}\sin 3\psi + \dots,
\end{aligned} \tag{8.3.7}$$

[dots denote terms of the order of $O(\varepsilon^2)$ and higher].
Equating the coefficients of $\cos\psi$ and $\sin\psi$ on both sides of (8.3.7), we obtain the so-called variational equations

$$\begin{aligned}
&\frac{d^2A}{dt^2} + A\frac{d\theta}{dt}\left(2\sigma_1 - \frac{d\theta}{dt}\right) \\
&= \varepsilon\frac{P_1}{D_1} - \varepsilon A\left\{\frac{3}{4}A\frac{d^2A}{dt^2} + \frac{3}{4}\left(\frac{dA}{dt}\right)^2 - \frac{1}{2}A^2\left(\frac{d\theta}{dt} - \sigma_1\right)^2\right\}, \\
&A\frac{d^2\theta}{dt^2} + 2\frac{dA}{dt}\left(\frac{d\theta}{dt} - \sigma_1\right) = -\varepsilon A^2\left\{\frac{A}{4}\frac{d^2\theta}{dt^2} + \frac{dA}{dt}\left(\frac{d\theta}{dt} - \sigma_1\right)\right\}.
\end{aligned} \tag{8.3.8}$$

The remaining terms in (8.3.7) give the equation for perturbation

$$\begin{aligned}
\frac{d^2x_1}{dt^2} + \sigma_1^2 x_1 &= -\frac{1}{4}A\left\{A\frac{d^2A}{dt^2} + \left(\frac{dA}{dt}\right)^2 - 2A^2\left(\frac{d\theta}{dt} - \sigma_1\right)^2\right\}\cos 3\psi \\
&\quad -A^2\left\{\frac{1}{4}A\frac{d^2\theta}{dt^2} + \frac{dA}{dt}\left(\frac{d\theta}{dt} - \sigma_1\right)\right\}\sin 3\psi.
\end{aligned} \tag{8.3.9}$$

The variational equations (8.3.8) are valid up to the terms of the order of $\varepsilon$ if

$$A = A_1, \quad \theta = \varepsilon \frac{P_1}{2A_1 d_1 \sigma_1} t + \theta_0, \tag{8.3.10}$$

where $A_1$ and $\theta_0$ are constants. Up to the terms of the first order, we write a solution of (8.3.9) in the form

$$x_1 = -\tfrac{1}{16} A^3 \cos 3\psi. \tag{8.3.11}$$

Thus, a solution of the first order to (8.3.5) has the form

$$r_1(t) = A_1 \cos(\sigma_1 t - \theta) - \varepsilon \tfrac{1}{16} A_1^3 \cos 3(\sigma_1 t - \theta) + O(\varepsilon^2), \tag{8.3.12}$$

where $\theta$ is determined by equality (8.3.10).

Thus, as a result of construction of a solution of the third order by the averaging method, we obtain relation (8.3.12) that coincides with the solution obtained in the previous item.

C. Equation (8.3.1) can be reduced [up to $r_1^3(t)$] to the quasilinear form

$$\ddot{r}_1 + \omega^2 r_1 = \varepsilon \left[ \omega^2 \frac{P_1}{d_1} \cos\tau - a r_1 - \omega^2 P_1 r_1^2 \cos\tau + \sigma_1^2 r_1^3 - r_1 \dot{r}_1^2 \right], \tag{8.3.13}$$

where $\varepsilon = d_1, \tau = \omega t, a = (\sigma_1^2 - \omega^2)/d_1$. By the Poincaré theorem, possible periodic solutions of quasilinear systems are located in the vicinity of the corresponding periodic solutions of the generating linear system provided that the parameter $\varepsilon$ is sufficiently small. In accordance with this result, we seek a solution of (8.3.13) in the form

$$r_1(t) = \varphi_0(t) + \varepsilon \varphi_1(t) + \varepsilon^2 \varphi_2(t) + \dots . \tag{8.3.14}$$

We substitute decomposition (8.3.14) into (8.3.13) and group terms of the same powers of $\varepsilon$. For $\varepsilon^0$, we obtain the generating equation

$$\frac{d^2 \varphi_0}{dt^2} + \omega^2 \varphi_0 = 0, \tag{8.3.15}$$

which has the periodic solution

$$\varphi_0(t) = A \cos\tau + A_1 \sin\tau. \tag{8.3.16}$$

Accounting the $O(\varepsilon)$-order terms, we obtain the following equation for $\varphi_1(t)$:

$$\frac{d^2 \varphi_1}{dt^2} + \omega^2 \varphi_1 = \omega^2 \left\{ A \left[ (3\bar{\sigma}_1^2 - 1) \frac{A^2 + A_1^2}{4} - \bar{a} \right] + P_1 \left( \frac{1}{d_1} - \frac{3A^2 + A_1^2}{4} \right) \right\} \cos\tau$$

$$+\omega^2 A_1\left\{(3\bar{\sigma}_1^2-1)\frac{A^2+A_1^2}{4}-\bar{a}-\frac{1}{2}P_1 A\right\}\sin\tau$$

$$+\omega^2\left\{P_1\frac{A_1^2-A^2}{4}+(\bar{\sigma}_1^2+1)A\frac{A^2-3A_1^2}{4}\right\}\cos 3\tau$$

$$+\omega^2 A_1\left\{(\bar{\sigma}_1^2+1)\frac{3A^2-A_1^2}{4}-\frac{1}{2}P_1 A\right\}\sin\tau, \qquad (8.3.17)$$

where

$$\bar{\sigma}_1=\frac{\sigma_1}{\omega},\quad \bar{a}=\frac{a}{\omega^2}=\frac{(\bar{\sigma}_1^2-1)}{d_1}.$$

By using the condition for the existence of periodic solutions (i.e., the absence of secular terms), we obtain

$$A\left[(3\bar{\sigma}_1^2-1)\frac{A^2+A_1^2}{4}-\bar{a}\right]+P_1\left(\frac{1}{d_1}-\frac{3A^2+A_1^2}{4}\right)=0,$$

$$A_1\left[(3\bar{\sigma}_1^2-1)\frac{A^2+A_1^2}{4}-\bar{a}-\frac{1}{2}P_1 A\right]=0. \qquad (8.3.18)$$

It follows from these equations that $A_1=0$ and $A$ is a root of the algebraic equation

$$\frac{3\bar{\sigma}_1^2-1}{4}A^3-\frac{3}{4}P_1 A^2+\frac{1-\bar{\sigma}_1^2}{d_1}A+\frac{P_1}{d_1}=0. \qquad (8.3.19)$$

Due to these results, we present the periodic solution of equation (8.3.16) in the form

$$\varphi_1(t)=\tfrac{1}{32}A^2[P_1-(\bar{\sigma}_1^2+1)A]\cos 3\tau+B\cos\tau. \qquad (8.3.20)$$

A new arbitrary constant $B$ here is determined from the periodicity condition of $\varphi_2(t)$. To find $B$, we consider the $O(\varepsilon^2)$-order terms at $\cos\tau$ and obtain the following relation:

$$B=\frac{2P_1+(5-3\bar{\sigma}_1^2)A}{3(3\bar{\sigma}_1^2-1)A^2-4\bar{a}-6P_1 A}\frac{A^3}{32}[P_1-(\bar{\sigma}_1^2+1)A]. \qquad (8.3.21)$$

Thus, in view of relations (8.3.14), (8.3.16), (8.3.20), and (8.3.21), a periodic solution of (8.3.13) in a neighborhood of resonance has the form

$$r_1(t)=A\cos\omega t+d_1\frac{A^2}{32}[P_1-(\bar{\sigma}_1^2+1)A]\cos 3\omega t$$

$$+d_1\frac{2P_1+(5-3\bar{\sigma}_1^2)A}{3(3\bar{\sigma}_1^2-1)A^2-4\bar{a}-6P_1 A}\frac{A^3}{32}[P_1-(\bar{\sigma}_1^2+1)A]\cos\omega t+\dots. \qquad (8.3.22)$$

This relation is similar to solution (8.3.12) obtained by the averaging method.

In conclusion, note the following: Applying formally the above-described procedure to (8.3.5), we obtain

$$r_1(t) = A\cos\omega t - d_1\frac{A^3}{16}\cos\omega t + \frac{3d_1^2A^5}{48d_1A^2 + 32(1-\bar{\sigma}_1^2)}\cos\omega t + \dots , \quad (8.3.23)$$

where $A$ is determined by the same relation as in the case of the Bubnov–Galerkin method for a single generalized coordinate, i.e., by (8.2.23) for $B = 0$. Keeping the terms up to the order of $O(A^3)$ inclusively, we can transform relation (8.3.19) into relation (8.2.23) (for $B = 0$).

II. We proceed to the generalized coordinates $r_1(t)$ and $p_1(t)$ that satisfy the system of nonlinear differential equations

$$\begin{gathered}\ddot{r}_1 + \sigma_1^2 r_1 + d_1(r_1^2\ddot{r}_1^2 + r_1\dot{r}_1^2 + r_1p_1\ddot{p}_1 + r_1\dot{p}_1^2) + d_2(p_1^2\ddot{r}_1 + 2p_1\dot{r}_1\dot{p}_1 \\ -r_1p_1\ddot{p}_1 - 2r_1\dot{p}_1^2) = \omega^2 P_1\cos\omega t, \\ \ddot{p}_1 + \sigma_1^2 p_1 + d_1(p_1\ddot{p}_1 + p_1\dot{r}_1^2 + p_1\dot{p}_1^2 + r_1p_1\ddot{r}_1) \\ +d_2(r_1^2\ddot{p}_1 + 2r_1\dot{r}_1\dot{p}_1 - r_1p_1\ddot{r}_1 - 2p_1\dot{r}_1^2) = 0.\end{gathered} \quad (8.3.24)$$

Since $1 + (d_1 + d_2)(r_1^2 + \dot{p}_1^2) > 0$, keeping the terms of the order of $O(r_1^3)$ inclusively, we can reduce system (8.3.24) to the normal form

$$\frac{dx}{dt} = X(x,t). \quad (8.3.25)$$

Here, $x$ and $X$ are four-dimensional vectors of the form

$$\begin{gathered} x^{\mathrm{T}} = (r_1, p_1, \dot{r}_1, \dot{p}_1), \quad X_1 = x_3, \quad X_2 = x_4, \\ X_3 = \omega^2 P_1\cos\omega t - \sigma_1^2 x_1 - \omega^2 P_1(d_1x_1^2 + d_2x_2^2)\cos\omega t \\ +d_1\sigma_1^2(x_1^3 + x_1x_2^2) - d_1x_1x_3^2 + (2d_2 - d_1)x_1x_4^2 - 2d_2x_2x_3x_4, \\ X_4 = -\sigma_1^2 x_2 + (d_2 - d_1)\omega^2 P_1x_1x_2\cos\omega t + d_1\sigma_1^2(x_1^2x_2 + x_2^3) \\ -d_1x_2x_4^2 + (2d_2 - d_1)x_2x_3^2 - 2d_2x_1x_3x_4. \end{gathered} \quad (8.3.26)$$

In [219], M. Urabe applies the Bubnov–Galerkin method to nonlinear periodic systems of the form (8.3.25) under some additional conditions for the right-hand sides of these systems. In particular, it is assumed that the vector function $X(x,t)$ and its first partial derivatives in $x$ are continuously differentiable with respect to $x$ and $t$ and, in addition, $X(x,t)$ is a $2\pi/\omega$-periodic function of $t$. It follows from relations (8.3.26) that these conditions are satisfied in the considered case.

In the considered case, in accordance with [219], if there exists an isolated periodic solution $\hat{x}(t)$ of system (8.3.25) that belongs to the domain of definition of $X(x,t)$, then, necessarily, (i) there exist Galerkin approximations $x_n(t)$

**Table 8.1**

| $t$ | $r_1(t) = A\cos\omega t$ | $p_1(t) = B\sin\omega t$ | $r_1(t)$ | $p_1(t)$ |
|---|---|---|---|---|
| | **Galerkin method** | | **Runge-Kutta method** | |
| 0 | 0.2741 | 0 | 0.2741 | 0 |
| 0.1 | 0.2577 | 0.1098 | 0.2576 | 0.1098 |
| 0.2 | 0.2104 | 0.2065 | 0.2101 | 0.2061 |
| 0.3 | 0.1379 | 0.2784 | 0.1376 | 0.2775 |
| 0.4 | 0.0489 | 0.3170 | 0.0489 | 0.3157 |
| 0.5 | −0.0460 | 0.3176 | −0.0456 | 0.3167 |
| 0.6 | −0.1353 | 0.2801 | −0.1347 | 0.2793 |
| 0.7 | −0.2085 | 0.2091 | −0.2079 | 0.2088 |
| 0.8 | −0.2567 | 0.1131 | −0.2564 | 0.1131 |
| 0.9 | −0.2741 | 0.0035 | −0.2741 | 0.0036 |
| 1.0 | −0.2587 | −0.1065 | −0.2587 | −0.1064 |
| 1.1 | −0.2123 | −0.2037 | −0.2122 | −0.2033 |
| 1.2 | −0.1405 | −0.2766 | −0.1405 | −0.2757 |
| 1.3 | −0.0519 | −0.3163 | −0.0522 | −0.3150 |
| 1.4 | 0.0430 | −0.3181 | 0.0422 | −0.3169 |
| 1.5 | 0.1327 | −0.2819 | 0.1316 | −0.2810 |
| 1.6 | 0.2065 | −0.2118 | 0.2056 | −0.2115 |
| 1.7 | 0.2556 | −0.1164 | 0.2551 | −0.1165 |
| 1.8 | 0.2741 | −0.0070 | 0.2739 | −0.0072 |

of all sufficiently large orders $n$ and (ii) certain operators related to the Jacobi matrix for $X(x,t)$ with respect to $x$ are bounded. By using these statements, one can prove that the Galerkin approximations $x_n(t)$ converge to $\hat{x}(t)$ uniformly as $n \to \infty$. Further, the existence of Galerkin approximations $x_n(t)$ for all sufficiently large $n$ implies the existence of the exact solution presented above. Using this result, M. Urabe formulates a criterion for the existence of the periodic solutions $\hat{x}(t)$.

Thus, due to the results stated in [219], we can consider the problem of the existence of an isolated periodic solution of system (8.3.25). In addition, it is of interest to estimate the error of approximation of the isolated periodic solution in dependence on the Galerkin approximation number. However, this procedure leads to rather awkward calculations. In [220], only simple second-order equations are considered.

In view of these difficulties, we investigate the nonlinear system (8.3.25) numerically. As an example, one can construct a periodic solution of (8.3.25) and (8.3.26) by the Runge–Kutta method, which gives the Cauchy solution with initial conditions $x^{\mathrm{T}} = (0.274145; 0; 0; 1.120516)$ that corresponds to the asymptotic periodic solution

$$r_1(t) = A\cos\omega t, \quad p_1(t) = B\sin\omega t. \tag{8.3.27}$$

Note the following specific feature of nonlinear oscillations: in a nonlinear system, different initial conditions can generate periodic solutions of different types. For the quantities in relations (8.3.26) and (8.3.27), we choose the values $d_1 = 0.577102$; $d_2 = -0.313260$; $\omega = 3.478513$; $A = 0.274145$; $B = 0.322125$. For the considered values of parameters, a solution in the form of (8.3.27) describes spatial sloshing (with period $T \approx 1.8$ sec) in a container in the form of two coaxial cylinders (with $R_0/R_1 = 0.5$ and $h/R_1 = 1$) under the influence of forced harmonic oscillations with $H/R_1 = 0.01$ and relative frequency $\omega/\sigma_1 = 1.02$. The computational results are shown in Table 8.1 where the second and third columns display the values obtained by using relations (8.3.27) and the fourth and fifth columns display the results of numerical integration of systems (8.3.25) and (8.3.26) by the Runge–Kutta method.

Values of the generalized coordinates $r_1(t)$ and $p_1(t)$ in Table 8.1 are given for a single period only. It is easy to see that the values of the (periodic) solution of system (8.3.25), (8.3.26) obtained by the Runge–Kutta method and the values obtained by relations (8.3.27) are close (as a rule, the difference is at most 0.001). From this, it is clear that the first harmonic component with the period of external excitation makes the major contribution to the periodic oscillations under consideration. Therefore, under the assumptions accepted in order to derive this system of differential equations, the other (higher) harmonics can be neglected.

## 8.4 Periodic Bubnov–Galerkin solution of (8.1.3)

In order to use the Bubnov–Galerkin method for finding periodic solutions of nonlinear system (8.1.3), we present the generalized coordinates $r_1(t)$ and $p_1(t)$ also in the form of a finite Fourier series with unknown coefficients [184]:

$$\begin{aligned} r_1(t) &= \alpha_0 + \sum_{k=1}^{n}(\alpha_k \cos k\omega t + \bar{\alpha}_k \sin k\omega t), \\ p_1(t) &= \beta_0 + \sum_{k=1}^{n}(\bar{\beta}_k \cos k\omega t + \beta_k \sin k\omega t). \end{aligned} \tag{8.4.1}$$

Choosing in these representations only first terms

$$r_1(t) = A\cos\omega t + \bar{A}\sin\omega t, \quad p_1(t) = \bar{B}\cos\omega t + B\sin\omega t, \tag{8.4.2}$$

we obtain the following explicit relations for the generalized coordinates $p_0(t)$, $p_2(t)$, and $r_2(t)$ from the last three equations of system (8.1.3) [since they are linear in $p_0(t)$, $p_2(t)$, and $r_2(t)$]:

$$\begin{aligned} p_0(t) &= c_0 + c_1 \cos 2\omega t + c_2 \sin 2\omega t, \\ p_2(t) &= s_0 + s_1 \cos 2\omega t + s_2 \sin 2\omega t, \\ r_2(t) &= e_0 + e_1 \cos 2\omega t + e_2 \sin 2\omega t, \end{aligned} \tag{8.4.3}$$

where

$$\begin{gathered} h_0 = \frac{d_{10}+d_8}{2(\bar\sigma_0^2-4)}, \quad h_2 = \frac{d_9+d_7}{2(\bar\sigma_2^2-4)}, \quad l_0 = \frac{d_{10}-d_8}{2\bar\sigma_0^2}, \quad l_2 = \frac{d_9-d_7}{2\bar\sigma_2^2}, \\ \bar\sigma_m = \frac{\sigma_m}{\omega}, \quad m=0,1,2, \quad c_0 = l_0(A^2+\bar A^2+B^2+\bar B^2), \\ c_1 = h_0(A^2-\bar A^2-B^2+\bar B^2), \quad c_2 = 2h_0(A\bar A+B\bar B), \\ s_0 = l_2(A^2+\bar A^2-B^2-\bar B^2), \quad s_1 = h_2(A^2-\bar A^2+B^2-\bar B^2), \\ s_2 = 2h_2(A\bar A-B\bar B), \quad e_0 = -2l_2(A\bar B+\bar A B), \\ e_1 = 2h_2(\bar A B-A\bar B), \quad e_2 = -2h_2(AB+\bar A\bar B). \end{gathered} \tag{8.4.4}$$

Substituting (8.4.2) and (8.4.3) into the Fredholm alternative

$$\begin{aligned} &\int_0^{\frac{2\pi}{\omega}} L_k(r_1,p_1,p_0,p_2,r_2)\cos\omega t dt = 0, \\ &\int_0^{\frac{2\pi}{\omega}} L_k(r_1,p_1,p_0,p_2,r_2)\sin\omega t dt = 0, \quad k=1,2, \end{aligned} \tag{8.4.5}$$

we get the following equations coupling the amplitudes $A$, $\bar A$, $B$, and $\bar B$:

$$\begin{aligned} A[\bar\sigma_1^2 - 1 + m_1(A^2+\bar A^2+\bar B^2) + m_2 B^2] + m_4 \bar A B\bar B &= P_1, \\ \bar A[\bar\sigma_1^2 - 1 + m_1(\bar A^2+B^2+A^2) + m_2 \bar B^2] + m_4 B\bar B A &= 0, \\ B[\bar\sigma_1^2 - 1 + m_1(B^2+\bar B^2+\bar A^2) + m_2 A^2] + m_4 \bar B A\bar B &= 0, \\ \bar B[\bar\sigma_1^2 - 1 + m_1(\bar B^2+A^2+B^2) + m_2 \bar A^2] + m_4 A\bar A B &= 0. \end{aligned} \tag{8.4.6}$$

Here,

$$\begin{gathered} m_1 = d_5\left(\tfrac12 h_0 - l_0\right) - d_3\left(\tfrac12 h_2 - l_2\right) - 2d_6 h_0 - 2d_4 h_2 - \tfrac12 d_1, \\ m_2 = -d_3\left(l_2 + \tfrac32 h_2\right) - d_5\left(l_0 + \tfrac12 h_0\right) + 2d_6 h_0 - 6d_4 h_2 + \tfrac12 d_1 - 2d_2, \\ m_4 = m_1 - m_2, \end{gathered} \tag{8.4.7}$$

Relations similar to (8.4.3) and, hence, (8.4.6) can also be obtained in a more general case if we keep a finite number of terms in the decompositions of

$r_1(t)$ and $p_1(t)$. By analyzing system (8.4.6) under the condition that $P_1 \not\equiv 0$ and $m_1 \neq m_2$ [54], one can obtain the following relations between the coefficients of decompositions (8.4.2):

$$\bar{A} = \bar{B} = 0, \quad A \neq 0, \quad |A| \neq |B|.$$

The amplitudes $A$ and $B$ are roots of the following system of algebraic equations associating the amplitudes $A$ and $B$ of forced oscillations with the frequency of the perturbation force $\omega$ and its amplitude $H$:

$$A(\bar{\sigma}_1^2 - 1 + m_1A^2 + m_2B^2) = P_1, \quad B(\bar{\sigma}_1^2 - 1 + m_1B^2 + m_2A^2) = 0. \quad (8.4.8)$$

By analyzing relations (8.4.2), (8.4.3), and (8.4.8), in view of the results presented in Sec. 8.2, one can see that there exist two types of approximate solutions of the nonlinear system (8.1.3):

(a) Planar regime, i.e., forced liquid sloshing in the plane of the excitation force (in the plane $xOz$):

$$r_1(t) = A\cos\omega t, \quad p_1(t) = 0, \quad r_2(t) \equiv 0,$$
$$p_k(t) = A^2(l_k + h_k\cos 2\omega t), \quad k = 0, 2, \quad (8.4.9)$$

where the amplitude $A$ is the root of the cubic equation

$$m_1A^3 + (\bar{\sigma}_1^2 - 1)A - P_1 = 0; \quad (8.4.10)$$

(b) Swirling regime formed in the domain of dynamic instability of planar regime as a result of parametric excitation of oscillations in the plane $xOy$:

$$r_1(t) = A\cos\omega t, \quad p_1(t) = B\sin\omega t, \quad r_2(t) = -2h_2AB\sin 2\omega t,$$
$$p_0(t) = l_0(A^2 + B^2) + h_0(A^2 - B^2)\cos 2\omega t,$$
$$p_2(t) = l_2(A^2 - B^2) + h_2(A^2 + B^2)\cos 2\omega t, \quad (8.4.11)$$

where $A$ and $B$ are determined by the equations

$$\left(m_1 - \frac{m_2^2}{m_1}\right)A^3 + (\bar{\sigma}_1^2 - 1)\left(1 - \frac{m_2}{m_1}\right)A - P_1 = 0,$$

$$B^2 = \frac{1 - \bar{\sigma}_1^2}{m_1} - \frac{m_2}{m_1}A^2. \quad (8.4.12)$$

Expressions for the response curves (8.2.23) and (8.2.24), which are derived in Sec. 8.2 under the condition that only two dominant generalized coordinates $r_1(t)$ and $p_1(t)$ are taken into account, correspond to a special case of the above-presented relations. To obtain these relations, we have to set $m_1 = -\frac{1}{2}d_1$ and $m_2 = \frac{1}{2}d_1 - 2d_2$ in (8.4.10) [or in (8.4.12)], i.e., to neglect the influence of the generalized coordinates of higher orders. Note that the amplitudes $A$ and $B$ are dimensionless. To obtain actual dimensions, one has to replace $A$ and $B$ by $AR_1$ and $BR_1$, respectively, where $R_1$ is the radius of the exterior cylinder.

## 8.5 Stability of periodic solutions of (8.1.3)

As it was already noted, depending on the amplitude $H$ and frequency $\omega$ of the external excitation, planar and swirling steady-state regimes are possible. The planar regime occurs for the frequencies outside a certain neighborhood of the lowest natural sloshing frequency. In the vicinity of this resonant zone, the planar sloshing becomes unstable and swirling may be observed. Periodic solutions of (8.1.3) in the form of (8.4.9) and (8.4.11) provide a possibility to describe these phenomena in a more complete way than it is suggested in Sec. 8.2. Planar steady-state sloshing corresponds to a periodic solution (8.4.9) and, in this case, some generalized coordinates are not excited. Swirling corresponds to solution (8.4.11) and, in this case, all generalized coordinates are excited.

Only stable solutions defined by relations (8.4.9) and (8.4.11) can be realized. We study the stability of the planar and swirling steady-state regimes by using the first Lyapunov method. To this end, we derive variational equations for planar motions under the condition that the Lyapunov unperturbed motion of system (8.1.3) is described by relation (8.4.9). Along with the unperturbed motion, we consider the perturbed motion of the form

$$\tilde{r}_1(t) = r_1(t) + \alpha(t), \quad \tilde{p}_1(t) = \beta(t), \quad \tilde{p}_0(t) = p_0(t) + \gamma(t),$$
$$\tilde{r}_2(t) = \delta(t), \quad \tilde{p}_2(t) = p_2(t) + \varepsilon(t), \tag{8.5.1}$$

which is close to the unperturbed motion, and whose initial conditions differ slightly enough from the initial conditions of the unperturbed motion. Here, $\alpha$, $\beta$, $\gamma$, $\delta$, and $\varepsilon$ denote differences between the perturbed solution (8.5.1) and unperturbed solution (8.4.9).

We now deduce variational equations for the perturbations $\alpha$, $\beta$, $\gamma$, $\delta$, and $\varepsilon$. To this end, we substitute the unperturbed solution (8.5.1) into system (8.1.3) and take into account the fact that the unperturbed solution (8.4.9) satisfies this system. Linearizing the obtained system with respect to perturbations, we obtain the following system of variational equations:

$$\begin{aligned}
&(1+d_1r_1^2+d_5p_0-d_3p_2)\ddot{\alpha} + d_6r_1\ddot{\gamma} + d_4r_1\ddot{\varepsilon} + (2d_1r_1\dot{r}_1+d_5\dot{p}_0-d_3\dot{p}_2)\dot{\alpha} + d_5\dot{r}_1\dot{\gamma}\\
&\quad -d_3\dot{r}_1\dot{\varepsilon} + [\sigma_1^2 + d_1(2r_1\ddot{r}_1 + \dot{r}_1^2) + d_6\ddot{p}_0 + d_4\ddot{p}_2]\alpha + d_5\ddot{r}_1\gamma - d_3\ddot{r}_1\varepsilon = 0,\\
&(1+d_2r_1^2+d_5p_0+d_3p_2)\ddot{\beta} - d_4r_1\ddot{\delta} + (2d_2r_1\dot{r}_1+d_5\dot{p}_0+d_3\dot{p}_2)\dot{\beta} + d_3\dot{r}_1\dot{\delta}\\
&\quad +[\sigma_1^2 + (d_1-d_2)r_1\ddot{r}_1 + (d_1-2d_2)\dot{r}_1^2 + d_6\ddot{p}_0 - d_4\ddot{p}_2]\beta + d_3\ddot{r}_1\delta = 0,\\
&\ddot{\gamma} + \sigma_0^2\gamma + d_{10}(\alpha\ddot{r}_1 + \ddot{\alpha}r_1) + 2d_8\dot{\alpha}\dot{r}_1 = 0,\\
&\ddot{\varepsilon} + \sigma_2^2\varepsilon + d_9(\alpha\ddot{r}_1 + \ddot{\alpha}r_1) + 2d_7\dot{\alpha}\dot{r}_1 = 0,\\
&\ddot{\delta} + \sigma_2^2\delta - d_9(\beta\ddot{r}_1 + \ddot{\beta}r_1) - 2d_7\dot{\beta}\dot{r}_1 = 0,
\end{aligned} \tag{8.5.2}$$

where $r_1(t)$, $p_0(t)$, and $p_2(t)$ are defined by relations (8.4.9).

By analogy, for spatial oscillations corresponding to solution (8.4.11), we consider the perturbed solution of system (8.1.1)

$$\tilde{r}_1(t) = r_1(t) + \alpha(t), \quad \tilde{p}_1(t) = p_1(t) + \beta(t), \quad \tilde{p}_0(t) = p_0(t) + \gamma(t),$$
$$\tilde{r}_2(t) = r_2(t) + \delta(t), \quad \tilde{p}_2(t) = p_2(t) + \varepsilon(t). \tag{8.5.3}$$

Substituting the perturbed solution (8.5.3) into system (8.1.1) and linearizing the obtained system with respect to perturbations, we obtain the system of variational equations

$$\begin{aligned}
&(1 + d_1 r_1^2 + d_2 p_1^2 + d_5 p_0 - d_3 p_2)\ddot{\alpha} + [(d_1 - d_2) r_1 p_1 + d_3 r_2]\ddot{\beta} + d_6 r_1 \ddot{\gamma} \\
&\quad + d_3 p_1 \ddot{\delta} + d_4 r_1 \ddot{\varepsilon} + (2 d_1 r_1 \dot{r}_1 + 2 d_2 p_1 \dot{p}_1 + d_5 \dot{p}_0 - d_3 \dot{p}_2)\dot{\alpha} \\
&\quad + [2 d_1 r_1 \dot{p}_1 + 2 d_2 (p_1 \dot{r}_1 - 2 r_1 \dot{p}_1) + d_3 \dot{r}_2]\dot{\beta} + d_5 \dot{r}_1 \dot{\gamma} + d_3 \dot{p}_1 \dot{\delta} - d_3 \dot{r}_1 \dot{\varepsilon} \\
&\quad + [\sigma_1^2 + d_1 (2 r_1 \ddot{r}_1 + \dot{r}_1^2 + p_1 \ddot{p}_1 + \dot{p}_1^2) - d_2 (p_1 \ddot{p}_1 + 2 \dot{p}_1^2) + d_6 \ddot{p}_0 + d_4 \ddot{p}_2]\alpha \\
&\quad + [d_1 r_1 \ddot{p}_1 + d_2 (2 p_1 \ddot{r}_1 + 2 \dot{r}_1 \dot{p}_1 - r_1 \ddot{p}_1) - d_4 \ddot{r}_2]\beta + d_5 \ddot{r}_1 \gamma + d_3 \ddot{p}_1 \delta + d_4 \ddot{r}_1 \varepsilon = 0, \\
&[(d_1 - d_2) r_1 p_1 + d_3 r_2]\ddot{\alpha} + (1 + d_1 p_1^2 + d_2 r_1^2 + d_5 p_0 + d_3 p_2)\ddot{\beta} + d_6 p_1 \ddot{\gamma} \\
&\quad - d_4 r_1 \ddot{\delta} - d_4 p_1 \ddot{\varepsilon} + [2 d_1 p_1 \dot{r}_1 + 2 d_2 (r_1 \dot{p}_1 - 2 p_1 \dot{r}_1) + d_3 \dot{r}_2]\alpha \\
&\quad + (2 d_1 p_1 \dot{p}_1 + 2 d_2 r_1 \dot{r}_1 + d_5 \dot{p}_0 + d_3 \dot{p}_2)\dot{\beta} + d_5 \dot{p}_1 \dot{\gamma} \\
&\quad + d_3 \dot{r}_1 \dot{\delta} + d_3 \dot{p}_1 \dot{\varepsilon} + [d_1 p_1 \ddot{r}_1 + d_2 (2 r_1 \ddot{p}_1 + 2 \dot{r}_1 \dot{p}_1 - p_1 \ddot{r}_1) - d_4 \ddot{r}_2]\alpha \\
&\quad + [\sigma_1^2 + d_1 (2 p_1 \ddot{p}_1 + r_1 \ddot{r}_1 + \dot{r}_1^2 + \dot{p}_1^2) - d_2 (r_1 \ddot{r}_1 + 2 \dot{r}_1^2) + d_6 \ddot{p}_0 - d_4 \ddot{p}_2]\beta \\
&\quad + d_5 \ddot{p}_1 \gamma + d_3 \ddot{r}_1 \delta + d_3 \ddot{p}_1 \varepsilon = 0, \\
&\ddot{\gamma} + \sigma_0^2 \gamma + d_{10} (\alpha \ddot{r}_1 + \ddot{\alpha} r_1 + \beta \ddot{p}_1 + \ddot{\beta} p_1) + 2 d_8 (\dot{\alpha} \dot{r} + \dot{\beta} \dot{p}_1) = 0, \\
&\ddot{\varepsilon} + \sigma_2^2 \varepsilon + d_9 (\alpha \ddot{r}_1 + \ddot{\alpha} r_1 - \beta \ddot{p}_1 - \ddot{\beta} p_1) + 2 d_7 (\dot{\alpha} \dot{r}_1 - \dot{\beta} \dot{p}_1) = 0, \\
&\ddot{\delta} + \sigma_2^2 \delta - d_9 (\alpha \ddot{p}_1 + \ddot{\alpha} p_1 + \beta \ddot{r}_1 + \ddot{\beta} r_1) - 2 d_7 (\dot{\alpha} \dot{p}_1 + \dot{\beta} \dot{r}_1) = 0,
\end{aligned} \tag{8.5.4}$$

where $r_1(t)$, $p_1(t)$, $p_0(t)$, $p_2(t)$, and $r_2(t)$ are defined by relations (8.4.11).

Thus, the stability analysis of periodic solutions (8.4.9) [or (8.4.11)] is reduced to the solution of system (8.5.2) [or (8.5.4)]. The obtained system consists of equations with periodic coefficients. By the Floquet–Lyapunov theorem, a fundamental set of solutions to this system takes the form

$$\alpha(t) = e^{\lambda t}\varphi_1(t), \quad \beta(t) = e^{\lambda t}\varphi_2(t), \quad \gamma(t) = e^{\lambda t}\varphi_3(t), \tag{8.5.5}$$
$$\delta(t) = e^{\lambda t}\varphi_4(t), \quad \varepsilon(t) = e^{\lambda t}\varphi_5(t),$$

where $\lambda$ is the characteristic exponent of the system and $\varphi_i$ are $\frac{2\pi}{\omega}$-periodic functions.

It follows from relations (8.5.1) and (8.5.3) that stability of solutions (8.4.9) and (8.4.11) depends on values of the characteristic exponent $\lambda$. If real parts of all characteristic exponents are negative, then periodic solutions are stable. If there exists at least one exponent with positive real part among the characteristic exponents, then periodic solutions (8.4.9) or (8.4.11) are unstable. The case where a characteristic exponent has real part equal to zero is more complicated. If all characteristic exponents are simple or multiple components have prime elementary divisors, then solutions (8.5.5) of the system of variational equations are bounded in time. If these conditions are not satisfied, then a secular instability occurs. This instability is caused by terms of the form

$$e^{\lambda_i t}\left[\frac{t^{m_i-1}}{(m_i-1)!}\varphi_{i1}(t) + \ldots + \varphi_{im_i}(t)\right]$$

in solution (8.5.5).

Let us deduce an equation to determine the characteristic exponents. To this end, we present the periodic functions $\varphi_i(t)$ in the following form of a truncated Fourier series by keeping only the first harmonics in the decompositions:

$$\varphi_i(t) = \tilde{a}_{i0} + \tilde{a}_i \cos\omega t + \tilde{b}_i \sin\omega t, \quad i = 1, 2, \ldots 5, \tag{8.5.6}$$

where $\tilde{a}_{i0}$, $\tilde{a}_i$, and $\tilde{b}_i$ are certain constant coefficients. We substitute (8.5.5) and (8.5.6) into the system of variational equations (8.5.2) [or (8.5.4)]. The perturbations $\gamma(t)$, $\delta(t)$, and $\varepsilon(t)$ are defined by the last three equations of the system and can be expressed in terms of $\tilde{a}_1$, $\tilde{b}_1$, $\tilde{a}_2$, and $\tilde{b}_2$. We obtain the following homogeneous system of linear algebraic equations for finding the constants $\tilde{a}_1$, $\tilde{b}_1$, $\tilde{a}_2$, and $\tilde{b}_2$:

$$\begin{aligned}
&A_{11}\tilde{b}_1 + \bar{\lambda}B_{12}\tilde{a}_1 + \bar{\lambda}B_{13}\tilde{b}_2 + A_{14}\tilde{a}_2 = 0,\\
&\bar{\lambda}B_{21}\tilde{b}_1 + A_{22}\tilde{a}_1 + A_{23}\tilde{b}_2 + \bar{\lambda}B_{24}\tilde{a}_2 = 0,\\
&\bar{\lambda}B_{31}\tilde{b}_1 + A_{32}\tilde{a}_1 + A_{33}\tilde{b}_2 + \bar{\lambda}B_{34}\tilde{a}_2 = 0,\\
&A_{41}\tilde{b}_1 + \bar{\lambda}B_{42}\tilde{a}_1 + \bar{\lambda}B_{43}\tilde{b}_2 + A_{44}\tilde{a}_2 = 0.
\end{aligned} \tag{8.5.7}$$

Here, $\bar{\lambda}$ denotes the ratio $\frac{\lambda}{\omega}$, the coefficients $A_{11}$, $B_{12}$, ..., $A_{44}$ of the linear algebraic system (8.5.7) are expressed in terms of the coefficients of nonlinear differential equations (8.1.3), the quantity $\bar{\lambda}$, and amplitudes $A$ and $B$ of the generalized coordinates $r_1(t)$ and $p_1(t)$ as follows:

$$\begin{aligned}
A_{11} &= \delta_1 + \delta_2 + \bar{\gamma}_0\bar{\alpha}_{01}A + \bar{\alpha}^2 d_3\bar{\alpha}_{11}B - \bar{\gamma}_2\bar{\alpha}_{12}A,\\
B_{12} &= 2A_0 + \bar{\gamma}_0\bar{\beta}_{01}A + d_3\bar{\beta}_{11}B - \bar{\gamma}_2\bar{\beta}_{12}A,\\
B_{13} &= C_2 + \bar{\gamma}_0\bar{\alpha}_{02}A + d_3\bar{\alpha}_{12}B + \bar{\gamma}_2\bar{\alpha}_{11}A,
\end{aligned}$$

$$
\begin{aligned}
A_{14} &= \delta_3 + C_2 + \bar{\gamma}_0\bar{\beta}_{02}A + \bar{\lambda}^2 d_3\bar{\beta}_{12}B + \bar{\gamma}_2\bar{\beta}_{11}A,\\
B_{21} &= -2A_0 - d_5\bar{\alpha}_{01}A - \bar{\gamma}_2\bar{\alpha}_{11}B - d_3\bar{\alpha}_{12}A,\\
A_{22} &= \delta_1 - \delta_2 - \bar{\lambda}^2 d_5\bar{\beta}_{01}A - \bar{\gamma}_2\bar{\beta}_{11}B - \bar{\lambda}^2 d_3\bar{\beta}_{12}A,\\
A_{23} &= \delta_3 - C_2 - \bar{\lambda}^2 d_5\bar{\alpha}_{02}A - \bar{\gamma}_2\bar{\alpha}_{12}B + \bar{\lambda}^2 d_3\bar{\alpha}_{11}A,\\
B_{24} &= C_2 - d_5\bar{\beta}_{02}A - \bar{\gamma}_2\bar{\beta}_{12}B + d_3\bar{\beta}_{11}A,\\
B_{31} &= -C_2 + d_5\bar{\alpha}_{01}B - \bar{\gamma}_2\bar{\alpha}_{11}A - d_3\bar{\alpha}_{12}B,\\
A_{32} &= \delta_3 - C_2 + \bar{\lambda}^2 d_5\bar{\beta}_{01}B - \bar{\gamma}_2\bar{\beta}_{11}A - \bar{\lambda}^2 d_3\bar{\beta}_{12}B,\\
A_{33} &= \bar{\delta}_1 + \bar{\delta}_2 + \bar{\lambda}^2 d_5\bar{\alpha}_{02}B - \bar{\gamma}_2\bar{\alpha}_{12}A + \bar{\lambda}^2 d_3\bar{\alpha}_{11}B,\\
B_{34} &= 2\bar{A}_0 + d_5\bar{\beta}_{02}B - \bar{\gamma}_2\bar{\beta}_{12}A + d_3\bar{\beta}_{11}B,\\
A_{41} &= \delta_3 + C_2 + \bar{\gamma}_0\bar{\alpha}_{01}B - \bar{\lambda}^2 d_3\bar{\alpha}_{11}A + \bar{\gamma}_2\bar{\alpha}_{12}B,\\
B_{42} &= -C_2 + \bar{\gamma}_0\bar{\beta}_{01}B - d_3\bar{\beta}_{11}A + \bar{\gamma}_2\bar{\beta}_{12}B,\\
B_{43} &= -2\bar{A}_0 + \bar{\gamma}_0\bar{\alpha}_{02}B - d_3\bar{\alpha}_{12}A - \bar{\gamma}_2\bar{\alpha}_{11}B,\\
A_{44} &= \bar{\delta}_1 - \bar{\delta}_2 + \bar{\gamma}_0\bar{\beta}_{02}B - \bar{\lambda}^2 d_3\bar{\beta}_{12}A - \bar{\gamma}_2\bar{\beta}_{11}B.
\end{aligned}
\tag{8.5.8}
$$

In (8.5.8), the following notation is used:

$$
\begin{aligned}
&\delta_1 = (\bar{\lambda}^2 - 1)A_0 + F_1, \quad \bar{\delta}_1 = (\bar{\lambda}^2 - 1)\bar{A}_0 + \bar{F}_1,\\
&\delta_2 = \tfrac{1}{2}[(\bar{\lambda}^2 + 1)A_1 + G_1], \quad \bar{\delta}_2 = \tfrac{1}{2}[(\bar{\lambda}^2 + 1)\bar{A}_1 + \bar{G}_1],\\
&\delta_3 = \tfrac{1}{2}[(\bar{\lambda}^2 + 1)B_1 + H_2], \quad \bar{\gamma}_0 = \bar{\lambda}^2 d_6 - d_5, \quad \bar{\gamma}_2 = \bar{\lambda}^2 d_4 + d_3,\\
&\bar{\alpha}_{01} = \frac{d_{10} - d_8 - \frac{1}{2}\bar{\lambda}^2 d_{10}}{\bar{\lambda}^2 + \bar{\sigma}_0^2}A, \quad \bar{\beta}_{01} = \frac{d_8 - d_{10}}{\bar{\lambda}^2 + \bar{\sigma}_0^2}A,\\
&\bar{\alpha}_{02} = \frac{d_{10} - d_8}{\bar{\lambda}^2 + \bar{\sigma}_0^2}B, \quad \bar{\beta}_{02} = \frac{d_{10} - d_8 - \frac{1}{2}\bar{\lambda}^2 d_{10}}{\bar{\lambda}^2 + \bar{\sigma}_0^2}B,\\
&\bar{\alpha}_{11} = \frac{d_7 - d_9}{\bar{\lambda}^2 + \bar{\sigma}_2^2}B, \quad \bar{\beta}_{11} = \frac{d_7 - d_9 + \frac{1}{2}\bar{\lambda}^2 d_9}{\bar{\lambda}^2 + \bar{\sigma}_2^2}B,\\
&\bar{\alpha}_{12} = \frac{d_7 - d_9 + \frac{1}{2}\bar{\lambda}^2 d_9}{\bar{\lambda}^2 + \bar{\sigma}_2^2}A, \quad \bar{\beta}_{12} = \frac{d_9 - d_7}{\bar{\lambda}^2 + \bar{\sigma}_2^2}A,
\end{aligned}
\tag{8.5.9}
$$

and

$$
\begin{aligned}
A_0 &= 1 + \tfrac{1}{2}(d_1 A^2 + d_2 B^2) + d_5 c_0 - d_3 s_0,\\
\bar{A}_0 &= 1 + \tfrac{1}{2}(d_1 B^2 + d_2 A^2) + d_5 c_0 + d_3 s_0,\\
A_1 &= \tfrac{1}{2}(d_1 A^2 - d_2 B^2) + d_5 c_1 - d_3 s_1,\\
\bar{A}_1 &= \tfrac{1}{2}(d_2 A^2 - d_1 B^2) + d_5 c_1 + d_3 s_1,
\end{aligned}
$$

$$
\begin{gathered}
B_1 = (d_1 - d_2)\tfrac{1}{2}AB + d_3 e_2, \quad C_2 = (d_1 - 3d_2)AB, \\
F_1 = \bar{\sigma}_1^2 - \tfrac{1}{2}(d_1 A^2 + d_2 B^2), \quad \bar{F}_1 = \bar{\sigma}_1^2 - \tfrac{1}{2}(d_1 B^2 + d_2 A^2), \\
G_1 = d_1 B^2 - \tfrac{3}{2}(d_1 A^2 + d_2 B^2) - 4(d_6 c_1 + d_4 s_1), \\
\bar{G}_1 = -d_1 A^2 + \tfrac{3}{2}(d_1 B^2 + d_2 A^2) - 4d_6 c_1 - d_4 s_1, \\
H_2 = 4d_4 e_2 - (d_1 + 3d_2)\tfrac{1}{2}AB,
\end{gathered} \tag{8.5.10}
$$

where $c_0$, $c_1$, $s_0$, $s_1$, and $e_2$ are defined by relations (8.4.4) (in view of the condition $\bar{A} = \bar{B} = 0$).

Note that relations (8.5.8), (8.5.9), and (8.5.10) correspond to the case of three-dimensional sloshing (8.4.11). In order to investigate the case of planar sloshing (8.4.9), we have to set $B = 0$ in all relations.

Since it is supposed that the system of linear homogeneous algebraic equations (8.5.7) with respect to constants $\tilde{a}_1$, $\tilde{b}_1$, $\tilde{a}_2$, and $\tilde{b}_2$ has a nonzero solution (otherwise, we have $\tilde{a}_i = \tilde{b}_i = 0$), the determinant of this system should be equal to zero:

$$
D(\lambda) = \begin{vmatrix} A_{11} & \bar{\lambda}B_{12} & \bar{\lambda}B_{13} & A_{14} \\ \bar{\lambda}B_{21} & A_{22} & A_{23} & \bar{\lambda}B_{24} \\ \bar{\lambda}B_{31} & A_{32} & A_{33} & \bar{\lambda}B_{34} \\ A_{41} & \bar{\lambda}B_{42} & \bar{\lambda}B_{43} & A_{44} \end{vmatrix} = 0. \tag{8.5.11}
$$

Expanding determinant (8.5.11), we obtain the characteristic equation in the form a polynomial of degree 12 in $\bar{\lambda}^2$. With regard to relations (8.5.8)–(8.5.10), it is rater difficult to present this equation in an explicit form. For this reason, the problem of finding values of the characteristic exponents $\lambda$ is reduced to computing all roots of the characteristic determinant (8.5.11) by numerical methods. It should be noted that if $\lambda$ is a root of the characteristic equation, then $-\lambda$ is a root, too, since this equation is even. Thus, unstable motions are associated with the presence of characteristic exponents with real parts ($\mathbf{Re}\,\lambda \neq 0$). In the present case, stable oscillations are also associated with imaginary roots ($\mathbf{Re}\,\lambda = 0$) of the characteristic determinant (8.5.11). According to the classification suggested in Sec. 8.2, these are presented by neutral solutions of the corresponding variational equations.

The expressions for coefficients of the system of linear algebraic equations (8.5.7) are much simpler in the case where we consider only two nonsymmetric harmonics associated with the generalized coordinates $r_1(t)$ and $p_1(t)$ (see Sec. 7.2). In this case, it is easy to present the characteristic determinant (8.5.11) in the form

$$
D(\lambda) = \alpha_4\bar{\lambda}^8 + \alpha_3\bar{\lambda}^6 + \alpha_2\bar{\lambda}^4 + \alpha_1\bar{\lambda}^2 + \alpha_0. \tag{8.5.12}
$$

The coefficients of this polynomial are defined by the relations

$$
\begin{aligned}
\alpha_4 &= (\tilde{B}_{11}\tilde{B}_{44} - \tilde{B}_{14}^2)(\tilde{B}_{22}\tilde{B}_{33} - \tilde{B}_{14}^2),\\
\alpha_3 &= \tilde{B}_{33}\tilde{B}_{44}(\tilde{A}_{11}\tilde{B}_{22} + \tilde{A}_{22}\tilde{B}_{11} + \tilde{B}_{12}) + \tilde{B}_{11}\tilde{B}_{22}(\tilde{A}_{22}\tilde{B}_{44} + \tilde{A}_{44}\tilde{B}_{33} + \tilde{B}_{34}^2)\\
&\quad -\tilde{B}_{14}\tilde{B}_{44}(\tilde{A}_{11}\tilde{B}_{14} + \tilde{A}_{23}\tilde{B}_{11} + \tilde{B}_{12}\tilde{B}_{13}) - \tilde{B}_{11}\tilde{B}_{14}(\tilde{A}_{23}\tilde{B}_{44} + \tilde{A}_{44}\tilde{B}_{14}\\
&\quad +\tilde{B}_{13}\tilde{B}_{34}) + (\tilde{B}_{11}\tilde{B}_{13} + \tilde{B}_{12}\tilde{B}_{14})(\tilde{B}_{13}\tilde{B}_{33} - \tilde{B}_{14}\tilde{B}_{34}) + (\tilde{B}_{13}\tilde{B}_{22}\\
&\quad -\tilde{B}_{12}\tilde{B}_{14})(\tilde{B}_{13}\tilde{B}_{44} + \tilde{B}_{14}\tilde{B}_{34}) - \tilde{B}_{14}\tilde{B}_{33}(\tilde{A}_{14}\tilde{B}_{22} + \tilde{A}_{22}\tilde{B}_{14} - \tilde{B}_{12}\tilde{B}_{13})\\
&\quad +\tilde{B}_{14}\tilde{B}_{22}(\tilde{B}_{13}\tilde{B}_{34} - \tilde{A}_{14}\tilde{B}_{33} - \tilde{A}_{22}\tilde{B}_{14}) - 2\tilde{B}_{13}^2\tilde{B}_{14}^2 + 2(\tilde{A}_{14} + \tilde{A}_{23})\tilde{B}_{14}^3,\\
\alpha_2 &= \tilde{A}_{11}\tilde{A}_{22}\tilde{B}_{33}\tilde{B}_{44} + \tilde{A}_{22}\tilde{A}_{44}\tilde{B}_{11}\tilde{B}_{22} + (\tilde{A}_{11}\tilde{B}_{22} + \tilde{A}_{22}\tilde{B}_{11} + \tilde{B}_{12}^2)(\tilde{A}_{22}\tilde{B}_{44}\\
&\quad +\tilde{A}_{44}\tilde{B}_{33} + \tilde{B}_{34}^2) - \tilde{A}_{11}\tilde{A}_{23}\tilde{B}_{14}\tilde{B}_{44} - \tilde{A}_{23}\tilde{A}_{44}\tilde{B}_{11}\tilde{B}_{14} - (\tilde{A}_{11}\tilde{B}_{14}\\
&\quad +\tilde{A}_{23}\tilde{B}_{11} + \tilde{B}_{12}\tilde{B}_{13})(\tilde{A}_{23}\tilde{B}_{44} + \tilde{A}_{44}\tilde{B}_{14} + \tilde{B}_{13}\tilde{B}_{34}) + (\tilde{A}_{11}\tilde{B}_{13}\\
&\quad +\tilde{A}_{14}\tilde{B}_{12})(\tilde{B}_{13}\tilde{B}_{33} - \tilde{B}_{14}\tilde{B}_{34}) + (\tilde{B}_{11}\tilde{B}_{13} + \tilde{B}_{12}\tilde{B}_{14})(\tilde{A}_{22}\tilde{B}_{13} - \tilde{A}_{23}\tilde{B}_{34})\\
&\quad +(\tilde{A}_{22}\tilde{B}_{13} - \tilde{A}_{23}\tilde{B}_{12})(\tilde{B}_{13}\tilde{B}_{44} + \tilde{B}_{14}\tilde{B}_{34}) + (\tilde{A}_{14}\tilde{B}_{34} + \tilde{A}_{44}\tilde{B}_{13})(\tilde{B}_{13}\tilde{B}_{22}\\
&\quad -\tilde{B}_{12}\tilde{B}_{14}) - \tilde{A}_{14}\tilde{A}_{22}\tilde{B}_{14}(\tilde{B}_{33} + \tilde{B}_{22}) + (\tilde{A}_{14}\tilde{B}_{22} + \tilde{A}_{22}\tilde{B}_{14}\\
&\quad -\tilde{B}_{12}\tilde{B}_{13})(\tilde{B}_{13}\tilde{B}_{34} - \tilde{A}_{14}\tilde{B}_{33} - \tilde{A}_{22}\tilde{B}_{14}) + 2\tilde{A}_{14}\tilde{A}_{23}\tilde{B}_{14}^2\\
&\quad +[\tilde{B}_{13}^2 - (\tilde{A}_{14} + \tilde{A}_{23})\tilde{B}_{14}]^2, \qquad (8.5.13)\\
\alpha_1 &= \tilde{A}_{11}\tilde{A}_{22}(\tilde{A}_{22}\tilde{B}_{44} + \tilde{A}_{44}\tilde{B}_{33} + \tilde{B}_{34}^2) + \tilde{A}_{22}\tilde{A}_{44}(\tilde{A}_{11}\tilde{B}_{22} + \tilde{A}_{22}\tilde{B}_{11} + \tilde{B}_{12}^2)\\
&\quad -\tilde{A}_{11}\tilde{A}_{23}(\tilde{A}_{23}\tilde{B}_{44} + \tilde{A}_{44}\tilde{B}_{14} + \tilde{B}_{13}\tilde{B}_{34}) - \tilde{A}_{23}\tilde{A}_{44}(\tilde{A}_{11}\tilde{B}_{14} + \tilde{A}_{23}\tilde{B}_{11}\\
&\quad +\tilde{B}_{12}\tilde{B}_{13}) + (\tilde{A}_{11}\tilde{B}_{13} + \tilde{A}_{14}\tilde{B}_{12})(\tilde{A}_{22}\tilde{B}_{13} - \tilde{A}_{23}\tilde{B}_{34}) + (\tilde{A}_{22}\tilde{B}_{13}\\
&\quad -\tilde{A}_{23}\tilde{B}_{12})(\tilde{A}_{14}\tilde{B}_{34} + \tilde{A}_{44}\tilde{B}_{13}) + \tilde{A}_{14}\tilde{A}_{22}(\tilde{B}_{13}\tilde{B}_{34} - \tilde{A}_{14}\tilde{B}_{33} - \tilde{A}_{22}\tilde{B}_{14})\\
&\quad -\tilde{A}_{14}\tilde{A}_{22}(\tilde{A}_{14}\tilde{B}_{22} + \tilde{A}_{22}\tilde{B}_{14} - \tilde{B}_{12}\tilde{B}_{13}) - 2\tilde{A}_{14}\tilde{A}_{23}\tilde{B}_{13}^2\\
&\quad +2\tilde{A}_{14}\tilde{A}_{23}(\tilde{A}_{14} + \tilde{A}_{23})\tilde{B}_{14},\\
\alpha_0 &= (\tilde{A}_{11}\tilde{A}_{44} - \tilde{A}_{14}^2)(\tilde{A}_{22}^2 - \tilde{A}_{23}^2),
\end{aligned}
$$

where $\tilde{A}_{11}$, $\tilde{B}_{11}$, …, $\tilde{B}_{44}$ are expressed in terms of the coefficients of motion of equations (8.2.1) and amplitudes $A$ and $B$ of the generalized coordinates $r_1(t)$ and $p_1(t)$ as follows:

$$
\begin{gathered}
\tilde{A}_{11} = \bar{\sigma}_1^2 - 1 - \tfrac{3}{2}d_1A^2 + \left(\tfrac{1}{2}d_1 - 2d_2\right)B^2, \quad \tilde{B}_{11} = 1 + \tfrac{3}{4}d_1A^2 + \tfrac{1}{4}d_2B^2,\\
\tilde{B}_{12} = 2 + d_1A^2 + d_2B^2, \quad \tilde{B}_{13} = (d_1 - 3d_2)AB; \quad \tilde{A}_{14} = (d_1 - 4d_2)AB,\\
\tilde{B}_{14} = \tfrac{1}{4}(d_1 - d_2)AB, \quad \tilde{A}_{22} = \bar{\sigma}_1^2 - 1 - \tfrac{1}{2}d_1(A^2 + B^2), \qquad (8.5.14)\\
\tilde{B}_{22} = 1 + \tfrac{1}{4}d_1A^2 + \tfrac{3}{4}d_2B^2, \quad \tilde{A}_{23} = (2d_2 - d_1)AB,\\
\tilde{B}_{33} = 1 + \tfrac{3}{4}d_2A^2 + \tfrac{1}{4}d_1B^2, \quad \tilde{B}_{34} = 2 + d_2A^2 + d_1B^2,\\
\tilde{A}_{44} = \bar{\sigma}_1^2 - 1 + \left(\tfrac{1}{2}d_1 - 2d_2\right)A^2 - \tfrac{3}{2}d_1B^2; \quad \tilde{B}_{44} = 1 + \tfrac{1}{4}d_2A^2 + \tfrac{3}{4}d_1B^2.
\end{gathered}
$$

Note that all roots of the characteristic polynomial (8.5.12) can be found fairly simply, e.g., by the Hitchcock method, which is not true for finding all roots of the characteristic determinant (8.5.11). This reasoning is used in the numerical realization of the method.

Thus, the study of stability of periodic solutions (8.4.9) and (8.4.12) to system (8.1.3) that describe planar (or three-dimensional) sloshing in a cavity is reduced to the problem of finding the roots of (8.5.11). In the case where only two main nonsymmetric modes [$r_1(t)$ and $p_1(t)$] are taken into account, the problem of stability is much simpler and can be reduced to the problem of computing the roots of the characteristic polynomial (8.5.12).

Note in conclusion that one can use also the method of slowly varying amplitudes whose essence is described in Sec. 8.2. This method can be applied to the problem of finding approximate expressions for the generalized coordinates $r_1(t)$, $p_1(t)$, $p_0(t)$, $p_2(t)$, and $r_2(t)$ that could be used in a subsequent study of stability of the revealed stationary regimes.

## 8.6 Forced sloshing in a vessel under angular excitations

Nonlinear sloshing inside a container subjected to angular harmonic motions with forcing frequency in a vicinity of the lowest natural sloshing frequency can also be studied by using the general nonlinear equations (4.7.8)–(4.7.16). We are going to construct steady-state harmonic solutions and investigate their stability. The results presented below can be regarded as a certain generalization of the corresponding results suggested in [173]. We also follow [173, 186]. In addition, among the works related to this topic, we note [164, 169].

Assume that an upright circular cylindrical container is subjected to angular harmonic motions governed by the law $\varphi = H\cos\omega t$. The container is moving as a physical pendulum with the angular velocity $\omega = (0, \dot\varphi, 0)$ defined in the coordinate system $Oxyz$, which is rigidly-fixed with respect to the body, i.e., $\varphi$ is the angle between the gravitational acceleration vector $\mathbf{g}$ and the $Ox$-axis that coincides with the axis of symmetry of the container. By using the results presented in Chap. 4, we obtain the modal equations

$$\begin{aligned}
L_1(r_1, p_1, p_0, r_2, p_2) &\equiv \mu_1(\ddot r_1 + \sigma_1^2 r_1\cos\varphi) + d_1 r_1(r_1\ddot r_1 + p_1\ddot p_1 + \dot r_1^2 + \dot p_1^2) \\
&+ d_2(p_1^2\ddot r_1 - r_1p_1\ddot p_1 + 2p_1\dot r_1\dot p_1 - 2r_1\dot p_1^2) + d_3(r_2\ddot p_1 + \dot p_1\dot r_2 - p_2\ddot r_1 - \dot r_1\dot p_2) \\
&+ d_4(r_1\ddot p_2 - p_1\ddot r_2) + d_5(\dot p_0\dot r_1 + p_0\ddot r_1) + d_6\ddot p_0 r_1 + c_1r_1^2\ddot\varphi - (c_1 + 3c_2)p_1\dot p_1\dot\varphi \\
&- c_2p_1^2\ddot\varphi + (c_3 - c_5)\dot p_0\dot\varphi + c_3p_0\ddot\varphi + (c_6 - c_4)\dot p_2\dot\varphi - c_4p_2\ddot\varphi - \lambda_0\ddot\varphi \\
&- G_{22}^3 r_1\dot\varphi^2 + \lambda g\sin\varphi = 0; \qquad (8.6.1)
\end{aligned}$$

$$L_2(r_1, p_1, p_0, r_2, p_2) \equiv \mu_1(\ddot p_1 + \sigma_1^2 p_1\cos\varphi) + d_1p_1(r_1\ddot r_1 + p_1\ddot p_1 + \dot r_1^2 + \dot p_1^2)$$

$$+d_2(r_1^2\ddot{p}_1 - r_1p_1\ddot{r}_1 + 2r_1\dot{r}_1\dot{p}_1 - 2p_1\dot{r}_1^2) + d_3(p_2\ddot{p}_1 + \dot{p}_1\dot{p}_2 + r_2\ddot{r}_1 + \dot{r}_1\dot{r}_2)$$
$$-d_4(p_1\ddot{p}_2 + r_1\ddot{r}_2) + d_5(\dot{p}_0\dot{p}_1 + p_0\ddot{p}_1) + d_6p_1\ddot{p}_0 + (c_1 + c_2)r_1p_1\ddot{\varphi}$$
$$+(c_1 + 3c_2)p_1\dot{r}_1\dot{\varphi} + (c_4 - c_6)\dot{r}_2\dot{\varphi} + c_4r_2\ddot{\varphi} - G_{22}^4p_1\dot{\varphi}^2 = 0; \qquad (8.6.2)$$
$$L_3(r_1, p_1, p_0) \equiv \mu_0(\ddot{p}_0 + \sigma_0^2p_0\cos\varphi) + d_6(r_1\ddot{r}_1 + p_1\ddot{p}_1) + d_8(\dot{r}_1^2 + \dot{p}_1^2)$$
$$+(c_5 - c_3)\dot{r}_1\dot{\varphi} + c_5r_1\ddot{\varphi} - \frac{1}{2}G_{22}^1\dot{\varphi}^2 = 0; \qquad (8.6.3)$$
$$L_4(r_1, p_1, r_2) \equiv \mu_2(\ddot{r}_2 + \sigma_2^2r_2\cos\varphi) - d_4(\ddot{r}_1p_1 + r_1\ddot{p}_1) - 2d_7\dot{r}_1\dot{p}_1$$
$$+(c_6 - c_4)\dot{p}_1\dot{\varphi} + c_6p_1\ddot{\varphi} = 0; \qquad (8.6.4)$$
$$L_5(r_1, p_1, p_2) \equiv \mu_2(\ddot{p}_2 + \sigma_2^2p_2\cos\varphi) + d_4(r_1\ddot{r}_1 - p_1\ddot{p}_1) + d_7(\dot{r}_1^2 - \dot{p}_1^2)$$
$$+(c_4 - c_6)\dot{r}_1\dot{\varphi} - c_6r_1\ddot{\varphi} - \frac{1}{2}G_{22}^2\dot{\varphi}^2 = 0, \qquad (8.6.5)$$

which form a system similar to system (8.1.3).

To describe the steady-state wave regimes, we present the generalized coordinates $r_1(t), \ldots, p_2(t)$ by the Fourier sums

$$r_l(t) = \alpha_0 + \sum_{k=1}^{n}[\alpha_k\cos k\omega t + \bar{\alpha}_k\sin k\omega t], \quad l = 1, 2,$$
$$p_m(t) = \beta_0 + \sum_{k=1}^{n}[\bar{\beta}_k\cos k\omega t + \beta_k\sin k\omega t], \quad m = 0, 1, 2, \qquad (8.6.6)$$

and apply the Bubnov–Galerkin method. Following the above-presented analysis, we keep only the dominant terms

$$r_1(t) = A\cos\omega t, \;\; p_1(t) = B\sin\omega t, \;\; p_0(t) = A_0 + B_0\cos 2\omega t,$$
$$r_2(t) = B_1\sin 2\omega t, \;\; p_2(t) = A_2 + B_2\cos 2\omega t \qquad (8.6.7)$$

in (8.6.6). Substituting (8.6.7) into the Fredholm alternative

$$\int_0^{\frac{2\pi}{\sigma}} L_1(r_1, p_1, p_0, r_2, p_2)\cos\omega t dt = 0,$$
$$\int_0^{\frac{2\pi}{\sigma}} L_2(r_1, p_1, p_0, r_2, p_2)\sin\omega t dt = 0, \qquad (8.6.8)$$
$$\cdots\cdots\cdots\cdots\cdots\cdots$$
$$\int_0^{\frac{2\pi}{\sigma}} L_5(r_1, p_1, p_2)\cos 2\omega t dt = 0,$$

we obtain the following equations with respect to $A$ and $B$:

$$a_0A^3 + a_1A^2 + a_2A + a_3 + B^2(a_4A + a_5) = 0,$$
$$B(b_0A^2 + b_1A + b_2 - b_3B^2) = 0; \tag{8.6.9}$$

and the following relations with respect to $A_0, \ldots, B_2$:

$$A_0 = \alpha_{01}A^2 + \alpha_{02}B^2 + \alpha_{03}A + \alpha_{04}, \;\; B_0 = \beta_{01}A^2 + \beta_{02}B^2 + \beta_{03}A + \beta_{04},$$
$$B_1 = B(\beta_{11}A + \beta_{12}), \tag{8.6.10}$$
$$A_2 = \alpha_{21}A^2 + \alpha_{22}B^2 + \alpha_{23}A + \alpha_{24}, \quad B_2 = \beta_{21}A^2 + \beta_{22}B^2 + \beta_{23}A + \beta_{24}.$$

Here,

$$\begin{aligned}
a_0 &= d_3\alpha_{21} - \tfrac{1}{2}d_1 - x_1\beta_{21} + x_2\beta_{01} - d_5\alpha_{01},\\
a_1 &= d_3\alpha_{23} + x_2\beta_{03} - x_1\beta_{23} - d_5\alpha_{03} + \big(x_4\beta_{01} - \tfrac{3}{4}c_1\\
&\quad - c_3\alpha_{01} + c_4\alpha_{21} + x_5\beta_{21}\big)H,\\
a_2 &= \mu_1\left[\bar{\sigma}_1^2\left(1 - \tfrac{3}{8}H^2\right) - 1\right] + d_3\alpha_{24} - x_1\beta_{24} + x_2\beta_{04} - d_5\alpha_{04}\\
&\quad + \left(x_4\beta_{03} - c_3\alpha_{03} + c_4\alpha_{23} + x_5\beta_{23} - \tfrac{1}{4}G_{22}^3H\right)H,\\
a_3 &= \left(x_4\beta_{04} - c_3\alpha_{04} + c_4\alpha_{24} + x_5\beta_{24} + \lambda_0 + \lambda\frac{g}{\omega^2}\right)H,\\
a_4 &= \tfrac{1}{2}d_1 - 2d_2 + d_3\alpha_{22} + x_1(\beta_{11} - \beta_{22}) + x_2\beta_{02} - d_5\alpha_{02},\\
a_5 &= x_1\beta_{12} + \left(c_2 + \tfrac{1}{4}c_1 + x_4\beta_{02} - c_3\alpha_{02} + c_4\alpha_{22} + x_5\beta_{22}\right)H,\\
b_0 &= \tfrac{1}{2}d_1 - 2d_2 - d_3\alpha_{21} - x_1\beta_{21} - d_5\alpha_{01} - x_6\beta_{01} + x_3\beta_{11},\\
b_1 &= \left(\tfrac{1}{2}c_1 + 2c_2 - x_5\beta_{11}\right)H + x_3\beta_{12} - d_3\alpha_{23} - x_1\beta_{23} - d_5\alpha_{03} - x_6\beta_{03},\\
b_2 &= \mu_1\left[\bar{\sigma}_1^2\left(1 - \tfrac{1}{8}H^2\right) - 1\right] - d_3\alpha_{24} - x_1\beta_{24} - d_5\alpha_{04}\\
&\quad - x_6\beta_{04} - x_5\beta_{12}H - \tfrac{3}{4}G_{22}^4H^2,\\
b_3 &= \tfrac{1}{2}d_1 + d_3\alpha_{22} + x_1\beta_{22} + d_5\alpha_{02} + x_6\beta_{02},
\end{aligned} \tag{8.6.11}$$

where

$$x_1 = \tfrac{1}{2}d_3 + 2d_4, \;\; x_2 = \tfrac{1}{2}d_5 - 2d_6, \;\; x_3 = 2d_4 - \tfrac{1}{2}d_3, \;\; x_4 = \tfrac{1}{2}c_3 - c_5,$$
$$x_5 = c_6 - \tfrac{1}{2}c_4, \;\; x_6 = \tfrac{1}{2}d_5 + 2d_6, \;\; \bar{\sigma}_m = \sigma_m/\omega, \quad m = 0, 1, 2,$$
$$\tilde{\alpha}_{11} = \bar{\sigma}_0^2\left(1 - \tfrac{1}{4}H^2\right), \;\; \tilde{\alpha}_{12} = -\tfrac{1}{8}\bar{\sigma}_0^2H^2, \;\; \tilde{\alpha}_{21} = 2\tilde{\alpha}_{12},$$
$$\tilde{\alpha}_{22} = \tilde{\alpha}_{11} - 4, \;\; \triangle_0 = 2\mu_0(\tilde{\alpha}_{11}\tilde{\alpha}_{22} - \tilde{\alpha}_{12}\tilde{\alpha}_{21}),$$
$$\alpha_{01} = \frac{1}{\triangle_0}[\tilde{\alpha}_{22}(d_6 - d_8) - \tilde{\alpha}_{12}(d_6 + d_8)],$$

$$\alpha_{02} = \frac{1}{\triangle_0}[\tilde{\alpha}_{22}(d_6 - d_8) + \tilde{\alpha}_{12}(d_6 + d_8)],$$

$$\alpha_{03} = \frac{H}{\triangle_0}(\tilde{\alpha}_{22}c_3 + 2\tilde{\alpha}_{12}x_4), \quad \alpha_{04} = \frac{G^1_{22}}{2\triangle_0}(\tilde{\alpha}_{12} + \tilde{\alpha}_{22})H^2,$$

$$\beta_{01} = \frac{1}{\triangle_0}[\tilde{\alpha}_{11}(d_6 + d_8) - \tilde{\alpha}_{21}(d_6 - d_8)],$$

$$\beta_{02} = \frac{1}{\triangle_0}[-\tilde{\alpha}_{11}(d_6 + d_8) - \tilde{\alpha}_{21}(d_6 - d_8)],$$

$$\beta_{03} = -\frac{H}{\triangle_0}(2\tilde{\alpha}_{11}x_4 + \tilde{\alpha}_{21}c_3), \quad \beta_{04} = -\frac{G^1_{22}}{2\triangle_0}(\tilde{\alpha}_{11} + \tilde{\alpha}_{21})H^2,$$

$$\hat{\alpha}_{11} = \bar{\sigma}_2^2\left(1 - \tfrac{1}{4}H^2\right), \quad \hat{\alpha}_{12} = -\tfrac{1}{8}\bar{\sigma}_2^2 H^2, \quad \hat{\alpha}_{21} = 2\hat{\alpha}_{12}, \tag{8.6.12}$$

$$\hat{\alpha}_{22} = \hat{\alpha}_{11} - 4, \quad \triangle_2 = 2\mu_2(\hat{\alpha}_{11}\hat{\alpha}_{22} - \hat{\alpha}_{12}\hat{\alpha}_{21}),$$

$$\beta_{11} = -\frac{1}{\mu_2\hat{\alpha}_{22}}(d_4 + d_7), \quad \beta_{12} = \frac{x_5 H}{\mu_2\hat{\alpha}_{22}},$$

$$\alpha_{21} = \frac{1}{\triangle_2}[\hat{\alpha}_{22}(d_4 - d_7) - \hat{\alpha}_{12}(d_4 + d_7)],$$

$$\alpha_{22} = \frac{1}{\triangle_2}[-\hat{\alpha}_{22}(d_4 - d_7) - \hat{\alpha}_{12}(d_4 + d_7)],$$

$$\alpha_{23} = \frac{H}{\triangle_2}[-\hat{\alpha}_{22}c_4 + 2\hat{\alpha}_{12}x_5), \quad \alpha_{24} = \frac{G^2_{22}}{2\triangle_2}(\hat{\alpha}_{12} + \hat{\alpha}_{22})H^2,$$

$$\beta_{21} = \frac{1}{\triangle_2}[\hat{\alpha}_{11}(d_4 + d_7) - \hat{\alpha}_{12}(d_4 - d_7)],$$

$$\beta_{22} = \frac{1}{\triangle_2}[\hat{\alpha}_{11}(d_4 + d_7) + \hat{\alpha}_{21}(d_4 - d_7)],$$

$$\beta_{23} = \frac{H}{\triangle_2}(\hat{\alpha}_{21}c_4 - 2\hat{\alpha}_{11}x_5), \quad \beta_{24} = -\frac{G^2_{22}}{2\triangle_2}(\hat{\alpha}_{11} + \hat{\alpha}_{21})H^2.$$

It follows from (8.6.9) that there exist two possible types of periodic solutions of (8.6.1)–(8.6.5) that imply

(a) planar steady-state regimes ($A \neq 0$ and $B = 0$),

$$r_1(t) = A\cos\omega t, \quad p_1(t) \equiv 0, \quad r_2(t) \equiv 0,$$

$$p_k(t) = A_k + B_k\cos 2\omega t, \quad k = 0, 2; \tag{8.6.13}$$

(b) swirling ($A \neq 0$ and $B \neq 0$), which is observed in the domain of dynamic instability of planar sloshing,

$$r_1(t) = A\cos\omega t, \quad p_1(t) = B\sin\omega t,$$

$$p_k(t) = A_k + B_k\cos 2\omega t, \quad k = 0, 2, \quad r_2(t) = B_1\sin 2\omega t. \tag{8.6.14}$$

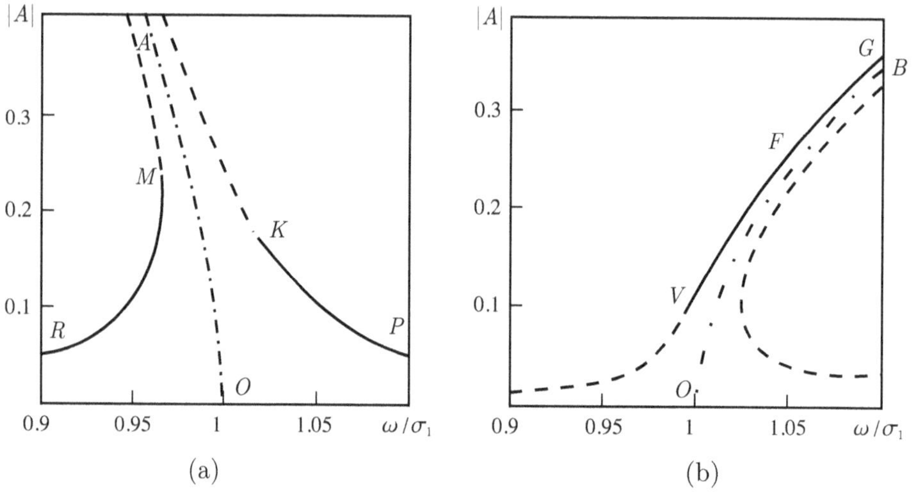


**Figure 8.2**

The response curves can be constructed by using (8.6.9); they are similar to those shown in Figs. 8.2. To clarify which of the possible steady-state regimes [by virtue of relations (8.6.13) and (8.6.14)] is actually observed, we have to study stability of the corresponding periodic solutions. To this end, we use the first Lyapunov method.

Along with the unperturbed motion defined by relations (8.6.13) and (8.6.14), we consider a perturbed motion of the form

$$\tilde{r}_1(t) = r_1(t) + \alpha(t), \quad \tilde{p}_1(t) = p_1(t) + \beta(t), \tag{8.6.15}$$
$$\tilde{p}_0(t) = p_1(t) + \gamma(t), \quad \tilde{r}_2(t) = r_2(t) + \delta(t), \quad \tilde{p}_2(t) = p_2(t) + \varepsilon(t),$$

which is close to the unperturbed motion, and whose initial conditions differ slightly enough from the initial conditions of the unperturbed motion.

We substitute the perturbed solution (8.6.15) into system (8.6.1)–(8.6.5) and take into account that the unperturbed solution satisfies this system. Linearizing the obtained system with respect to perturbations, we obtain the following system of variational equations:

$$q_{11}(t)\ddot{\alpha} + q_{12}(t)\ddot{\beta} + q_{13}(t)\ddot{\gamma} + q_{14}(t)\ddot{\delta} + q_{15}(t)\ddot{\varepsilon} + q_{16}(t)\dot{\alpha} + q_{17}(t)\dot{\beta}$$
$$+q_{18}(t)\dot{\gamma} + q_{19}(t)\dot{\delta} + q_{110}(t)\dot{\varepsilon} + q_{111}(t)\alpha + q_{112}(t)\beta$$
$$+q_{113}(t)\gamma + q_{114}(t)\delta + q_{115}(t)\varepsilon = 0; \tag{8.6.16}$$

$$q_{21}(t)\ddot{\alpha} + q_{22}(t)\ddot{\beta} + q_{23}(t)\ddot{\gamma} + q_{24}(t)\ddot{\delta} + q_{25}(t)\ddot{\varepsilon} + q_{26}(t)\dot{\alpha} + q_{27}(t)\dot{\beta}$$
$$+q_{28}(t)\dot{\gamma} + q_{29}(t)\dot{\delta} + q_{210}(t)\dot{\varepsilon} + q_{211}(t)\alpha + q_{212}(t)\beta$$
$$+q_{213}(t)\gamma + q_{214}(t)\delta + q_{215}(t)\varepsilon = 0; \tag{8.6.17}$$

$$q_{31}(t)\ddot{\alpha} + q_{32}(t)\ddot{\beta} + q_{33}(t)\ddot{\gamma} + q_{36}(t)\dot{\alpha} + q_{37}(t)\dot{\beta} + q_{311}(t)\alpha + q_{312}(t)\beta + q_{313}(t)\gamma = 0; \tag{8.6.18}$$

$$q_{41}(t)\ddot{\alpha} + q_{42}(t)\ddot{\beta} + q_{44}(t)\ddot{\delta} + q_{46}(t)\dot{\alpha} + q_{47}(t)\dot{\beta} + q_{411}(t)\alpha + q_{412}(t)\beta + q_{414}(t)\delta = 0; \tag{8.6.19}$$

$$q_{51}(t)\ddot{\alpha} + q_{52}(t)\ddot{\beta} + q_{55}(t)\ddot{\varepsilon} + q_{56}(t)\dot{\alpha} + q_{57}(t)\dot{\beta} + q_{511}(t)\alpha + q_{512}(t)\beta + q_{515}(t)\varepsilon = 0. \tag{8.6.20}$$

The coefficients of system (8.6.16)–(8.6.20) take the form

$$
\begin{gathered}
q_{11} = \mu_1 + d_1 r_1^2 + d_2 p_1^2 - d_3 p_2 + d_5 p_0, \quad q_{12} = (d_1 - d_2) r_1 p_1 + d_3 r_2, \\
q_{13} = d_6 r_1, \quad q_{14} = -d_4 p_1, \quad q_{15} = d_4 r_1, \\
q_{16} = 2 d_1 r_1 \dot{r}_1 + 2 d_2 p_1 \dot{p}_1 - d_3 \dot{p}_2 + d_5 \dot{p}_0, \\
q_{17} = 2 d_1 r_1 \dot{p}_1 + 2 d_2 (\dot{r}_1 p_1 - 2 r_1 \dot{p}_1) + d_3 \dot{r}_2 - (c_1 + 3c_2)\dot{\varphi} p_1, \\
q_{18} = d_5 \dot{r}_1 + (c_3 - c_5)\dot{\varphi}, \quad q_{19} = d_3 \dot{p}_1, \quad q_{110} = (c_6 - c_4)\dot{\varphi} - d_3 \dot{r}_1, \\
q_{111} = \mu_1 \sigma_1^2 \cos\varphi + d_1 (2 r_1 \ddot{r}_1 + p_1 \ddot{p}_1 + \dot{r}_1^2 + \dot{p}_1^2) \\
- d_2 (p_1 \ddot{p}_1 + 2\dot{p}_1^2) + d_4 \ddot{p}_2 + d_6 \ddot{p}_0 + 2 c_1 \ddot{\varphi} r_1 - G_{22}^3 \dot{\varphi}^2, \\
q_{112} = d_1 r_1 \ddot{p}_1 + d_2 (2 \ddot{r}_1 p_1 + 2 \dot{r}_1 \dot{p}_1 - r_1 \ddot{p}_1) - d_4 \ddot{r}_2 - 2 c_2 \ddot{\varphi} p_1 - (c_1 + 3c_2)\dot{\varphi}\dot{p}_1, \\
q_{113} = d_5 \ddot{r}_1 + c_3 \ddot{\varphi}, \quad q_{114} = d_3 \ddot{p}_1, \quad q_{115} = -d_3 \ddot{r}_1 - c_4 \ddot{\varphi}, \\
q_{21} = (d_1 - d_2) r_1 p_1 + d_3 r_2, \quad q_{22} = \mu_1 + d_1 p_1^2 + d_2 r_1^2 + d_3 p_2 + d_5 p_0, \\
q_{23} = d_6 p_1, \quad q_{24} = -d_4 r_1, \quad q_{25} = -d_4 p_1, \\
q_{26} = 2 d_1 \dot{r}_1 p_1 + 2 d_2 (r_1 \dot{p}_1 - 2 \dot{r}_1 p_1) + d_3 \dot{r}_2 + (c_1 + 3c_2)\dot{\varphi} p_1, \\
q_{27} = 2 d_1 p_1 \dot{p}_1 + 2 d_2 r_1 \dot{r}_1 + d_3 \dot{p}_2 + d_5 \dot{p}_0, \\
q_{29} = d_3 \dot{r}_1 + (c_4 - c_6)\dot{\varphi}, \quad q_{28} = d_5 \dot{p}_1, \quad q_{210} = d_3 \dot{p}_1, \\
q_{211} = d_1 \ddot{r}_1 p_1 + d_2 (2 r_1 \ddot{p}_1 - \ddot{r}_1 p_1 + 2 \dot{r}_1 \dot{p}_1) + (c_1 + c_2)\ddot{\varphi} p_1 - d_4 \ddot{r}_2, \\
q_{212} = \mu_1 \sigma_1^2 \cos\varphi + d_1 (r_1 \ddot{r}_1 + 2 p_1 \ddot{p}_1 + \dot{r}_1^2 + \dot{p}_1^2) - d_2 (r_1 \ddot{r}_1 + 2\dot{r}_1^2) \\
- d_4 \ddot{p}_2 + d_6 \ddot{p}_0 + (c_1 + c_2)\ddot{\varphi} r_1 + (c_1 + 3c_2)\dot{\varphi}\dot{r}_1 - G_{22}^4 \dot{\varphi}^2, \\
q_{213} = d_5 \ddot{p}_1, \quad q_{214} = d_3 \ddot{r}_1 + c_4 \ddot{\varphi}, \quad q_{215} = d_3 \ddot{p}_1, \\
q_{31} = d_6 r_1, \quad q_{32} = d_6 p_1, \quad q_{33} = \mu_0, \quad q_{36} = 2 d_8 \dot{r}_1 + (c_5 - c_3)\dot{\varphi}, \\
q_{37} = 2 d_8 \dot{p}_1, \quad q_{311} = d_6 \ddot{r}_1 + c_5 \ddot{\varphi}, \quad q_{312} = d_6 \ddot{p}_1, \\
q_{313} = \mu_0 \sigma_0^2 \cos\varphi, \quad q_{41} = -d_4 p_1, \quad q_{42} = -d_4 r_1, \quad q_{44} = \mu_2, \\
q_{46} = -2 d_7 \dot{p}_1, \quad q_{47} = (c_6 - c_4)\dot{\varphi} - 2 d_7 \dot{r}_1, \quad q_{411} = -d_4 \ddot{p}_1,
\end{gathered}
\tag{8.6.21}
$$

$$q_{412} = c_4\ddot{\varphi} - d_4\ddot{r}_1, \quad q_{414} = \mu_2\sigma_2^2 \cos\varphi, \quad q_{51} = d_4 r_1, \quad q_{52} = -d_4 p_1,$$

$$q_{55} = \mu_2, \quad q_{56} = 2d_7\dot{r}_1 + (c_4 - c_6)\dot{\varphi}, \quad q_{57} = -2d_7\dot{p}_1,$$

$$q_{511} = d_4\ddot{r}_1 - c_6\ddot{\varphi}, \quad q_{512} = -d_4\ddot{p}_1, \quad q_{515} = \mu_2\sigma_2^2 \cos\varphi.$$

Here, with regard to the restrictions imposed on the vector of angular velocity $\omega$, we set

$$\cos\varphi \cong 1 - \tfrac{1}{2}\varphi^2 = 1 - \tfrac{1}{2}H^2\cos^2\omega t. \tag{8.6.22}$$

Thus, the problem of stability of solutions (8.6.13) and (8.6.14) to the nonlinear system (8.6.1)–(8.6.5) is reduced to studying the solutions of the system of linear differential equations (8.6.16)–(8.6.20) with periodic coefficients (8.6.21). In order to investigate the behavior of these solutions, one can use the method presented in Sec. 8.5.

## 8.7 Analysis of the response curves

The response curves of the steady-state regimes are defined by relations (8.4.12) and (8.4.10). For both steady-state regimes [planar (8.4.9) and swirling (8.4.11)], the free surface patterns can be visualized by using the following relation:

$$x = f + h = p_0(t)Y_0(k_0\xi) + [r_1(t)\sin\eta + p_1(t)\cos\eta]Y_1(k_1\xi)$$

$$+[r_2(t)\sin 2\eta + p_2(t)\cos 2\eta]Y_2(k_2\xi) + h. \tag{8.7.1}$$

Consider a container with $R_1 = 1$, $R_0 = 0.5$, and $h = 2$, and the forcing amplitude $H = 0.01$. The response curves for planar sloshing plotted by using (8.4.10) are depicted in Fig. 8.2a. The solid curve (parts $RM$ and $KP$) denotes stable regimes and the dashed lines mark unstable regimes. The points $M$ and $K$ separate stable parts of the response curves from unstable ones. The response curves for swirling plotted by using (8.4.12) are depicted in Fig. 8.2b. As in the previous figure, the solid curve $VFG$ denotes the stable sloshing and the dashed curve denotes the unstable behavior. The point $V$ is a boundary point that separates the stable part from the unstable one. Analyzing the curves in Figs. 8.2, we can conclude that the resonance frequency depends nonlinearly on the amplitude of oscillations. The dash-dotted (skeleton) curve in these figures illustrates the dependence of frequency on the amplitude (for free sloshing).

The response curves for planar waves (the curves $RM$ and $KP$) and swirling (the curve $VFG$) are shown in Fig. 8.3a. Two cases are considered here: (1) the free surface (8.7.1) is approximated by only two dominant generalized coordinates $r_1(t)$ and $p_1(t)$, (2) all five generalized coordinates are taken into account, i.e., the equation of the free surface is chosen to be (8.7.1) (the solid line). The figure show that, in the case where five generalized coordinates

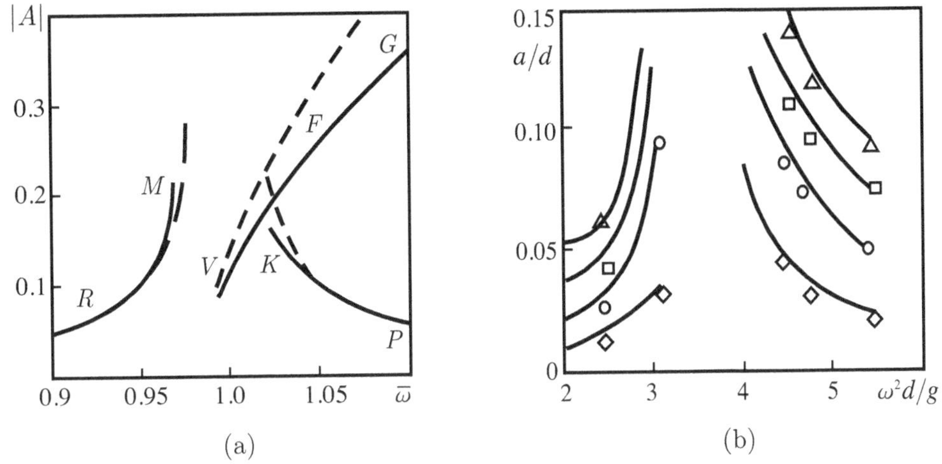


**Figure 8.3**

are taken into account, a considerable decrease of the maximum amplitude is observed (e.g., the maximum amplitude corresponding to plane oscillations decreases approximately by 25%) [184].

For a cylindrical container with parameters $R_1 = 1$ ($R_0 = 0$) and $h = 2$, Fig. 8.3b shows the response curves of planar sloshing, which are plotted by using relation (8.4.10), and the experimental data obtained in [5] ($\triangle$ — $H = 0.0454$; $\square$ — $H = 0.0344$; $\bigcirc$ — $H = 0.023$; $\diamond$ — $H = 0.0112$). In the figure, $a$ denotes the mean amplitude equal to the half-sum of two peaks measured near the wall in the plane of action of perturbation force, $d = 2R_1 = 2$, and g is the gravitational acceleration. In the case under consideration, the mean amplitude $a$ is computed by

$$a = \tfrac{1}{2}d(|p_0(t) + r_1(t) - p_2(t)| + |p_0(t) - r_1(t) - p_2(t)|).$$

Note that the theoretical values are close enough to experimental data.

We now consider the free surface patterns for planar steady-state regimes. Typical instant shapes of the free surface $\Sigma$ in a neighborhood of the main resonance in a cylindrical container were observed in experiments [147, 150].

For a cylindrical container with $R_1 = 1$ ($R_0 = 0$) and $h = 2$, for $H = 0.01$ and $\omega/\sigma_1 = 0.958$, the plots in Fig. 8.4 present several types of the free-surface profile for a planar wave obtained by using (8.4.9) at certain instants of time $t$. The plots show that the nodal line (point) in the middle of the free surface is mobile. Moreover, a kind of nonsymmetry of the perturbed free surface is observed and, usually, the elevation height of a "hump" exceeds the depth of a "hollow." This is easily seen for maximal amplitude values especially; for example, in the case under consideration, these amplitudes can reach, respectively,

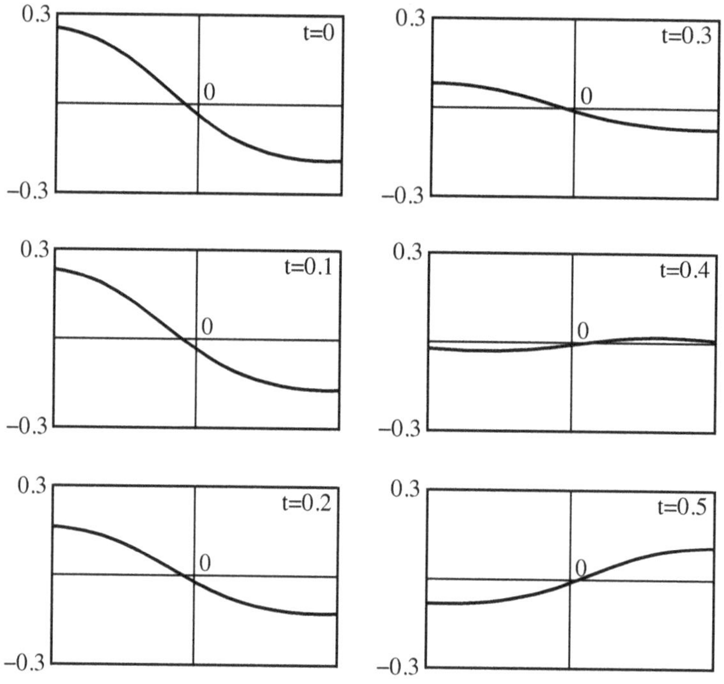


**Figure 8.4**

values 0.26 and 0.18 where the ratio of the "hump" height to the "hollow" depth is equal to 1.44. The difference between the "hump" height and the "hollow" depth increases when the perturbation amplitude $H$ increases. Note that this cannot be handled by the free-surface approximation (8.7.1) where only two dominant generalized coordinates $r_1(t)$ and $p_1(t)$ are used. These results are not changed considerably in values of the amplitude-frequency characteristics if we use nonlinear modal systems of higher dimensions. A detailed analysis of these results is given in [136, 142, 137, 138].

Specific features of this oscillatory behavior and their spatial visualization are presented in detail in [69] where it is shown that nonlinear phenomena do not disappear even in the case where the wave amplitude is small. Those studies are both experimental and theoretical. The pictures in Fig. 8.5 are obtained by using an animation grid program based on a five-dimensional modal system. These pictures present wave patterns that confirm the nonlinearity of sloshing behavior. It is caused by presence of the second order components that becomes important within a short time ($\Delta t \approx 0.01\,\text{sec}$) and visible in the form of a winding of nodal curve, which is the cross-line of the free surface and mean free surface $x = 0$. In this case, it is not a fixed straight line intersection as predicted by the linear theory. The curvature of the nodal curve is changing when the anti-nodal points (maximum and minimum surface elevations) change

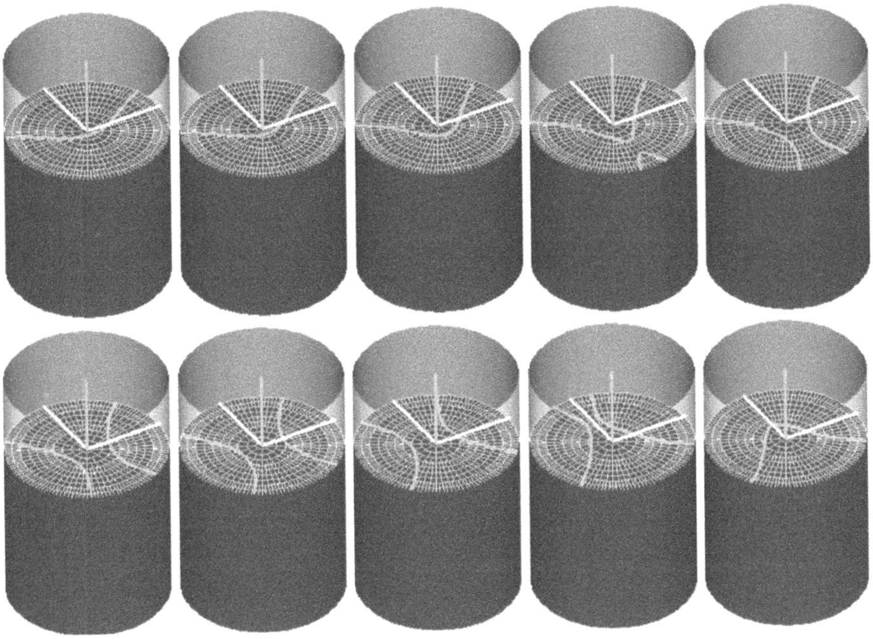

**Figure 8.5.** Steady-state sloshing, free planar oscillations, $H \equiv 0$, $h/R_1 = 2$, $A/R_1 = 0.0602626$; $\omega = 4.24$[rad/sec]. The non-zero initial conditions are $r_1(0) = 0.0602626$; $p_0(0) = 0.0011363446$; $p_2(0) = -0.00157382$.

their positions. The maximum elevation point (crest) is moving between diametrically opposite points of the vertical wall and forms a *"traveling"* wave. The contribution of the secondary modal components into the wave elevation grows along with the growth of $A$. Unlike the small-amplitude sloshing, in this case, the "traveling" wave is observed during the whole period $2\pi/\sigma$. We have also found that asymptotic approximations of initial conditions give incorrect results for $A/R_1 > 0.38$. This is caused by a jump to aperiodic solutions. This critical value is induced by the relative wave amplitude elevation over 0.45.

Each column in Fig. 8.6 presents an instant free-surface shape and its parametric plot $(r_1(t), p_1(t))$. It has a circular shape. This confirms the theoretical prediction $|A| = |B|$ for the free swirling wave. The free surface has a "frozen" shape that is "rotating" together with nodal line. Such a motion resembles the free surface spinning as a rigid body. The nodal line has nonzero curvature and does not pass through the cylinder axis.

The modal functions $(r_1, p_2)$ and $(p_1, r_2)$ correspond to longitudinal and transversal wave components, respectively. The modal function $p_0$ describes axial-symmetric deviations. The transients in Fig. 8.7 are calculated for $p_1 = r_2 \equiv 0$ that means that the transversal waves are not excited. The initial test values are chosen to be zero except for $r_1(0) = \text{const}$ (it is an asymmetric initial surface shaped by the primary natural mode).

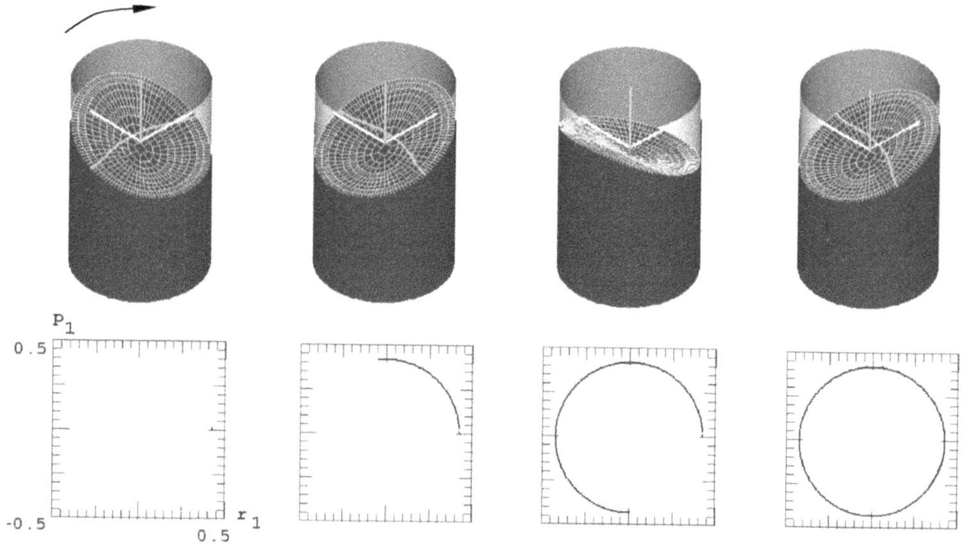


**Figure 8.6.** Steady-state sloshing, free swirling wave with $H \equiv 0$, $h/R_1 = 2$, $A/R_1 = 0.425$. The non-zero initial conditions are $r_1(0) = 0.425$; $p_0(0) = 0.056676$; $p_2(0) = -0.04$; $\dot{p}_1(0) = 1.8828$; $\dot{r}_2(0) = 0.347646$.

In Fig. 8.7, one can see a series of columns gathering three windows. The first window shows the plot $(r_1(t), p_0(t))$. The plot $r_1 = r_1(t)$ and the camera pictures displaying the free surface shapes are shown in the second and third windows. The graph of $r_1(t)$ shows that the transient motions imply the "beating" phenomenon (slow fluctuations of the amplitude). The parametric curve $(r_1(t), p_0(t))$ confirms an irregular behavior of sloshing.

In addition to a "traveling" wave, Fig. 8.7 reveals two other nonlinear surface wave phenomena. The first one is observed as a pair of symmetric crests sliding around the circular wall in the "clockwise" and "counterclockwise" directions, respectively. Their interference gives rise to a new surface shape with a single peak. In what follows, we treat this wave as a "traveling hollow" since its minimum elevation runs along the diameter of the mean free surface. The second wave can be treated as a nonlinear analogue of a standing wave since no crest run is observed. The minimum wave elevation (hollow) oscillates near the axis of the circular cylinder. Nodal curves are not connected. There are two surface peaks that co-exist and are located at diametrically opposite points of the free surface. The elevation of one of these peaks increases while the other one decreases. This wave can be physically treated as a flow beneath the mean free surface $\Sigma_0$.

These three wave motions imply the three possible modal energy redistributions. Denoting the "traveling" wave by $\mathcal{T}$, "traveling" hollow by $\mathcal{H}$, and "standing" wave by $\mathcal{S}$, for the case presented in Fig. 8.7, we observe the follow-

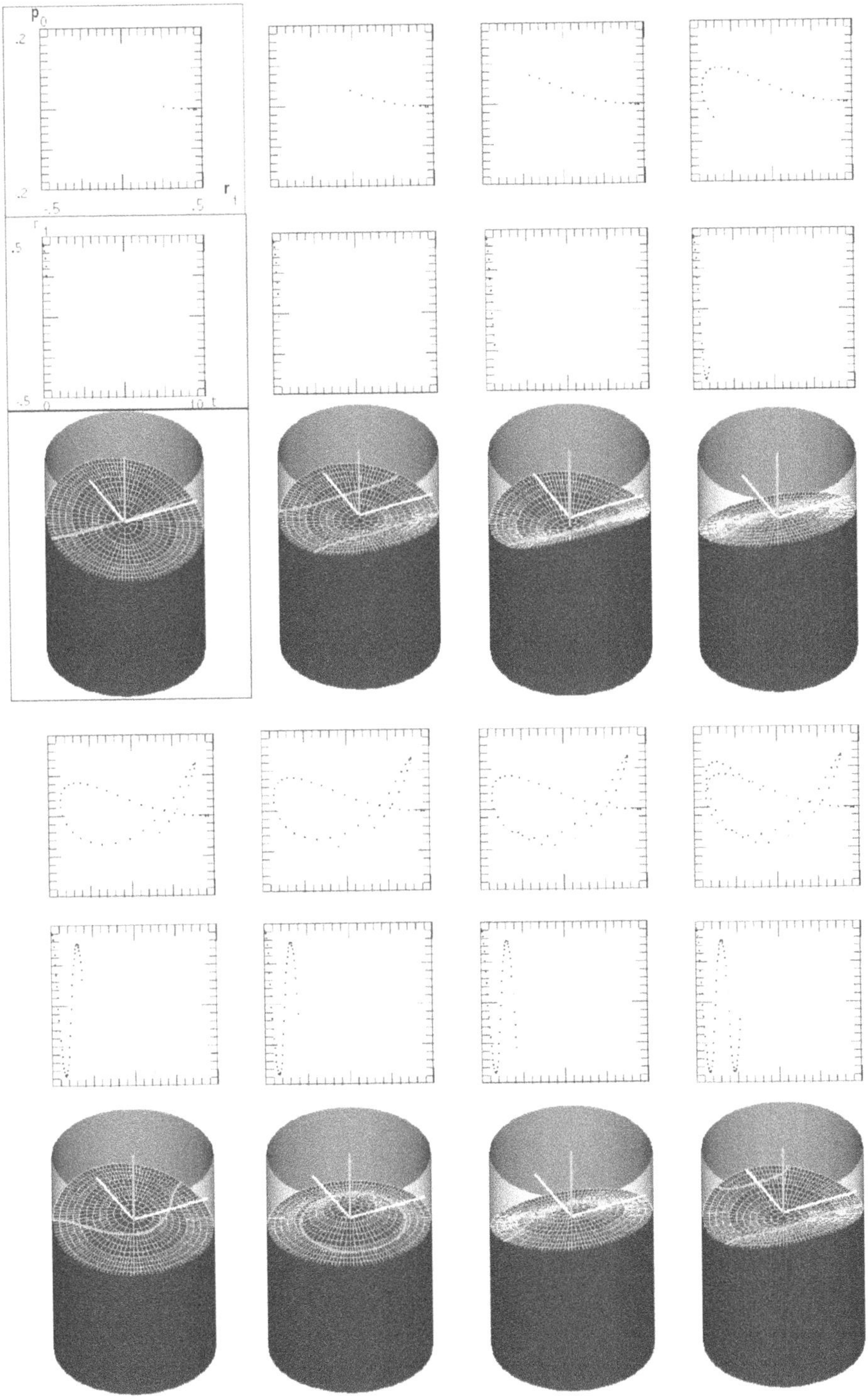

**Figure 8.7.** Free planar transient sloshing with $H \equiv 0$, $h/R_1 = 2$. Nonzero initial conditions are supposed for the case $r_1(0) = 0.4695$ only.

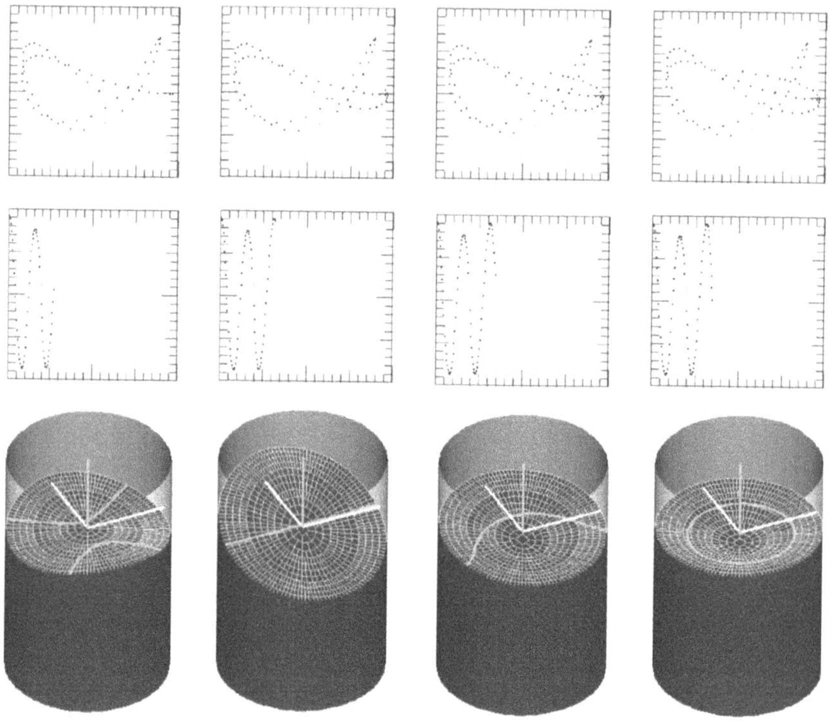

**Figure 8.7** (continued).

ing sequence of wave motions:

$$\mathcal{S}\ \mathcal{T}\ \mathcal{T}\ \mathcal{S}\ \mathcal{H}\ \mathcal{T}\ \mathcal{T}\ \mathcal{H}\ \mathcal{H}\ \mathcal{T}\ \mathcal{T}\ \mathcal{H}\ \mathcal{S}\ \mathcal{S}\ \mathcal{T}\ \mathcal{T}\ \mathcal{S}\ \mathcal{H}\ \mathcal{T}\ \mathcal{T}\ \mathcal{S}\ldots.$$

This sequence can be treated as a visual characterization of sloshing and liquid response in transients.

Our potential model neglects viscosity. However, on a long time scale, the damping is important because, for example, it leads to the free surface oscillations decaying. In case of waves excited resonantly, the damping can "kill" the transients. This means that after sufficiently long time the actual surface wave coincides with a stable steady-state wave. When analyzing sloshing response, we found out a frequency domain where no stable steady-state waves are observed. In Fig. 8.8, this frequency domain is bounded by the ordinates of points $M$ and $V$. In this domain, "beating" and "chaotic" waves do not vanish completely. We studied an experimental case in this frequency domain in order to obtain a visualization of sloshing phenomena associated with the breakdown of steady-state "planar" and "swirling" waves. The amplitude (force) fluid response have not been measured. It was discovered that effect of higher (secondary) modes cannot be neglected because there exist exotic wave patterns and nodal curve shapes. The simulation results presented below

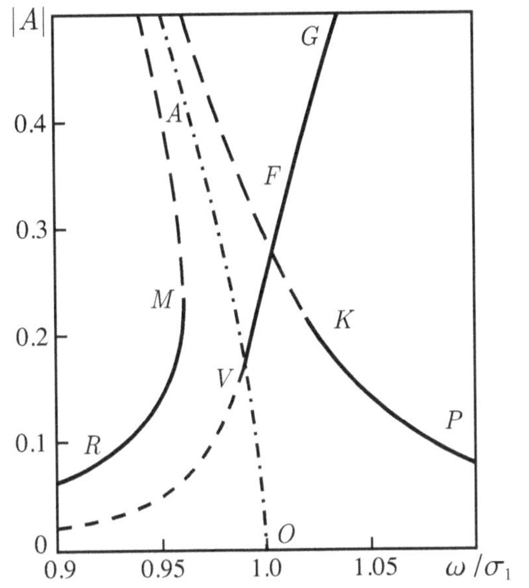


**Figure 8.8**

describe the breakdown of a swirling wave and are close to the corresponding experimental data.

Since the point $M$ is a turning point, the steady-state swirling wave corresponding to $M$ can be found by using our modal system. For $h/R_1 = 2$, $H/R_1 = 0.01$, this implies the excitation frequency $\omega = 4.1332231$ and nonzero initial conditions $r_1(0) = 0.091064535$; $p_1'(0) = 1.3292305$; $r_2'(0) = 0.057897668$.

When changing $\omega$ to 4.1, we drift the excitation frequency into the mentioned frequency domain between $P$ and $F$. The results of a numerical simulation are shown in Fig. 8.9. The parametric plot $(r_1(t), p_1(t))$ illustrates the contribution of primary components, the longitudinal $r_1$ and transversal $p_1$. In the photos we show grid-animated surface shapes. Initially, the surface waves are close to a swirling wave, for which its transversal component exceeds the longitudinal component ($B > A$ in the context of our approximation). Physically, this means that the center of the fluid mass oscillates with large magnitude near the $Oxz$-plane mainly. (Such a motions cannot be stable and, as it seems, this causes the breaking down.) The breakdown transients are presented by $(r_1, p_1)$-trajectories. For two periods, the $(r_1, p_1)$-trajectories have almost elliptic shape with a vertical focal axis. This axis rotates counterclockwise (see the plots in the last column). Since one of the semi-axes of the ellipse becomes smaller when rotating, $(r_1, p_1)$-trajectories in Fig. 8.9 correspond to waves, which are similar to a "planar" azimuthal rotation.

The grid visualization results shown in Fig. 8.9 confirm also an observable contribution of secondary modes in the wave patterns. Unlike the case of two-

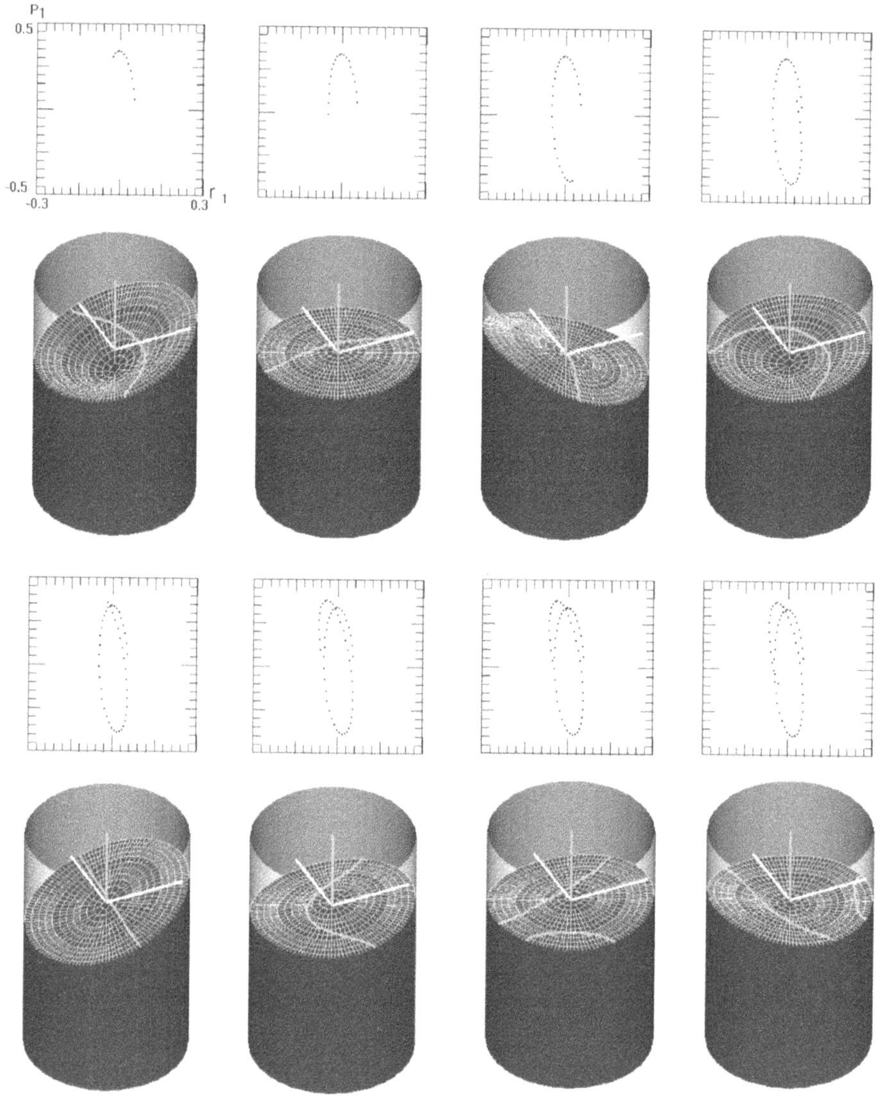


**Figure 8.9.** Breakdown of swirling wave; $h/R_1 = 2$, $H/R_1 = 0.01$.

dimensional approximation, in the present case, the nodal curve has no fixed shape and is changing together with surface. The observed dynamics of the nodal curve enables us to suggest a good explanation of the wave breakdown: the swirling wave breaks down since a traveling wave appears. This is confirmed by picture 6 and 7.

The examples considered in the present section enables us to conclude that the kinematics of nonlinear sloshing in a neighborhood of the main resonance can be studied qualitatively and quantitatively by the methods suggested in

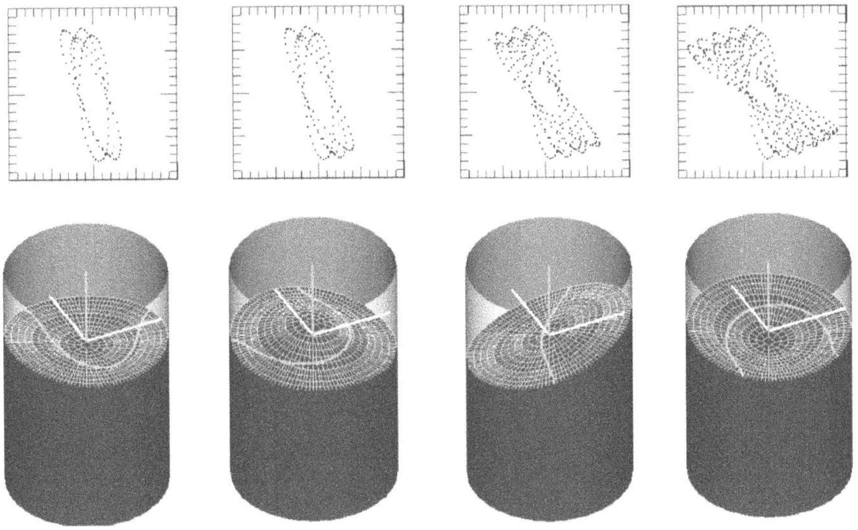

**Figure 8.9** (continued).

the present chapter. The results obtained by these methods meet the experimental data presented in [5]. When five generalized coordinates are taken into account, i.e., we keep the first five (two asymmetric and three symmetric) natural sloshing modes in the decomposition of the free surface, this enables us to suggest a more complete (qualitative and quantitative) description of the nonlinear wave phenomena.

In a similar way, we studied the case of sloshing in an upright circular cylindrical tank. The response curves for the inverse cone with the half-angle $\theta_0 = 20^o$ ($H = 0.01$ and $r_0 = 1.0$) is shown in Fig. 8.8. Other important forced nonlinear sloshing in conical tanks are described in [125, 126, 14, 73].

## 8.8 Forced liquid sloshing with damping effects

Liquid sloshing in containers is accompanied by dissipation of energy, and this leads to a damping of free oscillations and boundedness of amplitudes of forced sloshing in the resonance case. These phenomena can be studied only by using proper models of a viscous liquid. As a rule, in any approach that takes into account the damping of oscillatory processes in real liquid-filled mechanical systems, it is suggested to use a set of dissipative terms artificially added into discrete mathematical models, which are similar to models considered in the present monograph (see, e.g., [147, 149, 150]).

However, proper mathematical models can also be derived by using more rigorous arguments related to specific features of the oscillatory modes. In this section, we use the nonlinear modal system of equations of motion of a viscous liquid, which is suggested in Sec. 4.2. It has the form

$$
\begin{aligned}
L_1(r_1,p_1,p_0,r_2,p_2) &\equiv \mu_1(\ddot{r}_1+\alpha\dot{r}_1+\sigma_1^2 r_1)+d_1 r_1(r_1\ddot{r}_1+\dot{r}_1^2+p_1\ddot{p}_1+\dot{p}_1^2)\\
&+d_2(p_1^2\ddot{r}_1+2p_1\dot{r}_1\dot{p}_1-2r_1\dot{p}_1^2-r_1p_1\ddot{p}_1)+d_3(r_2\ddot{p}_1-p_2\ddot{r}_1+\dot{p}_1\dot{r}_2-\dot{r}_1\dot{p}_2)\\
&+d_4(r_1\ddot{p}_2-p_1\ddot{r}_2)+d_5(p_0\ddot{r}_1+\dot{p}_0\dot{r}_1)+d_6 r_1\ddot{p}_0\\
&+\alpha[(d_1r_1^2+d_2p_1^2)\dot{r}_1+(d_1-d_2)r_1p_1\dot{p}_1+d_3(r_2\dot{p}_1-p_2\dot{r}_1)\\
&d_4(r_1\dot{p}_2-p_1\dot{r}_2)+d_5p_0\dot{r}_1+d_6r_1\dot{p}_0]-\lambda H\omega^2\cos(\nu t)=0,\\
L_2(r_1,p_1,p_0,r_2,p_2) &\equiv \mu_1(\ddot{p}_1+\alpha\dot{p}_1+\sigma_1^2 p_1)+d_1 p_1(r_1\ddot{r}_1+\dot{r}_1^2+p_1\ddot{p}_1+\dot{p}_1^2)\\
&+d_2(r_1^2\ddot{p}_1-r_1p_1\ddot{r}_1+2r_1\dot{r}_1\dot{p}_1-2p_1\dot{r}_1^2)+d_3(r_2\ddot{r}_1+p_2\ddot{p}_1+\dot{r}_1\dot{r}_2+\dot{p}_1\dot{p}_2)\\
&-d_4(p_1\ddot{p}_2+r_1\ddot{r}_2)+d_5(p_0\ddot{p}_1+\dot{p}_0\dot{p}_1)+d_6p_1\ddot{p}_0\\
&+\alpha[(d_1p_1^2+d_2r_1^2)\dot{p}_1+(d_1-d_2)r_1p_1\dot{r}_1+d_3(p_2\dot{p}_1+r_2\dot{r}_1)\\
&-d_4(p_1\dot{p}_2+r_1\dot{r}_2)+d_5p_0\dot{p}_1+d_6p_1\dot{p}_0]=0,\\
L_3(r_1,p_1,p_0) &\equiv \mu_0(\ddot{p}_0+\alpha\dot{p}_0+\sigma_0^2p_0)+d_6(r_1\ddot{r}_1+p_1\ddot{p}_1)\\
&+d_8(\dot{r}_1^2+\dot{p}_1^2)+\alpha d_6(r_1\dot{r}_1+p_1\dot{p}_1)=0, \qquad (8.8.1)\\
L_4(r_1,p_1,r_2) &\equiv \mu_2(\ddot{r}_2+\alpha\dot{r}_2+\sigma_2^2r_2)-d_4(p_1\ddot{r}_1+r_1\ddot{p}_1)\\
&-2d_7\dot{r}_1\dot{p}_1-\alpha d_4(p_1\dot{r}_1+r_1\dot{p}_1)=0,\\
L_5(r_1,p_1,p_2) &\equiv \mu_2(\ddot{p}_2+\alpha\dot{p}_2+\sigma_2^2p_2)+d_4(r_1\ddot{r}_1-p_1\ddot{p}_1)\\
&+d_7(\dot{r}_1^2-\dot{p}_1^2)+\alpha d_4(r_1\dot{r}_1-p_1\dot{p}_1)=0.
\end{aligned}
$$

This system of equations describes forced nonlinear sloshing in a cylindrical container subjected to horizontal harmonic oscillations by the law $u(t) = H\cos\omega t$. The damping ratio $\alpha$ is associated with the logarithmic decrement of liquid sloshing for the first natural sloshing modes corresponding to the generalized coordinate $r_1(t)$ or $p_1(t)$.

We obtain the following approximate analytic solutions of the nonlinear system of equations (8.8.1) corresponding to steady-state forced sloshing [119, 139, 140]:

$$
\begin{aligned}
r_1(t) &= A\cos(\omega t+\varphi),\quad p_1(t)=B\sin(\omega t+\psi),\\
p_0(t) &= l_0(A^2+B^2)+h_0(A^2\cos[2(\omega t+\varphi)]-B^2\cos[2(\omega t+\psi)])\\
&+g_0(A^2\sin[2(\omega t+\varphi)]-B^2\sin[2(\omega t+\psi)]), \qquad (8.8.2)\\
r_2(t) &= 2AB[g_2\cos(2\omega t+\varphi+\psi)-h_2\sin(2\omega t+\varphi+\psi)+l_2\sin(\varphi-\psi)],\\
p_2(t) &= l_2(A^2-B^2)+h_2(A^2\cos[2(\omega t+\varphi)]+B^2\cos[2(\omega t+\psi)])\\
&+g_2(A^2\sin[2(\omega t+\varphi)]+B^2\sin[2(\omega t+\psi)]).
\end{aligned}
$$

For these relations, we use the notation

$$l_0 = \frac{d_6 - d_8}{2\mu_0 \sigma_0^2}, \quad h_0 = \frac{a_0(d_6 + d_8) - \bar{\alpha} b_0 d_6}{2(a_0^2 + b_0^2)}, \quad g_0 = \frac{\bar{\alpha} a_0 d_6 + b_0(d_6 + d_8)}{2(a_0^2 + b_0^2)},$$

$$l_2 = \frac{d_4 - d_7}{2\mu_2 \sigma_2^2}, \quad h_2 = \frac{a_2(d_4 + d_7) - \bar{\alpha} b_2 d_4}{2(a_2^2 + b_2^2)}, \quad g_2 = \frac{\bar{\alpha} a_2 d_4 + b_2(d_4 + d_7)}{2(a_2^2 + b_2^2)},$$

$$\bar{\sigma}_m = \frac{\sigma_m}{\omega}, \quad m = 0, 1, 2; \quad \bar{\alpha} = \frac{\alpha}{\omega},$$

$$a_k = \mu_k(\bar{\sigma}_k^2 - 4), \quad b_k = 2\bar{\alpha}\mu_k, \quad k = 0, 2,$$

and $A$, $B$, $\varphi$, and $\psi$ are unknown constants to be found.

After we choose the form of solutions for the generalized coordinates $r_1(t)$ and $p_1(t)$, the other generalized coordinates $p_0(t), r_2(t)$, and $p_2(t)$ can be found from system (8.8.1) as exact solutions by using constants $A$, $B$, $\varphi$, and $\psi$. As above, to find these constants , we use the Bubnov–Galerkin method. In this case, the corresponding equations have the form

$$\begin{aligned}
&\int_0^{\frac{2\pi}{\omega}} L_1(r_1, p_1, p_0, r_2, p_2) \cos(\omega t + \varphi) dt = 0, \\
&\int_0^{\frac{2\pi}{\omega}} L_1(r_1, p_1, p_0, r_2, p_2) \sin(\omega t + \varphi) dt = 0, \\
&\int_0^{\frac{2\pi}{\omega}} L_2(r_1, p_1, p_0, r_2, p_2) \cos(\omega t + \psi) dt = 0, \qquad (8.8.3) \\
&\int_0^{\frac{2\pi}{\omega}} L_2(r_1, p_1, p_0, r_2, p_2) \sin(\omega t + \psi) dt = 0.
\end{aligned}$$

As a result, we obtain the following system of equations with respect to $A$, $B$, $\varphi$, and $\psi$:

$$\begin{aligned}
&A\{m_2 + m_3 A^2 + [m_1 \cos 2(\varphi - \psi) + m_5 \sin 2(\varphi - \psi) - m_4] B^2\} = \lambda H \cos \varphi, \\
&A\{-m_0 + m_6 A^2 + [-m_5 \cos 2(\varphi - \psi) \\
&\qquad + m_1 \sin 2(\varphi - \psi) + m_7] B^2\} = \lambda H \sin \varphi, \\
&B\{m_0 + [m_5 \cos 2(\varphi - \psi) + m_1 \sin 2(\varphi - \psi) - m_7] A^2 - m_6 B^2\} = 0, \qquad (8.8.4) \\
&B\{m_2 + [m_1 \cos 2(\varphi - \psi) - m_5 \sin 2(\varphi - \psi) - m_4] A^2 + m_3 B^2\} = 0,
\end{aligned}$$

where

$$m_0 = \bar{\alpha}\mu_1, \;\; m_1 = \tfrac{1}{2}d_1 - d_2 - \left(\tfrac{1}{2}h_2 + l_2\right) d_3 - 2h_2 d_4 - \tfrac{1}{2}h_0 d_5 + 2h_0 d_6$$
$$+\bar{\alpha}\mathrm{g}_0 \left(\tfrac{1}{2}d_5 - d_6\right) + \bar{\alpha}\mathrm{g}_2\left(\tfrac{1}{2}d_3 + d_4\right),$$
$$m_2 = \mu_1(\bar{\sigma}_1^2 - 1), \;\; m_4 = d_2 + h_2 d_3 + 4h_2 d_4 + l_0 d_5 - \bar{\alpha}\mathrm{g}_2(d_3 + 2d_4),$$
$$m_5 = \mathrm{g}_0 \left(\tfrac{1}{2}d_5 - 2d_6\right) + \mathrm{g}_2 \left(\tfrac{1}{2}d_3 + 2d_4\right) + \bar{\alpha}\left[-\tfrac{1}{4}d_1 + \tfrac{1}{2}d_2\right. \tag{8.8.5}$$
$$\left. + \left(\tfrac{1}{2}h_2 + l_2\right) d_3 + h_2 d_4 + \tfrac{1}{2}h_0 d_5 - h_0 d_6\right],$$
$$m_7 = \bar{\alpha}\left[-\tfrac{1}{2}d_2 - h_2 d_3 - 2h_2 d_4 - l_0 d_5\right] - \mathrm{g}(d_3 + 4d_4),$$
$$m_3 = -m_1 - m_4, \;\; m_6 = m_5 + m_7.$$

This system is a generalization of the corresponding system presented in [16].

An analysis of (8.8.4) and the above-obtained relations (8.8.2) shows that two types of motions described by the nonlinear system (8.8.1) are possible, namely:

(1) Planar liquid sloshing in the plane $xOz$, i.e., in the plane of oscillations of the container. In this case, $A \neq 0$ and $B = 0$ and, hence,

$$r_1(t) = A\cos(\omega t + \varphi), \;\; p_1(t) \equiv 0, \;\; r_2(t) \equiv 0,$$

$$p_k(t) = [l_k + h_k \cos 2(\omega t + \varphi) + \mathrm{g}_k \sin 2(\omega t + \varphi)]A^2, \;\; k = 0, 2. \tag{8.8.6}$$

(2) Swirling, i.e., spatial sloshing motions arising in the domain where the dynamic stability of "plane" oscillations is lost due to excitation of parametric oscillations in the plane $xOy$, which is perpendicular to the plane of motion of the container. In this case, $A \neq 0$ and $B \neq 0$ and, hence, an approximate solution of the system of equations (8.8.1) takes the form (8.8.2).

We can not find an analytic solution of the algebraic system (8.8.4). To find amplitudes $A$ and $B$ and the initial phases $\varphi$ and $\psi$, we have to use various numerical methods for each value $\bar{\omega} = \frac{\omega}{\sigma_1}$. In this case, the procedure of numerical analysis is realized in three stages.

1. In the first stage, the amplitudes $A$ and $B$ are computed by using (8.8.4) for $\varphi = 0$, $\psi = 0$, and $\bar{\alpha} = 0$, i.e., by using only the first and fourth equations of the system.

2. The obtained values $A$ and $B$ are used as the first approximation of the solution of (8.8.4) in the case where the friction coefficient is included only in linear terms of system (8.8.1).

3. In the third stage, we find the amplitudes $A$ and $B$ and the initial phases $\varphi$ and $\psi$ by using equations (8.8.4), which are obtained for the complete model (8.8.1). In this case, for the initial values of the required parameters, we use the values obtained in the second stage for the model with linear friction.

Following this scheme and varying the parameters $H$, $\alpha$ and $\bar{\omega}$, A.M.Pil'kevich performed a series of numerical computations and a comparison of the computed results with the corresponding results for systems without friction ($\alpha = 0$) and the experimental data given in [5]. Parameters of the mathematical model were taken for a cylindrical container with the free surface radius $R_1 = 1.0$ m and the mean liquid depth $h = 2.0$ m. For liquids, water and motor oil (*castrol*) were chosen. Their friction coefficients are found from experimental data for the logarithmic decrement by using the method suggested in [147]. In this case, we have $\alpha = 0.0038\,\text{sec}^{-1}$ for water and $\alpha = 0.1491\,\text{sec}^{-1}$ for oil (*castrol*).

Table 8.2 contains numerical values of amplitudes and phases of forced oscillations for planar sloshing ($\alpha = 0.0038$) in dependence of the frequency of perturbation force $\bar{\omega} = \frac{\omega}{\sigma_1}$. Analogous computational results for spatial oscillations are presented in Table 8.3, where the friction is taken into account within the framework of the complete model. Some values of parameters in Tables 8.2 and 8.3 correspond to unstable solutions.

Tables 8.4 and 8.5 contain numerical values of the generalized coordinates $r_1(t)$, $p_1(t)$, $p_0(t)$, $r_2(t)$, and $p_2(t)$ obtained by using analytic solutions (8.8.1) and the numerical Runge–Kutta method for $\bar{\omega}$=1.07 [the initial conditions are chosen in accordance with (8.8.2)]. In Table 8.4, the comparison is performed for a single period ($T = 1.382573$, $A = 0.5694$, $B = 0.5814$, $\varphi = -0.06514$, and $\psi = -0.06376$), whereas Table 8.5 shows values of these parameters for the time intervals $t = n \cdot T$, where $n = 0, 10, 20, \ldots, 100$ [the initial conditions are chosen in accordance with relations (8.8.2)].

The results in Table 8.6 illustrate variations in the parameters $A$, $\varphi$, $B$, and $\psi$ in dependence of the forcing amplitude $H$ for the constant value $\bar{\omega} = 0.9$ in the case of a liquid that is more viscous than water [here, machine oil (*castrol*) with the friction coefficient $\alpha = 0.1491$ is used].

The results in Table 8.7 characterize the time dependence of generalized coordinates obtained by integration of the nonlinear system (8.8.1) by using the Runge–Kutta method for a viscous liquid (oil *castrol*) with the friction coefficient $\alpha = 0.1491$.

The results in Table 8.8 show the time dependence of the law of motion with respect to the generalized coordinate $r_1(t)$ in modes of plane oscillations for values $\alpha = 0.1491$ and $H = 0.05$. Typical response curves in mode $r_1(t)$ for the case of a viscous liquid are shown in Fig. 8.10. Parts *I* are stable branches of the response curves for liquid oscillations and *II* is the stable branch of the response curves for the rotation motion of the free surface.

The values presented above and related to the validation of the nonlinear model for forced sloshing of a viscous liquid confirm high efficiency of the model. Numerical values of oscillation amplitudes of the free surface, up to their limit values ($A \approx 0.69$), are close to the results obtained in the case of an ideal liquid and in experiments.

**Table 8.2**

| Ideal ($\alpha = 0$) | | Linear friction | | Complete model | |
|---|---|---|---|---|---|
| $\overline{\omega}$ | $A$ | $A$ | $\varphi$ | $A$ | $\varphi$ |
| 0.90 | 0.06622 | 0.06622 | −0.004285 | 0.06622 | −0.004291 |
| | 0.6504 | 0.6504 | −0.04826 | 0.6504 | −0.05408 |
| | −0.7166 | −0.7165 | 0.05464 | −0.7165 | 0.06242 |
| 0.91 | 0.07519 | 0.07519 | −0.004814 | 0.07519 | −0.004822 |
| | 0.5926 | 0.5926 | −0.04278 | 0.5926 | −0.046832 |
| | −0.6677 | −0.6677 | 0.04970 | −0.6677 | 0.05549 |
| 0.92 | 0.08672 | 0.08672 | −0.005496 | 0.08672 | −0.005508 |
| | 0.5341 | 0.5342 | −0.03749 | 0.5342 | −0.04024 |
| | −0.6208 | −0.6208 | 0.04510 | −0.62079 | 0.04941 |
| 0.93 | 0.1023 | 0.1023 | −0.006423 | 0.1023 | −0.006441 |
| | 0.4732 | 0.4732 | −0.03227 | 0.4732 | −0.03404 |
| | −0.5755 | −0.5754 | 0.04080 | −0.5754 | 0.04399 |
| 0.94 | 0.1253 | 0.1253 | −0.007802 | 0.1253 | −0.007833 |
| | 0.4060 | 0.4061 | −0.02686 | 0.4061 | −0.02791 |
| | −0.5313 | −0.5313 | 0.03678 | −0.5313 | 0.03912 |
| 0.95 | 0.1683 | 0.1682 | −0.01042 | 0.1682 | −0.01049 |
| | 0.3199 | 0.3200 | −0.0205 | 0.3200 | −0.02094 |
| | −0.4882 | −0.4882 | 0.03300 | −0.4882 | 0.03470 |
| 0.954245 | 0.2351 | 0.2331 | −0.01455 | 0.2331 | −0.01472 |
| | 0.2351 | 0.2371 | −0.01481 | 0.2371 | −0.01500 |
| | −0.4701 | −0.4701 | 0.03146 | −0.4701 | 0.03294 |
| 0.96 | −0.4459 | −0.4458 | 0.02944 | −0.4458 | 12.5970 |
| 0.97 | −0.4043 | −0.4042 | 0.02609 | −0.4042 | −6.2562 |
| 0.98 | −0.3634 | −0.3634 | 0.02294 | −0.3634 | 0.02353 |
| 0.99 | −0.3236 | −0.3236 | 0.03236 | −0.3236 | 0.02040 |
| 1.00 | −0.2853 | −0.2852 | 0.01728 | −0.2852 | 0.01754 |
| 1.01 | −0.2491 | −0.2491 | 0.01481 | −0.2491 | 0.01498 |
| 1.02 | −0.2161 | −0.2161 | 0.01263 | −0.2161 | 0.01274 |
| 1.03 | −0.1872 | −0.18721 | 0.01078 | −0.1872 | 0.01084 |
| 1.04 | −0.1627 | −0.1627 | 0.009236 | −0.1627 | 0.009276 |
| 1.05 | −0.1426 | −0.1426 | 0.007989 | −0.1426 | 0.008014 |
| 1.06 | −0.1262 | −0.1262 | 0.006987 | −0.1262 | 0.007004 |
| 1.07 | −0.1129 | −0.1129 | 0.006182 | −0.1129 | 0.006194 |
| 1.08 | −0.1021 | −0.1021 | 0.005530 | −0.1021 | 0.005539 |
| 1.09 | −0.09314 | −0.09314 | 0.004996 | −0.09314 | 0.005002 |
| 1.10 | −0.08573 | −0.08573 | 0.004553 | −0.08573 | 0.004558 |

**Table 8.3**

| Ideal ($\alpha = 0$; $H = 0.01$) | | | Water ($\alpha = 0.0038$; $H = 0.01$) | | | |
|---|---|---|---|---|---|---|
| $\overline{\omega}$ | $A$ | $B$ | $A$ | $\varphi$ | $B$ | $\psi$ |
| 0.90 | −0.02027 | 0.6865 | 0.04495 | -1.1056 | 0.6854 | −0.002594 |
| 0.91 | −0.02338 | 0.6344 | 0.04014 | -0.9521 | 0.6336 | −0.002966 |
| 0.92 | −0.02724 | 0.5837 | 0.03793 | -0.7729 | 0.5831 | −0.003426 |
| 0.93 | −0.03218 | 0.5337 | 0.03850 | -0.5850 | 0.5334 | −0.004010 |
| 0.94 | −0.03871 | 0.4842 | 0.04214 | -0.4114 | 0.4839 | −0.004779 |
| 0.95 | −0.04770 | 0.43467 | 0.04941 | −0.2693 | 0.43448 | −0.005833 |
| 0.96 | −0.06078 | 0.3853 | 0.06154 | −0.1645 | 0.3852 | −0.007360 |
| 0.97 | −0.08096 | 0.3376 | 0.08125 | −0.09491 | 0.3376 | −0.009708 |
| 0.98 | −0.1135 | 0.2978 | 0.1136 | −0.05518 | 0.2978 | −0.01347 |
| 0.99 | −0.1628 | 0.2807 | 0.1628 | −0.03870 | 0.2806 | −0.01914 |
| 1.00 | −0.2234 | 0.2960 | 0.2233 | −0.03611 | 0.2959 | −0.02602 |
| 1.01 | −0.2842 | 0.3319 | 0.2842 | −0.03892 | 0.3318 | −0.03284 |
| 1.02 | −0.3408 | 0.3746 | −0.3408 | 9.3816 | −0.3746 | 3.1025 |
| 1.03 | −0.3927 | 0.41838 | 0.3927 | −0.04780 | 0.4183 | −0.04474 |
| | 0.1123 | $0.2454 \cdot i$ | −0.1123 | −0.02203 | $0.2454 \cdot i$ | 0.01266 |
| | 0.2805 | 0.2225 | −0.2805 | 0.02602 | 0.2226 | 0.03178 |
| 1.04 | −0.4408 | 0.4612 | 0.4408 | −0.05232 | 0.4611 | −0.04993 |
| | 0.08082 | $0.3064 \cdot i$ | −0.08082 | -3.1970 | $0.3065 \cdot i$ | 22.0001 |
| | 0.3601 | 0.3274 | −0.3602 | 0.03712 | 0.3275 | 0.04058 |
| 1.05 | −0.4859 | 0.5026 | 0.4858 | −0.05670 | 0.5025 | −0.05477 |
| | 0.06485 | $0.3457 \cdot i$ | −0.06515 | 3.0532 | $0.3457 \cdot i$ | 3.1487 |
| | 0.4210 | 0.3977 | −0.4211 | 0.04470 | 0.3978 | 0.04720 |
| 1.06 | −0.5286 | 0.5426 | 0.5285 | −0.06097 | 0.5425 | −0.05935 |
| | 0.05491 | $0.3765 \cdot i$ | −0.05536 | −0.1221 | $0.3765 \cdot i$ | 0.005990 |
| | 0.4737 | 0.4556 | −0.4737 | 0.05094 | 0.4557 | 0.05289 |
| 1.07 | 0.5695 | 0.5815 | 0.5694 | −0.06514 | 0.5814 | −0.06376 |
| | 0.04800 | $0.4022 \cdot i$ | −0.04863 | 9.2690 | $0.4023 \cdot i$ | 9.4300 |
| | 0.5215 | 0.5068 | −0.5215 | 0.05646 | 0.5069 | 0.05805 |
| 1.08 | −0.6090 | 0.6194 | 0.6089 | −0.06926 | 0.6194 | −0.06807 |
| | 0.04290 | $0.4245 \cdot i$ | −0.04371 | −0.1890 | $0.4245 \cdot i$ | 0.004578 |
| | 0.5661 | 0.5538 | −0.5661 | 0.06155 | −0.5538 | 0.06289 |
| 1.09 | −0.6475 | 0.6566 | 0.6474 | −0.07336 | 0.6566 | −0.072325 |
| | 0.03897 | $0.4439 \cdot i$ | −0.03998 | −0.2213 | $0.4440 \cdot i$ | 0.004113 |
| | 0.6085 | 0.5980 | −0.6085 | 0.06639 | 0.5980 | 0.06754 |
| 1.10 | −0.6852 | 0.6933 | 0.6851 | −0.07750 | 0.6932 | −0.07658 |
| | 0.03584 | $0.4612 \cdot i$ | −0.03758 | −0.3019 | $0.4614 \cdot i$ | 0.003789 |
| | 0.6493 | 0.64019 | −0.6494 | 0.07110 | 0.6403 | 0.07210 |

**Table 8.4.** $\alpha$=0.0038; $H$=0.01; $\bar{\omega}$=1.07

| $t$ | Numerical / Analytic | | | | |
|---|---|---|---|---|---|
| | $r_1$ | $p_1$ | $p_0$ | $r_2$ | $p_2$ |
| 0.0 | 0.5682 | −0.03705 | 0.1072 | −0.008574 | −0.06426 |
| 0.1 | 0.5264 | 0.2212 | 0.1079 | 0.04949 | −0.04466 |
| | 0.5268 | 0.2214 | 0.1079 | 0.04952 | −0.04458 |
| 0.2 | 0.3777 | 0.4336 | 0.1095 | 0.06968 | 0.01298 |
| | 0.3785 | 0.4349 | 0.1096 | 0.06969 | 0.01335 |
| 0.3 | 0.1535 | 0.5575 | 0.1105 | 0.03670 | 0.06417 |
| | 0.1533 | 0.5602 | 0.1111 | 0.03641 | 0.06488 |
| 0.4 | −0.1012 | 0.5691 | 0.1098 | −0.02391 | 0.06964 |
| | −0.1030 | 0.5717 | 0.1111 | −0.02468 | 0.07028 |
| 0.5 | −0.3363 | 0.4657 | 0.1078 | −0.06555 | 0.02518 |
| | −0.3384 | 0.4671 | 0.1097 | −0.06649 | 0.02540 |
| 0.6 | −0.5040 | 0.2669 | 0.1061 | −0.05595 | −0.03508 |
| | −0.5051 | 0.2678 | 0.1080 | −0.05680 | −0.03517 |
| 0.7 | −0.5690 | 0.01266 | 0.1062 | −0.002286 | −0.06448 |
| | −0.5692 | 0.01405 | 0.1072 | −0.003065 | −0.06475 |
| 0.8 | −0.5174 | −0.2438 | 0.1083 | 0.05402 | −0.04018 |
| | −0.5179 | −0.2425 | 0.1080 | 0.05329 | −0.04053 |
| 0.9 | −0.3612 | −0.4495 | 0.1111 | 0.06961 | 0.01900 |
| | −0.3613 | −0.4499 | 0.1098 | 0.06882 | 0.01882 |
| 1.0 | −0.1334 | −0.5639 | 0.1123 | 0.03282 | 0.06755 |
| | −0.1315 | −0.5659 | 0.1111 | 0.03156 | 0.06754 |
| 1.1 | 0.1208 | −0.5654 | 0.1108 | −0.02800 | 0.06844 |
| | 0.1251 | −0.5670 | 0.1110 | −0.02977 | 0.06810 |
| 1.2 | 0.3517 | −0.4529 | 0.1076 | −0.06619 | 0.0210 |
| | 0.3563 | −0.4531 | 0.1095 | −0.06790 | 0.02005 |
| 1.3 | 0.5123 | −0.2471 | 0.1048 | −0.05214 | −0.03843 |
| | 0.5151 | −0.2471 | 0.1078 | −0.0534 | −0.03957 |
| $T$ | 0.5666 | −0.03585 | 0.1044 | −0.007492 | −0.06310 |
| | 0.5682 | −0.03705 | 0.1072 | −0.008574 | −0.06426 |

For planar steady-state waves with higher values of the friction coefficient $\alpha$, the difference between analytic and numerical solutions increases as $\bar{\omega}$ tends to $\bar{\omega}^*$ ($\bar{\omega}^*$ is the point, at which plane oscillations lose their stability). In these cases, one-term approximations are not sufficient for finding the generalized coordinates $r_1(t)$ and $p_1(t)$ (see Table 8.8). It is worth noting that the increase in $\alpha$ and $H$ strongly affects the variations of the amplitude of forced oscillations $C$ ($C = \sqrt{r_1^2(T) + p_1^2(T)}$). For $\bar{\omega} = 0.9$ (see Table 8.6) and $H < 0.025$, there are no steady-state motions in the form (8.8.6) (the amplitude $B$ is imaginary).

**Table 8.5**

| $t$ | $r_1$ | $p_1$ | $p_0$ | $r_2$ | $p_2$ |
|---|---|---|---|---|---|
| $0 \cdot T$ | 0.5682 | −0.03705 | 0.1072 | −0.008574 | −0.06426 |
| $10 \cdot T$ | 0.5693 | −0.04284 | 0.1076 | −0.009724 | −0.06437 |
| $20 \cdot T$ | 0.5706 | −0.03297 | 0.1075 | −0.007206 | −0.06466 |
| $30 \cdot T$ | 0.5682 | −0.02975 | 0.1058 | −0.006230 | −0.06391 |
| $40 \cdot T$ | 0.5669 | −0.03963 | 0.1050 | −0.008416 | −0.06311 |
| $50 \cdot T$ | 0.5685 | −0.03903 | 0.1052 | −0.008432 | −0.06312 |
| $60 \cdot T$ | 0.5681 | −0.03101 | 0.1045 | −0.006876 | −0.06294 |
| $70 \cdot T$ | 0.5661 | −0.03649 | 0.1037 | −0.008416 | −0.06244 |
| $80 \cdot T$ | 0.5671 | −0.04387 | 0.1045 | −0.01033 | −0.06269 |
| $90 \cdot T$ | 0.5690 | −0.03752 | 0.1055 | −0.009026 | −0.06344 |
| $100 \cdot T$ | 0.5681 | −0.03425 | 0.1054 | −0.008322 | −0.06350 |

**Table 8.6**

| Ideal/ Linear friction / Complete model | | | |
|---|---|---|---|
| $\overline{\omega}$ | $\omega^2 \cdot d/g$ | $A$ | $100 \cdot \lvert P_z \rvert / d^3$ |
| 0.90 | 2.9789 | 0.06622 | 0.05453 |
| | | 0.06622 | 0.05453 |
| | | 0.06622 | 0.05453 |
| 0.91 | 3.0455 | 0.07519 | 0.2607 |
| | | 0.07519 | 0.2607 |
| | | 0.07519 | 0.2607 |
| 0.92 | 3.1128 | 0.08672 | 0.6822 |
| | | 0.08672 | 0.6822 |
| | | 0.08672 | 0.6822 |
| 0.93 | 3.1808 | 0.1023 | 1.2713 |
| | | 0.1023 | 1.2712 |
| | | 0.1023 | 1.2712 |
| 0.94 | 3.2496 | 0.1253 | 2.1652 |
| | | 0.1253 | 2.1650 |
| | | 0.1253 | 2.1650 |
| 0.95 | 3.3191 | 0.1683 | 3.8626 |
| | | 0.1682 | 3.8621 |
| | | 0.1682 | 3.8621 |
| 0.9542 | 3.3489 | 0.2351 | 6.4888 |
| | | 0.2346 | 6.4699 |
| | | 0.2348 | 6.4797 |

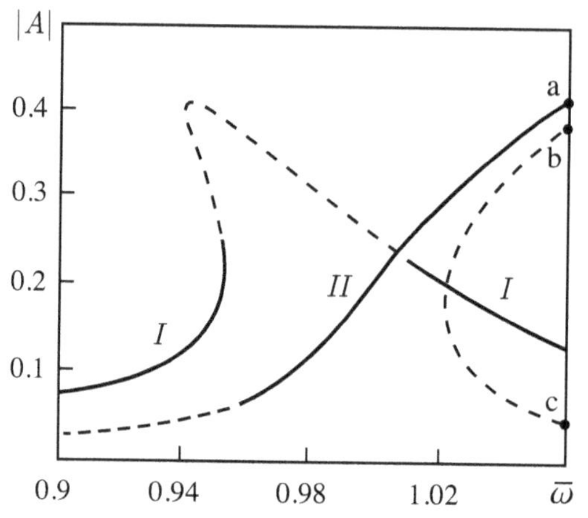


**Figure 8.10**

**Table 8.7**

| $H$ | $A$ | $\varphi$ | $B$ | $\psi$ | $C$ |
|---|---|---|---|---|---|
| 0.015 | −1.0519 | 1.4980 | $-0.7980 \cdot i$ | −3.2432 | |
| 0.02 | −0.7914 | −4.7628 | $0.3928 \cdot i$ | −0.1014 | |
| 0.023 | −0.6900 | −4.7465 | $0.4861 \cdot i$ | −0.1013 | |
| 0.025 | 0.6361 | −1.5928 | 0.2652 | −3.2428 | 0.03023 |
| 0.03 | 0.5337 | −1.5587 | 0.4401 | −3.2425 | 0.04479 |
| 0.035 | 0.4617 | 4.7640 | 0.5191 | 3.0411 | 0.05728 |
| 0.04 | 0.4090 | 4.8083 | −0.5657 | −0.1002 | 0.06879 |
| 0.045 | −0.3694 | −4.5681 | 0.5968 | −0.9973 | 0.07970 |
| 0.05 | −0.3391 | −4.5162 | 0.6191 | −0.9927 | 0.09018 |

**Table 8.8**

| $\overline{\omega}$ | $T$ | $t = n \cdot T$ | $r_1(t)$ (analytic) | $r_1(t)$ (numerical) |
|---|---|---|---|---|
| 0.882 | 1.6773 | 166.0498 | 0.3121 | 0.3000 |
| 0.884 | 1.6735 | 170.6946 | 0.3254 | 0.3103 |
| 0.886 | 1.6697 | 170.3093 | 0.3416 | 0.3217 |
| 0.887 | 1.6678 | 168.4494 | 0.3514 | 0.3280 |
| 0.89 | 1.6622 | 167.8816 | 0.4055 | 0.3499 |
| $\overline{\omega}^*$ | 1.6618 | 169.5083 | 0.4238 | 0.3514 |

Other specific features of forced oscillations of a viscous liquid in a cylindrical container (including finding the hydrodynamic force amplitude) are considered in [140]. Examples of applications of the Floquet theory to the stability analysis of solutions in some special cases are given in [119].

## 8.9 Hydrodynamic forces

We now consider the problem of hydrodynamic loads on the tank walls, which is important from the viewpoint of applications. As an example, we consider forced sloshing in an oscillating container. As it is known, the hydrodynamic load on a container can be associated with the hydrodynamic force

$$\boldsymbol{P} = \int_S p\boldsymbol{n}dS, \tag{8.9.1}$$

where $\boldsymbol{n}$ is the unit vector of the outer normal to the wetted surface $S$; $p$ is the pressure computed by using the Bernoulli equation

$$\frac{\partial \Phi}{\partial t} + \frac{1}{2}(\nabla\Phi)^2 - \nabla\Phi \cdot \dot{\boldsymbol{u}} - \mathbf{g} \cdot \boldsymbol{r} + \frac{p}{\rho} = 0, \tag{8.9.2}$$

were $\Phi$ is the potential of velocities, $\boldsymbol{r}$ is the radius vector of the liquid domain $Q$, $\rho$ is the liquid density, $\boldsymbol{u} = (0, 0, H\cos\omega t)$, and $\mathbf{g} = (-g, 0, 0)$.

In practice, the direct use of (8.9.1) is not possible. To derive an expression for the hydrodynamic force $\boldsymbol{P}$, we use the results presented in Sec. 2.8 and obtain

$$\boldsymbol{P} = -m(\ddot{\mathbf{u}} - \mathbf{g}) - \frac{d\boldsymbol{K}}{dt}, \tag{8.9.3}$$

where $m$ is the liquid mass and $\boldsymbol{K}$ is the momentum vector of liquid mass defined by the formula

$$\boldsymbol{K} = \rho \int_Q (\nabla\Phi)dQ.$$

In the general case where the equation of the perturbed free surface $\Sigma$ has the form

$$x = h + f = h + \sum_i \beta_i(t) f_i(y, z),$$

we obtain the following relations for the projections of the momentum vector $\boldsymbol{K}$ onto the axes of the Cartesian system:

$$K_x = \frac{1}{2}\sum_i \lambda_{i1}\beta_i(t)\dot{\beta}_i(t), \quad K_y = \sum_i \lambda_{i2}\dot{\beta}_i(t), \quad K_z = \sum_i \lambda_{i3}\dot{\beta}_i(t), \tag{8.9.4}$$

where

$$\lambda_{i1} = \rho \int_{\Sigma_0} f_i^2(y, z)dS, \ \lambda_{i2} = \rho \int_{\Sigma_0} y f_i(y, z)dS, \ \lambda_{i3} = \rho \int_{\Sigma_0} z f_i(y, z)dS. \tag{8.9.5}$$

In the present case, the quantities $p_0(t)$, $r_1(t)$, …, $p_2(t)$ play the role of the generalized coordinates $\beta_i(t)$. Up to the terms of the third order of smallness, we write the following expressions for the projections of the hydrodynamic force onto the axes of the body-fixed coordinate system:

$$P_x = -mg - \tfrac{1}{2}\lambda_1(r_1\ddot{r}_1 + p_1\ddot{p}_1 + \dot{r}_1^2 + \dot{p}_1^2), \tag{8.9.6}$$

$$P_y = -\lambda\ddot{p}_1, \quad P_z = -m\ddot{u} - \lambda\ddot{r}_1,$$

where

$$\lambda_1 = \pi\rho\int\limits_{R_0}^{R_1} Y_1^2(k_1\xi)\xi d\xi, \quad \lambda = \pi\rho\int\limits_{R_0}^{R_1} Y_1(k_1\xi)\xi^2 d\xi = \pi\rho i. \tag{8.9.7}$$

For the planar sloshing regime, the component $P_z$ of the hydrodynamic force on the $Oz$-axis is of the main interest (the cavity performs forced oscillations along this axis). Substituting the expressions for $u(t)$ and $r_1(t)$ into the last relation of system (8.9.6), we obtain the following relation for finding the force amplitude

$$|P_z| = \pi\rho\omega^2[Hh(R_1^2 - R_0^2) - iA]. \tag{8.9.8}$$

By using (8.9.8), we can estimate the contribution of inertia forces

$$|P_z^{\mathrm{f}}| = \pi\rho\omega^2 h(R_1^2 - R_0^2)H$$

and the wave motion of the free surface

$$|P_z^{\mathrm{w}}| = \pi\rho\omega^2 iA$$

to the projection of the total hydrodynamic force. For example, for a cylindrical container with sizes $R_1 = 1$ $(R_0 = 0)$ and $h = 2$ subjected to harmonic oscillations along the $Oz$-axis with $H = 0.01$, relative frequency $\omega/\sigma_1 = 0.958$, and the amplitude $A = 0.225$ of the main generalized coordinate (planar regime), the contribution of $P_z^b$ to the quantity $|P_z|$ is approximately 80%. This indicates that it is important to calculate the response curves $A(\omega/\sigma_1)$ of the generalized coordinate $r_1(t)$ more accurately when we compute the projection of the amplitude of the total hydrodynamic force onto the $Oz$-axis.

For a cylindrical container of radius $R_1 = 1$ and liquid depth $h = 2$, the behavior of the dimensionless amplitude of the force $P_z$ computed by using relation (8.9.8) and some experimental data obtained in [5] are shown in Fig. 8.11 by using the same notation as in Fig. 8.3b. Comparing the results displayed in Figs. 8.3b and 8.11, we see that the experimental data meet in a better accuracy the theoretical results for the projection of the total hydrodynamic force (see Fig. 8.11) than the mean amplitude (see Fig. 8.3b).

Within the framework of the described theory, we consider the problem of influence of a coaxial partition on the distribution of hydrodynamic loads on

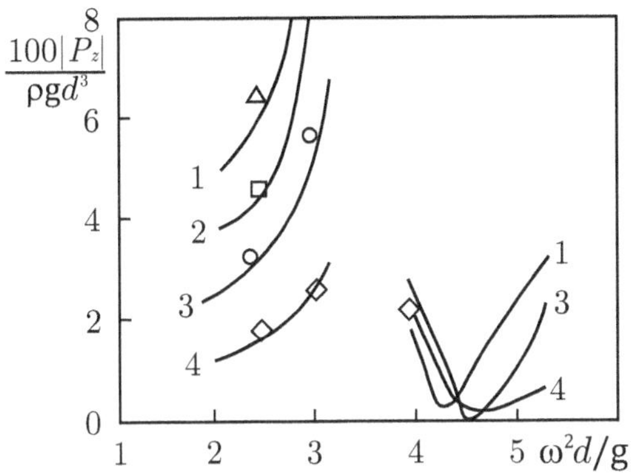


**Figure 8.11**

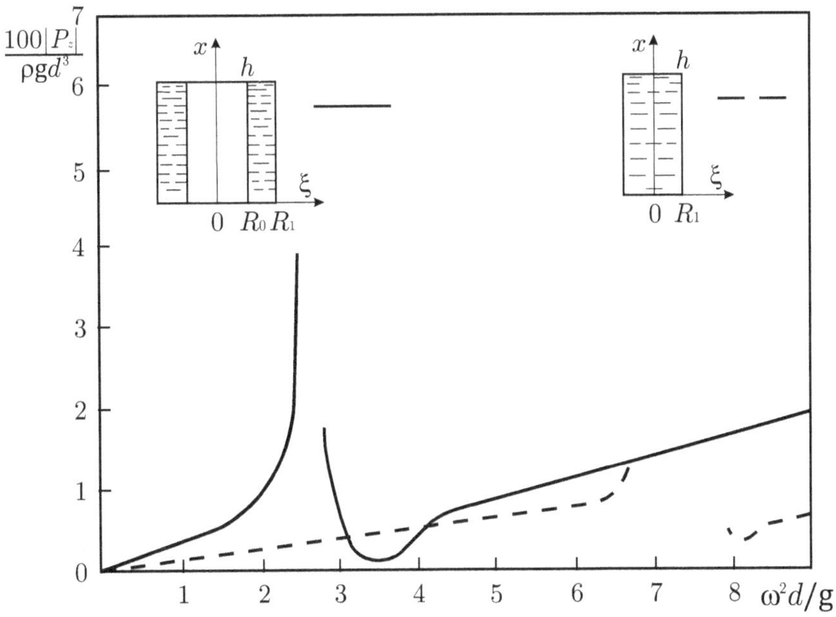


**Figure 8.12**

the container wall. It is known [5] that an advantage of a container formed by two coaxial right circular cylindrical surfaces is a great difference in the sloshing frequencies observed in internal and external cavities and also a decrease in the total hydrodynamic force caused by the difference of oscillation phases of liquid masses in these two cavities.

The response curves of the force $P_z$ for different containers are depicted in Figs. 8.12–8.14 (the forcing amplitude $H = 0.01$ and the liquid depth $h = 2$). The curves in Fig. 8.12 correspond to the cavity between two coaxial circular

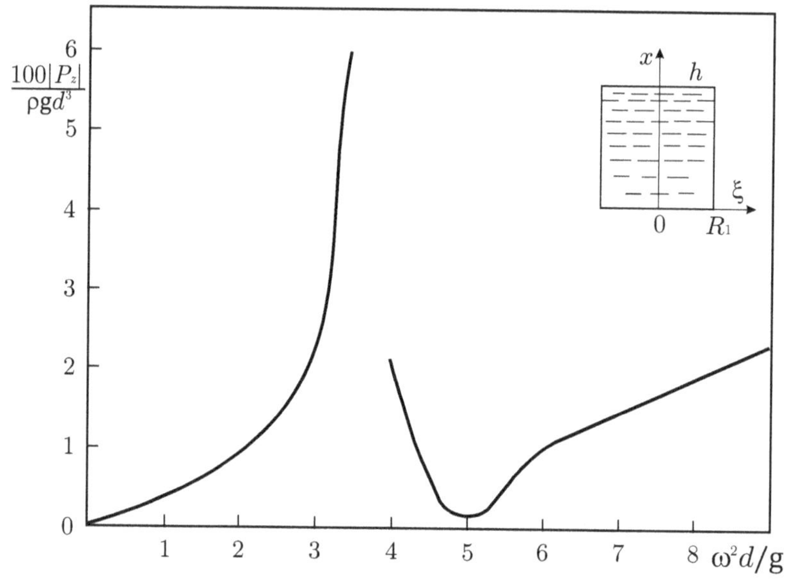


**Figure 8.13**

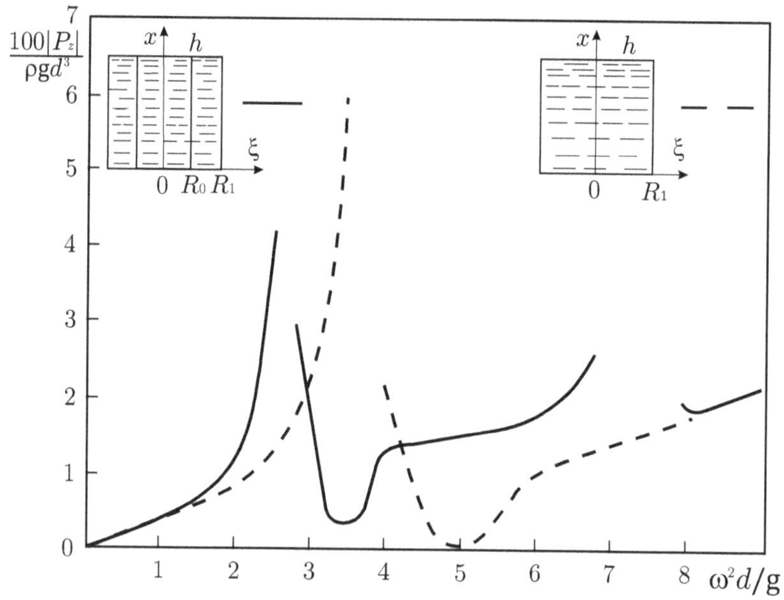


**Figure 8.14**

cylindrical surfaces of radii $R_1 = 1$ and $R_0 = 0.5$ (solid curve) and a cylindrical container of radius $R_1 = 0.5$ (dashed curves). The maximum value of the amplitude of the force $P_z$ for the cavity is a factor 1.5 smaller than the corresponding value for the cylindrical container of radius $R_1 = 1$ (Fig. 8.13). This is caused by a decrease in the resonance frequency (approximately by a factor of 1.15) and the liquid mass (by a factor of 0.25).

We now consider a container formed by two coaxial right circular cylinders of radii $R_1 = 1$ and $R_0 = 0.5$ whose both cavities are filled with liquid. Sloshing in both containers differ in phase. Therefore, the amplitude of the total force $P_z$ is at most the sum of the amplitudes of forces $P_z$ caused by the liquid motion in each container. To take into account the influence of the difference in phases of sloshing in the interior cylinder and in the annular cavity, it is necessary to solve the corresponding Cauchy problem for system (1.1.1). However, in view of the fact that the maximum amplitude of the force $P_z$ for the annular cavity with an unperturbed free surface is observed for $\omega^2 d/\mathrm{g} \approx 2.5$, the dashed curve in Fig. 8.12 makes it possible to conclude that the influence of the difference in the phases of sloshing in the internal and external cavities on the maximum value of the amplitude of the total force $P_z$ is not important. Therefore, the value of the projection of the amplitude of the total hydrodynamic force onto the $Oz$-axis can be replaced by the total amplitude of forces $P_z$ caused by the hydrodynamic loads on each cavity separately.

The response curves of the force $P_z$ for a cylindrical container of radius $R_1 = 1$ (dashed curve) and the response curves of the total force $P_z$ for the container separated by a coaxial partition of radius $R_0 = 0.5$ (solid line) are shown in Fig. 8.14. The plots in the figure show that the separation of the cylindrical container into two compartments leads to a considerable decrease (approximately by 30%) in the maximum value of the projection of the amplitude of the hydrodynamic force $P_z$ onto the $Oz$-axis.

## 8.10 Transient sloshing

The experimental results presented, e.g., in [6, 147, 149, 196] show that, in the case of sloshing in a neighborhood of the main resonance, a kind of beating is sometimes observed. Unlike the problem of finding steady-state regimes, studying the beating is more complicated because, in this case, the amplitude of oscillations is a function of time. In the present section, we show that the nonlinear equations (8.1.3) make it possible to describe the mode of beating observed in experiments. In what follows, we describe studies performed by A. M. Pil'kevich by using the Krylov–Bogolyubov method, the method of slowly varying amplitudes, and the Runge–Kutta method.

Consider a planar sloshing regime, i.e., the case where the generalized coordinates $p_1(t)$ and $r_2(t)$ are equal to zero and the generalized coordinates $r_1(t)$, $p_0(t)$, and $p_2(t)$ satisfy the system of nonlinear equations

$$\begin{aligned}&\ddot{r}_1 + \sigma_1^2 r_1 + d_1(r_1^2\ddot{r}_1 + r_1\dot{r}_1^2) + d_6 r_1\ddot{p}_0 + d_5(p_0\ddot{r}_1 + \dot{p}_0\dot{r}_1)\\&\quad + d_4 r_1\ddot{p}_2 - d_3(p_2\ddot{r}_1 + \dot{r}_1\dot{p}_2) - \omega^2 P_1\cos\omega t = 0;\end{aligned} \tag{8.10.1}$$

$$\ddot{p}_0 + \sigma_0^2 p_0 + d_{10} r_1 \ddot{r}_1 + d_8 \dot{r}_1^2 = 0; \tag{8.10.2}$$

$$\ddot{p}_2 + \sigma_2^2 p_2 + d_0 r_1 \ddot{r}_1 + d_7 \dot{r}_1^2 = 0. \tag{8.10.3}$$

As in the case of steady-state motions, we write a solution of (8.10.1) in the form

$$r_1(t) = A \cos \varphi, \tag{8.10.4}$$

where $\varphi = \omega t + \theta$. Then the generalized coordinates $p_0(t)$ and $p_2(t)$, which are solutions of (8.10.2) and (8.10.3), have the form

$$p_k(t) = A^2(l_k + h_k \cos 2\varphi), \quad k = 0, 2, \tag{8.10.5}$$

where $l_k$ and $h_k$ are constants defined by the equalities

$$h_0 = \frac{d_{10} + d_8}{2\bar{\sigma}_0^2 - 8}, \quad h_2 = \frac{d_9 + d_7}{2\bar{\sigma}_2^2 - 8}, \quad l_0 = \frac{d_{10} - d_8}{2\bar{\sigma}_0^2},$$

$$l_2 = \frac{d_9 - d_7}{2\bar{\sigma}_2^2}, \quad \bar{\sigma}_k = \frac{\sigma_k}{\omega}, \quad k = 0, 2.$$

We transform (8.10.1) to the quasilinear form

$$\ddot{r}_1 + \sigma_1^2 r = \varepsilon f(t, r_1, \dot{r}_1, \ldots, \ddot{p}_2), \tag{8.10.6}$$

where

$$f = \omega^2 \bar{P}_1 \cos \omega t - \bar{d}_1(r_1^2 \ddot{r}_1 + r_1 \dot{r}_1^2) - \bar{d}_6 r_1 \ddot{p}_0$$
$$- \bar{d}_5(p_0 \ddot{r}_1 + \dot{p}_0 \dot{r}_1) - \bar{d}_4 r_1 \ddot{p}_2 + \bar{d}_3(p_2 \ddot{r}_1 + \dot{r}_1 \dot{p}_2),$$

and the quantities $\bar{P}_1$, $\bar{d}_1$, ..., $\bar{d}_6$ are defined by the relations

$$P_1 = \varepsilon \bar{P}_1, \quad d_1 = \varepsilon \bar{d}_1, \ \ldots, \ d_6 = \varepsilon \bar{d}_6.$$

We seek a solution of (8.10.6) by applying the Krylov–Bogolyubov method [20]. To this end, we represent the original equation (8.10.6) in the form

$$\ddot{r}_1 + \omega^2 r_1 = \varepsilon \{ f - r_1 \triangle \}, \tag{8.10.7}$$

where $\varepsilon\triangle$ is the squared difference between the eigenfrequency and the external frequency. In the first approximation, we seek a solution of (8.10.7) in the form (8.10.4) where $A$ and $\theta$ are supposed to be certain functions of time defined as a solution of differential equations of the type

$$\frac{dA}{dt} = \varepsilon X(A, \theta), \quad \frac{d\varphi}{dt} = \omega + \varepsilon Y(A, \theta). \tag{8.10.8}$$

We substitute (8.10.4) and (8.10.5) into (8.10.7) and use system (8.10.8). Taking into account quantities of the order of $\varepsilon$ and equating the coefficients of

$\cos(\omega t+\theta)$ and $\sin(\omega t+\theta)$ in (8.10.7), we obtain the following equations with respect to $A$ and $\theta$:

$$\frac{dA}{dt}=\alpha\sin\theta,\quad \frac{d\theta}{dt}=\frac{\alpha}{A}\cos\theta+\beta+\gamma A, \tag{8.10.9}$$

where

$$\alpha=-\frac{\omega}{2}P_1,\quad \beta=\frac{\omega}{2}(\bar{\sigma}_1^2-1),$$

$$\gamma=\omega\left(\frac{h_0-2l_0}{4}d_5-\frac{h_2-2l_2}{4}d_3-\frac{d_1}{4}-h_0d_6-h_2d_4\right).$$

To obtain stationary modes, we set the right-hand sides of the equations of system (8.10.9) equal to zero because, for steady-state oscillations, the amplitude $A$ and the phase difference $\theta$ are constants. Thus, we obtain

$$\sin\theta=0,\quad \frac{\alpha}{A}\cos\theta+\beta+\gamma A^2=0$$

or

$$\left[\left(\frac{h_0}{2}-l_0\right)d_5-\left(\frac{h_2}{2}-l_2\right)d_3-\frac{d_1}{2}-2h_0d_6-2h_2d_4\right]A^3+(\sigma_1^2-1)A=P_1.$$

The last expression coincides with the response curves (8.10.11) of the planar regime.

Apparently, system (8.10.9) of nonlinear ordinary differential equations of the first order cannot be integrated analytically. In case of small phase angles $\theta$ ($\sin\theta\approx\theta$ and $\cos\theta\approx 1$), the system takes the form

$$\frac{dA}{dt}=\alpha\theta,\quad \frac{d\theta}{dt}=\frac{\alpha}{A}+\beta+\gamma A^2, \tag{8.10.10}$$

and admits the integration by quadrature

$$t=\pm\int\frac{dA}{\sqrt{2\alpha\left(\beta A+\alpha\ln|A|+\frac{\gamma}{3}A^3\right)+C_1}}+C_2,$$

where $C_1$ and $C_2$ are constants of integration. In this case, the behavior of the system is described by phase trajectories defined by the equation

$$A=\pm\sqrt{2\alpha\left(\beta A+\alpha\ln|A|+\frac{\gamma}{3}A^3\right)+C_1}. \tag{8.10.11}$$

The graph of the function $F(A)=2\alpha\left(\beta A+\alpha\ln|A|+\frac{1}{3}\gamma A^3\right)$ and phase trajectories computed for two different values of the constant $C_1$ — 0.005541 (1) and 0.005637 (2) — are depicted in Fig. 8.15; the point $S$ corresponds to the

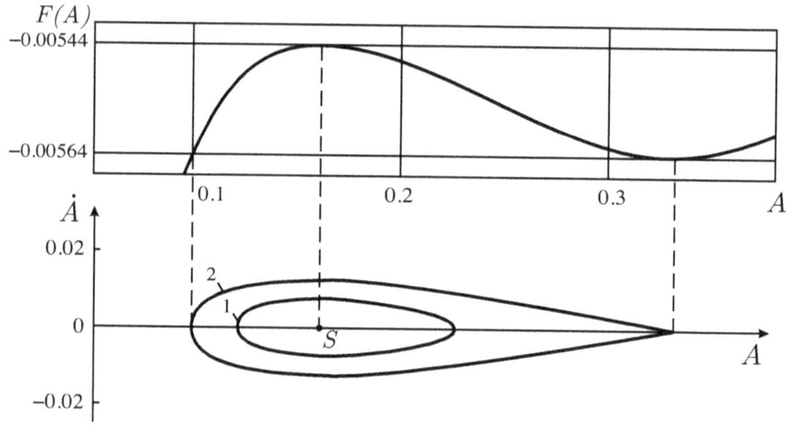


**Figure 8.15**

amplitude $A = 0.165$ of the steady-state motion. (The case of a cylindrical container of radius $R_1 = 1$ and depth of filling $h = 2$ subjected to harmonic oscillations with $H = 0.01$ and relative frequency $\omega/\sigma_1 = 0.95$ is considered.) We see that beating-type oscillations are possible in a neighborhood of stationary oscillations.

The system of nonlinear differential equations (8.10.9) can be studied with the use of characteristics, i.e., integral curves of the equation

$$\frac{d\theta}{dA} = \frac{\frac{\alpha}{A}\cos\theta + \beta + \gamma A^2}{\alpha \sin\theta} \tag{8.10.12}$$

that defines a functional dependence of quantities $A$ and $\theta$. To proceed the study, we represent (8.10.12) in the differential form

$$-\alpha A \sin\theta d\theta + (\alpha\cos\theta + \beta A + \gamma A^3)dA = 0. \tag{8.10.13}$$

We see that (8.10.13) is a total differential equation and its complete integral has the form

$$U(A,\theta) = \tfrac{1}{4}\gamma A^4 + \tfrac{1}{2}\beta A^2 + \alpha A\cos\theta = C, \tag{8.10.14}$$

where $C$ is a constant of integration. Knowing the complete integral (8.10.14) and using the first equation of system (8.10.9), we obtain the following equation for phase trajectories:

$$A = \alpha \sin\left\{\arccos\left[\frac{C - \frac{\gamma}{4}A^4 - \frac{\beta}{2}A^2}{\alpha A}\right]\right\}. \tag{8.10.15}$$

Phase trajectories computed by using this relation under the same conditions as in Fig. 8.15 are shown in Fig. 8.16. Here, $C = 0.0022\,(1)$, $0.0020\,(2)$,

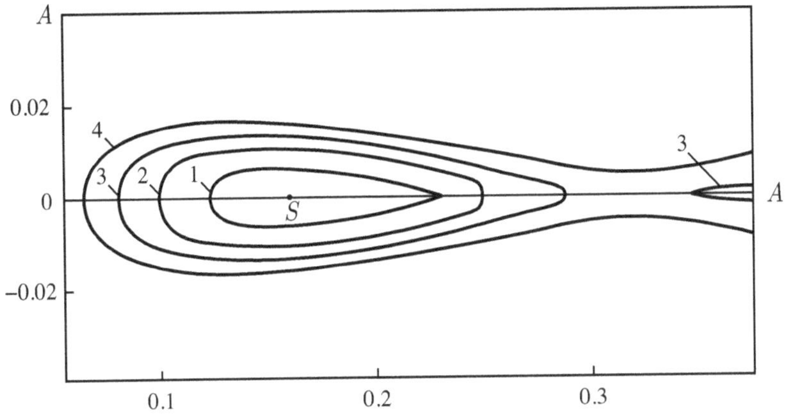


**Figure 8.16**

0.0018 (3), 0.0016 (4), the point $S$ denotes steady-state regimes with amplitude $A = 0.165$. Digit 3 in the figure denotes a separatrix, i.e., a closed curve that corresponds to maximum values of beating.

To clarify the behavior of solutions of system (8.10.1), we perform a numerical experiment. To this end, we transform system (8.10.1) to its normal form and integrate it by using the Runge–Kutta method. As initial conditions, we choose values obtained by relations (8.10.4) and (8.10.5) for $A = 0.1$.

First, we consider a cylindrical container of radius $R_1 = 1$ and depth of filling $h = 2$ subjected to harmonic oscillations with $H = 0.01$ and forcing frequency $\omega/\sigma_1 = 1.07$. An analysis shows that $r_1(t)$ varies by the sinusoidal law with period of oscillations $T = 1.6$ and, furthermore, oscillations reveal the presence of beating. The time dependence of the generalized coordinate over a period of beating is shown in Fig. 8.17. We can see that the envelope of beating satisfies the harmonic law rather well.

We now consider the case of a container formed by coaxial cylindrical surfaces of radii $R_0 = 0.5$ and $R_1 = 1$ and subjected to harmonic oscillations with relative frequency $\omega/\sigma_1 = 0.96$ (here, as in the previous case, $H = 0.01$ and $h = 2$). The results of computation enable us to conclude that $r_1(t)$ varies by the sinusoidal law (with period $T = 1.8$) and 51 oscillations occur for a period of beating with variations in the amplitude from $A_{\text{min}} = 0.075$ to $A_{\text{max}} = 0.26$. In this case, the envelope of beating is described by the harmonic law in a good approximation.

The beating mode can also be described analytically. First, consider the approximate system (8.10.10). Suppose that a stationary beating obeys the law [25]

$$A = \bar{A}\cos\nu t, \tag{8.10.16}$$

where $\bar{A}$ and $\nu$ are certain constants to be found. Substituting (8.10.16)

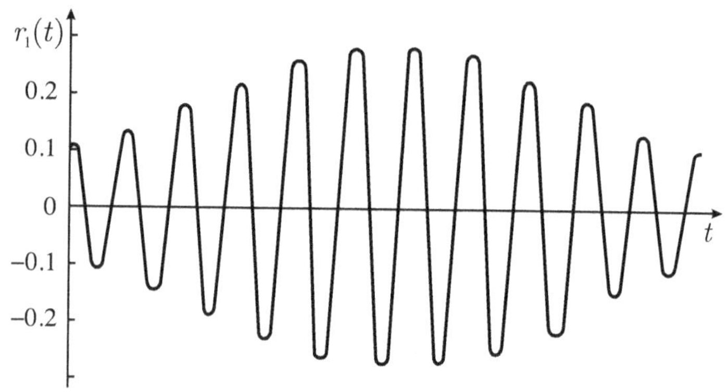


**Figure 8.17**

into (8.10.10), we obtain the following equation that relates the beat amplitude $\bar{A}$ and the beat frequency $\nu$:

$$\gamma\bar{A}^3\cos^3\nu t+\frac{\nu^2}{\alpha}\cos^2\nu t+\beta\bar{A}\cos\nu t+\alpha=0. \tag{8.10.17}$$

As a result, we obtain one equation with two unknowns $\bar{A}$ and $\nu$. If the quantity $\bar{A}$ (or $\nu$) can be defined, then we can obtain the value of $\nu$ (or $\bar{A}$) by using (8.10.17).

To perform an analysis of beating, one can use also the method of slowly varying amplitudes [25]. According to this method, the functions $r_1(t)$, $p_0(t)$, and $p_2(t)$ are presented in the form

$$\begin{aligned} r_1(t)&=A\cos\omega t+B\sin\omega t,\\ p_0(t)&=c_0+c_1\cos 2\omega t+c_2\sin 2\omega t,\\ p_2(t)&=s_0+s_1\cos 2\omega t+s_2\sin 2\omega t. \end{aligned} \tag{8.10.18}$$

Here, $A$ and $B$ are regarded as slowly varying functions of time, i.e., increments of these functions over a period are small relative to their mean values for a period. Thus, the following inequalities are true:

$$\left|\frac{\dot{A}}{A}\right|\ll 1,\quad \left|\frac{\dot{B}}{B}\right|\ll 1,\quad \left|\frac{\ddot{A}}{A}\right|\ll 1,\quad \left|\frac{\ddot{B}}{B}\right|\ll 1. \tag{8.10.19}$$

The quantities $c_0, \ldots, s_2$ in relations (8.10.18) are defined as follows:

$$\begin{aligned} c_0&=l_0(A^2+B^2),\quad c_1=h_0(A^2-B^2),\quad c_2=2h_0AB,\\ s_0&=l_2(A^2+B^2),\quad s_1=h_2(A^2-B^2),\quad s_2=2h_2AB. \end{aligned}$$

Substituting (8.10.18) into the differential equation (8.10.1) in view of inequalities (8.10.19), we obtain

$$\frac{1}{\omega^2}(\ddot{A}\cos\omega t+\ddot{B}\sin\omega t)+\frac{2}{\omega}(-\dot{A}\sin\omega t+\dot{B}\cos\omega t)+(\bar{\sigma}_1^2-1)(A\cos\omega t+B\sin\omega t)$$

$$+(A^2+B^2)\Bigg[-\frac{d_1^2}{2}(A\cos\omega t+B\sin\omega t)-2d_6h_0(A\cos\omega t+B\sin\omega t)$$

$$+d_5\left(\frac{h_0}{2}-l_0\right)(A\cos\omega t+B\sin\omega t)-d_4h_2(A\cos\omega t+B\sin\omega t)$$

$$-d_3\left(\frac{h_2}{2}-l_2\right)(A\cos\omega t+B\sin\omega t)\Bigg]-P_1\cos\omega t+\ \ldots\ =0, \qquad (8.10.20)$$

where dots denote the terms containing higher-order harmonics.

Equating the coefficients at $\cos\omega t$ and $\sin\omega t$ to zero and taking relations (8.10.19) into account, we obtain the following equations with respect to $A$ and $B$:

$$\frac{dA}{dt}=B[\beta+\gamma(A^2+B^2)],\quad \frac{dB}{dt}=\frac{\omega}{2}P_1-A[\beta+\gamma(A^2+B^2)]. \qquad (8.10.21)$$

Thus, we reduced the system of three differential equations of the second order to a system of equations of the first order.

Note that the steady-state mode (8.4.10) can be obtained from system (8.10.21) by equating the right-hand sides of equations to zero. As usual, we set

$$A=A_0\cos\nu t,\quad B=B_0\sin\nu t, \qquad (8.10.22)$$

where $A_0$ and $B_0$ are certain constants, and then substitute (8.10.22) into (8.10.21). We take into account the fact that the beat frequency should be small as compared with the excitation frequency ($\nu\ll\omega$) because, otherwise, $A$ and $B$ are not slowly varying functions of time. Then

$$-\nu A_0\sin\nu t=B_0\left[\beta+\frac{\gamma}{4}(A_0^2+3B_0^2)\right]\sin\nu t+\ \ldots,$$

$$\nu B_0\cos\nu t=\frac{\omega}{t}P_1-A_0\left[\beta+\frac{\gamma}{4}(3A_0^2+B_0^2)\right]\cos\nu t+\ \ldots, \qquad (8.10.23)$$

where dots denote terms containing higher-order harmonics.

Collecting the coefficients at $\sin\nu t$ and $\cos\nu t$, we obtain

$$-\nu A_0=B_0\left[\beta+\frac{\gamma}{4}(A_0^2+3B_0^2)\right],$$

$$\nu B_0=-A_0\left[\beta+\frac{\gamma}{4}(3A_0^2+B_0^2)\right]. \qquad (8.10.24)$$

Equations (8.10.24) contain three unknowns: the beat amplitudes $A_0$ and $B_0$ and the beat frequency $\nu$. It is clear that system (8.10.24) is undetermined. Solving this system, we can only express two unknowns in terms of the third one, which remains to be undetermined. To solve the system, we have to use some additional physical reasoning [25].

By using (8.10.24), we obtain the square of the beat amplitude

$$A_0^2 + B_0^2 = -\frac{4}{3}\frac{\beta}{\gamma} \tag{8.10.25}$$

and express the beat frequency in terms of the product of beat amplitudes as follows:

$$\nu = \frac{\gamma}{2} A_0 B_0. \tag{8.10.26}$$

Unlike relation (8.10.17), relations (8.10.24) and, hence, (8.10.25) and (8.10.26) were derived directly from system (8.10.1)–(8.10.3). The beat amplitude determined from (8.10.24) is independent of $H$. In this sense, relation (8.10.17) is preferable because it takes into account a dependence of the beat amplitude on the excitation amplitude.

Thus, on the basis of three approaches (the Krylov–Bogolyubov method, numerical integration, and the method of slowly varying amplitudes), which are presented above and used to investigate a beating mode of the considered mechanical system, we can conclude that the system of equations (8.10.1)–(8.10.3) provides a possibility to describe a beat mode qualitatively. For a quantitative estimation of the obtained results, a comparison with experimental data and subsequent theoretical study are required.

## 8.11 Forced liquid sloshing in a tapered conical tank

We consider resonant steady-state behavior in the case of forced liquid sloshing in a tank subjected to harmonic translational excitation. For the sake of brevity, we assume that excitation is directed along the $Oz$-axis in the notation of Fig. 5.7 and this implies $\eta_i = 0$, $i \neq 3$, and $\eta_3 = H\cos(\sigma t)$. Our goal is to find a time-periodic solution of (5.4.5)–(5.4.11) corresponding to steady-state wave regimes. The low-order generalized coordinates $r_1(t)$ and $p_1(t)$ are presented by the Fourier series

$$r_1(t) = \sum_{k=1}^{\infty} \left(A_{2k-1}\cos(k\sigma t) + A_{2k}\sin(k\sigma t)\right),$$

$$p_1(t) = \sum_{k=1}^{\infty} \left(B_{2k-1}\cos(k\sigma t) + B_{2k}\sin(k\sigma t)\right),$$

where, according to Moiseev's asymptotics, the contribution of the leading asymptotic is associated with the first harmonics, i.e.,

$$r_1(t) = A_1 \cos \sigma t + A_2 \sin \sigma t + o\left(\varepsilon^{1/3}\right);$$
$$p_1(t) = B_1 \cos \sigma t + B_2 \sin \sigma t + o\left(\varepsilon^{1/3}\right) \tag{8.11.1}$$

and $A_1 \sim A_2 \sim B_1 \sim B_2 = O(\varepsilon^{1/3})$, $\varepsilon = H$.

Substituting (8.11.1) into the modal equations (5.4.5), (5.4.8) and (5.4.9), we can see that the generalized coordinates $p_0(t)$, $r_2(t)$, and $p_2(t)$ are functions of dominant amplitude parameters $A_1$, $A_2$, $B_1$ and $B_2$, i.e.

$$p_0(t) = -\left(A_1^2 + A_2^2 + B_1^2 + B_2^2\right) o_0^{(0)} - \frac{1}{2}\left(A_1^2 - A_2^2 + B_1^2 - B_2^2\right) o_0^{(2)} \cos 2\sigma t$$
$$- \left(A_1 A_2 + B_1 B_2\right) o_0^{(2)} \sin 2\sigma t + o\left(\varepsilon^{2/3}\right), \tag{8.11.2}$$

$$r_2(t) = -2\left(A_1 B_1 + A_2 B_2\right) o_2^{(0)} - \left(A_1 B_1 - A_2 B_2\right) o_2^{(2)} \cos 2\sigma t$$
$$- \left(A_1 B_2 + A_2 B_1\right) o_2^{(2)} \sin 2\sigma t + o\left(\varepsilon^{2/3}\right), \tag{8.11.3}$$

$$p_2(t) = \left(A_1^2 + A_2^2 - B_1^2 - B_2^2\right) o_2^{(0)} + \frac{1}{2}\left(A_1^2 - A_2^2 - B_1^2 + B_2^2\right) o_2^{(2)} \cos 2\sigma t$$
$$+ \left(A_1 A_2 - B_1 B_2\right) o_2^{(2)} \sin 2\sigma t + o\left(\varepsilon^{2/3}\right). \tag{8.11.4}$$

Similarly, one can find

$$r_3(t) = (A_1(A_1^2 + A_2^2 - 3B_1^2 - B_2^2) - 2A_2 B_1 B_2) o_3^{(1)} \cos \sigma t$$
$$+(A_2(A_1^2 + A_2^2 - B_1^2 - 3B_2^2) - 2A_1 B_1 B_2) o_3^{(1)} \sin \sigma t$$
$$+(A_1(A_1^2 - 3A_2^2 - 3B_1^2 + 3B_2^2) + 6A_2 B_1 B_2) o_3^{(3)} \cos 3\sigma t$$
$$+(A_2(3A_1^2 - A_2^2 - 3B_1^2 + 3B_2^2) - 6A_1 B_1 B_2) o_3^{(3)} \sin 3\sigma t + o(\varepsilon), \tag{8.11.5}$$

$$p_3(t) = (B_1(3A_1^2 + A_2^2 - B_1^2 - B_2^2) + 2A_1 A_2 B_2) o_3^{(1)} \cos \sigma t$$
$$+(B_2(A_1^2 + 3A_2^2 - B_1^2 - B_2^2) + 2A_1 A_2 B_1) o_3^{(1)} \sin \sigma t$$
$$+(B_1(3A_1^2 - 3A_2^2 - B_1^2 + 3B_2^2) - 6A_1 A_2 B_2) o_3^{(3)} \cos 3\sigma t$$
$$+(B_2(3A_1^2 - 3A_2^2 - 3B_1^2 + B_2^2) + 6A_1 A_2 B_1) o_3^{(3)} \sin 3\sigma t + o(\varepsilon), \tag{8.11.6}$$

where coefficients $o_m^k$ are defined in [68].

Substituting (8.11.1), (8.11.2)–(8.11.4) into the modal equations (5.4.7) and (5.4.6) and gathering the low-order terms at the first harmonics, we obtain the system of algebraic equations

$$\begin{cases}\left(m_1\left(A_1^2+A_2^2+B_1^2\right)+m_2B_2^2\right)A_1+m_3A_2B_1B_2+\left(\bar\sigma_1^2-1\right)A_1=H\Lambda,\\ \left(m_1\left(A_1^2+A_2^2+B_2^2\right)+m_2B_1^2\right)A_2+m_3A_1B_1B_2+\left(\bar\sigma_1^2-1\right)A_2=0,\\ \left(m_1\left(A_1^2+B_1^2+B_2^2\right)+m_2A_2^2\right)B_1+m_3A_1A_2B_2+\left(\bar\sigma_1^2-1\right)B_1=0,\\ \left(m_1\left(A_2^2+B_1^2+B_2^2\right)+m_2A_1^2\right)B_2+m_3A_1A_2B_1+\left(\bar\sigma_1^2-1\right)B_2=0\end{cases} \tag{8.11.7}$$

with respect to the dominant amplitude parameters, where coefficients $m_i$ depend on hydrodynamic coefficients of the modal equations.

The algebraic system (8.11.7) is similar to the corresponding system suggested for a non-truncated tank where we showed that its solvability condition is $A_2 = B_1 = 0$ and, therefore, there are only two nonzero amplitude parameters that can be found from the system

$$\begin{aligned}&m_1A_1^3+m_2A_1B_2^2+\left(\bar\sigma_1^2-1\right)A_1=H\Lambda;\\ &m_1B_2^3+m_2A_1^2B_2+\left(\bar\sigma_1^2-1\right)B_2=0,\end{aligned} \tag{8.11.8}$$

whose solution, obviously, depends on $m_i$ and, in turn, on the nondimensional ratio of the bottom and free surface radii $\mathtt{r}_1$, $\bar\sigma_1(\mathtt{r}_1)$, and $\theta_0$ ($m_i = m_i(\bar\sigma_1,\mathtt{r}_1,\theta_0)$).

As shown for non-truncated tank, one can distinguish two types of solutions of (8.11.8) and the corresponding steady-state wave regimes. The solution of the first type, $A_1 \neq 0$, $B_2 = 0$, implies the so-called planar waves. The solution of the second type, $A_1 \neq 0$, $B_2 \neq 0$, leads to the so-called swirling. The planar waves are described by the asymptotic solution

$$\begin{aligned}p_1=r_2=p_3=0;\quad &r_1=A_1\cos\sigma t+o(\varepsilon);\\ p_2=A_1^2o_2^{(0)}&+\frac12A_1^2o_2^{(2)}\cos2\sigma t+o(\varepsilon^2),\\ r_3=A_1^3o_3^{(1)}&\cos\sigma t+A_1^3o_3^{(3)}\cos3\sigma t+o(\varepsilon^3),\\ p_0=-A_1^2o_0^{(0)}&-\frac12A_1^2o_0^{(2)}\cos2\sigma t+o(\varepsilon^2),\end{aligned} \tag{8.11.9}$$

where the amplitude parameter $A_1$ is a root of the cubic equation

$$m_1A_1^3+(\bar\sigma_1^2-1)A_1-H\Lambda=0. \tag{8.11.10}$$

Swirling implies

$$\begin{aligned}r_1(t)=A_1\cos\sigma t+o(\varepsilon);\quad &p_1(t)=B_2\sin\sigma t+o(\varepsilon);\\ r_2(t)=-A_1B_2o_2^{(2)}&\sin2\sigma t+o(\varepsilon^2),\end{aligned}$$

$$\begin{aligned} p_0(t) &= -\left(A_1^2 + B_2^2\right) o_0^{(0)} - \frac{1}{2}\left(A_1^2 - B_2^2\right) o_0^{(2)} \cos 2\sigma t + o(\varepsilon^2), \\ p_2(t) &= \left(A_1^2 - B_2^2\right) o_2^{(0)} + \frac{1}{2}\left(A_1^2 + B_2^2\right) o_2^{(2)} \cos 2\sigma t + o(\varepsilon^2), \\ r_3(t) &= A_1(A_1^2 - B_2^2) o_3^{(1)} \cos \sigma t + A_1(A_1^2 + 3B_2^2) o_3^{(3)} \cos 3\sigma t + o(\varepsilon^3), \\ p_3(t) &= B_2(A_1^2 - B_2^2) o_3^{(1)} \sin \sigma t + B_2(3A_1^2 + B_2^2) o_3^{(3)} \sin 3\sigma t + o(\varepsilon^3), \end{aligned} \tag{8.11.11}$$

where the amplitude parameters $A_1$ and $B_2$ are computed from the equations

$$m_6 A_1^3 + m_5(\bar{\sigma}_1^2 - 1)A_1 - H\Lambda = 0; \quad B_2^2 = (m_1 A_1^2 - (\bar{\sigma}_1^2 - 1))/m_2 > 0, \tag{8.11.12}$$

$m_5 = m_3/m_1$ and $m_6 = m_4 m_5$. The inequality in (8.11.12) is a solvability condition.

When constructing the time-periodic solution, we assume that the forcing frequency $\sigma$ is close to the lowest natural frequency $\sigma_{11}$ of sloshing, i.e.

$$\sigma \approx \sigma_{11}. \tag{8.11.13}$$

The constructed solution is valid if and only if the coefficients of the polynomial terms in the amplitude parameters are of the order $O(1)$. However, these coefficients can be large when $2\sigma$ is close to one of the natural sloshing frequencies $\sigma_{2i}$ or $\sigma_{0i}$, $i \geq 1$, or, alternatively, when $3\sigma$ tends to one of the natural sloshing frequencies $\sigma_{3i}$, $i \geq 1$, or $\sigma_{1i}$, $i \geq 2$. These possibilities are related to the so-called *secondary resonances*. A necessary condition for secondary resonance is provided by the relations

$$2\sigma \approx \sigma_{0n}, \quad 2\sigma \approx \sigma_{2n}, \quad 3\sigma \approx \sigma_{3n}, \quad 3\sigma \approx \sigma_{1(n+1)}, \quad n \geq 1, \tag{8.11.14}$$

in addition to (8.11.13). The secondary resonance is unavoidable when equalities in (8.11.14) and (8.11.13) are strong.

To analyze the secondary resonances with strong equalities in (8.11.14) and (8.11.13), we plot the graphs of $i_{m,n}(\theta_0, \mathtt{r}_1)$ as functions of the nondimensional parameter $\mathtt{r}_1$ ($\mathtt{r}_1$ is the ratio of the bottom and free surface radii) for a fixed value of the semi-apex angle

$$\begin{gathered} i_{0,n}(\theta_0, \mathtt{r}_1) = \frac{\sigma_{0n}}{2\sigma_{11}} = \frac{1}{2}\sqrt{\frac{\varkappa_{0n}}{\varkappa_{11}}}; \quad i_{2,n}(\theta_0, \mathtt{r}_1) = \frac{\sigma_{2n}}{2\sigma_{11}} = \frac{1}{2}\sqrt{\frac{\varkappa_{2n}}{\varkappa_{11}}}, \\ i_{3,n}(\theta_0, \mathtt{r}_1) = \frac{\sigma_{3n}}{3\sigma_{11}} = \frac{1}{3}\sqrt{\frac{\varkappa_{3n}}{\varkappa_{11}}}; \quad i_{1(n+1)}(\theta_0, \mathtt{r}_1) = \frac{\sigma_{1(n+1)}}{3\sigma_{11}} = \frac{1}{3}\sqrt{\frac{\varkappa_{1(n+1)}}{\varkappa_{11}}}, \\ n \geq 1; \end{gathered}$$

the plots are shown in Figs. 8.18–8.20. The functions $i_{m,n} = i_{m,n}(\theta_0, \mathtt{r}_1)$ are independent of the forcing frequency $\sigma$, and it is easily seen that the condition

$$i_{m,n} = 1 \tag{8.11.15}$$

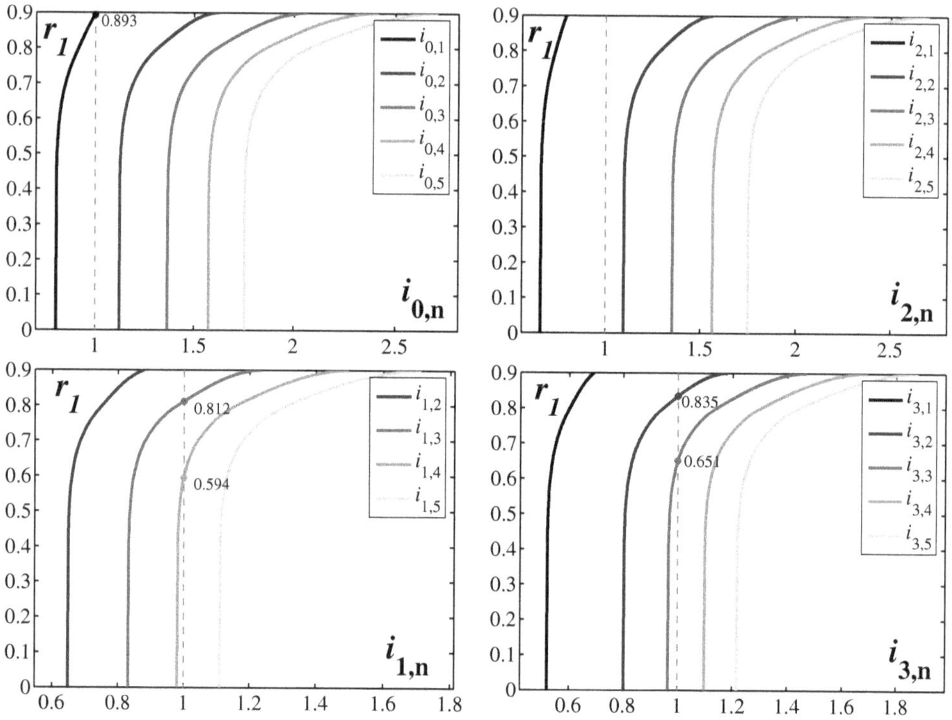


**Figure 8.18.** The graphs of $\mathtt{r}_1 = \mathtt{r}_1(i_{m,n})$ ($\mathtt{r}_1$ is the ratio of the bottom and free surface radii) for the semi-apex angle $\theta_0 = 30°$. The secondary resonance is expected at $\mathtt{r}_1 = 0.8116$, $\mathtt{r}_1 = 0.5939$, $\mathtt{r}_1 = 0.8926$, $\mathtt{r}_1 = 0.835$, and $\mathtt{r}_1 = 0.651$.

for a fixed pair of $m$ and $n$ is equivalent to the condition that the corresponding $m, n$-equation in (8.11.14) and (8.11.13) is strong. The case $\mathtt{r}_1 = 0$ corresponds to a V-shaped conical tank while the limit $\mathtt{r}_1 \to 1$ implies the shallow water condition.

The calculations were performed for the three semi-apex angles $\theta_0 = 30°$, $\theta_0 = 45°$, and $\theta_0 = 60°$. The strong equality $i_{0,1} = 1$ is observed for $\mathtt{r}_1 = 0.8926$ and this means that the first axisymmetric mode is the subject of the secondary resonance for larger $\mathtt{r}_1$, and then the double harmonics $2\sigma$ can be amplified. As for the triple harmonics $3\sigma$, they can occur for the modes (1,3), (1,4), (3,2) and (3,3). In particular, for $\mathtt{r}_1 = 0.651$, the modes (3,3) are the subject to the secondary resonance while the modes (3,2) are resonantly excited at $\mathtt{r}_1 = 0.835$. Finally, the modes (1,3) are exposed to the secondary resonance at $\mathtt{r}_1 = 0.8116$ and the modes (1,4) at $\mathtt{r}_1 = 0.5939$. The strong secondary resonances for the semi-apex angle $\theta_0 = 30°$ are not expected for the nondimensional ratio values $\mathtt{r}_1 \lesssim 0.5$.

As follows from Fig. 8.19, the secondary resonances exist also for $\theta_0 = 45°$ at $\mathtt{r}_1 = 0.6386$ [modes (1,3)], $\mathtt{r}_1 = 0.7972$ [mode (0,1)] and $\mathtt{r}_1 = 0.7$ [modes (3,2)].

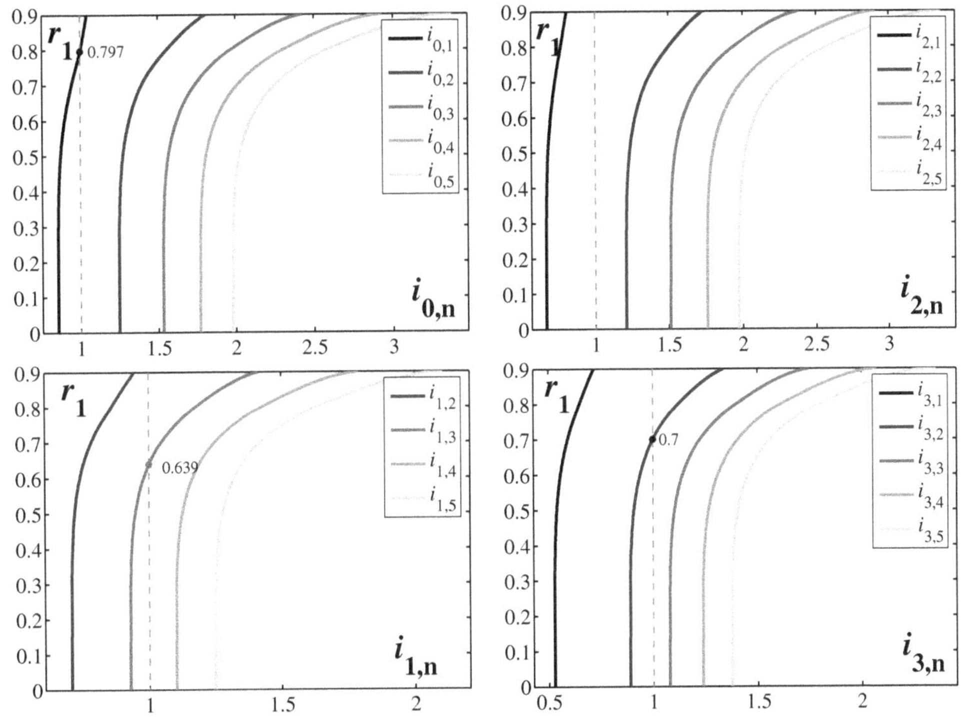


**Figure 8.19.** The graphs of $\mathtt{r}_1 = \mathtt{r}_1(i_{m,n})$ ($\mathtt{r}_1$ is the ratio of the bottom and free surface radii) for the semi-apex angle $\theta_0 = 45°$. The secondary resonance is expected at $\mathtt{r}_1 = 0.6386$, $\mathtt{r}_1 = 0.7972$, and $\mathtt{r}_1 = 0.7$.

This implies that the constructed Moiseev-type modal equations can be applicable for the nondimensional ratios $\mathtt{r}_1 \lesssim 0.6$. The plots in Fig. 8.20 show two critical values of $\mathtt{r}_1$ for $\theta_0 = 60°$. These are $\mathtt{r}_1 = 0.67$ [the secondary resonance for the mode (0,1)] and $\mathtt{r}_1 = 0.3196$ [the secondary resonance for the modes (3,2)]. Moreover, $i_{3,2}$ is close to 1 for $\mathtt{r}_1 \lesssim 0.5$ while $i_{0,1} \approx 1$ for $0.55 \lesssim \mathtt{r}_1$. This means that it may be necessary to revise the derived modal equations in order to account for secondary resonances for this value of the semi-apex angle.

The hydrodynamic instability of time-periodic solutions (8.11.9) and (8.11.11) was studied by using the first Lyapunov method. Figures 8.21 and 8.22 present (in terms of the amplitude parameters $A_1$ and $B_2$) the response curves associated with the steady-state wave motions. With regard to the secondary resonance analysis and the related limitations of the Moiseev-type modal equations, we consider mainly the semi-apex angles $\theta_0 = 30°$ and $\theta_0 = 45°$ and, respectively, the ratios $\mathtt{r}_1 = 0.5$ and 0.4. By analyzing the response curves, it is possible to estimate the effective frequency ranges for planar and/or swirling sloshing. The amplitude parameter $A_1$ measures the longitudinal wave component while $B_2$ corresponds to the cross-wave component. The solid lines mark

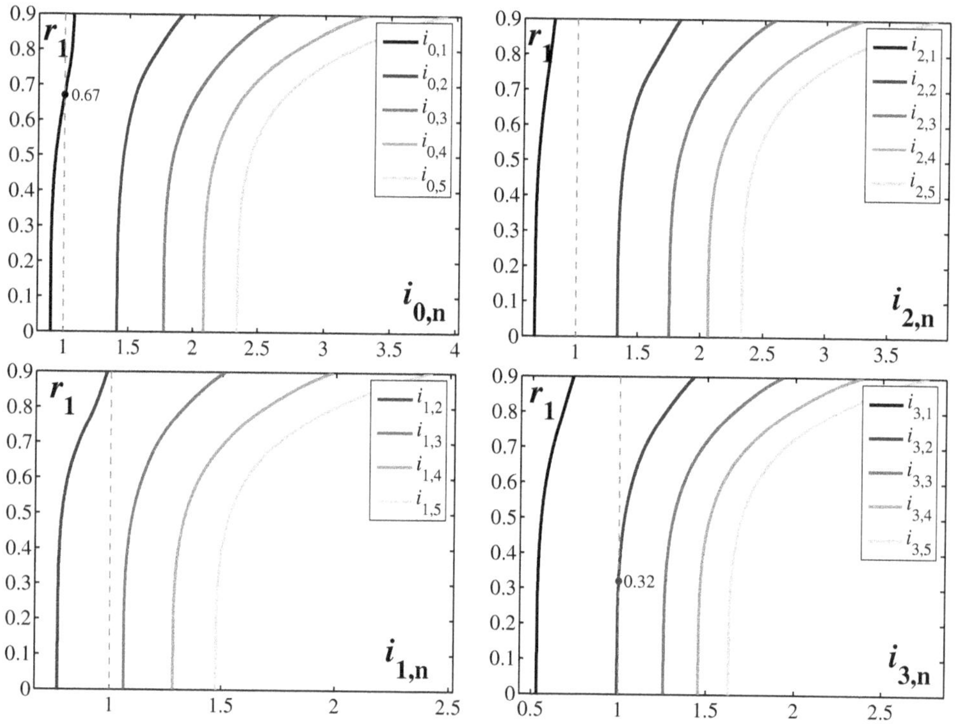


**Figure 8.20.** The graphs of $\mathtt{r}_1 = \mathtt{r}_1(i_{m,n})$ ($\mathtt{r}_1$ is the ratio of of the bottom and free surface radii) for the semi-apex angle $\theta_0 = 60°$. The secondary resonance is expected as $\mathtt{r}_1 = 0.67$ and $\mathtt{r}_1 = 0.3196$.

stable steady-state wave regimes while dashed lines are used to denote their instability.

The figures show that the response curves are qualitatively similar to those known for non-truncated conical tanks and upright circular cylindrical tanks. By considering these figures, we can come to the following conclusions:

*Firstly*, the planar sloshing (branches $K^1K^2$ and $M^1M^2$) is always unstable in a neighborhood of the linear resonance ($\sigma/\sigma_1 = 1$); the instability is expected for the forcing frequencies between the abscissas of $K$ and $M$, where $K$ is the turning point while $M$ is the Poincaré bifurcation point at which the branch $MM^3$ of unstable swirling emerges.

*Secondly*, stable swirling exists at $\sigma/\sigma_1 = 1$. The "swirling" branch, $N^2N^1$, is divided by the Hopf bifurcation point $N$; the subbranch $NN^1$ corresponds to stable steady-state wave motions (the abscissa of $N$ is less than 1) while the subbranch $NN^2$ implies unstable steady-state swirling.

*Thirdly*, the interval between the abscissas of $N$ and $M$ is the frequency range where both planar and swirling steady-state wave regimes are unstable; in this frequency range, irregular (chaotic) waves are expected.

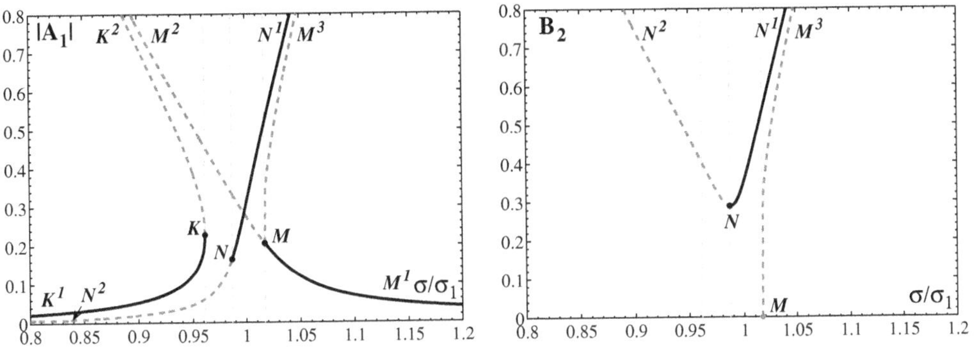


**Figure 8.21.** The response curves of planar and swirling resonant steady-state sloshing for the semi-apex angle $\theta_0 = 30°$, nondimensional ratio $\mathtt{r}_1 = 0.5$, and nondimensional excitation amplitude $H = 0.01$. The amplitude parameters $A_1$ and $B_2$ imply longitudinal and transverse wave components.

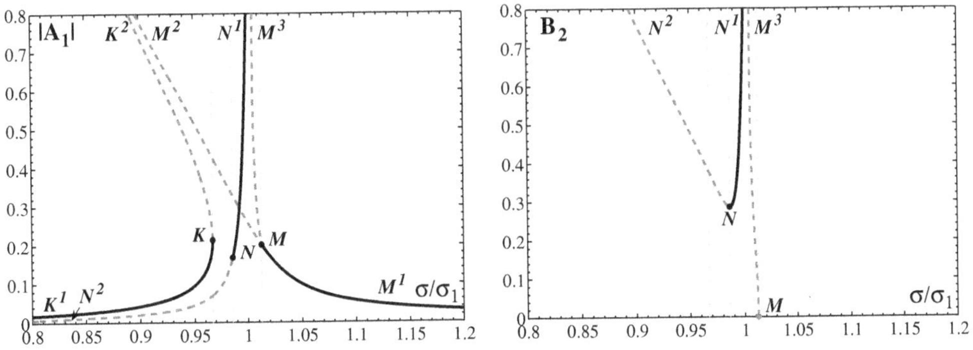


**Figure 8.22.** The same curves as in Fig. 8.21 for $\theta_0 = 45°$, $\mathtt{r}_1 = 0.4$ and $H = 0.01$.

The list of publications on experimental studies of nonlinear resonant sloshing in a truncated conical tank is very short. Being interested in such experimental data to validate our theoretical results, we paid attention to [28] where appropriate experimental results were mentioned in the context of the Tuned Liquid Dampers equipped with a conical tank. Thanks to Prof. Fabio Casciati and Dr. Emiliano Matta (Politecnico di Torino, Italy), we got a more detailed report on these experiments documented in the PhD Thesis [145]. In the Thesis, for an experimental study, a tank with the semi-apex angle $\theta_0 = 30°$ is used and the hydrodynamic force in such a tank subjected to horizontal harmonic excitation is measured. The Thesis contains a set of the hydrodynamic force recordings and, moreover, suggests a kind of classification of the liquid motions based on both measurements and observations. Because the experimental series were conducted on a relatively short time scale, the classification is successful to a certain extent only. In some cases, it is possible to reveal an almost steady-state wave regime (planar or swirling), while a lot of Matta's experimental data indicate strong breaking waves and irregular motions, which may be explained as either continuing transients or hydrodynamic instability. However, these irregular and almost steady-state liquid motions can be found as they follow from our analysis: irregular waves were detected for $\sigma/\sigma_1 < 1$, a few stable model series reveal swirling for $\sigma/\sigma_1 > 1$, and stable planar waves are observed far away from the linear resonance $\sigma/\sigma_1 = 1$. These results support our theory qualitatively. In this case it is not easy to obtain a quantitative validation of the theory because the experimental time series are too short. Additionally, these experiments were performed in the case of relatively small (almost shallow) depths of liquid where the ratio $\mathtt{r}_1 = 0.852$. An analysis of the secondary resonances shows that this ratio is too close to $\mathtt{r}_1 = 0.8926$ [the secondary resonance by the mode (0,1)] and $\mathtt{r}_1 = 0.835$ [the secondary resonance by the mode (3,2)]. Unfortunately, this means that due to these secondary resonances the Moiseev-type modal equations are most likely inapplicable to shallow water sloshing in this case and cannot provide a quantitative agreement of data to confirm the theory. The secondary resonances are implicitly confirmed in observations [145]. Breaking waves and overturning are almost always detected for forcing frequencies in a neighborhood of $\sigma/\sigma_1 = 1$ where our theory predicts irregular waves or swirling. As was discussed in Chaps. 8 and 9 of [56], these phenomena are typical attributes of multiple secondary resonances, especially, for the case of shallow water.

# Bibliography


[1] K. A. Abgayan and I. M. Rapoport, *Rockets dynamics*, Mashinostroenie, Moscow, 1969, in Russian.

[2] H. N. Abramson, Dynamic behavior of liquids in moving containers, *Appied Mechanics Reviews* **16** (1966), 501–506.

[3] ______, *The dynamic behavior of liquids in moving containers, with applications to space vehicle technology*, NASA, Report no. NASA, SP-106, Washington, 1966.

[4] ______, Dynamics of contained liquids: A personal odyssey, *Applied Mechanics Reviews* **56** (2003), R1–R7.

[5] H. N. Abramson, W. Chu and D. Kana, Some studies of nonlinear lateral sloshing in rigid containers, *Journal of Applied Mechanics* **33** (1966), 66–74.

[6] Y. N. Anosov, On nonlinear vibrations of liquid in a cylindrical cavity, *Soviet Applied Mechanics* **2** (1966), 14–17.

[7] M. Arai, Experimental and numerical studies of sloshing in liquid cargo tanks with internal structures, *IHI Engineering Review* **19** (1986), 1–9.

[8] T. Y. Azizov, V. Hardt, N. Kopachevsky and R. Mennicken, On the problem of small motions and normal oscillations of a viscous fluid in a partially filled container, *Mathematische Nachrichten* **248–249** (2003), 3–39.

[9] A. M. Basin, I. O. Velednicky and A. G. Lahovicky, *Hydrodynamics of Ships in Shallow Water*, Sudostroenie, Leningrad, 1976, in Russian.

[10] H. Bateman, *Partial differential equations of mathematical physics*, Cambridge University Press, 1932.

[11] H. F. Bauer, Nonlinear mechanical model for the description of propellant sloshing, *AIAA Journal* **4** (1966), 1662–1668.

[12] ______, *Nonlinear propellant sloshing in a rectangular container of infinite length*, Developments in theoretical and applied mechanics (W. Shaw, ed.), 3, Pregamon Press, New York, 1967, pp. 725–759.

[13] ______, Flüssigkeitsschwingungen in Kegelbehälterformen, *Acta Mechanica* **43** (1982), 185–200.

[14] H. F. Bauer and W. Eidel, Non-linear liquid motion in conical container, *Acta Mechanica* **73** (1988), 11–31.

[15] T. B. Benjamin and G. Ursell, The stability of a plane free surface of a liquid in vertical periodic motion, *Proceedings of the Royal Society, London A* **225** (1954), 505–515.

[16] V. L. Berdichevsky, *Variational principles of continuum mechanics. I. Fundamentals*, Interaction of mechanics and mathematics, Springer-Verlag, Heidelberg, Dordrecht, London, New York, 2009.

[17] R. R. Berlot, Production of rotation in confined liquid through translational motions of the boundaries, *Journal of Applied Mechanics, Transactions ASME* **26** (1959), 513–516.

[18] J. Billingham, Nonlinear sloshing in zero gravity, *Journal of Fluid Mechanics* **464** (2002), 365–391.

[19] ———, On a model for the motion of a contact line on a smooth solid surface, *European Journal of Applied Mathematics* **17** (2006), 347–382.

[20] N. Bogolyubov and Y. Mitropolski, *Asymptotic methods in non-linear mechanics*, Gordon and Breach, New York, 1961.

[21] G. I. Bogomaz and S. A. Sirota, *Oscillations of a liquid in containers: methods and results of experimental studies*, National Space Agency of Ukraine, Dnepropetrovsk, 2002, in Russian.

[22] I. B. Bogoryad, *Oscillations of a viscous liquid in a cavity of a rigid body*, Tomsk University, Tomsk, 1999, in Russian.

[23] I. B. Bogoryad, I. A. Druzhinin and S. V. Chakhlov, *Study of transient processes with large disturbances of the free surface of a liquid in a closed compartment*, Dynamics of spacecraft apparatus and investigation of the space, Mashinostroenie, Moscow, 1986, in Russian, pp. 194–203.

[24] I. B. Bogoryad, I. A. Druzhinin, G. Z. Druzhinina and E. E. Libin, *Introduction to the dynamics of vessels with a liquid*, Tomsk University, Tomsk, 1977, in Russian.

[25] V. V. Bolotin, *The dynamic stability of elastic systems*, Holden-Day, San Francisco, 1964.

[26] J. Boussinesq, Théorie de l'intumescence liquide, appelée onde solitaire onde translation, se propageant dans un canal rectangulaire, *Comptes Rendus Mathematique, Paris* **72** (1871), 755–759.

[27] A. Cariou and G. Casella, Liquid sloshing in ship tanks: a comparative study of numerical simulation, *Marine Structures* **12** (1999), 183–198.

[28] F. Casciati, A. D. Stefano and E. Matta, Simulating a conical tuned liquid damper, *Simulation Modelling Practice and Theory* **11** (2003), 353–370.

[29] K. M. Case and W. C. Parkinson, Damping of surface waves in an incompressible liquid, *Journal of Fluid Mechanics* **2** (1957), 172–184.

[30] A.-L. Cauchy, M´emoire sur la th´eorie de la propagation des ondes ‘a la surface d‘un fluide pesant d‘une profondeur ind´efinie, *M´em. Pr´esent´es Divers Savans Acad. R. Sci. Inst. France (Prix Acad. R. Sci., concours de 1815 et de 1816)* **I** (1827), 3–312.

[31] F. Chernous'ko, *Motion of a rigid body with cavities containing viscous liquid*, Nauka, 1968.

[32] F. L. Chernous'ko, On free oscillations of a viscous fluid in a vessel, *Journal of Applied Mathematics and Mechanics* **30** (1966), 990–1003.

[33] ______, Oscillations of a vessel containing a viscous fluid, *Fluid Dynamics* **2** (1967), 39–43.

[34] ______, *Motion of a rigid body with cavities containing viscous liquid*, VTs AN SSSR, 1968, in Russian.

[35] W.-H. Chu, Subharmonic oscillations in an arbitrary tank resulting from axial excitation, *Journal of Applied Mechanics, Transactions ASME* **35** (1968), 148–154.

[36] W.-H. Chu and D. D. Kana, A theory for nonlinear transverse vibrations of a partially filled ellastic tank, *AIAA Journal* **5** (1967), 1828–1835.

[37] R. M. Cooper, Dynamic of liquid in moving containers, *ARS Journal* **30** (1960), 725–729.

[38] A. D. D. Craik, The origin of water wave theory, *Annual Review of Fluid Mechanics* **36** (2004), 1–28.

[39] ______, George Gabriel Stokes water wave theory, *Annual Review of Fluid Mechanics* **37** (2005), 23–42.

[40] F. T. Dodge, D. D. Kana and H. N. Abramson, Liquid surface oscillations in longitudinally excited rigid cylindrical containers, *AIAA Journal* **3** (1965), 685–695.

[41] L. V. Dokuchaev, *On composition of nonlinear equations of motion of a body with a liquid patially filled by a liquid*, Oscillations of elastic constructions with a liquid, Volna, Moscow, 1976, pp. 149–156.

[42] L. V. Dokuchaev and I. A. Lukovskii, Methods for determining the hydrodynamic characteristics of a moving vessel with partitions, *Fluid Dynamics* **3** (1968), 143–148.

[43] I. A. Druzhinin and S. V. Chakhlov, *Computation of nonlinear oscillations of liquid in a vessel (the Cauchy problem)*, Dynamics of Elastic and Rigid Vodies interacting with a Liquid, Tomsk University, Tomsk, 1981.

[44] M. Eastham, An eigenvalue problem with parameter in the boundary condition, *Quarterly Journal of Mathematics* **13** (1962), 304–320.

[45] C. R. Easton and I. Catton, Initial value technique in free-surface hydrodynamics, *Journal of Computational Physics* **9** (1972), 424–439.

[46] R. D. Edge and G. Walters, The period of standing gravity waves of largest amplitude on water, *Journal of Geophysical Research* **69** (1964), 1674–1675.

[47] M. Efroimsky, Relaxation of wobbling asteroids and comets – theoretical problems, perspectives of experimental observation, *Planetary and Space Science* **9** (2001), 937–965.

[48] T. Eguchi and O. Niho, A numerical simulation of 2-dimensional sloshing problem, *Mitsui Zosen Technical Review* (1989), 1–19.

[49] L. Euler, Continuation des recherches sur la th´eorie du mouvement des fluides, *Acad. Sci. Berlin* **11** (1757), 316–361.

[50] ———, Principes g´eneraux du mouvement des fluides, *M´em. Acad. Sci. Berlin* (1757), 271–315.

[51] ———, Principia motus fluidorum, *Novi Commentarii Acad. Sci. Petropolitanae* **6** (1761), 271–311.

[52] O. M. Faltinsen, A nonlinear theory of sloshing in rectangular tanks, *Journal of Ship Research* **18** (1974), 224–241.

[53] O. M. Faltinsen, O. F. Rognebakke, I. A. Lukovsky and A. N. Timokha, Multidimensional modal analysis of nonlinear sloshing in a rectangular tank with finite water depth, *Journal of Fluid Mechanics* **407** (2000), 201–234.

[54] O. M. Faltinsen, O. F. Rognebakke and A. N. Timokha, Resonant three-dimensional nonlinear sloshing in a square base basin, *Journal of Fluid Mechanics* **487** (2005), 1–42.

[55] ———, Two-dimensional resonant piston-like sloshing in a moonpool, *Journal of Fluid Mechanics* **575** (2007), 359–397.

[56] O. M. Faltinsen and A. N. Timokha, *Sloshing*, Cambridge University Press, Cambridge, 2009.

[57] ———, Multimodal analysis of weakly nonlinear sloshing in a spherical tank, *Journal of Fluid Mechanics* **719** (2013), 129–164.

[58] M. Faraday, On a peculiar class of acoustical figures. On certain forms assumed by groups of particles upon vibrating elastic surfaces, *Philosophical Transactions of the Royal Society of London* **121** (1831), 299–340.

[59] S. F. Feschenko, I. A. Lukovsky, B. I. Rabinovich and L. V. Dokuchaev, *Methods of determining the added liquid mass in mobile cavities*, Naukova Dumka, Kiev, 1969, in Russian.

[60] L. L. Fontenot, *Dynamic stability of space vehicles. Vol. 7. The dynamics of liquid in fixed and moving containers*, NASA, Report no. NASA, CR-941, Washington, 1968.

[61] K. O. Friedrichs, On the derivation of the shallow water theory, Appendix to "The formation of breakers and bores" by J. J. Stoker, *Communications on Pure and Applied Mathematics* **1** (1948), 81–87.

[62] D. Fultz and T. S. Murty, Experiments on the frequency of finite amplitude axisymmetric waves in circular cylinder, *Journal of Geophysical Research* **68** (1963), 1457–1862.

[63] M. Funakoshi and S. Inoue, Surface waves due to resonant oscillation, *Journal of Fluid Mechanics* **192** (1988), 219–247.

[64] ———, Bifurcations in resonantly forced water waves, *European Journal of Mechanics. B/Fluids* **10** (1991), 31–36.

[65] R. F. Ganiev and V. V. Kholopova, Nonlinear oscillations of a fluid-containing body moving in space, *Soviet Applied Mechanics* **11** (1975), 1187–1196.

[66] I. G. Gataullin and V. I. Stolbetsov, On the determination of nonlinear oscillations of a liquid in a circular cylinder sector, *Fluid Dynamics* **5** (1970), 820–825.

[67] I. Gavrilyuk, M. Hermann, I. Lukovsky, O. Solodun and A. Timokha, Natural sloshing frequencies in rigid truncated conical tanks, *Engineering Computations* **25** (2008), 518–540.

[68] ______, Weakly-nonlinear sloshing in a truncated circular conical tank, *Fluid Dynamics Research* **45** (2013), 1–30.

[69] I. Gavrilyuk, I. Lukovsky and A. N. Timokha, A multimodal approach to nonlinear sloshing in a circular cylindrical tank, *Hybrid Methods in Engineering* **2** (2000), 463–483.

[70] ______, Sloshing in circular conical tank, *Hybrid Methods in Engineering* **3** (2001), 322–378.

[71] I. Gavrilyuk, I. Lukovsky, Y. Trotsenko and A. Timokha, Sloshing in a vertical circular cylindrical tank with an annular baffle. Part 2. Nonlinear resonant waves, *Journal of Engineering Mathematics* **57** (2007), 57–78.

[72] I. P. Gavrilyuk, I. A. Lukovsky, V. L. Makarov and A. N. Timokha, *Evolutional problems of the contained fluid*, Institute of Mathematics of NASU, 2006.

[73] I. P. Gavrilyuk, I. A. Lukovsky and A. N. Timokha, Linear and nonlinear sloshing in a circular conical tank, *Fluid Dynamics Research* **37** (2005), 399–429.

[74] H. P. Greenspan, On the non-linear interaction of inertial modes, *Journal of Fluid Mechanics* **36 (Part 2)** (1969), 257–264.

[75] L. Guo and K. Morita, Numerical simulation of 3D sloshing in a liquid-solid mixture using particle methods, *International Journal for Numerical Methods in Engineering* **95** (2013), 771–790.

[76] M. D. Haskind, *Hydrodynamic theory of ship rolling and pitching*, Nauka, Moscow, 1973, in Russian.

[77] C. Hayashi, *Forced oscillations in non-linear systems*, Nippon Printing and Publishing Co, 1952.

[78] D. M. Henderson and J. W. Miles, Faraday waves in 2:1 resonance, *Journal of Fluid Mechanics* **222** (1991), 449–470.

[79] P. Holmes, Chaotic motions in a weakly nonlinear model for surface waves, *Journal of Fluid Mechanics* **162** (1986), 365–388.

[80] R. E. Hutton, *An investigation of nonlinear, nonplanar oscillations of fluid in cylindrical container*, NASA, Report, NASA; D-1870, 1963.

[81] ______, Fluid-particle motion during rotary sloshing, *Journal of Applied Mechanics, Transactions ASME* **31** (1964), 145–153.

[82] R. A. Ibrahim, V. N. Pilipchuk and T. Ikeda, Recent advances in liquid sloshing dynamics, *Applied Mechanics Reviews* **54** (2001), 133–199.

[83] N. Joukowski, On motions of a rigid body with cavity filled by homogeneous liquid, *Journal of Russian Physical-Mathematical Society* **XVI** (1885), 30–85, in Russian.

[84] D. D. Kana, U. S. Lindholm and H. N. Abramson, An experimental study of liquid instability in a vibrating elastic tank, *AIAA Symposium on Structural Dynamics and Aeroelasticity* (1965), 162–168.

[85] G. H. Keulegan and L. H. Carpenter, Forces on cylinders and plates in a oscillating fluid, *Journal of Research of the National Bureau of Standards* **60** (1958), 423–440.

[86] K. S. Kolesnikov, *Propellant rocket as a control object*, Mashinostroenie, Moscow, 1969, in Russian.

[87] A. N. Komarenko and I. A. Lukovskii, Stability of nonlinear fluid oscillations in a shell performing harmonic oscillations, *Soviet Applied Mechanics* **10** (1974), 1118–1122.

[88] N. D. Kopachevsky and S. G. Krein, *Operator approach to linear problems of hydrodynamics. Volume 1: Self-adjoint problems for an ideal fluid*, Birkhauser Verlag, Basel – Boston – Berlin, 2003.

[89] ———, *Operator approach to linear problems of hydrodynamics. Volume 2: Nonself-adjoint problems for viscous fluid*, Birkhauser Verlag, Basel – Boston – Berlin, 2003.

[90] D. Korteweg and G. de Vries, On the change of form of long waves advancing in a rectangular canal and on a new type of long stationary waves, *Philosophical Magazine* **39** (1895), 422–443.

[91] T. S. Krasnopolskaya and A. Y. Shvets, Dynamical chaos for a limited power supply for fluid oscillations in cylindrical tanks, *Journal of Sound and Vibration* **322** (2009), 532–553.

[92] S. G. Krein, Oscillations of a viscous fluid in a container, *Doklady Akademii Nauk SSSR* **159** (1964), 262–265, in Russian.

[93] A. N. Krylov, *Ship rolling and pitching. Collection of works*, 1, Academiya Nauk SSSR, 1951, in Russian.

[94] M. La Rocca, M. Scortino and M. A. Boniforti, A fully nonlinear model for sloshing in a ritating container, *Fluid Dynamics Research* **27** (2000), 225–229.

[95] J. L. Lagrange, Sur la mani'ere de rectifier deux entroits des Principes de Newton relatifs 'a la propagation du son et au mouvement des ondes, *Nouv. M´em. Acad., Berlin* **1889** (1786), 591–609.

[96] H. Lamb, *Hydrodynamics*, Cambridge University Press, 1932.

[97] E. E. Libin and N. G. Yakutova, *On rotary wave in a cylindical tank*, Oscillations of elastic constructions with a liquid, Volna, Novosibirsk, 1974, in Russian.

[98] O. S. Limarchenko, Direct method for solution of nonlinear dynamic problem on the motion of a tank with fluid, *Dopovidi Akademii Nauk Ukrains'koi RSR, Series A* **11** (1978), 99–1002, in Ukrainian.

[99] ———, Variational-method investigation of problems of nonlinear dynamics of a reservoir with a liquid, *Soviet Applied Mechanics* **16** (1980), 74–79.

[100] ______, Application of a variational method to the solution of nonlinear problems of the dynamics of combined motions of a tank with fluid, *Soviet Applied Mechanics* **19** (1983), 1021–1025.

[101] ______, Direct method of solving problems on the combined spatial motions of a body-fluid system, *Soviet Applied Mechanics* **19** (1983), 715–721.

[102] ______, Specific features of application of perturbation techniques in problems of nonlinear oscillations of a liquid with free surface in cavities of noncylindrical shape, *Ukrainian Mathematical Journal* **59** (2007), 45–69.

[103] O. S. Limarchenko and V. V. Yasinskii, *Nonlinear dynamics of constructions with a fluid*, Kiev Polytechnical University, Kiev, 1996, in Russian.

[104] L. G. Loitsianskii, *Fluid and gas mechanics*, Nauka, Moscow, 1973, in Russian.

[105] D. O. Lomen and L. L. Fontenot, Fluid behavior in parabolic containers under going vertical excitation, *Journal of Mathematical Physics* **46** (1967), 43–53.

[106] V. V. Lugovskii, *Hydrodynamics of nonlinear rolling and pitching of ships*, Sudostroenie, Leningrad, 1980, in Russian.

[107] J. G. Luke, A variational principle for a fluid with a free surface, *Journal of Fluid Mechanics* **27** (1967), 395–397.

[108] I. A. Lukovskii, An investigation into the motion of a rigid body with liquid performing nonlinear vibrations, *Soviet Applied Mechanics* **3** (1967), 70–74.

[109] ______, Approximate method of solution of nonlinear problems in the dynamics of a liquid in a vessel executing a prescribed motion, *Soviet Applied Mechanics* **17** (1981), 172–178.

[110] ______, Variational methods of solving dynamic problems for fluid-containing bodies, *International Applied Mechanics* **40** (2004), 1092–1128.

[111] I. A. Lukovsky, On constructing a solution of a nonlinear problem on free oscillations of a liquid in basins of arbitray shape, *Dopovidi AN UkrSSR, Ser. A* (1969), 207–210, in Ukrainian.

[112] ______, *Nonlinear sloshing in tanks of complex geometrical shape*, Naukova Dumka, Kiev, 1975, in Russian.

[113] ______, *Variational method in the nonlinear problems of the dynamics of a limited liquid volume with free surface*, Oscillations of elastic constructions with liquid, Volna, Moscow, 1976, in Russian, pp. 260–264.

[114] ______, *Studying the nonlinear sloshing in a cicular cylindical tank with arbitrary bottom shape by a variational method*, Dynamics and stability of managed systems, Institute of Mathematics, Kiev, 1977, in Russian, pp. 32–44.

[115] ______, *Variational method for solving the nonlinear problem on liquid sloshing in tanks of complicated shape*, Dynamics and stability of managed systems, Institute of Mathematics, Kiev, 1977, in Russian, pp. 45–61.

[116] ______, *Implementation of a variational principle of the Ostrogradsky type to solution of nonlinea problems of the rigid body dynamics with cavities containing a liquid*, Dynamics and stability of mechanical systems, Institute of Mathematics, Kiev, 1980, in Russian, pp. 3–14.

[117] ———, *Nonlinear mathematical modesl in the rigid body dynamics with cavities containing an ideal liquid*, Studies in the theory of complex variable with applications to the continuum mechanics, Institute of Mathematics, Kiev, 1986, in Russian, pp. 102–119.

[118] ———, *Usage of variational principle to derivations of equations of the body-liquid dynamics*, Dynamics of spacecraft apparatus and study of the space, Moscow, 1986, in Russian, pp. 182–194.

[119] ———, *Introduction to nonlinear dynamics of rigid bodies with the cavities partially filled by a fluid*, Naukova Dumka, Kiev, 1990.

[120] ———, *To the problem on composing finite-dimensional models in the dynamics of a limited liquid volume by a variational method*, Modeling the dynamic processes of interaction in the body-liquid systems, Institute of Mathematics, Kiev, 1990, in Russian, pp. 5–18.

[121] ———, On the theory of nonlinear sloshing of a weakly-viscous liquid, *Dopovidi of National Academy of Sciences of Ukraine* (1997), 80–84, in Ukrainian.

[122] ———, Linearization and delinearization in the hydrodynamic-type problems by using the Ostrogradsky action, in: *Proceedings of the Ukrainian Mathematical Congress, Section 8 "Numerical Mathematics and Mathematical Problems of Mechanics"*, pp. 71–80, Institite of Mathematics, Kiev, 2002, in Russian.

[123] ———, On solving spectral problems on linear sloshing in conical tanks, *Dopovidi NANU* (2002), 53–58, in Ukrainian.

[124] I. A. Lukovsky, M. Y. Barnyak and A. N. Komarenko, *Approximate Methods of Solving the Problems of the Dynamics of a Limited Liquid Volume*, Naukova Dumka, Kiev, 1984, in Russian.

[125] I. A. Lukovsky and A. N. Bilyk, *Forced sloshing in movable axisymmetic conical tanks*, Numerical-analytical studies of the dynamics and stability of multidimensional systems, Institute of Mathematics, Kiev, 1985, in Russian, pp. 12–26.

[126] ———, *Study of forcced nonlinear sloshing in conical tanks with a small apex angle*, Applied problems in the dynamics and stability of mechanical systems, Institute of Mathematics, Kiev, 1987, in Russian, pp. 5–14.

[127] I. A. Lukovsky, D. V. Ovchynnykov and A. N. Timokha, Asymptotic nonlinear multimodal method for liquid sloshing in an upright circular cylindrical tank. Part 1: Modal equations, *Nonlinear Oscillations* **14** (2012), 512–525.

[128] I. A. Lukovsky and A. M. Pilkevich, *Study of nonlinear sloshing in coaxial cylinder by variational method*, Bounday problems of the mathematiucal physics, Institute of Mathematics, Kiev, 1978, in Russian, pp. 79–91.

[129] ———, Determining the added mass of a limited liquid volume by variational method and perturbation technique, in: *Dynamics of elastic and rigid bodies interacting with a liquid*, Proceedings of V Seminar, pp. 102–112, Tomsk, 1984, in Russian.

[130] ———, *On liquid motions in an upright oscillating circular tank*, Numerical-analytical studies of the dynamics and stabnility of multidimensional systems, Institute of Mathematics, Kiev, 1985, in Russian, pp. 3–11.

[131] I. A. Lukovsky and G. A. Shvets, *Nonlinear equations of wave motions in axisymmetric tanks and computation of the hydrodynamic coefficients*, Applied methods for studying the physic-and-mechanic processes, Institute of Mathematics, Kiev, 1979, in Russian, pp. 36–48.

[132] I. A. Lukovsky, O. V. Solodun and A. N. Timokha, Eigen oscillations of a liquid sloshing in truncated conical tanks, *Acoustic Bulletin* **9** (2006), 42–61, in Russian.

[133] I. A. Lukovsky and A. N. Timokha, *Variational methods in nonlinear problems of the dynamics of a limited liquid volume*, Institute of Mathematics of NASU, 1995, in Russian.

[134] ______, *Nonlinear theory of sloshing in mobile tanks: classical and non-classical problems*, Problems of analytical mechanics and applications, Institute of Mathematics, Kiev, 1999, in Russian, pp. 169–201.

[135] ______, Modal modeling of nonlinear sloshing in tanks with non-vertical walls. Non-conformal mapping technique, *International Journal of Fluid Mechanics Research* **29** (2002), 216–242.

[136] I. O. Lukovsky and D. V. Ovchynnykov, *Nonlinear mathematical model of the fifith order in the sloshing problem for a cylindrical tank*, Problems of the dynamics and stability of multidimensional systems. Transactions of Institute of Mathematics of NASU, 47, Institite of Mathematics, Kiev, 2003, in Ukrainian, pp. 119–160.

[137] ______, *An optimal third-order model in the problem on nonlinear liquid sloshing in a cylindrical tank*, Problems of the dynamics and stability of multidimensional systems. Proceedings of Institute of Mathematics of NASU, 2, Institite of Mathematics of NASU, Kiev, 2005, in Ukrainian, pp. 254–265.

[138] ______, *Analysis of a force interation for a mobile cylindrical reseivor by using a ten-mode nonlinear model*, Analytical mechanics and its applications. Proceedings of Institute of Mathematics of NASU, 5, Institite of Mathematics of NASU, Kiev, 2008, in Ukrainian, pp. 204–244.

[139] I. O. Lukovsky and A. M. Pilkevich, *Studies on nonlinear sloshing of a viscous liquid in a cylindrical tank*, Problems of the dynamics and stability of multidimensional systems. Proceedings of Institute of Mathematics of NASU, 2, Institite of Mathematics of NASU, Kiev, 2005, in Ukrainian, pp. 266–275.

[140] ______, *Some properties of forced sloshing of a viscous liquid in circular cylindrical tanks*, Problems of the dynamics and stability of multidimensional systems. Proceedings of Institute of Mathematics of NASU, 4, Institite of Mathematics of NASU, Kiev, 2007, in Ukrainian, pp. 147–164.

[141] I. O. Lukovsky and O. V. Solodun, A nonlinear model of liquid motions in cylindrical compartment tanks, *Dopovidi NANU* (2001), 51–55, in Ukrainian.

[142] ______, *Study of the forced liquid sloshing in circular tanks by using a seven-mode model of the third order*, Problems of the dynamics and stability of multidimensional systems. Transactions of Institute of Mathematics of NASU, 47, Institite of Mathematics, Kiev, 2003, in Ukrainian, pp. 161–179.

[143] A. I. Lurye, *The analytical mechanics*, Fizmatgiz, Moscow, 1961, in Russian.

[144] O. D. D. Maggio and A. S. Rehm, Nonlinear free oscillations of a perfect fluid in a cylindical container, *AIAA Symposium on Structural Dynamics and Aeroelasticity* **30** (1965), 156–161.

[145] E. Matta, *Sistemi di attenuazione della risposta dinamica a massa oscillante solida e fluida*, Ph.D. thesis, Politecnico di Torino, Torino, 2002.

[146] L. Matthiessen, Akustische Versuche die Kleinsten Transversalwellen der Flussigkeiten betreffend, *Annalen der Physic und Chemie* **134** (1868), 107–177.

[147] G. N. Mikishev, *Experimental methods in the dynamics of spacecraft apparatus*, Mashinostroenie, 1978, in Russian.

[148] G. N. Mikishev and N. Y. Dorozhkin, Experimental study of free oscillations of liquid in vessels, *Izv. AN SSSR, OTN, Mekhanika i mashinostroenie* (1961), 48–53, in Russian.

[149] G. N. Mikishev and B. I. Rabinovich, *Dynamics of a Solid Body with Cavities Partially Filled with Liquid*, Mashinostroenie, 1968, in Russian.

[150] ———, *Dynamics of Thin-Walled Structures with Compartments Containing a Liquid*, Mashinostroenie, 1971, in Russian.

[151] J. W. Miles, Nonlinear surface waves in closed basins, *Journal of Fluid Mechanics* **75** (1976), 419–448.

[152] ———, Internally resonant surface waves in circular cylinder, *Journal of Fluid Mechanics* **149** (1984), 1–14.

[153] ———, Resonantly forces surface waves in circular cylinder, *Journal of Fluid Mechanics* **149** (1984), 15–31.

[154] ———, Parametrically excited, progressive cross-waves, *Journal of Fluid Mechanics* **186** (1988), 129–146.

[155] ———, Parametrically excited, standing cross-waves, *Journal of Fluid Mechanics* **186** (1988), 119–127.

[156] ———, On Faraday waves, *Journal of Fluid Mechanics* **248** (1993), 671–683.

[157] ———, Faraday waves: rolls versus squares, *Journal of Fluid Mechanics* **269** (1994), 353–371.

[158] ———, On Faraday waves, *Journal of Fluid Mechanics* **269** (1994), 372–372.

[159] L. M. Milne-Thomson, *Theoretical hydrodynamics*, Courier Dover Publications, 1968.

[160] G. A. Moiseev, Some problems of delinearization in dynamics of complex oscillatory systems, *Soviet Applied Mechanics* **8** (1972), 88–96.

[161] N. N. Moiseev, On the theory of nonlinear vibrations of a liquid of finite volume, *Journal of Applied Mathematics and Mechanics* **22** (1958), 860–872.

[162] ———, *Variational problems in the theory of oscillations of a liquid and a body with a liquid*, Variational methods in the problems of oscillations of a liquid and a body with a liquid, Moscow, 1962, in Russian, pp. 9–118.

[163] N. N. Moiseev and A. A. Petrov, *Numerical methods for computing the eigenoscillations of a limited liquid volume*, Computer Center of Academy of Sciences of USSR, 1966, in Russian.

[164] N. N. Moiseyev and V. V. Rumyantsev, *Dynamic Stability of Bodies Containing Fluid*, Applied Physics and Engineering 6, Springer, Berlin, Heidelberg, 1968.

[165] R. E. Moore and L. M. Perko, Inviscid fluid flow in an accelerating cylindrical container, *Journal of Fluid Mechanics* **22** (1964), 305–320.

[166] R. H. Multer, Exact nonlinear model of wave generation, *Journal of the Hydraulics Division, ASME* **99 (HY1)** (1973), 31–46.

[167] A. D. Myshkis, V. G. Babskii, A. D. Kopachavskii, L. A. Slobozhanin and A. D. Tiuptsov, *Low-gravity fluid mcehanics: Mathematical theory of capillary phenomena*, Springer-Verlag, Berlin and New York, 1987.

[168] A. H. Naifeh, *Introduction to perturbation techniques*, Wiley, 1981.

[169] T. Nakayama and K. Washizu, Nonlinear analysis of liquid motion in a container subjected to forced pitching oscillation, *International Journal for Numerical Methods in Engineering* **15** (1980), 1207–1220.

[170] G. S. Narimanov, Motions of a solid body whose cavity is partly filled by a liqud, *Applied Mathematics and Mechanics (PMM)* **20** (1956), 21–38.

[171] ______, Movement of a tank partly fled by a fluid: the taking into account of non-smallness of amplitude, *Prikl. Math. Mech.* **21** (1957), 513–524, in Russian.

[172] ______, On the oscillations of liquid in the mobile cavities, *Izv. of the AS USSR, OTN* (1957), 71–74, in Russian.

[173] G. S. Narimanov, L. V. Dokuchaev and I. A. Lukovsky, *Nonlinear dynamics of flying apparatus with liquid*, Mashinostroenie, Moscow, 1977, in Russian.

[174] A. I. Nekrasov, *The exact theory of steady waves on the surface of a heavy fluid*, Izdat. Akad. Nauk. SSSR, Moscow, 1951, in Russian.

[175] I. Newton, *Philosophiae Naturalis Principia Mathematica*, Jussu Societatis Regiae ac Typis J. Streater. Engl. transl. N Motte, 1687.

[176] J. R. Ockendon and H. Ockendon, Resonant surface waves, *Journal of Fluid Mechanics* **59** (1973), 397–413.

[177] T. Okamoto and M. Kawahara, Two-dimensional sloshing analysis by Lagrangian finite element method, *International Journal for Numerical Methods in Fluids* **11** (1990), 453–477.

[178] D. E. Okhozimskii, On the theory of a body motions when there is a cavity partlu filled by a liquid, *Applied Mathematics and Mechanics (PMM)* **20** (1956), 3–20.

[179] M.-A. Ostrogradsky, Memoire sur la propagation des ondes dans un bassin cylindrique, *Memoires a l'Academie Royale des Sciences, De lInstitut de France* **III** (1832), 23–44.

[180] L. V. Ovsyannikov, N. I. Makarenko, V. I. Nalimov, V. Y. Lyapidevski, P. I. Plotnikov, I. V. Sturova, V. I. Bukreev and V. A. Vladimirov, *Nonlinear problems of the theory of surface and internal waves*, Nauka, Novosibirsk, 1985.

[181] W. G. Penny and A. T. Price, Finite periodic stationary waves in a perfect liquid, *Philosophical Transactions of the Royal Society, A* **244** (1952), 254–284.

[182] L. M. Perko, Large-amplitude motions of liquid-vapour interface in an accelerating container, *Journal of Fluid Mechanics* **35** (1969), 77–96.

[183] A. S. Peters and J. J. Stoker, The motion of a ship, as a floating rigid body, in a seaway, *Communications on Pure and Applied Mathematics* **10** (1957), 399–490.

[184] A. M. Pilkevich, *Analysis of forced liquid sloshing in co-axial cylindical reseivors*, Applied methods of studying the physical-mechanical processes, Institute of Mathematics, Kiev, 1979, in Russian, pp. 49–63.

[185] ———, *Constructing the approximate solutions of nonlinear equations on wave motions of a liquid in a container*, Dynamics and stability of mechanical systems, Institute of Mathematics, Kiev, 1980, in Russian, pp. 16–21.

[186] ———, *On motions of a liquid in a cylindical tank peforming angular oscillations*, Numerical-analytical methods of study of the dynamics ans stability of complex systems, Institute of Mathematics, Kiev, 1984, in Russian, pp. 32–40.

[187] ———, *On liquid motions in an elliptical cylinder*, Mathematical modeling of dynamic processes in systems with a liquid, Institute of Mathematics, Kiev, 1988, in Russian, pp. 82–92.

[188] P.-S. Laplace, Marquis de, Suite des recherches sur plusieurs points du systeme du monde (XXV-XXVII), *M ´em. Pr ´esent ´es Divers Savans Acad. R. Sci. Inst. France* (1776), 525–552.

[189] B. I. Rabinovich, Equations of perturbed motions of a solid body with a cylindrical cavity partltly filled by a liquid, *Applied Mathematics and Mechanics (PMM)* **20** (1956), 39–49.

[190] ———, On the theory of small vibrations of a rigid body with a cavity partially filled with a viscous incompressible liquid, *Soviet Applied Mechanics* **4** (1968), 54–58.

[191] ———, *Introduction to dynamics of spacecraft*, Mashinostroenie, Moscow, 1975, in Russian.

[192] B. Ramaswamy, M. Kawara and T. Nakayama, Lagrangian finite element method for the analysis of two-dimensional sloshing problems, *International Journal for Numerical Methods in Fluids* **6** (1986), 659–670.

[193] Rayleigh, On the crispations of fluid resting upon a vibrating support, *Philosophical Magazine* **16** (1883), 50–58.

[194] ———, Deep water waves, progressive or stationary, to the third order of approximation, *Proceedings of the Royal Society A* **91** (1915), 345–353.

[195] S. Rebouillat and D. Liksonov, Fluid-structure interaction in partially filled liquid containers: A comparative review of numerical approaches, *Computers & Fluids* **39** (2010), 739–746.

[196] A. Royon-Lebeaud, E. J. Hopfinger and A. Cartellier, Liquid sloshing and wave breaking in circular and square-base cylindrical containers, *Journal of Fluid Mechanics* **577** (2007), 467–494.

[197] L. W. Schwartz and A. K. Whitney, A semi-analytic solution for nonlinear standing waves in deep water, *Journal of Fluid Mechanics* **107** (1981), 147–171.

[198] Y. I. Sekerzh-Senkovich, Three-dimensional problem on standing waves of finite amplitude on the free surface of a heavy liquid, *Doklady Academii Nauk SSSR* **86** (1952), 35–38, in Russian.

[199] R. L. Seliger and G. B. Whitham, Variational principles in continuum mechanics, *Proceedings of the Royal Society A* **305** (1968), 1–25.

[200] P. N. Shankar and R. Kidambi, A modal method for finite amplitude, nonlinear sloshing, *Pramana* **59** (2002), 631–651.

[201] V. M. Shashin, Dynamics of liquid in a conical tank during transition from small to appreciable gravity, *Fluid Dynamics* **3** (1968), 49–52.

[202] V. Y. Shkadov and Z. D. Zapryanov, *Viscous liquid flow*, Moscow University, Moscow, 1984, in Russian.

[203] A. Y. Shvets and V. A. Sirenko, Peculiarities of transition to chaos in nonideal hydrodynamics systems, *Chaotic Modeling and Simulation* **2** (2012), 303–310.

[204] J. M. Sicilian and J. R. Tegart, Comparison of FLOW-3D calculations with very large amplitude slosh data, in: *Proc. ASME/JSME Pressure Vessels and Piping Conference, Honolulu, HI, July 23-27, 1989 (A90-32339 13-31). New York, American Society of Mechanical Engineers*, pp. 23–30, 1989.

[205] R. Skalak and M. Yarymovych, Forced large amplitude surface waves, in: *Proc. 4-th U.S. Nat. Congr. Appl. Mech.*, 2, pp. 1411–1418, 1962.

[206] F. Solaas, *Analytical and numerical studies of sloshing in tanks*, Ph.D. thesis, Norwegian Institute of Technology, 1995.

[207] F. Solaas and O. M. Faltinsen, Combined numerical and analytical solution for sloshing in two-dimensional tanks of general shape, *Journal of Ship Research* **41** (1997), 118–129.

[208] A. J. Stofan and A. L. Armsted, *Analytical and experimental investigation of forces and frequencies resulting from liquid sloshing in a spherical tank*, NASA, Lewis Research Center, Clevelend, Ohio, Report no. D–1281, 1962.

[209] J. J. Stoker, *Water waves. The mathematical theory with applications*, Interscience Publishers, Inc., New York, Chichester, Brisbane, Toronto, Singapore, 1957.

[210] V. I. Stolbetsov, Nonsmall liquid oscillations in a right circular cylinder, *Fluid Dynamics* **2** (1967), 41–45.

[211] ______, Oscillations of liquid in a vessel in the form of a rectangular parallelepiped, *Fluid Dynamics* **2** (1967), 44–49.

[212] ______, Equations of nonlinear oscillations of a container partially filled with a liquid, *Fluid Dynamics* **4** (1969), 95–99.

[213] V. I. Stolbetsov and V. M. Fishkis, A mechanical model of a liquid performing small oscillations in a spherical cavity, *Fluid Dynamics* **3** (1968), 79–81.

[214] T. C. Su and Y. Wang, Numerical simulation of three-dimensional large amplitude liquid sloshing in cylindrical tanks subjected to arbitrary excitations, in: *Proc., 1990 Pressure Vessels and Piping Conf., ASME, New York, N.Y.*, pp. 127–148, 1990.

[215] I. E. Sumner, *Experimental investigations of stability boundaries for planar and nonplanar sloshing in spherical tanks*, NASA, Report no. TN D-3210, Washington, 1966.

[216] I. E. Sumner and A. J. Stofan, *An experimental investigation of the viscous damping of liquid sloshing in spherical tanks*, NASA Technical Note, TN D-1991, Report, 1963.

[217] I. Tadjbaksh and J. B. Keller, Standing surface waves of finite amplitude, *Journal of Fluid Mechanics* **8** (I960), 443–451.

[218] G. Taylor, An experimental study of standing waves, *Proceedings of the Royal Society, London. A* **210** (1953), 44–59.

[219] M. Urabe, Galerkin's procedure for nonlinear periodic systems, *Archive for Rational Mechanics and Analysis* **20** (1965), 120–152.

[220] M. Urabe and A. Reiter, Numerical computation of nonlinear forced oscillations by Galerkin's procedure, *Journal of Mathematical Analysis and Applications* **14** (1966), 107–140.

[221] J. H. G. Verhagen and L. Wijngaarden, Non-linear oscillations of fluid in a container, *Journal of Fluid Mechanics* **22** (1965), 737–752.

[222] L. Wang, Z. Wang and Y. Li, A SPH simulation on large-amplitude sloshing for fluids in a two-dimensional tank, *Earthquake Engineering and Engineering Vibration* **12** (2013), 135–142.

[223] H. J. Weiss, *A nonlinear analysis for sloshing for forces and moments on a cylindrical tanks*, NASA, Report no. NASA; CR-221, Washington, 1966.

[224] J. H. Woodward and H. F. Bauer, Fluid behavior in a longitudinally excited cylindrical tank of arbitrary sector-annular cross section, *AIAA Journal* **8** (1970), 713–719.

[225] V. A. Yakubovich and V. M. Starzhinski, *Linear differential equations with periodic coefficients*, 1 & 2, John Wiley, New York, 1975.

[226] G. C. Yeh, Free and forced oscillations of a liquid in an axisymmetric tank at low-gravity environments, *Journal of Applied Mechanics, Transactions ASME* **34** (1967), 23–28.

[227] H. C. Yuen and B. M. Lake, *Nonlinear dynamics of deep-water gravity waves*, Advances in Applied Mechanics (C.-S. Yih, ed.), 22, Academic Press, Inc., 1982, pp. 68–231.

[228] V. E. Zakharov, Stability of periodic waves of finite amplitude on the surface of a deep fluid, *Journal of Applied Mechanics and Technical Physics* **9** (1968), 190–194.

# Index

# De Gruyter Studies in Mathematical Physics

**Volume 27**
*Ivan A. Lukovsky*
Nonlinear Dynamics, 2015
ISBN 978-3-11-031655-1, e-ISBN (PDF) 978-3-11-031657-5,
e-ISBN (EPUB) 978-3-11-038973-9, Set-ISBN 978-3-11-031658-2

**Volume 26**
*Vincent Laude*
Phononic Crystals, 2015
ISBN 978-3-11-030265-3, e-ISBN(PDF) 978-3-11-030266-0,
e-ISBN (EPUB) 978-3-11-038791-9, Set-ISBN 978-3-11-030267-7

**Volume 25**
*Sergey D. Algazin, Igor A. Kijko*
Aeroelastic Vibrations and Stability of Plates and Shells, 2015
ISBN 978-3-11-033836-2, e-ISBN (PDF) 978-3-11-033837-9,
e-ISBN (EPUB) 978-3-11-038945-6, Set-ISBN 978-3-11-040491-3

**Volume 24**
*Igor Olegovich Cherednikov, Tom Mertens, Frederik F. van der Veken*
Wilson Lines in Quantum Field Theory, 2014
ISBN 978-3-11-030910-2, e-ISBN (PDF) 978-3-11-030921-8,
e-ISBN (EPUB) 978-3-11-038293-8, Set-ISBN 978-3-11-030922-5

**Volume 23**
*Boris A. Arbuzov*
Non-perturbative Effective Interactions in the Standard Model, 2014
ISBN 978-3-11-030292-9, e-ISBN(PDF) 978-3-11-030521-0,
e-ISBN (EPUB) 978-3-11-038805-3, Set-ISBN 978-3-11-030522-7

**Volume 22**
*Sergei M. Kopeikin (Ed.)*
Frontiers in Relativistic Celestial Mechanics, Volume 2 Applications and Experiments, 2014
ISBN 978-3-11-034545-2, e-ISBN (PDF) 978-3-11-034566-7,
e-ISBN (EPUB) 978-3-11-037953-2, Set-ISBN 978-3-11-034567-4

www.ingramcontent.com/pod-product-compliance
Lightning Source LLC
Chambersburg PA
CBHW050038220326
41599CB00041B/7202
*9783110555363*